Faraday Laser

Jingbiao Chen · Tiantian Shi · Duo Pan · Zheyi Ge ·
Jia Zhang · Zijie Liu · Xiaomin Qin · Yaqi Wang

Faraday Laser

A Frequency-Stabilized Diode Laser Based on Faraday Atomic Optical Filters

Science Press
Beijing

Springer

Jingbiao Chen
School of Electronics
Peking University
Beijing, China

Duo Pan
Peking University
Beijing, China

Jia Zhang
Peking University
Beijing, China

Xiaomin Qin
Peking University
Beijing, China

Tiantian Shi
Peking University
Beijing, China

Zheyi Ge
Peking University
Beijing, China

Zijie Liu
Peking University
Beijing, China

Yaqi Wang
Peking University
Beijing, China

ISBN 978-981-97-8022-8 ISBN 978-981-97-8023-5 (eBook)
https://doi.org/10.1007/978-981-97-8023-5

Jointly published with Science Press
The print edition is not for sale in China mainland. Customers from China mainland please order the print book from: Science Press.
ISBN of the Co-Publisher's edition: 978-7-03-079648-6

The original submitted manuscript has been translated into English. The translation was done using artificial intelligence. A subsequent revision was performed by the author(s) to further refine the work and to ensure that the translation is appropriate concerning content and scientific correctness. It may, however, read stylistically different from a conventional translation.

This Springer imprint is published by the registered company Springer Nature Singapore Pte Ltd.
The registered company address is: 152 Beach Road, #21-01/04 Gateway East, Singapore 189721, Singapore

This book is dedicated to Michael Faraday.

Foreword

In modern human activities, lasers have become indispensable facilities and tools akin to electrical appliances, indispensable in various aspects such as production, daily life, education, health care, and entertainment. Remarkable indeed, for an invention scarcely six decades old to have achieved such widespread adoption, showcasing a marvel of modern science and technology. The laser family boasts a diverse array, encompassing solid-state lasers, gas lasers, dye lasers, diode lasers (also known as laser diodes, LD), and free-electron lasers, among others. These lasers further distinguish themselves by their wavelength outputs, spanning the infrared, visible, ultraviolet, and X-ray spectra, as well as by their continuous or pulsed modes of operation. The diode laser, specifically, stands out as the most prevalent and practical member of the laser family due to its compactness, long lifespan, and ease of use, particularly in fields such as optical communication, information processing, scientific research, precision measurement, and consumer electronics. Despite its relatively affordable pricing, which contributes to a modest market share within the broader laser industry, the diode laser continuously evolves, enhancing its capabilities, with a promising future ahead.

The diode laser, or laser diode, operates efficiently through electrical current excitation. However, it inherently faces limitations such as short cavity length, low Q-factor, broad linewidth, poor monochromaticity, large divergence angle, limited power output, poor temperature stability, and a tendency for "mode-hopping." Nevertheless, users can tailor its performance to their specific applications by incorporating additional optical feedback to extend the cavity length. This extended path can incorporate highly stable frequency-selective devices (e.g., gratings, filters), transforming the system into an "External Cavity Diode Laser (ECDL)." Furthermore, the use of highly stable reference frequency sources, such as externally coupled high-Q resonators or atomic (including molecular and ionic) spectral lines with electrical feedback, enables linewidth compression and frequency stabilization. The Faraday laser, the subject of this book, belongs to this class of ECDLs. Initially, the design and integration of these performance-enhancing components were the responsibility of users, who would customize them with "bare" diodes (laser chips with packaging). Over time, however, specialized enterprises have taken on this role, facilitating users to focus solely on their research objectives.

The Faraday laser utilizes a Faraday anomalous dispersion optical filter (FADOF) as a frequency-selective element placed within the external cavity, imparting remarkable monochromaticity to the laser output. Additionally, it incorporates laser frequency stabilization techniques, including saturated absorption spectroscopy, modulation transfer spectroscopy, and Pound–Drever–Hall (PDH) technique, achieving exceptional frequency stability. This innovative technology stems from the research group led by Prof. Jingbiao Chen at Peking University, whose integrated laser system has been internationally recognized as the "Faraday laser." Professor Chen's team has built upon the legacy of Peking University's pioneering work in FADOF, initiated by the late Professors Lemin Zheng and Junxiong Tang in the 1980s, focusing on optical communication and later exploring its potential in external cavity diode laser.

This book presents a comprehensive and relatively accessible account of Faraday lasers, drawing on the rich experimental and theoretical endeavors of Prof. Chen's team at Peking University. It delves into the fundamental structure, manufacturing techniques, physical properties, and developmental trends of laser diodes. Furthermore, it examines the constituents of ECDLs, highlighting the selection and application of frequency-selective devices, including the Faraday anomalous dispersion optical filter (FADOF). The book meticulously details the Faraday atomic filter, contrasting various atomic filters, with a particular focus on the Faraday anomalous dispersion optical filter (FADOF) and the Voigt anomalous dispersion optical filter (VADOF), whose primary distinction lies in the orientation of the magnetic field relative to the optical axis. It also provides a thorough description of the overall system architecture, encompassing excitation power supplies, temperature control circuits, and even manufacturing processes. The book delves deeply into the frequency stabilization of Faraday lasers, introducing techniques such as saturated absorption, modulation transfer, and PDH stabilization, with a special emphasis on the research group's distinctive utilization of modulation transfer spectroscopy. Furthermore, it presents novel perspectives on the application of Faraday lasers, particularly in the field of atomic clocks. For instance, the use of Faraday lasers in laser-pumped cesium beam atomic clock has achieved stability exceeding that of magnetically selected state cesium clocks 5071A by more than half an order of magnitude and outperforms other commercial optically pumped cesium clocks. Furthermore, the innovative concept of an optical-microwave dual-frequency atomic clock is proposed and partially realized, along with suggestions for its application in cold-atom active optical clocks. Additionally, theoretical and practical recommendations for the Faraday laser's use in diverse fields such as atomic gravimeters, magnetometers, and optical communications are abundant. As a conclusion, the book offers a visionary outlook on the future development of this laser, envisioning possibilities such as miniaturization, smart integration, high power, and multi-channel operation, demonstrating a broad mindset and numerous innovative ideas anticipated to materialize in the future.

In summary, this is a monograph on Faraday lasers that offers both nearly accessible explanations and profound theoretical discussions, coupled with abundant practical demonstrations and meticulous descriptions of technological processes and techniques. It serves as an invaluable training material for scientific researchers, engineers, and technicians involved in the application, development, and production of these lasers. Furthermore, it holds significant reference value for general scientific researchers and university students working with lasers. For these reasons, I am delighted to recommend this book to all!

Beijing, China
June 2024

Yiqiu Wang

Preface

With the innovation of science and technology, laser technology, being one of the cornerstones of the modernized engineering tools, plays a vital role in plethora of fields including scientific research, industrial manufacturing, communication systems, national defense, and military applications. Among different aspects of laser technology, diode lasers have garnered widespread interest due to their development and applications. The Faraday laser, a diode laser realized through the Faraday magneto-optic rotation effect, exhibits advantages such as high frequency stability, narrow spectral linewidth, and strong resistance to temperature and current fluctuations. It has been widely applied in the field of quantum precision measurement, including atomic clocks, atomic gravimeters, magnetometers, and other related devices.

The compilation of this book, *Faraday Laser*, is based on in-depth research into diode laser technology and the widespread application of Faraday lasers in modern scientific and technological fields. It integrates the latest research findings, technological advancements and various applications, aiming to provide a comprehensive and systematic introduction to the principles, technologies, applications, and future development trends of Faraday lasers.

The distinctive features of this book are reflected in following aspects:

1. **Clarity and Precision**
   Starting from the development history and trends of semiconductor lasers, this book provides relevant information on the principles, structure, performance characteristics, application scenarios and requirements of Faraday lasers, as well as key components such as Faraday laser-specific semiconductor laser diodes and Faraday atomic filters. Additionally, it delves into the frequency stabilization techniques, promotional applications, and future development trends of Faraday lasers, thereby forming a comprehensive technical system.
2. **Practical Relevance**
   This book not only focuses on the introduction of theoretical knowledge but also emphasizes real-world applications and operations. Through a wealth of experimental data and case studies, it demonstrates the application effects and

prospects of Faraday lasers in various fields, providing readers with practical technical guidance and references.

3. **Future Insights**
   Based on the introduction of existing technologies, this book also provides insights and predictions on the future development trends of Faraday lasers. It discusses the future direction and application prospects of Faraday lasers from various aspects, including miniaturization into microchips, intelligent integration, high-power output, and multi-frequency capabilities, offering readers forward-thinking perspectives.

This book is primarily compiled by the research team at Peking University, led by Jingbiao Chen, Tiantian Shi, Duo Pan, and Zheyi Ge, who collaborated to shape the overall framework and key content of the book. The development history and trends of semiconductor lasers as well as frequency-stabilized semiconductor lasers and their applications are introduced by Duo Pan, Tianyu Liu, Jia Zhang and Zhiyang Wang, in Chap. 1. Tiantian Shi and Jia Zhang have authored Chap. 2, Overview of Faraday Lasers, providing the details of the basic principles, structural composition, and performance characteristics of Faraday lasers, along with their prospective applications and demands. Chapter 3, Laser Diodes for Faraday Lasers, authored by Zheyi Ge and Yaqi Wang, includes the basic principles, main classifications, and performance characteristics of semiconductor laser diodes, as well as the materials and preparation techniques specifically used for semiconductor laser diodes in Faraday lasers. Chapter 4, Faraday Atomic Filters, is completed by Tiantian Shi, Zheng Xiao, Suyang Wei, Xiaolei Guan and Xun Gao. It discusses the development history and main classifications of atomic filters, elucidates the basic principles and core preparation techniques of Faraday atomic filters, and presents related theoretical and simulation results. Tiantian Shi, Zijie Liu and Ziqi Lu have described the overall technical scheme, optical resonant cavity, circuit system, frequency-stabilized power supply, as well as the main performance characteristics and specifications in Chap. 5, Faraday Laser Scheme and Technology. Chapter 6, Frequency Stabilization Techniques for Faraday Lasers, is written by Duo Pan, Zhiyang Wang and Suyang Wei to analyze the saturated absorption spectroscopy and modulation transfer spectroscopy of Faraday lasers, as well as the Pound–Drever–Hall (PDH) frequency stabilization technique. As the name suggests, Chap. 7, Applications of Faraday Lasers, discusses the applications of Faraday lasers in atomic clocks, atomic gravimeters, magnetometers, optical communications, laser wavelength standards, ultra-narrowband weak coherent light amplification, Raman spectroscopy detection, and other scenarios. Extensive analysis regarding the applicability of these lasers is conducted by Duo Pan, Hangbo Shi, and Ziqi Lu. Lastly, Tiantian Shi, Xiaomin Qin, Xiaolei Guan, and Xun Gao conclude the book with Chap. 8, Future Trends, by providing insights and predictions on the future trends of Faraday lasers towards miniaturization and intelligent integration, as well as the development of high-power, multi-frequency, quantum mode-locked, cold-atom, Faraday-Michelson lasers, and other advancements.

The research and compilation process of this book would not have been possible without the extensive support and assistance from multiple institutions. These invaluable contributions not only accelerated the research progress but also greatly enriched the content and depth of this book.

We extend our heartiest gratitude to experts and scholars including Yiqiu Wang, Lemin Zheng, Donghai Yang, Taiqian Dong, Junxiong Tang, Xuzong Chen, Zhigang Zhang, Hong Guo, Xiaoji Zhou, Anhong Dang, Zhong Wang, Yanhui Wang, Xiang Peng, Teng Wu, Aimin Wang, Jianmin Wang, Weimin Wang, Xianghui Qi, Wei Xiong, Zhangyuan Chen, Chuanwen Zhu and Huifang Hou from Peking University; Yi Luo, Li You, Guilu Long, Hongbo Sun, Lijun Wang, Zhibiao Hao, Xue Wang, Qiang Liu, Yidong Tan, Bo Wang, Yanying Feng and Jianwei Zhang from Tsinghua University; Jianwei Pan, Haifeng Jiang, Qiang Zhang and Hanning Dai from University of Science and Technology of China; Bin Luo, Song Yu, Xiaoguang Zhang, Guanjun Gao and Xing Chen from Beijing University of Posts and Telecommunications; Kaikai Huang from Zhejiang University; Xiaojun Xu, Zining Yang, Yanhong Wang, Zhiguo Wang and Xiong Yang from National University of Defense Technology; Longsheng Ma, Yanyi Jiang and Xinye Xu from East China Normal University; Qiang Lin from Zhejiang University of Technology; Pengyuan Chang from Nanjing University of Posts and Telecommunications; Limin Li from Wenzhou University; Wei Zhou, Lina Bai, Wei Liang, Zhiqi Li and Miao Miao from Xidian University; Yuzhu Wang, Weibiao Chen, Liang Liu, Desheng Lv, Yuan Sun, Jinyin Wan and Xin Wang from Shanghai Institute of Optics and Fine Mechanics, Chinese Academy of Sciences; Zhaohui Ye, Mingsheng Zhan, Kelin Gao, Baolong Lv, Hua Guan, Xueren Huang, Tingyun Shi, Jiehua Chen, Lingxiang He and Faquan Li from Innovation Academy for Precision Measurement Science and Technology, CAS; Wenxin Wang, Ruquan Wang and Yanmei Yu from Institute of Physics CAS, for providing us with the guidance and assistance regarding fundamental exploration and theoretical research.

We are also thankful for the support and assistance from experts and technicians including Tianchu Li, Xiang Fang, Fang Fang, Zhanjun Fang, Wei Zhuang, Pingwei Lin, Yige Lin, Jifeng Qu, Shiying Cao, Yijie Pan, Shuo Jia, Fei Meng, Zhenfei Song, Fasong Zheng, Aimin Zhang, Yuzhuo Wang, Weiliang Chen and Yanfei Wang from National Institute of Metrology, China; Yu Wang and Dong Hu from the 304th Institute, Third Research Institute of China Aerospace Science and Industry Corporation Limited; Haijun Chen, Zhongzheng Liu and Yufei Yan from the 12th Research Institute of China Electronics Technology Group Corporation; Shougang Zhang, Hong Chang, Deshui Yu, Jun Ruan, Tao Liu, Ruifang Dong, Fuyu Sun, Enxue Yun and Zhijing Du from National Time Service Center Chinese Academy Of Sciences; Xiaoyu Ma from Institute of Semiconductors, Chinese Academy of Sciences; Xiaohua Chen and Jun Li from BWT Co., Ltd., for enlightening us throughout our experimental journey and device development.

We would also like to thank Detian Li, Jiqing Lian and Jiang Chen from Lanzhou Institute of Physics, CAST; Jun Ge, Yifei Hu, Shengkang Zhang, Xiaobo Xue, Liang Wang, Xiumei Wang, Tong Shen, Yabei Su and Yudong Ding from the 203rd Institute, Second Research Institute of China Aerospace Science and Industry Corporation Limited; Yinxiao Miao, Tieli Zhang, Ke Liu and Hao Liu from China Academy of Launch Vehicle Technology; Wenxi Zhang, Weihu Zhou and Zhihong Gao from Aerospace Information Research Institute, Chinese Academy of Sciences; Qiang Wei and Xingwen Zhao from Chengdu Spaceon Elecstronics Co.,Ltd.; Yongjie Xie, Haihui Zhou, Yelong Hong, Zhenfeng Liu, Guirong Sun, Peng Liu, Xiangqian Zhao, Jianfu Tian and Renjie Shan from Zhejiang Faraday Laser Technology Co., Ltd., for helping us in the industrial promotion of our book.

Last but certainly not the least, we express our sincere gratitude to Jyoti for her thorough and thoughtful review of the English version of our book. Her meticulous corrections, constructive criticism and valuable suggestions have greatly enhanced the clarity and coherence of our book, for which we are deeply appreciative.

Once again, we express our heartfelt gratitude to all the experts and scholars who have worked diligently on this book. At the same time, we would also like to thank the editors and staff of the Springer for their hard work, which has made the smooth publication of this book possible. We hope that this book will provide readers with useful references and inspirations, and contribute to the development of Faraday laser technology and its applications.

Beijing, China

Jingbiao Chen
Tiantian Shi
Duo Pan
Zheyi Ge
Jia Zhang
Zijie Liu
Xiaomin Qin
Yaqi Wang

Synopsis

High-precision, stable-frequency diode lasers are fundamental instruments for cutting-edge scientific research. They are currently the core components in rapidly developing international fields such as quantum precision measurement, time-frequency communication, and atomic physics and are of great significance to economic development and security construction. Starting from the development history and trends of diode lasers, this book systematically introduces the fundamental principles and fabrication techniques of Faraday lasers. It elaborates in detail the remarkable advantages of Faraday lasers, which include their ability to intrinsically align with atomic transition lines upon startup and their resistance to temperature and current perturbations. In addition, the book also introduces the practical applications and important value of Faraday lasers in devices such as cesium atomic clocks, atomic gravimeters, and underwater optical communication systems and discusses the future development trends of Faraday lasers.

This book is suitable for researchers and engineers in the field of frequency-stabilized diode lasers and can also serve as a textbook for advanced undergraduate and graduate courses in quantum precision measurement, precision spectroscopy, and other professional fields.

Contents

Chapter 1
Introduction

Diode lasers, initially developed in the 1960s, have evolved into essential core devices in the field of modern optoelectronic technology through decades of continuous innovation and development. Starting from the initial homojunction structure, advancements have progressed through single heterojunction, double heterojunction, quantum well, quantum wire, and quantum dot structures. Furthermore, the incorporation of technologies such as Distributed Feedback (DFB), Distributed Bragg Reflection (DBR), Vertical-Cavity Surface-Emitting Laser (VCSEL), Photonic-Crystal Surface-Emitting Laser (PCSEL), and External-Cavity Diode Laser (ECDL) has significantly enhanced the performance of diode lasers and expanded their applications in various fields. Currently, diode lasers are widely used in precision spectroscopy, communication, data storage, medical, industrial processing, and other fields, especially for the field of quantum precision measurement, diode lasers have become the core devices of atomic clocks, quantum interferometers, atomic magnetometers, and other scientific instruments. This chapter focuses on the development history and trends of diode lasers, especially the external cavity diode laser.

1.1 Development History and Trends of Diode Lasers

Since the first laser diode was introduced in 1962 [1–3], it has been rapidly developed and widely used due to its significant advantages such as small size, high power, high efficiency, simple structure, low cost, and easy tuning. It has been widely used in many aspects such as laser spectroscopy, quantum precision measurement, laser optical communication, laser storage, high-speed optical printing, optical information processing, optical computing, and optical sensing. In recent years, diode lasers have demonstrated significant value in quantum detection, cold atom physics, atomic spectroscopy research, gravitational wave detection, especially in the field

J. Chen et al., *Faraday Laser*, https://doi.org/10.1007/978-981-97-8023-5_1

of quantum precision measurement, diode lasers have become essential components in numerous quantum detection instruments, including atomic clocks, atomic gravimeters, atomic magnetometers, atomic gyroscopes.

Diode lasers, as a type of electromagnetic wave oscillator, exhibit frequency variations over time. Rapid frequency changes result in laser linewidth broadening, while slower changes are often induced by environmental influences on the laser cavity, pump source, and other factors, resulting in frequency drift and deterioration of medium and long-term stability. The linewidth and frequency stability of diode lasers are very important indicators for plethora of applications, especially in fundamental research in the quantum field (such as high-resolution spectroscopy, laser cooling and trapping of atoms, and new quantum frequency standards) and industrial applications (such as atomic gravimeters, atomic gyroscopes, atomic magnetometers, atomic clocks, and various quantum sensors), where narrow linewidth and superior frequency stability are particularly critical. There are mainly two methods to narrow the linewidth of diode lasers: electrical feedback and optical feedback. Given the sensitivity of diode lasers to optical feedback, the method of narrowing the laser linewidth by optical feedback can greatly simplify the structure of the frequency-stabilized laser. At the same time, using optical feedback and Pound–Drever–Hall (PDH) frequency stabilization technology, one can get significant linewidth narrowing effect, which have been widely used in optoelectronic technology and other fields. The improvement of laser frequency drift and medium and long-term stability is often achieved by external atomic frequency stabilization.

1.1.1 History of the Development of External Cavity Diode Lasers

The development of quantum precision measurement technology has greatly progressed physics. Among the various physical quantities measured, time–frequency measurement stands out with its excellent accuracy and stability, becoming the most accurate measurable physical quantity at present. In 1967, the 13th International Metrology Conference redefined the unit of time, "second", changing from the "astronomical second" based on the Earth's revolution to the "atomic second" based on atomic transitions. The duration of 9,192,631,770 cycles of radiation corresponding to the transition between the two hyperfine energy levels of the undisturbed ground state of cesium 133 (^{133}Cs) atoms at Kelvin absolute zero was defined as 1 s [4]. In 2018, the 26th International Metrology Conference revised the definition of the second, considering the transition frequency of the undisturbed ground state hyperfine transition of ^{133}Cs atoms as a constant $\Delta\nu_{Cs} = 9{,}192{,}631{,}770$ Hz, and converting this frequency into time, i.e., $1\ s = 9{,}192{,}631{,}770/\Delta\nu_{Cs}$; the revised definition has been effective since May 20, 2019 [5]. Devices that use quantum level transitions to generate accurate and stable time frequencies are called atomic clocks, which are widely used in frontier scientific exploration [6, 7], timekeeping [8], navigation and positioning [9] and other fields. With the continuous development of atomic physics, atomic vapor cells, high-performance microwave sources, laser cooling and trapping, new lasers and frequency stabilization technologies, and new atomic clock principle schemes, the performance of atomic clocks is improving. Currently, the ion optical clock and the neutral atom optical lattice clock have achieved the best systematic uncertainty of 10^{-18} and even 10^{-19}, internationally [10, 11]. For example, to accurately detect the narrow linewidth of quantum transitions, when a diode laser is used as the local oscillator light source of a passive optical frequency atomic clock, it is required to narrow the laser linewidth as much as possible. Therefore, the local oscillator laser linewidth directly determines the width of the atomic clock transition spectrum, which is crucial to the frequency stability and overall performance of the clock.

The external cavity diode laser (ECDL) has a narrow linewidth, is tunable, and has high frequency stability. It can meet the requirements of precision measurement fields for light sources and is one of the ideal laser sources for new atomic clocks, atomic magnetometers [12], atomic gravimeters [13, 14] and other research and applications. In addition to basic research such as quantum frequency standards [15], laser cooling [16] and high-resolution spectroscopy [17], narrow linewidth, tunable external cavity diode lasers are also widely used in optical storage [18], optical communication [19], laser sensing [20, 21], laser radar [22] and other fields.

The linewidth of the diode laser can be represented by the revised Schawlow–Townes formula [23]:

$$\Delta\nu_{\mathrm{LD}} = \frac{2\pi h\nu_0\left(\Delta\nu_{1/2}\right)^2\mu}{P}\left(1+\alpha^2\right) \tag{1.1}$$

where, P is the output power, $\mu \equiv N_2/(N_2 - N_1)$ characterizes the particle number inversion, N_1 and N_2 represent the particle numbers of the lower and upper energy levels, respectively, $\Delta v_{1/2}$ is the cavity mode linewidth of the resonant cavity, α is the Henry coupling parameter, proposed by Henry in 1982 [24], $(1 + \alpha^2)$ represents the Schawlow–Townes linewidth correction factor caused by spontaneous radiation.

In 1980, Lang and Fleming and others confirmed that external cavity feedback has the effect of narrowing the laser linewidth [25, 26], the method of using external cavity feedback limits the frequency band of longitudinal mode selection within the gain spectrum, suppresses spontaneous radiation noise, reduces the emission spectrum width, and effectively narrows the laser linewidth. The laser linewidth output by the external cavity diode laser can be represented as [27]:

$$\Delta\nu = \frac{\Delta\nu_{\mathrm{LD}}}{[1 + (L_D/nL_{\mathrm{LD}})]} \tag{1.2}$$

where, n is the refractive index of the light-emitting part of the laser diode, L_{D} is the external cavity length, L_{LD} is the length of the laser diode, and $\Delta\nu_{\mathrm{LD}}$ is the laser linewidth of the ordinary diode laser in formula (1.1). The above formula shows that the linewidth of the diode laser can be effectively narrowed by using an external cavity. However, at the same time, the increase in the length of the resonant cavity caused by the external cavity structure will reduce the interval between adjacent longitudinal modes, which is far less than the gain bandwidth of the laser diode, making it difficult to achieve stable single longitudinal mode output. Therefore, the external cavity laser needs to be equipped with an external cavity mode selector for mode selection [28]. Depending on the external cavity device, external cavity diode lasers are usually divided into grating-type, interference-filter-type, waveguide-type, and Faraday lasers. The first three types of lasers use macroscopic device characteristics for frequency selection, while Faraday lasers use atomic filtering effects for frequency selection. Further, we provide a detailed discussion of the structure, principle, research status, and application in quantum precision measurement of various types of external cavity diode lasers in upcoming sections of this chapter.

1. Grating type external cavity diode laser

The grating type external cavity diode laser usually uses a diffraction grating as the external cavity frequency selection device to achieve wavelength tuning and linewidth narrowing. The diffraction grating is composed of a medium with a periodically changing thickness and a constant refractive index, or a medium with a periodically changing refractive index and a constant thickness. Its wavelength selectivity can be expressed by the diffraction formula:

$$m\lambda = a(\sin\theta_i + \sin\theta_d) \tag{1.3}$$

where, λ is the output wavelength, m is the diffraction order, a is the diffraction constant, θ_i and θ_d are the incident angle and diffraction angle, respectively. Common diffraction gratings include reflection gratings, transmission gratings, and blaze gratings. In addition, with the development of micro-integration technology, volume holographic gratings (VHG) are also used as external cavity feedback elements in external cavity diode lasers. Due to their small size, narrow spectral response, small spatial acceptance angle, high diffraction efficiency, wide working band, and excellent thermal stability, they are of great significance in high-precision miniature lasers.

Grating-type external cavity diode lasers are primarily categorized into Littrow and Littman configurations, which use different structures to achieve laser frequency selection. In the Littrow external cavity diode laser, the diffraction grating serves as both the frequency selector and feedback device, as shown in Fig. 1.1. The light emitted from the laser diode overlaps with the first-order diffracted light after passing through the diffraction grating, forming feedback in the cavity, and the zero-order diffracted light serves as the output light. By adjusting the angle of the diffraction grating, the wavelength of the output laser can be adjusted, and the direction of the laser output is controlled. However, when selecting the wavelength by adjusting the grating angle, the direction of the output laser will also change, which cannot meet the practical application needs of certain specific fields. To overcome this problem, the structure is further improved to get the laser output from the back end of the diode [29], as shown in Fig. 1.2.

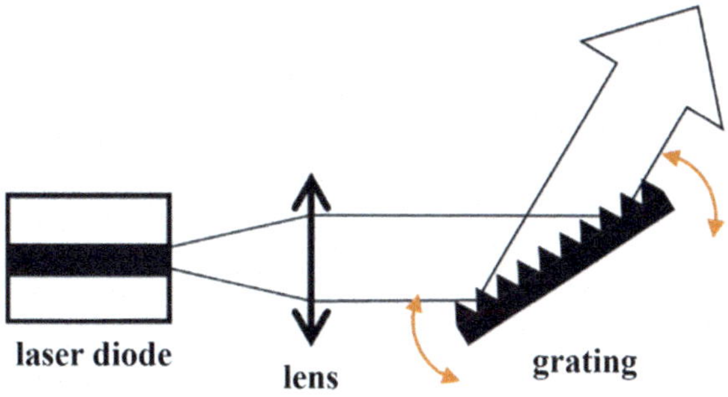

Fig. 1.1 Littrow type external cavity diode laser

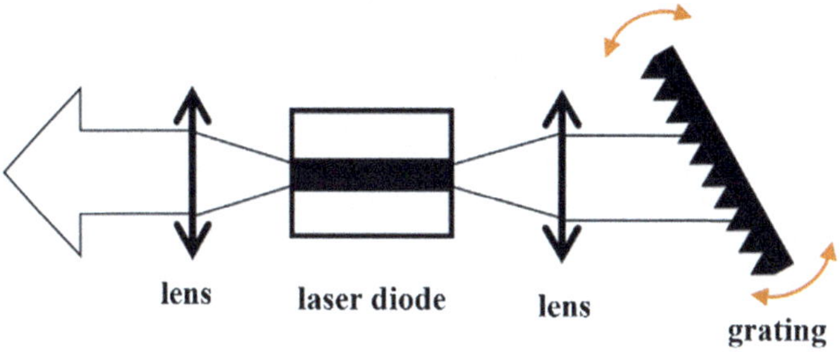

Fig. 1.2 Improved Littrow-type external cavity diode laser

The Littman diode laser adopts a folded cavity structure. Based on the Littrow external cavity diode laser, a mirror is added. The diffraction grating only serves as a frequency selection device, and the laser feedback is realized through the external cavity mirror, as shown in Fig. 1.3. The laser light emitted from the laser diode is collimated and incident on the grating, producing 0th order and + 1st order diffraction. The 0th order light serves as the output light, and the + 1st order light is reflected back to the grating by the mirror. According to formula (1.3), the + 1st order diffracted light of this incident light will return to the laser diode along the original path, forming external cavity feedback. A small rotation of the mirror can change the cavity length and select the feedback wavelength. The advantage of this structure is that after two diffraction frequency selections of the grating, better mode selection characteristics can be obtained; the grating remains fixed, and the direction of the laser output will not change during the process of adjusting the output laser wavelength. However, the introduction of the mirror increases the complexity of the system. In addition, secondary diffraction will increase the loss in the cavity, so a high reflectivity grating is needed.

(1) *Development history and current status of Littrow-type external cavity diode lasers*

In the 1960s, the Littrow configuration was proposed for laser wavelength selection [30]. After improvements and development, the currently implemented Littrow-type

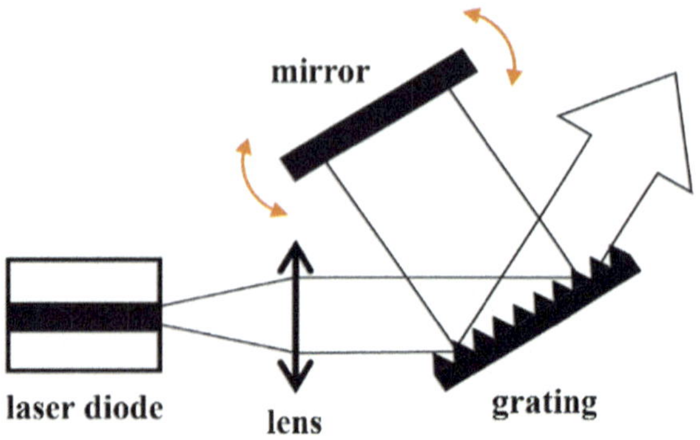

Fig. 1.3 Littman-type external cavity diode laser

external cavity diode laser has achieved kHz level narrow linewidth and a tuning range exceeding 100 nm.

In 2019, Chen and colleagues from National Cheng Kung University in Taiwan reported a 445 nm blue InGaN diode laser based on the traditional Littrow structure [31]. By adjusting the holographic grating with a piezoelectric ceramic transducer (PZT), the laser achieved a wavelength tuning range of over 4 nm, a maximum output power of 20 mW, and a single longitudinal mode laser output with a linewidth of 4.7 MHz.

To overcome the interference of ridge waveguide back reflection and intracavity modes and to achieve more stable laser output, researchers proposed the use of a single-angled-facet (SAF) gain chip to replace the traditional gain chip. In 2016, Shin and colleagues from the Australian National University implemented a 1080 nm single-mode external cavity diode laser using a single-angled facet [32]. The wavelength tuning range of this external-cavity diode laser exceeds 100 nm. In free-running mode, it achieves a Gaussian-fitted linewidth of 22 kHz within an integration time of 22.5 ms, a Lorentzian-fitted linewidth of 4.2 kHz, and demonstrates long-term frequency stability better than 40 kHz@11 h. In 2020, Kapasi and others reported a 2 μm narrow-band linewidth tunable laser for gravitational wave detection [33]. Using a commercial single-angled facet gain chip and a diffraction grating to form a Littrow type external cavity structure, the grating is controlled by a piezoelectric ceramic, and the linewidth reached 20 kHz within an integration time of 10 ms. The usable laser output power exceeded 9 mW, up to 15 mW, and the wavelength tuning range was 120 nm.

(2) *Development history and current status of Littman type external cavity diode laser*

In 1978, Littman and Metcalf first proposed the Littman–Metcalf external cavity structure (abbreviated as Littman type) and applied it to dye lasers [34]. In 1994, Fuh and colleagues applied this structure to external cavity diode lasers[35]. This structure has now become one of the classic configurations for grating-based external cavity diode lasers.

In 2018, Torcheboeuf and colleagues reported a 2.2 μm working wavelength Micro-electro-mechanical systems (MEMS) tunable Littman type external cavity diode laser [36].

The system consists of a laser diode with an anti-reflection coating, a collimating lens, a diffraction grating, and a MEMS mirror. The tunable range of the output laser is 222 nm, the power within this range is 8 ~ 24 mW, the side mode suppression ratio is more than 50 dB, and the output laser frequency tuning range is ~ 18 GHz.

In 2019, Hoppe and colleagues from the German company Sacher Lasertechnik proposed a new type of Littman-type external cavity diode laser based on MEMS [37]. The system mainly consists of a gain chip with a curved waveguide, a collimating lens, a MEMS mirror, and a reflection grating. After the laser forms oscillations in the cavity, it is coupled out from the back of the semiconductor gain chip. Using the MEMS rotating mirror can simultaneously change the cavity length and angle, and the wavelength can be tuned without mode hopping, with a tuning speed up to the kHz level. Based on this, in 2021, Hoppe and colleagues reported a MEMS-type external cavity diode laser covering a wavelength of 1980 ~ 2090 nm [38]. Using a GaSb-based laser diode with a curved waveguide and a resonant-driven MEMS driver, a fast-tuning, mode-hop-free external cavity diode laser was realized. Under different center wavelengths, it has a mode-hop-free frequency range from 8.4 to 34 GHz.

(3) *Development history and current status of micro-integrated external cavity diode laser*

Micro-integrated external cavity diode lasers usually use volume holographic gratings as external cavity frequency selectors. All micro-components are integrated on a single substrate, with a compact structure, small size, and insensitivity to mechanical vibrations. A miniature thermo-electric cooler (TEC) is used for precise temperature control, making it more adaptable to the environment [39]. It is expected to achieve narrow linewidth lasers with small volume, high stability, and high reliability. Volume holographic gratings have the advantages of small size, narrow spectral response, small spatial acceptance angle, and high diffraction efficiency, and are gradually being applied to miniature high-stability lasers.

In 2012, Kürbis and colleagues from the Ferdinand Braun Institute in Germany proposed the first generation of micro-integrated external cavity diode lasers [40], used for space Rb atom interference measurement experiments. This external cavity diode laser consists of a ridge waveguide gain chip, a collimating lens, and a volume holographic grating. All of these components are mounted on the same substrate. The laser output power can reach up to 120 mW, with a linewidth of less than 100 kHz at a 10 μs integration time, and a continuous tuning range of several GHz or more. Under a working environment of 28.5 °C and 164 mA, the short-term linewidth is 60 kHz, and the intrinsic linewidth is 3.6 kHz. Subsequently, the group further optimized the temperature control module and circuit module of this external cavity diode laser. By 2014, they reported the third generation of micro-integrated external cavity diode lasers [41], featuring an output power of over 35 mW, a short-term linewidth of less than 47 kHz, an intrinsic linewidth of 3 kHz, a continuously tunable range of over 27.5 GHz, and excellent mechanical stability and reliability. In 2020, the team developed a micro-integrated narrow linewidth external cavity diode laser [42], which uses the master oscillator power amplifier mechanism to amplify the optical power. The output center wavelength is 1064.490 nm, the optical power is 570 mW, and the 1 ms integrated linewidth is 13 kHz. This external cavity diode laser passed the $8.8g_{RMS}$ vibration testing and was used for iodine spectral analysis in space exploration rockets.

In summary, the grating-type external cavity diode laser uses a diffraction grating as an external cavity frequency selection device, which can simultaneously achieve frequency selection and optical feedback functions. The laser linewidth can be narrowed to several tens kHz level, and it can achieve a tuning range of 100 nm or even sub-micron level in the output laser wavelength micron band. However, it is difficult to further narrow the linewidth of the grating-type external cavity diode laser, and the mode and power of its output laser are sensitive to environmental noise, mechanical deformation, and temperature fluctuations. With the development of photonic integration and MEMS technology, realizing smaller volume, lower noise, and more stable external cavity diode lasers has become a future development trend.

2. Interference filter type external cavity diode laser

Traditional Littrow and Littman type external cavity diode lasers use diffraction gratings for wavelength selection, requiring precise fixation of the grating angle and the length of the external cavity, so the system is extremely sensitive to temperature and mechanical vibration. Narrowband filters provide another scheme for wavelength selection, as they have sufficiently narrow bandwidth (~ 0.3 nm) [43], which can ensure single-mode operation of the laser. In addition, this external cavity structure can use a cat's eye reflector, which can avoid instability caused by non-collimation and beam offset of light, thereby improving the mechanical robustness of the system. Commonly used narrowband filters include interference filter (IF), Fabry–Pérot (F-P) etalon, atomic filters, etc., among which interference filters and Fabry–Pérot etalons use multi-beam interference effects for frequency selection.

(1) *Interference filter type external cavity diode laser*

Interference filters are thin film filters that allow only a specific spectral range of light to pass through based on the principle of multi-beam interference, with the advantages of high transmittance and narrow bandwidth. The wavelengths of its transmitted and reflected light will change with the change of the direction of the incident light, and the relationship between the output wavelength of the laser $\lambda(\theta)$ and the incident light λ_0 and the angle θ of the interference filter is [43]

$$\lambda(\theta) = \lambda_0\sqrt{1 - \sin^2\theta / n_{eff}^2} \tag{1.4}$$

where, n_{eff} represents the effective refractive index. As can be seen from the above formula, by finely adjusting the angle between the incident light and the interference filter, the continuous tuning of the output laser wavelength can be achieved. This

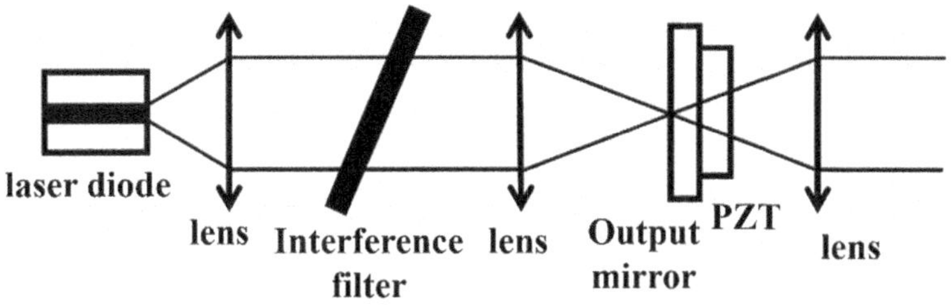

Fig. 1.4 Schematic diagram of the structure of the interference filter type external cavity diode laser

structure has a smaller angle sensitivity than the grating type, so it has better anti-vibration ability.

The structure of the interference filter type external cavity diode laser [43, 44] is shown in Fig. 1.4. The light emitted by the laser diode passes through the collimating lens and then through the interference filter, and then is reflected by the cat's eye mirror to form light feedback. By adjusting the angle of the interference filter, the laser output wavelength can be changed. Compared with the Littrow or Littman structure, the combination of the interference filter and the cat's eye mirror offers easier laser frequency adjustment and optimized light feedback. This design provides better stability and features both narrow linewidth and tunable characteristics.

In 2006, Baillard from the Paris Observatory in France used an interference filter as a frequency selection component to build an external cavity diode laser [43], and realized prototypes of different wavelengths in the visible light (698 nm) and near-infrared (852 nm) bands, with a linewidth as low as 14 kHz and a tuning range of more than 20 nm. This laser is used for atomic physics experiments such as laser cooling of Cs, Rb, and Sr.

In 2011, Schmidt and colleagues built a compact interference filter type external cavity diode laser [45], the laser cavity length is 80 mm, the output power is up to 50 mW, the intrinsic linewidth is less than 10 kHz, and a mode-hop-free range of about 9 GHz is achieved by controlling the gain ratio, which can cover ^{87}Rb atom D_2 line. Using this laser as a light source, the team built a laser system for high-precision atomic interference measurement experiments. In the same year, Wang and colleagues from Tsinghua University and Peking University realized an interference filter type external cavity diode laser with an output wavelength of 657 nm [46], the instantaneous linewidth of the Lorenz fit reached 7 kHz.

In 2012, Thompson and colleagues from the University of Melbourne reported a single-mode external cavity diode laser that uses a broadband (3 nm) interference filter for frequency selection [47]. The wavelength tuning range of this laser exceeds 14 nm, with a laser linewidth of 26 kHz. By employing a cat's eye structure as the reflector, the frequency noise and sensitivity to vibrations are significantly reduced.

In 2020, the research group of Zhang at the National Time Service Center developed a compact and highly stable interference filter type external cavity diode laser [48], with an output wavelength of 698 nm, specifically for space strontium atomic clocks. The laser linewidth is about 180 kHz, the output laser power is 35 mW, and it has a current control tuning range of over 40 GHz and a piezoelectric ceramic control tuning range of 3 GHz. This external cavity diode laser has been applied in the prototype of the space ultra-stable laser system.

(2) *Fabry–Pérot standard external cavity diode laser*

Fabry–Pérot standard frequency selection typically involves two forms: one places a tilted Fabry–Pérot standard in the cavity, and the other uses the standard formed by two reflectors directly as the resonant cavity. Due to the multi-beam interference effect, the Fabry–Pérot standard has selective transmittance to light of different frequencies in the cavity. Only light of certain specific frequencies can pass through the standard and oscillate back and forth in the cavity, while light of other frequencies cannot oscillate due to excessive loss. Using a high Q-value Fabry–Pérot etalon for frequency selection can achieve a very narrow laser linewidth. With the development of technology, the Q value of the Fabry–Pérot cavity is getting higher and higher, and the linewidth narrowing effect is becoming more and more significant.

In 1975, Voumard and colleagues from the University of Bern in Switzerland reported a scheme using a Fabry–Pérot standard as a laser frequency selection device [49]. The team used a diode laser without an antireflection coating combined with a Fabry–Pérot etalon to realize an external cavity diode laser. When using a Fabry–Pérot standard with a finesse of 20, the spectral linewidth is effectively narrowed to 0.04 nm without changing the total output power.

In 2014, Aoyama and colleagues from Tohoku University in Japan reported an external cavity diode laser realized by a coherent optical negative feedback system [50], as shown in Fig. 1.5. The system is composed of only one lens and one Fabry–Pérot etalon, and the output linewidth is narrowed from 6.4 MHz to 6.5 kHz. They also proved through experiments that this coherent optical negative feedback method can effectively reduce the linewidth and frequency modulation (FM) noise without affecting the relative intensity noise (RIN).

In 2021, Hao and colleagues from Nanjing University reported a self-injection locked diode laser utilizing a high-quality factor fiber Fabry–Pérot (FFP) resonator [51]. A distributed feedback laser is used as the light source, and the Q value of this

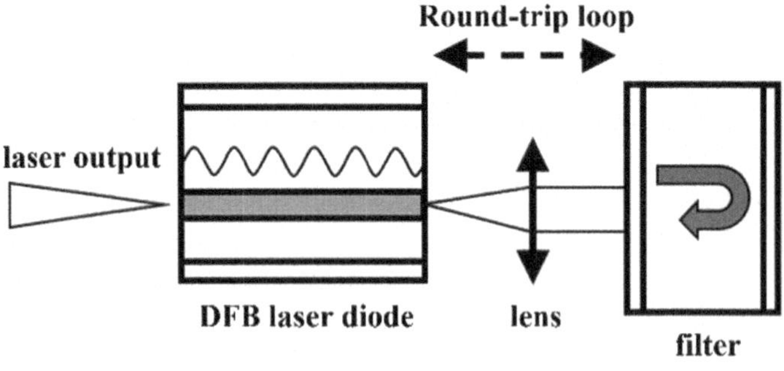

Fig. 1.5 Schematic diagram of the FP-external cavity diode laser structure [47]

fiber Fabry–Pérot resonator is 3.95×10^7, which is plug-and-play in the all-fiber feedback loop. The free-running linewidth of the laser is 145 Hz, and the white noise is 50 Hz^2/Hz, which is more than 42 dB lower than the unlocked distributed feedback laser.

In 2023, Liang and colleagues from the Suzhou Institute of Nanotechnology and Nanobionics, Chinese Academy of Sciences, achieved a small narrow linewidth laser with a volume of 67 mL by self-injecting the distributed feedback laser into a high Q value (7.7×10^8) micro Fabry–Pérot cavity with a volume of 0.5 mL [52]. The Lorentz linewidth reached 60 mHz, and the integrated linewidth was about 80 Hz. The high Q value micro Fabry–Pérot cavity is made by connecting a plane mirror and a concave mirror to a 7 mm-long hollow fused silica pad. The frequency noise performance of this laser exceeds that of commercial narrow linewidth lasers and self-injection locked integrated narrow linewidth lasers, and can be mass-produced and used in outdoor laboratory environments in the future.

In terms of narrowing the linewidth of lasers using external cavity technology, the team of Academician Tianchu Li has carried out a series of work. In 2011, they proposed using a monolithic folded Fabry–Pérot cavity (MFC) made of quartz glass to achieve optical feedback for a 689 nm external cavity diode laser [53]. The laser is based on the grazing diffraction configuration, and the external cavity frequency is locked to the resonance frequency of the folded Fabry–Pérot standard by a low-bandwidth Hänsch–Couillaud servo loop, and leading to the locked linewidth is about 6.8 kHz. The folded Fabry–Pérot etalon external cavity diode laser is locked to the reference laser using a phase-locked loop, and phase locking is achieved by feeding back to the folded Fabry–Pérot standard through a piezoelectric ceramic. The minimum unit gain frequency of the open-loop servo loop is 10.2 kHz. This optical feedback folded Fabry–Pérot etalon external cavity diode laser has the advantages of compact structure, better open-loop frequency and power stability, better low-frequency noise and narrow linewidth. In 2012, they proposed a narrow linewidth external cavity diode laser that achieves external optical feedback through a coupled cavity structure [54], using a high-precision (10,000) double-mirror non-confocal Fabry–Pérot standard, by adjusting the piezoelectric ceramic voltage to control the laser frequency of the external cavity diode laser to be consistent with the resonance frequency of the Fabry–Pérot standard, and to achieve feedback phase matching. The laser linewidth is narrowed to 100 Hz, and the instantaneous linewidth is reduced to 30 Hz. In the frequency range of 10 Hz–10 kHz, the laser phase noise is significantly reduced by more than 50 dB. In the same year, the research team used high-precision Fabry–Pérot standard external optical feedback technology to achieve the first hertz-level relative linewidth distributed feedback laser, and the laser phase noise was significantly suppressed in the entire measurement frequency range [55]. Specifically, at the Fourier frequency of 17 kHz, which is approximately the linewidth of the Fabry–Pérot reference cavity, the laser phase noise was significantly suppressed by more than 92 dB. In the Lorentz fit, the linewidth of a single laser was reduced from 7 MHz to 4.4 Hz. By using two identical lasers for beat frequency, an instantaneous linewidth of 220 mHz was achieved.

In addition, the team led by Academician Tianchu Li proposed a general analytical form of round-trip phase shift in grating feedback diode lasers in 2011 [56]. With this new form, when only one constraint is satisfied, the round-trip phase shift can be independent of the rotation angle under a first-order approximation. This condition is referred to as the Quasi Synchronous Tuning (QST) condition. In the QST region, a larger mode-hop-free tuning range can be obtained. The quasi-synchronous pivot is not strictly limited by the grating surface and its epitaxy, and only a one-degree-of-freedom adjustment structure is needed to accurately find and locate the quasi-synchronous pivot, making the design of the external cavity diode laser simpler and the output of the laser more stable and reliable. In 2014, they experimentally verified the QST theory [57]. Through QST, a Littman–Metcalf type external cavity diode laser was realized, with a mode-hop-free range of over 2 THz (6 nm), where the pivot point is 65 mm away from the traditional strictly synchronous rotation center. The accuracy of achieving a mode-hop-free tuning range of over 1 THz has been relaxed to more than 300 μm. Due to the simplification of the adjustment structure, compared with the traditional Rigorous Synchronous Tuning (RST), the design of the external cavity diode laser can be made easier, and the laser can be made more stable and reliable.

In summary, the interference filter type external cavity diode laser uses the principle of multi-beam interference for filtering, which is more stable than the grating type external cavity diode laser, and has a linewidth narrowing to kHz. In the future, while ensuring narrow linewidth performance, a larger tunable range can be achieved with the help of QST theory; Additionally, the use of integrated technology will facilitate the development of more compact and stable laser systems.

3. Waveguide type external cavity diode laser

With the development of photonic integration technology, the waveguide external cavity structure has become another optional scheme besides the discrete external cavity and the combined external cavity. While achieving narrow linewidth with the use of waveguide external cavity, it greatly reduces the size and volume of the system and can be integrated with other components, improving the stability and reliability of the system, and has become a research hotspot in recent years. The following mainly introduces the external cavity diode lasers based on Fiber Bragg Grating (FBG) and silicon-based waveguide structures.

(1) *Fiber Bragg grating type*

The structure of the Fiber Bragg Grating type external cavity diode laser is shown in Fig. 1.6. One end of the gain chip is coated with a high reflectivity (HR) film, and the other end is coated with an anti-reflectivity (AR) film. One end of the Fiber Bragg Grating is directly coupled to the anti-reflectivity film surface of the gain chip through a tapered fiber lens, and the other end serves as the end reflector of the external cavity. The laser wavelength is determined by selecting the appropriate Fiber Bragg Grating to determine its Bragg wavelength.

In 2016, Lynch and colleagues from the University of Southampton in the UK reported an external cavity diode laser based on integrated optical fiber (IOF) [58].

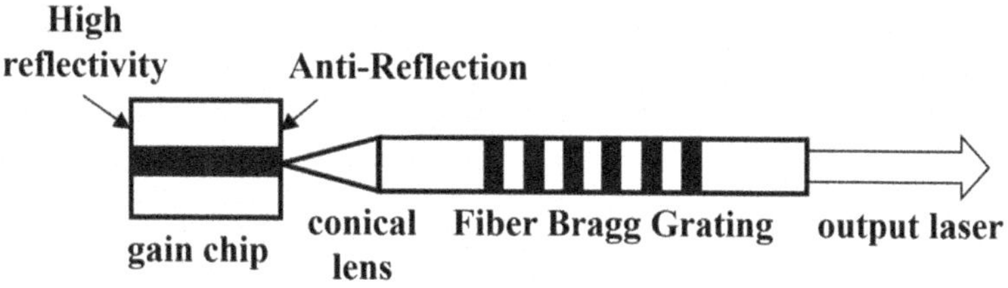

Fig. 1.6 Schematic diagram of the structure of a fiber Bragg grating type external cavity diode laser

They fused photosensitive fiber onto a silicon substrate, used a direct ultraviolet writing process to write Bragg gratings into the fiber, and docked this device with a semiconductor gain chip to form a laser cavity. The realized that external cavity diode laser operates in single mode on the acetylene P13 line (1532.83 nm), with an output power of 9 mW and a linewidth less than 14 kHz, exhibiting better relative intensity noise characteristics as compared to commercial external cavity diode lasers. By integrating planar glass waveguides onto silicon wafers, this laser achieved good mechanical and thermal stability characteristics, better photosensitivity, lower loss, and higher waveguide structure complexity and precision.

In 2016, Wei and colleagues from the Shanghai Institute of Optics and Precision Mechanics, Chinese Academy of Sciences, reported a 1550 nm diode laser with a linewidth of sub-kHz, combining the advantages of long external cavity and high Q-value Fabry–Pérot cavity optical feedback [59]. They locked a commercial distributed feedback laser onto the resonant transmission peak of the fiber Bragg grating type Fabry–Pérot cavity through a 4 m polarization-maintaining fiber loop, with a cavity Q-factor of 6.5×10^6, and the linewidth narrowed from 1 MHz to 1 kHz. Frequency noise of the laser above 1 kHz is as low as 40 Hz^2/Hz, and it provides quasi-continuous tunability over 0.8 nm.

In 2017, Zhang and colleagues from the Shanghai Institute of Optics and Precision Mechanics, Chinese Academy of Sciences, designed a mode-hop-free external cavity diode laser using a fiber Bragg grating with enhanced thermal sensitivity [60]. They used a gain diode with an antireflection coating and angled facet as the active part, packaged the fiber Bragg grating into a glue-filled V-groove, and fixed the gain diode and fiber Bragg grating on an aluminum nitride substrate, ensuring good thermal conductivity for the thermo-electric cooler (TEC). This laser has a narrow linewidth of 35 kHz and a high linear thermal tuning speed of 65 pm/°C (8.125 GHz/°C), with a continuous mode-hop-free range of 0.5 nm, which is 5 times the spacing between adjacent cavity modes.

In 2022, the team further improved the structure of the external cavity diode laser module [61], and integrated the external cavity into a 14-pin butterfly package, using a polarization-maintaining all-fiber ring resonator to achieve frequency-stable transfer technology. The intrinsic Lorentz linewidth of the output laser is 15 kHz, and the relative frequency stability reaches the 10^{-12} order of magnitude, which can be used as a CO_2 detection lidar light source.

In 2019, Huang and colleagues from the University of California, Santa Barbara, reported an extended distributed feedback (E-DBR) laser fully integrated on Si [62], integrating III–V gain materials and ultra-low loss Si waveguides on the same chip, and adding a 15 mm long low-loss, weakly perturbed Bragg grating reflector in the waveguide. The laser linewidth is about 1 kHz, with an output power exceeding 37 mW. A high Q-value ring is added to the cavity, realizing a high Q-value ring-assisted distributed feedback laser (Ring-assisted E-DBR, RAE-DBR) with a linewidth less than 500 Hz.

In summary, compared to traditional Bragg reflectors, fiber Bragg grating structures provide a longer cavity length, thus having a better linewidth narrowing effect, capable of reaching a few kHz or even less than 1 kHz. However, its tuning methods are limited, and the absorption loss of the material is relatively large, so its application range is somewhat restricted.

(2) *Silicon-based waveguide type*

Silicon-based waveguide external cavity diode lasers are devices that integrate active gain and passive waveguides, where the waveguide serves as an external resonant cavity. Its three key technologies are [63]: ① Using low-loss waveguide materials, including silicon oxynitride (SiON), silicon dioxide (SiO_2), silicon nitride (Si_3N_4), silicon on insulator (SOI), etc.; ② Narrowing the linewidth by increasing the cavity length, currently mainly using micro-ring resonators (MRR) and spiral waveguides to increase the effective optical path; ③ Implementing effective schemes for integrating active gain and passive cavities.

The performance of traditional fiber Bragg grating external cavity diode lasers is affected by the sensitivity of the fiber grating to vibration. In 2009, the American RIO company invented a 1550 nm external cavity diode laser [64], which etched a narrow-band Bragg grating on a silicon-based silicon dioxide planar waveguide, forming a cavity together with a semiconductor gain diode, and integrated into a 14-pin butterfly package. The laser linewidth is less than or equal to 2.6 kHz, the phase/frequency noise is comparable to that of long-cavity fiber lasers, the relative intensity noise is less than or equal to – 147 dBc/Hz at a frequency of 1 kHz, and the power is greater than 10 mW. This planar external cavity diode laser has high integration, strong shock resistance, and can work under vibration and harsh environmental conditions.

In 2018, Li from Tsinghua University and colleagues reported a new type of on-chip diode laser [65], which coupled a commercial Fabry–Pérot diode laser with an external high Q-value (2×10^5) micro-resonator, using its introduced Vernier effect and optical self-injection effect to achieve linewidth narrowing. The laser has a wide wavelength tuning range of 17 nm and a narrow linewidth of 8 kHz. In addition, the on-chip output power of this laser is 7 dBm, the typical side-mode suppression ratio (SMSR) is greater than 45 dB, and it is a new scheme to realize a narrow linewidth, wide tunable range on-chip diode laser.

In 2020, Fan from the University of Twente in the Netherlands reported a hybrid integrated InP-Si_3N_4 tunable wideband external cavity diode laser with an intrinsic linewidth of 40 Hz [66]. By using a wide bandgap Si_3N_4 waveguide circuit, the impact of two-photon absorption in narrow bandgap Si waveguides can be overcome. The laser includes an InP gain section and a low-loss Si_3N_4 waveguide feedback circuit to extend the cavity length. The extension of the optical cavity length is achieved through a 33 mm long spiral, and the increase in optical length is achieved through the resonant excitation of three cascaded micro-ring resonators in the cavity, ultimately extending the cavity length on the chip to 0.5 m. The laser's single-mode oscillation achieves a maximum fiber-coupled output power of 23 mW, with a spectral coverage exceeding 70 nm at a wavelength of 1.55 μm, and a side-mode suppression ratio greater than 60 dB. Since the loss of SiN waveguides is several orders of magnitude lower than that of the most advanced Si waveguides, it is an ideal choice to build a fully integrated low-noise photonic platform based on SiN. High-power, low-noise diode lasers are crucial in many applications such as coherent communication, laser radar, and remote sensing. Fully integrated lasers on SiN eliminate the need for free space or fiber coupling to SiN waveguides, improving the scalability and stability of the device.

In 2021, the University of California, Santa Barbara, and the California Institute of Technology's, Xiang and colleagues achieved a high-performance laser integrated on a silicon nitride (Si_3N_4) waveguide using a multi-layer hetero-integration method [67], with a power output of tens of mW through the waveguide and a basic linewidth of sub-kHz. Related work also shows that coupling this laser with a high-Q SiN ring resonator with ultra-low loss can achieve a hertz-level intrinsic linewidth. The same year, the team reported a hybrid integrated laser [68]. The traditional distributed feedback laser is self-injection locked into a Si_3N_4 micro-ring resonator with a high Q factor (over 2.6×10^9) and finesse (over 4.2×10^4), reducing noise by 5 orders of magnitude and achieving a frequency noise of 0.2 $Hz^2\ Hz^{-1}$, with a corresponding linewidth of 1.2 Hz.

In 2023, Zanarella and colleagues from Columbia University in the United States reported a chip-level visible laser platform [69], which can achieve tunable narrow linewidth lasers from near-ultraviolet to near-infrared wavelengths. The team designed a broadband low-loss optical feedback scheme based on a ring resonator, using high-Q micron-level Si_3N_4 resonators and commercial Fabry–Pérot laser diodes to achieve a coarse tuning range of 12.5 nm and a mode-hop-free fine tuning range of 33.9 GHz, with an intrinsic linewidth as low as a few kHz. In addition, the laser also has 267 GHz μs^{-1} micro-adjustment speed, 10 mW fiber-coupled power, and a typical side mode suppression ratio of over 35 dB.

In summary, for waveguide external cavity diode lasers, their greatest advantage lies in high integration, flexible selection of frequency selection components, and the ability to achieve long effective cavity lengths. They have excellent linewidth and noise performance and have good prospects for commercial production and application.

4. Faraday laser

In precision measurement systems such as atomic clocks, atomic magnetometers, and atomic gravimeters, lasers with wavelengths corresponding to atomic transition lines and stable frequencies are essential. Traditional external cavity diode lasers use macroscopic devices such as gratings, interference filters, and Fabry–Pérot etalon for frequency selection. These systems require wavelength meters and adjustments to the current, temperature, and the angle of the frequency selection device to align the wavelength with the atomic spectral line, making them susceptible to mechanical vibrations. By contrast, using a Faraday Anomalous Dispersion Optical Filter (FADOF) with a narrow transmission bandwidth as a frequency selection device offers significant advantages. With this approach, the quantum transition lines of atoms or molecules serve as a frequency reference allowing the output frequency to directly correspond to the atomic wavelength, without the need to adjust the working current, temperature, and cavity length of the laser. Besides using atomic filters for frequency selection can overcome the effects of mechanical vibrations, current fluctuations, and temperature variations on the transmission spectrum, and has better robustness.

Taking alkali metal atoms as an example, FADOF structurally includes two polarizers with orthogonal polarization directions and an atomic vapor cell of alkali metal atoms, as well as a magnetic field parallel to the direction of light propagation and a thermal insulation device. Through the magneto-optical effect of atoms and the selection of the polarization direction of transmitted light by two orthogonal polarizers, atomic filtering is achieved.

As an important indicator of FADOF performance, the transmission spectrum effectively characterizes its properties. The center frequency, transmission bandwidth, and peak transmission of the FADOF transmission spectrum are affected by the magnetic field and temperature, and the design parameters of FADOF can be determined by theoretically calculating the transmission spectrum. At present, theoretical research on the calculation of the FADOF transmission spectrum under small light intensity has advanced significantly [70–72], and the open-source software ElecSus, developed by Durham University in the UK, can calculate the transmission spectrum under small light intensity [73]. Currently, Peking University has developed theory and calculation software for the FADOF transmission spectrum under strong light, details of which are provided in Sect. 4.4 of this book.

In 1956, Öhman first realized the Faraday atomic filter [74], placing a gas medium between two orthogonal polarizers, while applying a strong magnetic field in a parallel direction. This setup, based on the atomic resonance Faraday rotation effect to filter out background light, was applied to the observation of solar spectra. In 1969, Sorokin and colleagues from IBM Watson Laboratory realized Na-FADOF [75], and first applied it to dye laser frequency stabilization, with the FADOF transmission peak corresponding to the sodium atom $3S_{1/2} \rightarrow 3P_{1/2}$ and $3S_{1/2} \rightarrow 3P_{3/2}$ transition.

In 1991, Dick and Shay reported the experimental demonstration results of Rb-FADOF [76], which studied the Rb atom $5^2S_{1/2} \rightarrow 5^2P_{3/2}$ transition. The atomic filter of hyperfine transition, with a measured bandwidth of 1 GHz and a transmission of 63%, has an out-of-band suppression rate of 10^5. In 2011, Miao and colleagues from Peking University achieved a transmission bandwidth of 1.3 GHz for the Rb-FADOF [71, 77], with a central wavelength of 780.241 nm, and applied it to the frequency selection of external cavity diode lasers to realize first Faraday laser, with AR coated laser diode. In 2012, Wang and others from Peking University used the Faraday magneto-optical effect and saturated absorption effect to achieve a FADOF working at the cesium atom $6S_{1/2}, F = 3 \rightarrow 7P_{3/2}, F' = 2, 3$ transition levels, with a bandwidth of 3.9 MHz [78]. In 2022, Yan and colleagues from Shanxi University experimentally demonstrated a dual-wavelength band-pass Faraday filter (DW-FADOF) working on the Rb D_1 and D_2 lines [79]. This FADOF provides a transmission band covering two typical atomic transitions, which can be used as a dual-signal carrier, and has great application potential in wavelength division multiplexing communication and weak signal extraction. In 2023, Guan and colleagues from Beijing University of Posts and Telecommunications, National Institute of Metrology, China, and Peking University used laser cooling technology to realize a FADOF based on ^{87}Rb cold atoms [80], which effectively overcomes the impact of Doppler broadening on the FADOF bandwidth, achieving a 420 nm narrow-band FADOF with a passband of 2.7(2) MHz at a peak transmission of 3.2%, which is the narrowest bandwidth FADOF known internationally. To improve the peak transmission, the team introduced circularly polarized light polarization pumping, achieving a peak transmission of 15.6% and a transmission bandwidth of 6.6(4) MHz for the 780 nm narrow-band FADOF [81].

According to the working mode classification, FADOF can be divided into active and passive types. The difference between them is that in passive atomic filters, the lower energy level of atomic transition is the ground state, while in active atomic filters, it is the excited state. The emergence of active atomic filters has enriched the working wavelength of FADOF and expanded its application range. The working atoms and their corresponding wavelengths of the realized FADOF mainly include the following categories: cesium atoms 455 nm [79, 82], 459 nm [83], 852 nm [84], 894 nm [18], 1470 nm [85], Rb atoms 420 nm [86], 532 nm [87], 776 nm [88], 780 nm [89], 795 nm [90], 1529 nm [19, 91], potassium atoms at 766 nm [92], 770 nm [93], sodium atoms at 589 nm [94], strontium atoms at 461 nm [95] and calcium atoms at 423 nm [96], with a typical transmission bandwidth on the order of GHz. FADOF has characteristics such as high transmission, narrow bandwidth, frequency stability, large field of view, and fast time response. Moreover, most atomic physics experiments require tuning the output wavelength of diode lasers to the atomic transition wavelength, so FADOF can be used as a frequency selector for external cavity diode lasers. As early as the 1960s, people began to use FADOF for laser frequency stabilization. It was found that Faraday filters can be used as frequency selectors for lasers, improving laser performance [97–109]. The research group at Peking University first proposed that a Faraday atomic filter combined with a laser diode coated with an anti-reflection film can be used to create a laser source that is immune to large fluctuations in temperature and current, and named it the Faraday laser. A detailed review can be found in Chap. 2 of this book.

In recent years, Peking University has further stabilized the frequency of the Faraday laser using a modulation transfer spectroscopy stabilization scheme, a key step towards extending the applications of Faraday lasers to areas such as atomic clocks. In 2022, Chang and colleagues at Peking University first achieved a cesium atomic optical frequency standard based on the Faraday laser [110]. The Faraday laser can operate in single-frequency and dual-frequency modes at 852 nm, with a free-running linewidth of 17 kHz. Using modulation transfer spectroscopy (MTS) for stabilization, the self-evaluated frequency stability under single-frequency conditions after stabilization is $3 \times 10^{-14}/\sqrt{\tau}$, and the beat frequency linewidth of the two modes under dual-frequency conditions after stabilization is narrowed to 85 Hz. In the same year, the team used a double-layer atomic cell as a frequency reference to reduce the impact of temperature fluctuations on the quantum reference. Using MTS technology, they achieved higher stability of laser frequency, with short-term self-evaluated stability reaching $5.8 \times 10^{-15}/\sqrt{\tau}$ [111].

Figure 1.7 summarizes the development trend of the linewidth of diode lasers and Faraday lasers using FADOF optical feedback for frequency stabilization. It can be found that using FADOF for external cavity frequency selection can directly narrow the laser linewidth to tens of kHz, without the need for an external locking loop, and it overcomes the influence of the internal cavity mode of the laser diode. It is worth mentioning that after the Faraday laser is stabilized with MTS technology, the linewidth of the laser will be further narrowed. Researchers aim to reduce the linewidth to the Hz level, which would eliminate the need for PDH ultra-stable cavities in specific applications. This advancement would enable simultaneous linewidth narrowing and frequency stabilization using atoms.

Faraday lasers use atomic transition frequencies for frequency selection, which have the advantages of narrow linewidth, stable frequency, strong robustness to current and temperature fluctuations, and a variety of selectable wavelengths. The characteristic of the output wavelength directly corresponding to the atomic spectral line means that the Faraday laser does not have the problem of large wavelength changes, so it has good adaptability to the environment and does not require manual wavelength adjustment. It has unique advantages in special fields such as satellite applications. This high-performance laser can be used in fields such as atomic clocks,

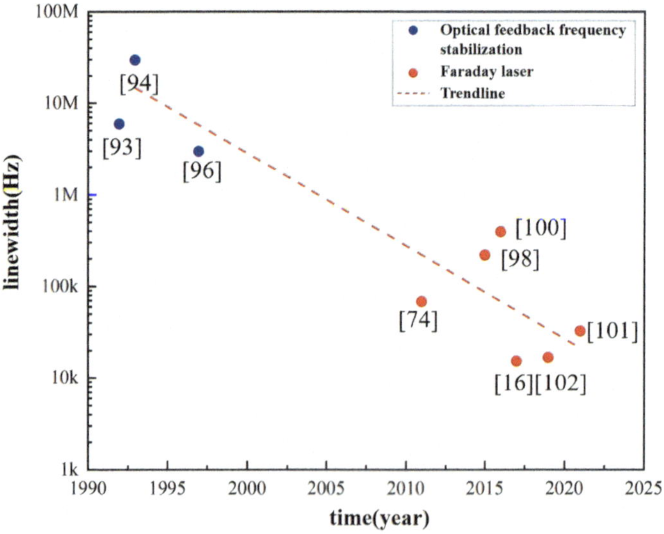

Fig. 1.7 Development trend of FADOF frequency selection laser linewidth

optical communication, laser radar, and navigation timing. In the future, by developing miniaturized FADOF or FADOF inside hollow fibers, and combining with integrated processes, further miniaturization, volume reduction, stability improvement, and application scenario expansion can be achieved.

1.1.2 Development Trend of Diode Lasers

The development trend of diode lasers can be detailed from the following aspects:

1. Technological innovation and material progress

With the progress of material science, new diode laser materials are being developed. These materials can achieve more efficient photoelectric conversion, higher output power, and longer life. For example, using quantum well structures, distributed feedback (DFB) structures, or vertical-cavity surface-emitting lasers (VCSELs), photonic crystal surface-emitting lasers (PCSELs), etc., can improve the performance of the laser.

2. High power and high efficiency

In high-power application fields such as industrial processing and laser weapon systems, the requirements for diode lasers are higher output power and better beam quality. At the same time, in order to improve the overall efficiency of the system and reduce energy consumption, universities, institutes, and companies are focusing on improving designs, using more efficient pump sources, and optimizing thermal conduction to reduce energy loss and obtain lasers with higher electro-optic conversion efficiency.

3. Miniaturization and integration

With the advancement of microelectronics technology, diode lasers are moving towards miniaturization and integration. This is especially important for portable devices and applications with limited space. For example, due to its structural characteristics, VCSEL is easy to achieve high integration and miniaturization, and therefore, has a wide range of application prospects in fields such as near-eye display and biometric recognition. In addition, integration also includes integrating lasers with detectors, optical switches, and other optoelectronic components to form a complete optical communication or sensor system.

4. Wavelength diversity and precise control

Different application fields have varying demands for laser wavelength. With the advancement of laser manufacturing technology, achieving a wide wavelength range and precise wavelength control has become one of the development trends. Through precise design of the laser structure and material selection, lasers can be tailored to output specific wavelengths to meet the needs of specific applications. With the development of wavelength control technology, such as tunable lasers, the wavelength can be adjusted in real time according to demand, increasing the flexibility and application range of lasers. This allows diode lasers to meet more diverse application needs, such as multi-wavelength transmission in optical fiber communication, and different diagnostic and treatment applications in biomedicine.

5. Intelligence and automation

With the advancement of intelligent manufacturing and automation, diode lasers also need to exhibit higher levels of intelligence, such as adaptive modulation, real-time monitoring, and fault self-diagnosis functions. Currently, lasers are gradually being equipped with sensors and controllers to achieve real-time monitoring and self-adjustment to adapt to different operating environments and needs. In terms of automation, the setup and adjustment of lasers can be done through software, reducing the dependence on professionals and improving production efficiency. Such development trends can improve the reliability and ease of use of laser systems and reduce maintenance costs.

6. Cost-effectiveness

With the maturity of production technology and mass production, the cost of diode lasers is gradually decreasing, which allows diode laser technology to be more widely applied in various fields. In addition, improving production efficiency and reducing energy consumption are also important aspects of improving cost-effectiveness.

7. Environment and energy efficiency

Environmental protection and energy saving have become important issues of global concern. The advantages of diode lasers in high energy efficiency and low energy consumption make them have important application potential in the fields of new energy, environmental protection technology, and sustainable development.

8. Expansion of application fields

With the development of technologies such as 5G communication, artificial intelligence, and the Internet of Things, the application fields of diode lasers are constantly expanding, including but not limited to cloud computing, big data, Internet of Things, autonomous driving, biotechnology, medical health, etc. The rapid development of these emerging fields puts forward new technical and performance requirements for diode lasers.

Overall, the development trend of diode lasers is diversified, covering technical innovation, cost control, application expansion, intelligence, and other aspects. With the advancement of technology and market demand, diode lasers will continue to develop towards high performance, low cost, and wide application. As the core light source for many applications, Narrow linewidth external cavity diode lasers, which offer high stability have important research value. In the future, with the continuous development of photonic integration technology, high-performance external cavity diode lasers with narrow linewidth will inevitably tend towards full automation and chipization, thereby stimulating greater application potential and driving the development of diode lasers and downstream applications.

The Faraday laser, which is the focus of this book, is compact, efficient, and reliable, with a wide range of applications in various fields. Due to the inevitable interference from environmental factors such as mechanical vibrations and temperature on macroscopic frequency selection devices, and in applications related to precise measurements such as atomic clocks, it is necessary to precisely control the angle or position of the frequency selection components to achieve laser output corresponding to atomic transition frequencies. Therefore, problems such as easy mode hopping, easy loss of lock, and difficulty in adjustment exist in practical applications. However, the Faraday laser uses the transition spectral lines of atoms in the atomic filter as a frequency reference, and the output laser frequency intrinsically aligns with the atomic transition frequency, which can be used immediately upon startup, and the frequency can be maintained in resonance with the atomic transition spectral line for a long time. Its significant advantages include: narrow linewidth, wavelength corresponding to atomic spectral lines, good robustness to temperature and current fluctuations of laser diode, and unique advantages in related applications. It is expected to become the mainstream technology solution for frequency-stabilized diode lasers in the future.

The currently implemented Faraday lasers are mainly based on passive Rb-FADOF and Cs-FADOF. In the future, Faraday lasers based on other alkali metal atom FADOFs and active FADOFs can be realized to further expand the working band of Faraday lasers and enrich their application scenarios. The transmission spectrum of FADOF is affected by the intensity of the laser light, and the current intensity of the light in the laser cavity cannot be accurately calculated. By conducting theoretical research on the internal light intensity of the laser, the situation of the FADOF transmission spectrum can be determined more accurately determined, thereby, controlling the performance of the laser with high accuracy. In addition, by preparing miniature FADOFs or fiber FADOFs, miniaturized Faraday lasers can be realized, which have important practical value and are currently being broken through. This kind of miniaturized, narrow linewidth Faraday laser is expected to be applied to optical communication, laser cooling, high-resolution spectroscopy, laser sensing, laser radar and other fields in the future, becoming a high-quality laser light source. It is also expected to be quickly applied to benchmark atomic clocks, atomic gravimeters and other large precision measurement devices, reducing system volume, improving system performance, and expanding a wider range of application scenarios.

1.2 Frequency-Stabilized Diode Lasers and Their Applications

1.2.1 Frequency-Stabilized Diode Lasers

As mentioned earlier, the instantaneous frequency fluctuation of a diode laser can be reduced by external cavity feedback, that is, the linewidth of the diode laser can be narrowed. Currently, better external cavity diode lasers can narrow the linewidth of the output laser to 1 MHz or even lower. However, for long-term frequency fluctuations, such as frequency drift caused by mechanical vibrations, operating temperature, operating current, etc., active frequency stabilization must be used to control the output frequency of the laser and achieve better frequency stability.

The basic principle of active frequency stabilization is to lock the frequency of the laser to a stable reference frequency. If the output frequency of the laser is not equal to the reference frequency, an error signal is generated based on the difference between this output frequency and the reference frequency. The error signal is fed back through a feedback circuit to lock the frequency of the laser, usually the error signal is fed back to the current end of the laser and the piezoelectric ceramic. This reference frequency source can be a super-stable passive resonator, such as a high-precision Fabry–Pérot cavity, atomic, molecular or ion transition lines, etc. The current frequency stabilization method based on the F-P cavity reference is mainly PDH frequency stabilization technology. Common frequency stabilization methods based on microscopic particle transition lines include saturated absorption spectroscopy frequency stabilization, Lamb dip frequency stabilization, modulation transfer spectroscopy frequency stabilization, polarization spectroscopy frequency stabilization, wavelength modulation frequency stabilization, two-color laser frequency stabilization, etc. In practical applications, frequency-stabilized diode lasers may need to combine multiple methods to achieve the required frequency stability. The following briefly introduces the common saturated absorption spectroscopy frequency stabilization, modulation transfer spectroscopy frequency stabilization, and PDH frequency stabilization.

1. Saturated absorption spectroscopy frequency stabilization

A strong laser field acting on an absorbing medium will cause saturation of absorption, so "strong light saturates weak light detection" can be used to obtain a saturated absorption spectroscopy signal. By making the frequency of the strong incident laser equal to the weak detection laser, the strong incident laser causes absorption saturation on passing through the absorption cell. Due to the Doppler effect, atoms of different speeds perceive different laser frequencies. When the weaker detection light passes through the absorption atomic vapor cell from the opposite direction, atoms with zero velocity interact with both the pump light and the detection light, reducing the absorption of the detection light. This interaction produces a sharp saturated absorption peak, which can be used to stabilize the frequency of the laser to the corresponding atom or molecule's saturated absorption peak.

The experimental equipment for saturated absorption spectroscopy frequency stabilization is relatively simple and the experimental cost is low, but there will be a Doppler background affecting the experiment, that is, the Doppler background is difficult to be completely eliminate. In addition, saturated absorption spectroscopy frequency stabilization belongs to the internal modulation frequency stabilization of diode lasers, that is, the modulation signal is directly added to the laser, which will introduce additional noise, such as frequency noise and intensity noise, etc. Saturated absorption spectroscopy frequency stabilization can be used for experiments that do not require high accuracy.

2. Modulation transfer spectroscopy frequency stabilization

When two beams of light recombine in the opposite direction and interact with atoms, the unmodulated probe light and the modulated pump light interact with the nonlinear medium together, and the probe light obtains the modulated sidebands. The modulation method for the pump light can be phase modulation and amplitude modulation, among which the modulated phase information can be extracted through phase-sensitive detection, obtaining a high-resolution modulation transfer spectrum.

The probe light, after modulation transfer, also exhibits phase modulation characteristics. However, the intensity of the modulation sidebands of the probe light (i.e., the $\pm$ 1st order sidebands) is related to the detuning of the laser relative to the central frequency of the atomic transition. This relationship with frequency detuning exhibits a distinct absorption line shape, where the intensity of the modulation sidebands is maximized when the laser is in resonance with the atoms. By demodulating the beat signal of the probe light sidebands and the main frequency in a phase-sensitive manner, the modulation transfer spectrum line is obtained. This process, through high-frequency modulation and demodulation, shifts the signal frequency to high frequencies for detection. According to the noise spectrum law, the high-frequency noise is much smaller than that at low frequencies, so extracting the laser frequency information at high frequencies can significantly improve the signal-to-noise ratio of the spectrum line and effectively reduce the impact of low-frequency noise in the loop. At the same time, when the detuning is too large, the intensity of the modulated sidebands of the probe light rapidly decreases, so the modulation transfer spectrum shows the characteristic of no Doppler background. The heterodyne detection method can effectively eliminate the impact of noise that is independent of the laser frequency and can achieve high-sensitivity spectral line detection.

3. PDH locking

In 1946, Pound used a microwave cavity to stabilize the microwave frequency. In 1983, Drever and Hall used the PDH method to lock the laser frequency on the optical cavity, obtaining a laser with a linewidth less than 100 Hz. PDH locking is a reference to this frequency stabilization principle, locking the laser frequency on the Fabry–Pérot resonator. Using PDH locking technology to lock the laser frequency on the Fabry–Pérot cavity can achieve ultra-narrow output linewidth at the Hz or sub-Hz level.

During the PDH locking process, the laser to be stabilized is first phase-modulated, which can use electro-optic modulation, acousto-optic modulation, etc. Taking electro-optic modulation as an example, the laser to be stabilized passes through an electro-optic modulator driven by a radio frequency source, and the phase-modulated laser is incident perpendicularly to the Fabry–Pérot cavity and resonates with the cavity. A quarter-wave plate can be used to convert the reflected light into linearly polarized light, which then enters the photodetector. The light signal can be converted into an electrical signal through the photodetector, and this electrical signal and the radio frequency source signal after phase shifting enter the mixer for mixing. The error signal after mixing is filtered and amplified and input to the servo feedback circuit, and then fed back to the frequency control part of the laser, thereby achieving the locking of the laser frequency on the highly stable Fabry–Pérot cavity.

PDH locking technology plays an important role in high-resolution precision measurement, high-performance cold atomic clocks, ultra-fine spectroscopy research, etc., due to its excellent short-term frequency stability and ultra-narrow output linewidth. Currently, obtaining ultra-narrow linewidth lasers requires the use of ultra-high precision, ultra-low loss optical resonators, which puts high requirements on the material of the cavity and the coating of the cavity mirror. Since the stability of the laser also has extremely strict requirements for vibration isolation and noise, reduction of thermal noise of the optical cavity also needs further research. In general, the conditions for implementing PDH locking technology are very demanding, requiring ultra-high precision optical resonators, even cooling them to low temperature, leading to a relatively complex and expensive experimental system.

In summary, a frequency-stabilized diode laser is a semiconductor device capable of outputting a laser at a specific frequency Its key feature is the ability to maintain the stability of the laser frequency over a long period. This is necessary in many scientific and engineering applications, such as quantum physics experiments, fiber optic communications, laser radar, and precision measurements, all of which require extremely high frequency stability. For diode lasers, frequency drift caused by thermal vibrations, mechanical vibrations, and electronic noise of the laser diode can reach the GHz level, so it is necessary to stabilize the frequency of the laser. The basic principle of frequency stabilization is to lock the frequency of the laser to a very stable reference frequency, and to lock the frequency to a super-stable resonator. Affected by the environment, the resonator itself will also have a certain frequency drift. In order to achieve long-term stability of the laser frequency, it is necessary to directly lock the laser frequency on the quantum transition spectrum line. Frequency stabilization using atomic or molecular spectrum lines is common, including saturated absorption spectroscopy stabilization, modulation transfer spectroscopy stabilization, polarization spectroscopy stabilization, etc. Detailed frequency stabilization schemes can be found in Chap. 6. With the development of technology and changes in application requirements, frequency-stabilized diode lasers will continue to develop and improve, offering greater value to scientific and engineering research.

1.2.2 Applications of Frequency-Stabilized Diode Lasers

Frequency-stabilized diode lasers, due to their high frequency stability and narrow linewidth characteristics, have a wide range of applications in scientific research and industry. In industry and medicine, there are usually high requirements for the instantaneous power of the frequency-stabilized laser to ensure sufficient energy for material processing.

1. Spectral analysis

In spectroscopy, frequency-stabilized lasers can interact with samples to analyze the absorption, emission, or scattering characteristics of light, thereby determining the composition and properties of the sample. For example, measuring the spectrum of laser interaction with cells to assist in the treatment and diagnosis of diseases, structural analysis of molecules and atoms, etc.

2. Quantum sensing

Using the high sensitivity of frequency-stabilized lasers to measure physical quantities, such as magnetic fields, gravity, etc. In the context of today's big data, precise real-time measurement of these physical quantities can establish real-time monitoring and data processing of a physical system, greatly enhancing our exploration of the physical world. For example, gravitational wave detection, real-time measurement of global terrain, etc.

3. Precision measurement

In the field of precision measurement, such as length, time, temperature, etc., frequency-stabilized lasers provide a stable reference frequency source for calibrating and improving the accuracy of measurement equipment.

4. Space–time benchmark

The application of frequency-stabilized lasers to achieve high-precision optical clocks, based on which timekeeping systems, ground benchmark systems, ocean benchmark systems, and deep space benchmark systems are established.

5. Fiber optic communication

In fiber optic communication systems, frequency-stabilized diode lasers are used to provide a stable light source to ensure the accuracy and reliability of data transmission, used for sending and receiving signals.

6. Atomic interferometer

In atomic interferometers, frequency-stabilized lasers are used to precisely control the frequency of the two beams in the interferometer.

In atomic interferometers, frequency-stabilized lasers are used to precisely control the frequency of the two beams in the interferometer, in order to observe the interference phenomenon of atoms, and further study atomic physics and quantum mechanics.

References

1. R.N. Hall, G.E. Fenner, J.D. Kingsley et al., Coherent light emission from GaAs junctions. Phys. Rev. Lett. **9**(9), 366 (1962)
2. T.M. Quist, R.H. Rediker, R.J. Keyes et al., Semiconductor maser of GaAs. Appl. Phys. Lett. **1**(4), 91–92 (1962)
3. M.I. Nathan, W.P. Dumke, G. Burns et al., Stimulated emission of radiation from GaAs p–n junctions. Appl. Phys. Lett. **1**(3), 62–64 (1962)
4. J. Terrien, News from the international bureau of weights and measures. Metrologia **4**(1), 41–45 (1968)
5. M. Stock, R. Davis, E. De Mirandés et al., The revision of the SI-the result of three decades of progress in metrology. *Metrologia* **56**(2), 022001 (2019)
6. A. Derevianko, M. Pospelov, Hunting for topological dark matter with atomic clocks. Nat. Phys. **10**(12), 933–936 (2014)
7. P. Delva, J. Lodewyck, S. Bilicki et al., Test of special relativity using a fiber network of optical clocks. Phys. Rev. Lett. **118**(22) (2017)
8. S.H. Zhao, S.W. Dong, S.S. Bai et al., Research on frequency control algorithm based on the combination of benchmark frequency standard and time-saving frequency standard. Chin. J. Sci. Instrum. **41**(8), 67–75 (2020)
9. L. Maleki, J. Prestage, Applications of clocks and frequency standards: from the routine to tests of fundamental models. Metrologia **42**(3), S145–S153 (2005)
10. S.M. Brewer, J.S. Chen, A.M. Hankin et al., $^{27}Al^+$ quantum-logic clock with a systematic uncertainty below 10^{-18}. Chin. J. Sci. Instrum. Phys. Rev. Lett. **123**(3), 033201 (2019)
11. Q.Z. Zhi, K.J. Arnold, R. Kaewuam et al., $^{176}Lu^+$ clock comparison at the 10^{-18} level via correlation spectroscopy. Sci. Adv. **9**(18), 1971 (2023)
12. V.I. Yudin, A.V. Taichenachev, Y.O. Dudin et al., Vector magnetometry based on electromagnetically induced transparency in linearly polarized light. Phys. Rev. A At. Mol. Opt. Phys. **82**(3), 033807 (2010)
13. S. Merlet, L. Volodimer, M. Lours et al., A simple laser system for atom interferometry. Appl. Phys. B Lasers Opt. **117**(2), 749–754 (2014)
14. M.J. Lv, G.M. Wang, S.H. Yan et al., Overview of integrated light source system for atomic interferometer gravimeter. J. Electron. Meas. Instrum. **35**(7), 1–10 (2021)
15. W. Zhuang, J. Chen, Active Faraday optical frequency standard. Opt. Lett. **39**(21), 6339–6342 (2014)
16. D. Sesko, C.G. Fan, C.E. Wieman, Production of a cold atomic vapor using diode-laser cooling, in *Collected Papers of Carl Wieman* (2008), pp. 267–269.
17. W. Lenth, Optical heterodyne spectroscopy with frequency- and amplitude-modulated diode lasers. Opt. Lett. **8**(11), 575–577 (1983)
18. S.L. Portalupi, M. Widmann, C. Nawrath et al., Simultaneous Faraday filtering of the Mollow triplet sidebands with the Cs-D_1 clock transition. Nat. Commun. **7**(1), 1–6 (2016)
19. P. Chang, H. Peng, S. Zhang et al., A Faraday laser lasing on Rb 1529 nm transition. Sci. Rep. **7**(1), 1–8 (2017)
20. S.N. Andreev, E.S. Mironchuk, I.V. Nikolaev et al., High precision measurements of the $^{13}CO_2/^{12}CO_2$ isotope ratio at atmospheric pressure in human breath using a 2 μm diode laser. Appl. Phys. B **104**(1), 73–79 (2011)
21. M.W. Kudenov, B. Pantalone, R. Yang, Dual-beam potassium Voigt filter for atomic line imaging. Appl. Opt. **59**(17) (2020)
22. S. Yu, Z. Zhang, M. Li et al., Multi-frequency differential absorption lidar incorporating a comb-referenced scanning laser for gas spectrum analysis. Opt. Express **29**(9), 12984–12995 (2021)
23. A. Yariv, *Optical Electronics in Modern Communications* (House of Electronics Industry, Beijing, 2002)
24. C. Henry, Theory of the linewidth of diode lasers. IEEE J. Quantum Electron. **18**(2), 259–264 (1982)

25. R. Lang, K. Kobayashi, External optical feedback effects on semiconductor injection laser properties. IEEE J. Quantum Electron. **16**(3), 347–355 (1980)
26. M. Fleming, A. Mooradian, Spectral characteristics of external-cavity controlled diode lasers. IEEE J. Quantum Electron. **17**(1), 44–59 (1981)
27. F. Riehle, *Frequency Standards: Basic and Applications* (WILEY-VCH, 2004)
28. G.J. Steckman, W. Liu, R. Platz et al., Volume holographic grating wavelength stabilized laser diodes. IEEE J. Sel. Top. Quantum Electron. **13**(3), 672–678 (2007)
29. M.S. Zediker, S. Stry, L. Hildebrandt et al., Compact tunable diode laser with diffraction-limited 1 Watt for atom cooling and trapping, in *Proceedings of SPIE 5336, High-Power Diode Laser Technology and Applications II* (2004). https://doi.org/10.1117/12.525059
30. T. Hard, Laser wavelength selector and output coupler. IEEE J. Quantum Electron. **5**(6), 321–321 (1969)
31. M.H. Chen, S.C. Hsiao, K.T. Shen et al., Single longitudinal mode external cavity blue InGaN diode laser. Opt. Laser Technol. **116**, 68–71 (2019)
32. D.K. Shin, B.M. Henson, R.I. Khakimov et al., Widely tunable, narrow linewidth external-cavity gain chip laser for spectroscopy between 1.0–1.1 μm. Opt. Express **24**(24), 27403–27414 (2016)
33. D.P. Kapasi, J. Eichholz, T. Mcrae et al., Tunable narrow-linewidth laser at 2 μm wavelength for gravitational wave detector research. Opt. Express **28**(3), 3280–3288 (2020)
34. M.G. Littman, H.J. Metcalf, Spectrally narrow pulsed dye laser without beam expander. Appl. Opt. **17**(14), 2224–2227 (1978)
35. N. Chen, H.C. Chen, C.P. Liang et al., A tunable diode laser, in *IEEE Instrumentation and Measurement Technology Conference*, Hamamatsu, Japan (1994)
36. N. Torcheboeuf, S. Droz, I. Simonyte et al., MEMS tunable Littman–Metcalf diode laser at 2.2 μm for rapid broadband spectroscopy in aqueous solutions, in *2018 IEEE International Diode Laser Conference* (2018)
37. M. Hoppe, H. Rohling, S. Schmidtmann et al., New wide tunable external cavity inter-band cascade laser based on a micro electro-mechanical system device, in *MOEMS and Miniaturized Systems XVIII* (2019). https://doi.org/10.1117/12.2507176
38. M. Hoppe, S. Schmidtmann, C. Aßmann et al., High speed external cavity diode laser concept based on a resonantly driven MEMS scanner for the mid-infrared region. Appl. Opt. **60**(15), C92–C97 (2021)
39. Z.Q. Qi, W.L. Hu, Z.Z. Shen et al., Research progress in narrow linewidth diode lasers with external cavities. Opt. Optoelectron. Technol. **21**(2), 1–10 (2023)
40. C. Kürbis, E. Luvsandamdin, A. Sahm et al., Micro-integrated, narrow linewidth extended cavity diode laser for atom interferometry in space, in *2013 Conference on Lasers & Electro-Optics Europe & International Quantum Electronics Conference*, Munich (2012)
41. E. Luvsandamdin, C. Kürbis, M. Schiemangk et al., Micro-integrated extended cavity diode lasers for precision potassium spectroscopy in space. Opt. Express **22**(7), 7790–7798 (2014)
42. C. Kürbis, A. Bawamia, M. Krüger et al., Extended cavity diode laser master-oscillator-power-amplifier for operation of an iodine frequency reference on a sounding rocket. Appl. Opt. **59**(2), 253–262 (2020)
43. X. Baillard, A. Gauguet, S. Bize et al., Interference-filter-stabilized external-cavity diode lasers. Opt. Commun. **266**(2), 609–613 (2006)
44. P. Zorabedian, W.R. Trutna, Interference-filter-tuned, alignment-stabilized, semiconductor external-cavity laser. Opt. Lett. **13**(10), 826–828 (1988)
45. M. Schmidt, M. Prevedelli, A. Giorgini et al., A portable laser system for high-precision atom interferometry experiments. Appl. Phys. B **102**(1), 11–18 (2010)
46. Z. Wang, X. Lv, J. Chen, A 657-nm narrow bandwidth interference filter-stabilized diode laser. Chin. Opt. Lett. **9**(4), 041402 (2011)
47. D.J. Thompson, R.E. Scholten, Narrow linewidth tunable external cavity diode laser using wide bandwidth filter. Rev. Sci. Instrum. **83**(2), 023107 (2012)
48. L. Zhang, T. Liu, L. Chen et al., Development of an interference filter-stabilized external-cavity diode laser for space applications. Photonics **7**(1), 12 (2020)

49. C. Voumard, R. Salathé, H. Weber, Mode selection by etalons in external diode laser cavities. Appl. Phys. **7**(2), 123–126 (1975)
50. K. Aoyama, R. Yoshioka, N. Yokota et al., Experimental demonstration of linewidth reduction of laser diode by compact coherent optical negative feedback system. Appl. Phys. Express **7**(12), 122701 (2014)
51. L. Hao, X. Wang, D. Guo et al., Narrow-linewidth self-injection locked diode laser with a high-Q fiber Fabry–Pérot resonator. Opt. Lett. **46**(6), 1397–1400 (2021)
52. W. Liang, Y. Liu, Compact sub-hertz linewidth laser enabled by self-injection lock to a sub-milliliter FP cavity. Opt. Lett. **48**(5), 1323–1326 (2023)
53. Y. Zhao, Y. Peng, T. Yang et al., External cavity diode laser with kilohertz linewidth by a monolithic folded Fabry–Pérot cavity optical feedback. Opt. Lett. **36**(1), 34–36 (2011)
54. Y. Zhao, Y. Li, Q. Wang et al., 100-Hz linewidth diode laser with external optical feedback. IEEE Photonics Technol. Lett. **24**(20), 1795–1798 (2012)
55. Y. Zhao, Q. Wang, F. Meng et al., High-finesse cavity external optical feedback DFB laser with hertz relative linewidth. Opt. Lett. **37**(22), 4729–4731 (2012)
56. E. Zang, S. Wang, Y. Zhao et al., Quasi synchronous tuning for grating feedback lasers. Appl. Opt. **50**(26), 5080–5084 (2011)
57. S. Wang, Y. Li, Y. Zhao et al., Mode-hopping-free scanning over 2 THz by means of quasi-synchronous tuning. Appl. Phys. B Lasers Opt. **114**(3), 381–384 (2014)
58. S.G. Lynch, C. Holmes, S.A. Berry et al., External cavity diode laser based upon an FBG in an integrated optical fiber platform. Opt. Express **24**(8), 8391 (2016)
59. F. Wei, F. Yang, X. Zhang et al., Subkilohertz linewidth reduction of a DFB diode laser using self-injection locking with a fiber Bragg grating Fabry–Pérot cavity. Opt. Express **24**(15), 17406–17415 (2016)
60. L. Zhang, F. Wei, G. Sun et al., Thermal tunable narrow linewidth external cavity laser with thermal enhanced FBG. IEEE Photonics Technol. Lett. **29**(4), 385–388 (2017)
61. Q. Su, F. Wei, G. Sun et al., Frequency-stabilized external cavity diode laser at 1572 nm based on frequency stability transfer. IEEE Photonics Technol. Lett. **34**(4), 203–206 (2022)
62. D. Huang, M.A. Tran, J. Guo et al., High-power sub-kHz linewidth lasers fully integrated on silicon. Optica **6**(6), 745–752 (2019)
63. C. Luo, R. Zhang, B. Qiu et al., Waveguide external cavity narrow linewidth diode lasers. J. Semicond. **42**(4), 041308 (2021)
64. M. Alalusi, P. Brasil, S. Lee et al., Low noise planar external cavity laser for interferometric fiber optic sensors, in *Fiber Optic Sensors and Applications VI* (2009). https://doi.org/10.1117/12.828849
65. Y. Li, Y. Zhang, H. Chen et al., Tunable self-injected Fabry–Pérot laser diode coupled to an external high-Q Si_3N_4/SiO_2 microring resonator. J. Lightwave Technol. **36**(16), 3269–3274 (2018)
66. Y. Fan, A. van Rees, P.J.M. van der Slot et al., Hybrid integrated InP–Si_3N_4 diode laser with a 40-Hz intrinsic linewidth. Opt. Express **28**(15), 21713–21728 (2020)
67. C. Xiang, J. Guo, W. Jin et al., High-performance lasers for fully integrated silicon nitride photonics. Nat. Commun. **12**, 6650 (2021)
68. W. Jin, Q.-F. Yang, L. Chang et al., Hertz-linewidth diode lasers using CMOS-ready ultra-high-Q microresonators. Nat. Photonics **15**(5), 346–353 (2021)
69. M. Corato-zanarella, A. Gil-molina, X. Ji et al., Widely tunable and narrow-linewidth chip-scale lasers from near ultraviolet to near-infrared wavelengths. Nat. Photonics **17**(2), 157–164 (2022)
70. B. Luo, *Principle and Technology of Atomic Filter* (Beijing University of Posts and Telecommunications Press, Beijing, 2018)
71. X. Miao, *An External-Cavity Diode Laser Immune to the Current and Temperature Fluctuations* (Peking University, Beijing, 2011)
72. B. Yin, T.M. Shay, Theoretical model for a Faraday anomalous dispersion optical filter. Opt. Lett. **16**(20), 1617–1619 (1991)

73. J. Keaveney, C.S. Adams, I.G. Hughes, ElecSus: extension to arbitrary geometry magneto-optics. Comput. Phys. Commun. **224**, 311–324 (2018)
74. Y. Öhman, On some new auxiliary instruments in astrophysical research VI. A tentative monochromator for solar work based on the principle of selective magnetic rotation. Stockh. Obs. Ann. **19**(4), 9–11 (1956)
75. P.P. Sorokin, J.R. Lankard, V.L. Moruzzi et al., Frequency-locking of organic dye lasers to atomic resonance lines. Appl. Phys. Lett. **15**(6), 179–181 (1969)
76. D.J. Dick, T.M. Shay, Ultrahigh-noise rejection optical filter. Opt. Lett. **16**(11), 867–869 (1991)
77. X. Miao, L. Yin, W. Zhuang et al., Note: demonstration of an external-cavity diode laser system immune to current and temperature fluctuations. Rev. Sci. Instrum. **82**(8), 086106 (2011)
78. Y. Wang, S. Zhang, D. Wang et al., Nonlinear optical filter with ultranarrow bandwidth approaching the natural linewidth. Opt. Lett. **37**(19), 4059–4061 (2012)
79. Y. Yan, J. Yuan, L. Wang et al., A dual-wavelength bandpass Faraday anomalous dispersion optical filter operating on the D_1 and D_2 lines of Rb. Opt. Commun. **509**, 127855 (2022)
80. X. Guan, W. Zhuang, T. Shi et al., 420-nm Faraday optical filter with 2.7-MHz ultranarrow bandwidth based on laser cooled ^{87}Rb atoms. IEEE Photonics Technol. Lett. **35**(12), 672–675 (2023)
81. X. Guan, W. Zhuang, T. Shi et al., Cold-atom optical filtering enhanced by optical pumping. Front. Phys. **10** (2022)
82. J. Menders, P. Searcy, K. Roff et al., Blue cesium Faraday and Voigt magneto-optic atomic line filters. Opt. Lett. **17**(19), 1388–1390 (1992)
83. X. Xue, D. Pan, X. Zhang et al., Faraday anomalous dispersion optical filter at ^{133}Cs weak 459 nm transition. Photonics Res. **3**(5), 275–278 (2015)
84. J. Xiong, B. Luo, L. Yin et al., The characteristics of Ar and Cs mixed Faraday optical filter under different signal powers. IEEE Photonics Technol. Lett. **30**(8), 716–719 (2018)
85. D. Pan, T. Shi, B. Luo, J. Chen, H. Guo, Atomic optical stimulated amplifier with optical filtering of ultra-narrow bandwidth. Sci. Rep. **8**, 6567 (2018)
86. L. Ling, G. Bi, Isotope ^{87}Rb Faraday anomalous dispersion optical filter at 420 nm. Opt. Lett. **39**(11), 3324–3327 (2014)
87. Y. Peng, W. Zhang, L. Zhang et al., Analyses of transmission characteristics of Rb, ^{85}Rb and ^{87}Rb Faraday optical filters at 532 nm. Opt. Commun. **282**(2), 236–241 (2009)
88. Q. Sun, W. Zhuang, Z. Liu et al., Electrodeless-discharges-vapor-lamp-based Faraday anomalous-dispersion optical filter. Opt. Lett. **36**(23), 4611–4613 (2011)
89. X. Xue, Z. Tao, Q. Sun et al., Faraday anomalous dispersion optical filter with a single transmission peak using a buffer-gas-filled Rb cell. Opt. Lett. **37**(12), 2274–2276 (2012)
90. J.A. Zielińska, F.A. Beduini, N. Godbout et al., Ultranarrow Faraday rotation filter at the Rb D_1 line. Opt. Lett. **37**(4), 524–526 (2012)
91. Q. Sun, Y. Hong, W. Zhuang et al., Demonstration of an excited-state Faraday anomalous dispersion optical filter at 1529 nm by use of an electrodeless discharge Rb vapor lamp. Appl. Phys. Lett. **101**(21), 211102 (2012)
92. Y. Zhang, X. Jia, Z. Ma et al., Potassium Faraday optical filter in line-center operation. Opt. Commun. **194**(1), 147–150 (2001)
93. E.T. Dressler, A.E. Laux, R.I. Billmers, Theory and experiment for the anomalous Faraday effect in potassium. J. Opt. Soc. Am. B Opt. Phys. **13**(9), 1849–1858 (1996)
94. H. Chen, C.Y. She, P. Searcy et al., Sodium-vapor dispersive Faraday filter. Opt. Lett. **18**(12), 1019–1021 (1993)
95. J.A. Gelbwachs, Y.C. Chan, Passive Fraunhofer-wavelength atomic filter at 460.7 nm. IEEE J. Quantum Electron. **28**(11), 2577–2581 (1992)
96. Y.C. Chan, J.A. Gelbwachs, A Fraunhofer-wavelength magnetooptic atomic filter at 422.7 nm. IEEE J. Quantum Electron. **29**(8), 2379–2384 (1993)
97. P. Wanninger, E.C. Valdez, Diode-laser frequency stabilization based on the resonant Faraday effect. IEEE Photonics Technol. Lett. **4**(1), 94–96 (1992)

98. K. Choi, J. Menders, P. Searcy et al., Optical feedback locking of a diode laser using a cesium Faraday filter. Opt. Commun. **96**(4–6), 240–244 (1993)
99. X.T. Jun, J.W. Qing, L. Yimin et al., Experimental study of a model digital space optical communication system with new quantum devices. Appl. Opt. **34**(15), 2619–2622 (1995)
100. X.J. Lu, J.X. Tang, L.M. Zheng, Experimental study on an external cavity diode laser using Rb Faraday anomalous dispersion filter. Chin. J. Electron. **25**(2), 79–82 (1997)
101. F.Q. Li, Y.P. Wang, X.W. Cheng et al., Faraday anomalous dispersion optical filter atomic frequency-stabilized diode laser through optical feedback. Chin. J. Lasers **32**(10), 1317–1320 (2005)
102. Z. Tao, Y. Hong, B. Luo et al., Diode laser operating on an atomic transition limited by an isotope ^{87}Rb Faraday filter at 780 nm. Opt. Lett. **40**(18), 4348–4351 (2015)
103. Z. Tao, X. Zhang, D. Pan et al., Faraday laser using 1.2 km fiber as an extended cavity. J. Phys. B At. Mol. Opt. Phys. **49**(13), 13LT01 (2016)
104. J. Keaveney, W.J. Hamlyn, C.S. Adams et al., A single-mode external cavity diode laser using an intra-cavity atomic Faraday filter with short-term linewidth $<$ 400 kHz and long-term stability of $<$ 1 MHz. Rev. Sci. Instrum. **87**(9), 095111 (2016)
105. T. Shi, X. Guan, P. Chang et al., A dual-frequency Faraday laser. IEEE Photonics J. **12**(4), 1–11 (2020)
106. P. Chang, Y. Chen, H. Shang et al., A Faraday laser operating on Cs 852 nm transition. Appl. Phys. B **125**(12), 230 (2019)
107. M.D. Rotondaro, B.V. Zhdanov, M.K. Shaffer et al., Narrowband diode laser pump module for pumping alkali vapors. Opt. Express **26**(8), 9792–9797 (2018)
108. H. Tang, H. Zhao, R. Wang et al., 18 W ultra-narrow diode laser absolutely locked to the Rb D_2 line. Opt. Express **29**(23), 38728–38736 (2021)
109. H. Tang, H. Zhao, D. Zhang et al., Polarization insensitive efficient ultra-narrow diode laser strictly locked by a Faraday filter. Opt. Express **30**(16), 29772–29780 (2022)
110. P. Chang, H. Shi, J. Miao et al., Frequency-stabilized Faraday laser with 10^{-14} short-term instability for atomic clocks. Appl. Phys. Lett. **120**(14), 141102 (2022)
111. H. Shi, P. Chang, Z. Wang et al., Frequency stabilization of a cesium Faraday laser with a double-layer vapor cell as frequency reference. IEEE Photonics J. **14**(6), 1–6 (2022)

Chapter 2
Overview of Faraday Lasers

A Faraday laser is a type of laser that uses a Faraday Anomalous Dispersion Optical Filter (abbreviated as FADOF) as selective frequency component, which is a new type of diode laser. In principle, the output wavelength of the Faraday laser directly corresponds to the atomic transition spectrum line, and the laser frequency is always restricted within the transmission bandwidth of the atomic filter. In addition, the Faraday laser has the ability to resist disturbances in the driving current and operating temperature of the laser diode, and has the advantages of output frequency automatically corresponding to atomic transitions, long-term continuous operation, and immediate use. This chapter provides a detailed introduction to the basic working mechanism, structure, and performance advantages of the Faraday laser, as well as its application in fields such as quantum precision measurement.

2.1 Basic Principles of Faraday Lasers

Faraday magneto-optical rotation effect is an extremely important discovery in the process of human scientific research. It was first discovered by the famous British physicist and chemist Michael Faraday (1791–1867) in 1845 [1–5]. When a polarized light passes through a glass under the action of a magnetic field, if the direction of light propagation is the same as the direction of the magnetic field, the polarization direction of this light will rotate, validating "magnetic force and light are proved to have a relation to each other" [6]. This indicates that under the action of a magnetic field, the interaction between light and the medium will cause the polarization direction of the light to rotate. Since its discovery, the Faraday magneto-optical rotation effect has attracted the attention of many scientists. The famous physicist James Clerk Maxwell (1831–1879) was inspired by the Faraday magneto-optical rotation effect and published his masterpiece *On Physical Lines of Force* [7], which played an important role in further establishing the basic theory of the electromagnetic field. In

J. Chen et al., *Faraday Laser*, https://doi.org/10.1007/978-981-97-8023-5_2

the 1950s and 1960s, with the rise of laser and optoelectronic technology, people used the Faraday magneto-optical rotation effect to make many optoelectronic devices, such as magneto-optical modulators, magneto-optical switches, and magneto-optical isolators. These devices based on the Faraday magneto-optical rotation effect have greatly promoted the development of lasers, optoelectronics, and related fields of science and technology, and further promoted the wide applications in the fields of quantum precision measurement and quantum information sensing.

FADOF is an atomic filter implemented based on the principle of the Faraday magneto-optical rotation effect (as shown in Fig. 2.1). It includes two polarization elements with orthogonal polarization directions and alkali metal atoms and other media, as well as a magnetic field parallel to the direction of light propagation, with the black arrow indicating the polarization direction of the light. The Faraday magneto-optical rotation effect can be understood as follows: when the direction of the external magnetic field is consistent with the direction of light propagation, the atomic magnetic sublevels undergo Zeeman splitting under the action of the longitudinal magnetic field. The direction of the quantization axis determined by the magnetic field is consistent with the direction of light propagation. The incident linearly polarized light can be decomposed into left-handed circularly polarized light and right-handed circularly polarized light. The dispersion curves of the two circularly polarized lights will split, as their refractive indices are different at the same frequency, causing their phase difference to change after propagating the same path in the atomic medium. Finally, when the light exits the atomic medium, its linear polarization direction rotates relative to the initial incident polarization direction. By setting an appropriate length of the atomic medium, the polarized light near a specific quantum transition frequency can undergo a nearly 90° angle deviation, and then output through the polarimeter, achieving atomic filtering.

There are primarily two methods for achieving atomic filtering. One is the commonly used FADOF based on the above-mentioned Faraday magneto-optical effect [8], which has high background noise suppression ratio [9], narrow bandwidth [10], high transmission [11], high response rate [12] and other features; the other is the "absorption-reemission type" [13], which uses the absorption-reemission process of atoms to the signal light to complete signal filtering, generally known as atomic

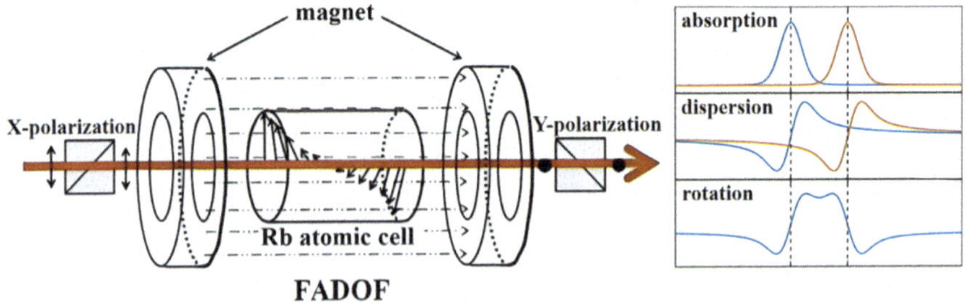

Fig. 2.1 Schematic diagram of FADOF structure and optical rotation principle

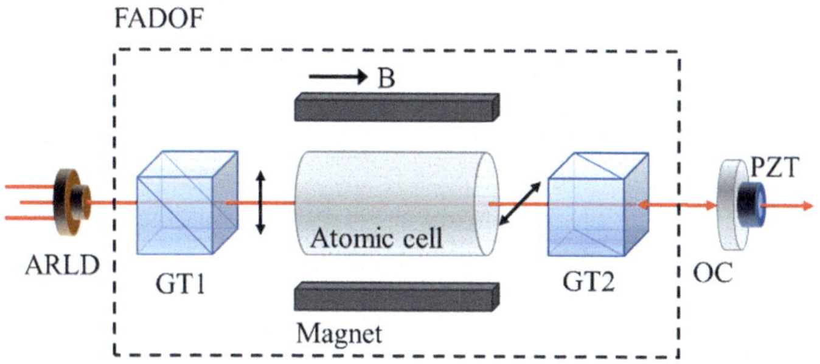

Fig. 2.2 Schematic diagram of Faraday laser principle

resonance filter, with narrow linewidth, large field of view angle, high background noise suppression and other features.

The Faraday laser is a type of external cavity diode laser that uses a FADOF as a frequency selection component. Its working principle is shown in Fig. 2.2, which mainly includes a wide-spectrum laser diode (Antireflection-coated laser diode, ARLD) as the gain medium, a FADOF providing frequency selection function, and optical coupling output mirror providing optical feedback to establish laser oscillation. After the Faraday laser forms oscillation and reaches a steady state, the light emitted by the antireflection-coated laser diode passes through the first polarizing element Glan–Taylor prism (GT-1). Due to the magneto-optical effect, while passing through the atomic medium, the laser beam will undergo Faraday rotation due to the presence of a magnetic field in the direction of light propagation, thus causing a certain deflection in the polarization direction. By setting specific temperature and magnetic field intensity for FADOF, only the light corresponding to the atomic transition frequency undergoes a 90° polarization rotation after passing through the atomic vapor cell. This portion of light then passes through the second polarizing element, the Glan–Taylor prism (GT-2), with maximum efficiency. After passing through GT-2, the light beam is incident vertically onto the output mirror, and a certain proportion of the light is reflected by the mirror, subsequently returning to the ARLD in a direction parallel but opposite to the incident light, thereby forming resonant feedback and enabling laser emission, which is named as Faraday laser.

2.2 Faraday Laser Structure Composition

The optical part of the Faraday laser mainly includes the laser source (laser diode), atomic filter, and laser resonator, as shown in Fig. 2.3. The Faraday laser uses ARLD to provide gain, the atomic filter is placed in the resonator as a frequency selector for frequency selection, the cavity mirror with a certain reflectivity is used to provide optical feedback, and finally, the laser output frequency can stably work within the transmission spectrum bandwidth of the atomic filter, and is immune to the current and temperature fluctuations of the laser diode. Compared with traditional external cavity diode lasers with frequency selection based on interference filters, gratings, and F-P cavities, the Faraday laser uses a narrow-bandwidth atomic filter for frequency selection, and the output laser frequency is determined by the cavity mode closest to the FADOF transmission spectrum peak frequency. Due to the superior frequency stability and robustness, it has significant application advantages in precision spectroscopy, atomic clocks, atomic magnetometers, atomic gravimeters, atomic interferometers, optical communication, laser radar, military defense and other fields in basic scientific research.

For the light source of the Faraday laser, ARLD is commonly used now. Considering the basic theory of resonator frequency selection, the diode laser utilizes the resonator formed between the back surface of the laser diode and the output coupling mirror for mode selection. However, the resonator formed by the front and back surfaces of the laser diode will affect the frequency selection, causing mode hopping and mode instability. By Using ARLD as the gain medium and applying an antireflection coating to the front surface of the laser diode, internal cavity modes are eliminated, thereby overcoming its impact on the laser output mode. In previous Faraday laser research, related research groups often used Toptica's laser diodes from Germany. With the continuous development of chip manufacturing and other processes in China, Peking University and related cooperation partners have now developed TO9 packaged tunable 780 nm/852 nm laser diodes, and have successfully coated the front surface of the laser diode with an antireflection film, meeting the needs of Faraday lasers.

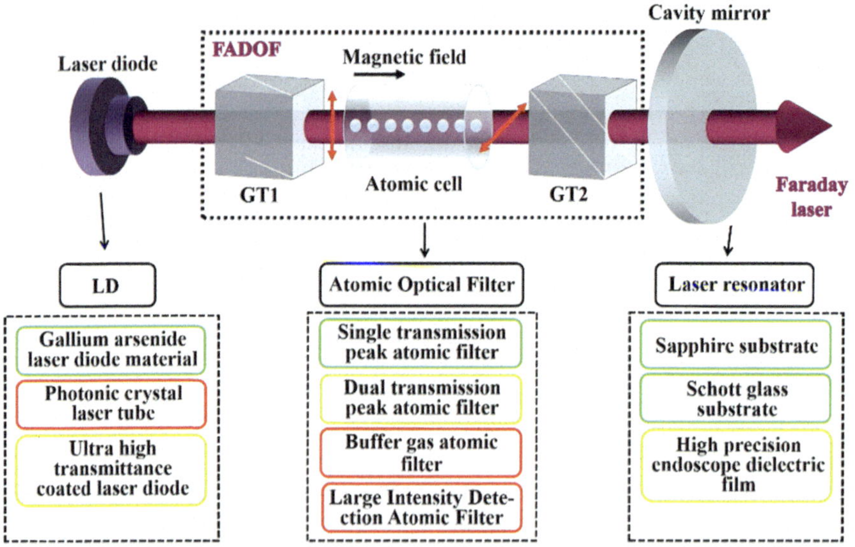

Fig. 2.3 Schematic diagram of the structure of the Faraday laser

As a core component of the Faraday laser, the atomic filter can be divided into single transmission peak atomic filters and multiple transmission peak atomic filters according to the transmission spectrum line shape. In addition, there are special atomic filters filled with buffer gas and atomic filters for high light intensity detection to meet different needs. In order to achieve excellent performance, precise temperature control and magnetic field control are required for the atomic cell. For its insulation, polytetrafluoroethylene can be used as the insulation material, and a heating sheet can be wrapped around the atomic cell for temperature adjustment. For the magnetic field, permanent magnets are mostly used, and coils also be used in some experiments. By applying appropriate temperature and magnetic field, only the specific light near the atomic resonance transition line in the wide spectrum fluorescence range can pass through the FADOF, oscillate and amplify in the resonator, and ultimately achieve laser emission once reaching the threshold. Since the transmission peak is near the atomic transition spectrum, the output frequency of the laser can directly correspond to the atomic transition.

Filling the atomic cell of the atomic filter with buffer gas can broaden the transmission spectrum and increase the continuous tuning range of the laser frequency. Buffer gas is filled into the atomic cell to reduce collisions between selective atoms, but at the same time, the collision frequency shift between selective atoms and buffer gas atoms is enhanced, which broadens the transmission spectrum of the Faraday anomalous dispersion atomic filter correspondingly. The transmission bandwidth will contain more cavity modes, increasing the continuous tuning range of the laser. There is always a stable longitudinal mode of the laser in the transmission spectrum of the Faraday anomalous dispersion atomic filter, effectively avoiding the problem of inability to oscillate. The Faraday laser is essentially a special type of diode laser, with an atomic frequency selector added to its resonant cavity. In addition to the Faraday anomalous dispersion atomic filter providing light feedback for specific frequencies, a resonant cavity is formed between the coupling output mirror having a specific reflectance and the back surface of the laser diode. Under the action of cavity feedback, the linewidth of the laser spectrum can also be narrowed.

2.3 Faraday Laser Performance Characteristics

2.3.1 Special Performance of Faraday Laser

Diode lasers usually use external cavity light feedback to narrow the laser linewidth, and employ macroscopic devices such as gratings, interference filters, Fabry–Pérot cavities (F-P cavities), and other macro devices to achieve frequency selection. In high-precision frequency stabilization systems, such as optical clock systems, the linewidth of the diode laser used as the clock transition laser needs to be further narrowed. The Pound–Drever–Hall (PDH) frequency stabilization technology is usually used to narrow the laser frequency on a high Q-value optical resonant cavity to achieve excellent coherent laser output. However, the resonance frequency of the optical resonant cavity is easily affected by external temperature, pressure, and mechanical vibration, leading to a deterioration in the long-term frequency stability of the laser. To improve the long-term stability of the laser system, a more stable frequency reference is needed, such as the quantum transition spectrum lines provided by atoms or molecules. Dichroic atomic vapor laser lock (DAVLL), saturated absorption spectroscopy (SAS), modulation transfer spectroscopy (MTS), and other methods are usually used for laser frequency stabilization. To further optimize laser performance, the center frequency stability and reproducibility of the quantum reference spectrum line, narrow linewidth, and high Q factor are required. Therefore, cold atoms cooled by lasers or trapped in optical lattices, or single ions trapped in ion traps can replace thermal atoms in the cell as quantum frequency references. By stabilizing the laser frequency on the transition spectrum line of atoms, molecules, or ions through electrical feedback, the frequency stability of the laser can be improved to 10^{-19} or even better. Compared to the above methods, a Faraday laser using FADOF as a frequency selection device can achieve laser frequency within the transmission bandwidth of the atomic filter (corresponding to atomic transition spectral lines) as soon as it is turned on. Unlike the electrical feedback methods in PDH frequency stabilization and SAS frequency stabilization schemes, this method uses optical feedback and has a simple structure. The laser output from the laser diode is reflected by the cavity mirror to form optical feedback after passing through the FADOF. Due to the Faraday magneto-optical effect, only laser modes with frequencies within the FADOF transmission bandwidth can form optical feedback. Therefore, the output wavelength of the Faraday laser can always be limited within the transmission bandwidth of the atomic filter, and it is immune to fluctuations in the driving current and operating temperature of the laser diode. Due to the significant advantages of Faraday lasers over traditional diode lasers, more and more research groups around the world have carried out theoretical and experimental research on Faraday lasers. Below we discuss several important developments and achievements in improving the performance parameters of Faraday lasers internationally.

2.3.2 Performance Improvement and Development of Faraday Lasers

Since the 1960s, scientists have been conducting research on atomic filters and applications. In 1969, Sorokin and colleagues from IBM's Watson Laboratory realized Na-FADOF [14], and for the first time applied it to dye laser frequency stabilization, with the FADOF transmission peak corresponding to the sodium atom $3S_{1/2} \rightarrow 3P_{1/2}$ and $3S_{1/2} \rightarrow 3P_{3/2}$ transition. In 1990, Lee and Campbell from the University of Texas at Austin used an Rb Faraday atomic filter to achieve frequency locking of the $Al_xGa_{1-x}As/GaAs$ material 780 nm diode laser through the construction of a narrowband optical feedback loop [15]. In 1995, Kitching and Yariv from the California

Institute of Technology [16] achieved amplitude squeezed states of quantum well diode lasers by combining pump suppression with weak optical feedback, where weak feedback provided by the Faraday atomic filter can also provide dispersive loss. Through experimental design, the combination of pump suppression and weak optical feedback with dispersive loss achieved a 0.9 dB amplitude squeeze of the diode laser under free-running conditions at room temperature, which is a significant improvement over the previous best of 0.33 dB [17]. The system uses a laser with an anti-reflection coating on the front surface combined with a Faraday atomic filter, which improves the feedback coupling efficiency and further narrows the internal noise of the laser.

In China, research teams from Peking University, the Institute of Precision Measurement Science and Technology Innovation of the Chinese Academy of Sciences (formerly the Wuhan Institute of Physics and Mathematics of the Chinese Academy of Sciences), Harbin Institute of Technology, and the National University of Defense Technology have conducted research on atomic filters. In 1997, Lu and others from Peking University conducted exploratory research on an external cavity diode laser using an Rb Faraday atomic filter [18]. The output frequency of the diode laser can be aligned with the four transmission peaks of the FADOF. Under specific current and temperature working conditions, single-mode, stable frequency, and narrow linewidth operation are achieved. When using a Faraday atomic filter to construct external cavity feedback, it is crucial to maintain the laser output power and avoid multi-mode output caused by strong feedback. To achieve this, the team has researched to make the optical power used for feedback account for three-thousandths of the total output power, which also avoids the effect of weak feedback on the laser's frequency stability and linewidth narrowing. Through the construction of the above weak feedback, the laser linewidth is narrowed from 30 to 3 MHz. It has been observed that when the magnetic field applied to the Faraday atomic filter is reduced from 500 to 400 G, the laser frequency can be continuously tuned over a range of 200 MHz. In 2005, Li and colleagues from the Institute of Precision Measurement Science and Technology Innovation of the Chinese Academy of Sciences [19], using GaAlAs material diode laser and Cs atomic Faraday anomalous dispersion atomic filter to make an external cavity feedback diode laser, under constant temperature, the wavelength no longer changes with the change of current within a certain range. When the current changes by 3 ~ 4 mA, the laser output wavelength is always limited within the transmission bandwidth of the atomic filter, and the frequency fluctuation range is 75 MHz/2 h.

In 2011, Miao and colleagues from Peking University [20], using Rb FADOF as a frequency selector, for the first time using ARLD to build a Faraday laser, under the condition of 6.4% feedback intensity, a diode laser immune to diode current and temperature fluctuations was realized, with a laser linewidth of 69 kHz. In this work, Miao and colleagues used ARLD as a gain to eliminate the influence of the diode internal cavity mode on the output laser mode. Most importantly, when the laser diode operates at different temperatures, as long as the operating current is greater than the threshold, the laser output frequency correspond to the transmission spectrum of the atomic filter. This laser diode, as a gain medium, has a fluorescence spectrum width of 60 nm, avoiding the need to heat or cool the laser diode to tune its emission spectrum characteristics.

For international research groups engaged in this direction, Zentile and colleagues from Durham University in the UK use Rb atomic Faraday anomalous dispersion atomic filter to achieve frequency selection [21], analyzed the advantages and disadvantages of using long resonant cavity and short resonant cavity to build Faraday lasers. Using a longer resonant cavity can reduce the free spectral range (FSR), so that there is always a cavity mode near the highest point of the transmission spectrum. In this way, it is not necessary to actively stabilize the resonant cavity length, no matter how the modes compete, the output laser frequency is always limited within the transmission spectrum of the atomic filter. For example, Peking University proposed using a 1.2 km fiber as the resonant cavity of the Faraday laser, and its laser output frequency change range is less than 100 kHz [22]. Using a shorter resonant cavity, by increasing the FSR, making it greater than the FADOF transmission spectrum profile, so that there is at most one cavity mode within the transmission spectrum profile. This method requires active stabilization of the resonant cavity length to ensure the existence of a cavity mode within the transmission spectrum, and to ensure long-term stability.

In the aforementioned related work, although atomic filters have been used to construct frequency-stable Faraday lasers, the term "Faraday laser" was first proposed internationally by Tao and colleagues from Peking University in 2015 [23], and subsequently received widespread recognition and use in academic journals internationally. This team used an Rb atomic Faraday anomalous dispersion atomic filter as a frequency selector, a laser diode with an anti-reflection coating as the laser gain medium, and achieved a Faraday laser that is immune to diode temperature and current fluctuations. In the work of Tao and colleagues, considering the issues associated with the use of natural Rb, namely the difficulty in restricting the laser frequency to resonate precisely with a specific atomic transition due to the presence of different isotopes, they chose to use ^{87}Rb atoms with a concentration of 96.5%. The length of the vapor cell was 5 cm, and the polarization extinction ratio of the two Glan–Taylor prisms was 1×10^{-5}. In the experiment, by analyzing the effects of the magnetic field and temperature on the highest transmission of the transmission spectrum, the optimal working parameters of the FADOF were determined to obtain the best laser performance. This Faraday laser maintain an output wavelength consistently around 780.2456 nm despite changing in the drive current of the anti-reflection coated laser diode from 55 to 140 mA and temperature variations from 15 to 35 °C. The output wavelength of the laser is always around 780.2456 nm, the wavelength standard deviation is less than or equal to 0.27 pm, which can correspond to the atomic transition line.

In 2017, Chang and colleagues from Peking University used an electrodeless discharge lamp to realize the excited state Faraday laser for the first time, which can directly output the 1529 nm Faraday laser corresponding to the Rb atom $5P_{3/2} \rightarrow 4D_{5/2}$ transition [24]. In comparison to the traditional excited state atomic filters, there is no need for an additional frequency-locked laser to excite the atom from the 5S state to the 5P state. This work utilizes the excited state atomic filter of an electrodeless discharge lamp for frequency selection, combined with the anti-reflection coated laser diode, when the drive current of the diode changes from 85 to 171 mA, and the diode temperature changes from 11 to 32 °C, the laser can work stably, and the long-term frequency fluctuation range of the laser within 24 h is maintained within 600 MHz.

Under the frequency selection effect of the atomic filter, although the output frequency of the Faraday laser can be limited near the atomic transition line, its frequency stability still needs to be further improved. In order to obtain a laser with ultra-high frequency stability, one method is to lock the output frequency of the laser to the atomic transition line. In 2022, Chang and others from Peking University first locked the output frequency of the cesium-atom 852 nm Faraday laser to the atomic transition line through modulation transfer spectroscopy. In this work, the Faraday laser can be adjusted to single-frequency or dual-frequency output. When working in single-frequency mode, the short-term stability measured by the residual error signal evaluation method reaches $3 \times 10^{-14}/\sqrt{\tau}$; when working in dual-frequency mode, the dual-frequency laser beat frequency linewidth can be narrowed to 85 Hz [25]. The experimental setup is shown in Fig. 2.4. The external cavity is formed by the back surface of the laser diode with an anti-reflection coating and the reflection cavity mirror. Frequency selection is achieved using a Faraday anomalous dispersion atomic filter. In single-frequency mode, by modulation transfer spectroscopy, the laser frequency is locked to the cesium atom $6^2S_{1/2}F = 4 \rightarrow 6^2P_{3/2}F = 5$ cyclic transition line, greatly improving the frequency stability of the Faraday laser. In dual-frequency mode, one of the laser frequencies is locked to the $6^2S_{1/2}F = 3 \rightarrow 6^2P_{3/2}F = 2$ transition line, effectively narrowing the linewidth of the beat signal between the dual-frequency lasers, reducing the noise of the microwave signal, and opening up a new path for optically generated microwaves.

The realized Faraday laser has a power output of approximately 20 mW, making it suitable for applications in quantum precision measurement such as laser-pumped cesium beam atomic clock, spectral measurement, and ultra-high stability PDH frequency-stabilized lasers. However, in scenarios involving laser weapons and other high-power laser requirements, the power of the above-mentioned laser may not be sufficient. Therefore, the development of high-power Faraday lasers is also very important and has a broad application prospect. The National University of Defense Technology has conducted an in-depth research in the direction of high-power Faraday lasers. In 2022, the National University of Defense Technology used a Faraday anomalous dispersion atomic filter based on ^{87}Rb as a frequency selection device, and a diode laser module with an output power of 47 W composed of 18 single emitters as a gain medium light source, achieving a Faraday laser output of 38.3 W under a drive current of 5 A, narrowing the 4 nm gain medium fluorescence spectrum to 0.005 nm (2.6 GHz) laser output spectrum [26]. To avoid mode competition under free-running conditions, the output surface of the diode laser module is coated with anti-reflection, with a residual reflectivity of less than 0.5%. The experimental setup is shown in Fig. 2.5. To improve the polarization frequency selection characteristics of the Faraday anomalous dispersion atomic filter, a ring resonator structure is used to increase the efficiency of the external cavity and to ensure high power output.

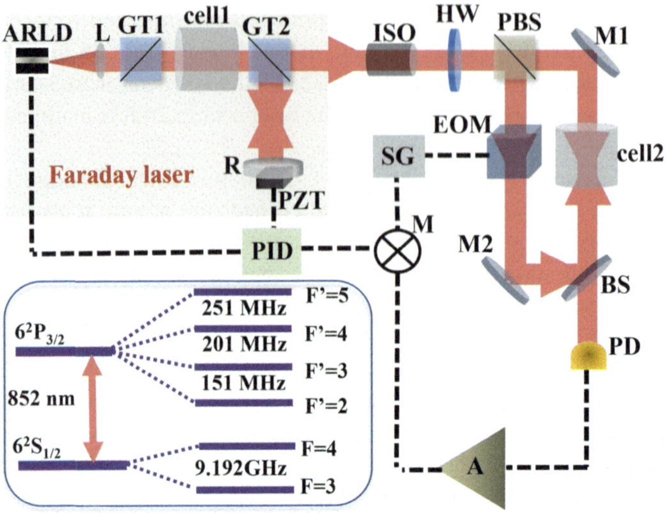

Fig. 2.4 The experimental setup. *ARLD* anti-reflection coated laser diode, *L* lens, *GT1* Glan–Taylor prism 1, *cell1* cell 1, *GT2* Glan–Taylor prism 2, *R* cavity mirror, *PZT* piezoelectric ceramic, *PID* proportional integral differential feedback circuit, *PBS* polarizing beam splitter, *EOM* electro-optic modulator, *SG* signal generator, *A* amplifier

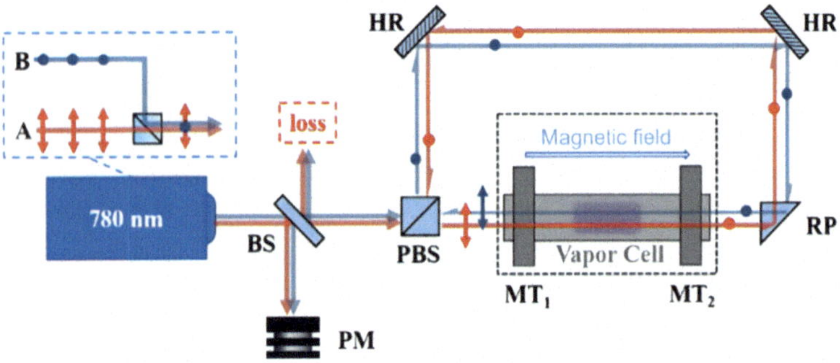

Fig. 2.5 Experimental setup diagram [26]

In order to further enhance the stability of the atomic cell temperature, in 2022, Shi and colleagues from Peking University improved the atomic reference cell by adopting a double-layer structure, with a vacuum between the two layers of glass, which can effectively avoid low-frequency temperature fluctuations in the reference atomic cell, thereby reducing the impact on frequency stability. Ultimately, they achieved a Faraday laser based on the cesium-atom Faraday anomalous dispersion atomic filter, and locked its output frequency using modulation transfer spectroscopy, achieving a short-term stability of the output laser frequency up to $5.8 \times 10^{-15}/\tau^{1/2}$, and dropping to 2.0×10^{-15} [27] at 50 s. This work greatly improved the stability of the Faraday laser output frequency by filling the cell of the Faraday anomalous dispersion atomic filter with buffer gas and using the double-layer atomic cell as the reference signal source in the modulation transfer spectroscopy. It has a wide range of applications in quantum optics, quantum precision measurement, and other fields. With the development of technology, it is possible to achieve Faraday lasers based on other alkali metal atom FADOFs and active FADOFs in the future, to further expand the working band of Faraday lasers; by conducting theoretical research on the internal light intensity of the laser, the transmission spectrum of FADOF can be more accurately determined, thereby more precisely controlling the performance of the laser. By using high-power Faraday lasers to excite atoms, the internal structure of atoms can be further studied, and high-power Faraday lasers can also be used directly as cooling light in cold atom experiments. In addition, by preparing micro FADOFs or fiber FADOFs, miniaturized Faraday lasers can be achieved. It is believed that in the near future, the performance of Faraday lasers can be further improved and widely used in quantum precision measurement, optical communication, basic research, and other fields.

2.4 Faraday Laser Application Scenarios and Requirements

As a new type of diode laser, the Faraday laser essentially uses FADOF as a frequency selector, with the output laser wavelength corresponding to the atomic transition line. At the same time, compared with traditional external cavity diode lasers, Faraday lasers have the ability to resist disturbances to diode drive current and operating temperature changes, and the output wavelength is always limited within the atomic filter transmission bandwidth. Due to the characteristics of FADOF, only wavelengths corresponding to atomic transition lines can pass through the filter, which has important application prospects in quantum frequency standards, laser cooling, and high-resolution spectroscopy, and also has unique advantages in new atomic clocks, atomic gravimeters, and other applications.

2.4.1 Application Scenarios of Faraday Lasers

The output laser frequency of the Faraday laser directly corresponds to the atomic transition spectrum line. After frequency stabilization, a 1 s sampling Allen deviation stability better than 10^{-13} can be achieved. This ultra-stable laser can further be used as a pump light and probe light applied to the optically pumped cesium clock. In the research of microwave atomic clocks, laser-pumped cesium beam atomic clock has high long-term stability and plays an important role in practical applications. Among them, the laser-pumped cesium beam atomic clock has the advantages of high atomic utilization rate and simple structure, and can achieve higher frequency stability compared to the magnetically selected cesium beam atomic clock. One of the most representative laser-pumped cesium beam atomic clock internationally is the American 5071A cesium clock, which has a high-quality stability of 5×10^{-12}/s and 8.5×10^{-13}/100 s, and long-term frequency stability exceeding 1×10^{-14}. Recently, Peking University and the 12th Research Institute of China Electronics Technology Group have cooperated to conduct research on applying Faraday lasers to optically pumped cesium clocks. Using a single Faraday laser as the light source, the output laser is locked to the cesium atom $6^2S_{1/2}F = 4 \rightarrow 6^2P_{3/2}F' = 5$ transition using modulation transfer spectroscopy, and is divided into two beams, one as a probe light, and the other after the Acoustic Optical Modulator (AOM) shifts the frequency by 251.4 MHz to correspond to the $6^2S_{1/2}F = 4 \rightarrow 6^2P_{3/2}F' = 4$ transition as the pump light. Due to its advantages of narrow linewidth, frequency stability, and short-term stability, it has now achieved a stability of 7.45×10^{-13}/s, and 2.9×10^{-13}/100 s.

Faraday lasers also hold significant applications in the field of active optical clocks. In 2005, Chen of Peking University first proposed the active optical clock technology scheme [28], by using an atomic ensemble as the gain medium, the stimulated radiation can be directly used as the clock laser signal. In 2014, Zhuang and colleagues from Peking University realized the active optical clock using a bad-cavity Faraday laser [29], the center frequency is determined by the cesium atom $6^2S_{1/2}F = 4 \rightarrow 6^2P_{3/2}F' = 4, 5$ cross transition spectrum line. Since the laser works in the bad-cavity regime, it has the advantages of cavity-pulling suppression and narrow linewidth, which can effectively overcome the problem of cavity length thermal noise. The linewidth obtained by optical heterodyne beating frequency using two sets of the same independent system is 281 Hz.

Faraday lasers can also be used in underwater optical communication. High-sensitivity underwater wireless optical communication is subjected to interference from solar noise and other background noise. In 2022, Zhang and colleagues from Beijing University of Posts and Telecommunications used an 852 nm Faraday laser and a 1 GHz bandwidth FADOF to achieve air–water and underwater wireless optical communication resistant to background noise [30]. The research results show that after using the Faraday laser, the Q factor of the system is 3.37 dB higher than when using an interference filter. By adding a FADOF filter at the receiving end, the Q factor is 14.67 dB higher than without FADOF, verifying that the implementation of Faraday lasers and FADOF can improve the performance of underwater optical communication under broadband background noise interference.

Currently, Faraday lasers are rapidly developing due to their excellent performance and huge application potential, and have made a series of important progress. However, there are still some research difficulties, such as how to accurately obtain the transmission spectrum of FADOF when forming a resonant cavity inside the laser, to avoid the influence of atomic depolarization in FADOF on the performance of Faraday lasers, whereas achieving high power, narrow linewidth, single longitudinal mode Faraday lasers are the major difficulties in current Faraday laser research that need to be conquered. In the future development of Faraday lasers, interdisciplinary integration will play an important role. Due to the performance characteristics of Faraday lasers, research combining them with precision spectroscopy and cold atom physics will become an important research direction for future Faraday lasers.

2.4.2 Faraday Laser Application Requirements

Traditional narrow linewidth diode lasers generally use gratings, interference filters, F-P cavities, and other devices for frequency selection, and have played an important role in various fields. However, the above macro frequency selection devices do not have quantum frequency references and are inevitably affected by environmental factors such as mechanical vibration and temperature. In addition, in applications related to precision measurement such as atomic clocks, it is necessary to precisely control the angle or position of the frequency selection device to achieve laser output at the corresponding atomic transition frequency. Faraday lasers use the atomic transition lines in atomic filters as the basis for frequency selection. The output laser frequency can intrinsically align with the atomic transition frequency, achieving better frequency stability. The wavelength corresponding to the atomic spectral line with narrow linewidth, and its good robustness to temperature and current fluctuations give Faraday lasers significant advantages over traditional narrow linewidth diode lasers. It is expected to become a research hotspot in the field of diode lasers in the future. Faraday lasers are expected to be applied to optical communication, laser cooling, high-resolution spectroscopy, laser sensing, laser radar, and other fields in the future, becoming a high-quality laser source.

Recently, Faraday lasers are also found to be advantageous in the field of quantum technology, particularly in quantum precision measurement. The realization of various quantum technologies largely utilizes the characteristics of laser interaction with atoms and ions (such as atomic clocks, magnetometers, gravimeters, gyroscopes, etc.). Diode lasers corresponding to atomic spectral lines are the foundation and important component of quantum precision measurement technology, and they hold an irreplaceable position in defense, scientific research, and civilian economy. Especially in the field of defense and military industry, diode lasers are widely used in quantum precision measurement, quantum computing and quantum communication, global satellite navigation, and radar detection. At present, in the above application needs, we mainly rely on commercial diode lasers, but these diode lasers based on grating feedback and other solutions, the frequency cannot correspond intelligently with atomic spectral lines, and the laser frequency is easily affected by current fluctuations and temperature fluctuations, there are unavoidable technical shortcomings. The lack of frequency stability and poor long-term continuous operation capability to some extent restrict the development of the quantum field.

To solve this problem, Peking University first proposed the use of Faraday atomic filters as intracavity frequency selection devices in 2011 [20], and developed a Faraday laser based on this principle. The output frequency of Faraday laser is immune to current and temperature fluctuations of the laser diode, which has an overwhelming advantage. The realized laser can maintain a stable output wavelength within a current fluctuation of ± 35 mA and a temperature fluctuation of ± 10 °C, and continuously work on the atomic spectral line for more than 6 months, which can solve the difficult problems of frequency stability and long-term continuous operation. The unique properties of the Faraday laser make it promising to be quickly applied to large-scale precision measurement devices such as reference atomic clocks and atomic gravimeters with reduced system volume, improved system performance, expanding a wider range of application scenarios, and further promoting the development of defense, economy, and scientific research fields such as the reproduction and traceability of the international system of units, the upgradation of the global satellite navigation system, the construction of timing and timekeeping systems, information network communication, and verification of physical principles.

References

1. M. Faraday, On the magnetization of light and the illumination of magnetic lines of force. Philos. Trans. R. Soc. Lond. **136**, 1–20 (1846)
2. M. Faraday, Experimental researches in electricity. Nineteenth series. Philos. Trans. R. Soc. Lond. **136**, 1–20 (1846)
3. J. Tyndall, *Faraday as a Discoverer* (D. Appleton, 1868)
4. G. Li, *Research on the Content of "Magnetic-Optic Rotation" in "Faraday's Diary"* (Inner Mongolia Normal University, Inner Mongolia, 2006)
5. P.N. Schatz, A.J. Mccaffery, The Faraday effect. Q. Rev. Chem. Soc. **23**(4), 552–584 (1969)
6. J.A. Crowther, *The Life and Discoveries of Michael Faraday* (Society for Promoting Christian Knowledge, 1918)
7. J.C. Maxweel, On physical lines of force. Lond. Edinb. Dubl. Philos. Mag. J. Sci. **21**(141), 338–348 (1861)
8. B. Luo, *Principle and Technology of Atomic Filter* (Beijing University of Posts and Telecommunications Press, Beijing, 2018)
9. D.J. Dick, T.M. Shay, Ultrahigh-noise rejection optical filter. Opt. Lett. **16**(11), 867–869 (1991)
10. J. Menders, K. Benson, S.H. Bloom et al., Ultranarrow line filtering using a Cs Faraday filter at 852 nm. Opt. Lett. **16**(11), 846–848 (1991)
11. B. Yin, T.M. Shay, Faraday anomalous dispersion optical filter for the Cs 455 nm transition. IEEE Photonics Technol. Lett. **4**(5), 488–490 (1992)
12. J.Y. Xiong, *Research on Key Technologies in Atomic Filters* (Peking University, Beijing, 2018)
13. J. Gelbwachs, C. Klein, J. Wessel, Infrared detection by an atomic vapor quantum counter. IEEE J. Quantum Electron. **14**(2), 77–79 (1978)
14. P.P. Sorokin, J.R. Lankard, V.L. Moruzzi et al., Frequency-locking of organic dye lasers to atomic resonance lines. Appl. Phys. Lett. **15**(6), 179–181 (1969)
15. W.D. Lee, J.C. Campbell, Optically stabilized $Al_xGa_{1-x}As$/GaAs laser using magnetically induced birefringence in Rb vapor. Appl. Phys. Lett. **58**(10), 995–997 (1991)
16. J. Kitching, A. Yariv, Y. Shevy, Room temperature generation of amplitude squeezed light from a diode laser with weak optical feedback. Phys. Rev. Lett. **74**(17), 3372 (1995)
17. S. Machida, Y. Yamamoto, Y. Itaya, Observation of amplitude squeezing in a constant-current–driven diode laser. Phys. Rev. Lett. **58**(10), 1000 (1987)
18. X.J. Lu, J.X. Tang, L.M. Zheng, Experimental study on external cavity diode laser using Rb-Faraday anomalous dispersion filter. Acta Electron. Sin. **02**, 79–82 (1997)

19. F.Q. Li, Y.P. Wang, X.W. Cheng et al., Frequency stabilization of diode laser using atomic Faraday anomalous dispersion optical filter for optical feedback. Chin. J. Lasers **10**, 15–18 (2005)
20. X.Y. Miao, L.F. Yin, W. Zhuang et al., Note: demonstration of an external-cavity diode laser system immune to current and temperature fluctuations. Rev. Sci. Instrum. **82**(8), 086106 (2011)
21. M. Zentile, *Applications of the Faraday Effect in Hot Atomic Vapours* (Durham University, 2015)
22. Z. Tao, X. Zhang, D. Pan et al., Faraday laser using 1.2 km fiber as an extended cavity. J. Phys. B At. Mol. Opt. Phys. **49**(13), 13LT01 (2016)
23. Z.M. Tao, Y.L. Hong, B. Luo et al., Diode laser operating on an atomic transition limited by an isotope ^{87}Rb Faraday filter at 780 nm. Opt. Lett. **40**(18), 4348–4351 (2015)
24. P.Y. Chang, H.F. Peng, S.N. Zhang et al., A Faraday laser lasing on Rb 1529 nm transition. Sci. Rep. **7**(1), 8995 (2017)
25. P.Y. Chang, H.B. Shi, J.X. Miao et al., Frequency-stabilized Faraday laser with 10^{-14} short-term instability for atomic clocks. Appl. Phys. Lett. **120**(14), 141102 (2022)
26. H. Tang, H.Z. Zhao, D. Zhang et al., Polarization insensitive efficient ultra-narrow diode laser strictly locked by a Faraday filter. Opt. Express **30**(16), 29772–29780 (2022)
27. H.B. Shi, P.Y. Chang, Z.Y. Wang et al., Frequency stabilization of a cesium Faraday laser with a double-layer vapor cell as frequency reference. IEEE Photonics J. **14**(6), 1–6 (2022)
28. J. Chen, Active optical clock. Chin. Sci. Bull. **54**(3), 348–352 (2009)
29. W. Zhuang, J. Chen, Active Faraday optical frequency standard. Opt. Lett. **39**(21), 6339–6342 (2014)
30. J. Zhang, G. Gao, B. Wang et al., Background noise resistant underwater wireless optical communication using Faraday atomic line laser and filter. J. Lightwave Technol. **40**(1), 63–73 (2022)

Chapter 3
Laser Diodes for Faraday Lasers

The laser diode (LD), also known as diode lasers, have witnessed significant improvements in its performance parameters and have made substantial progress since its inception in 1962. Compared to gas or solid-state lasers, diode lasers exhibit numerous advantages including small size, light weight, high efficiency, and low cost, along with the facility for easier pumping and modulation. Consequently, diode lasers have found extensive and profound applications in various fields including quantum precision measurement, laser spectroscopy analysis, cold atom physics, laser communication, as well as national defense and military affairs.

This chapter will focus on the fundamental principles and main classifications of laser diodes, along with a detailed account of the performance characteristics and manufacturing processes of laser diodes for Faraday laser. The objective of this chapter is to assist readers in achieving a comprehensive and profound understanding of high-performance, specialized laser diodes within the realm of precision measurement, thereby fostering the further development and application of diode laser technology.

3.1 Laser Diodes

Diode lasers, as a type of miniaturized light source device, have a core working material composed of PN junctions or PIN junctions made from direct bandgap semiconductor materials. These diode lasers working materials are diverse, each with a specific band structure and electron transition characteristics, corresponding to different laser wavelengths and spectral features. Currently, researchers have successfully used various semiconductor materials to create lasers with different performances, including gallium arsenide (GaAs), gallium nitride (GaN), indium arsenide (InAs), indium antimonide (InSb), cadmium sulfide (CdS), cadmium telluride (CdTe), lead selenide (PbSe), lead telluride (PbTe), aluminum gallium

J. Chen et al., *Faraday Laser*, https://doi.org/10.1007/978-981-97-8023-5_3

arsenide ($Al_xGa_{1-x}As$), and indium phosphide arsenide (InPAs), etc. The discovery and application of these materials not only enrich the types of diode lasers but also provide a solid foundation for their wide applications in scientific research, industry, military, and medical fields.

The excitation methods of diode lasers are mainly divided into three types: electric injection, optical pumping, and high-energy electron beam excitation. Among them, electric injection, due to its ease of implementation, has become the excitation method used by the vast majority of diode lasers. In the electric injection process, by applying a forward voltage to the PN junction, the carriers in the junction plane area recombine, thereby generating stimulated emission, that is, a forward-biased diode that forms self-sustaining laser oscillation under the action of the resonant cavity. That is why diode lasers are alternatively known as laser diodes. For the semiconductor gain medium, its electron transitions occur between wide energy bands, not between discrete narrow energy levels. This transition characteristic makes the transition energy not a fixed value, but shows a wide spectral feature. This wide spectral characteristic causes the output wavelength of the diode laser to be distributed within a certain range. The wavelengths emitted by different semiconductor gain materials cover multiple bands from 0.3 to 34 μm, and the specific wavelength band depends on the bandgap of the material used. For example, the output wavelength of the common $Al_xGa_{1-x}As$ double heterojunction laser is mainly concentrated between 750 and 890 nm. Figure 3.1 shows the basic structure of a GaAs laser. Its core part is a rectangular PN junction, with a length of about 250 μm and a width of about 100 μm [1].

The laser's volume is comparable to a pinhole. Its two ends are kept parallel, forming the front and back reflection surfaces of the laser resonator. The semiconductor gain medium, as the core component of the laser, forms the conduction band

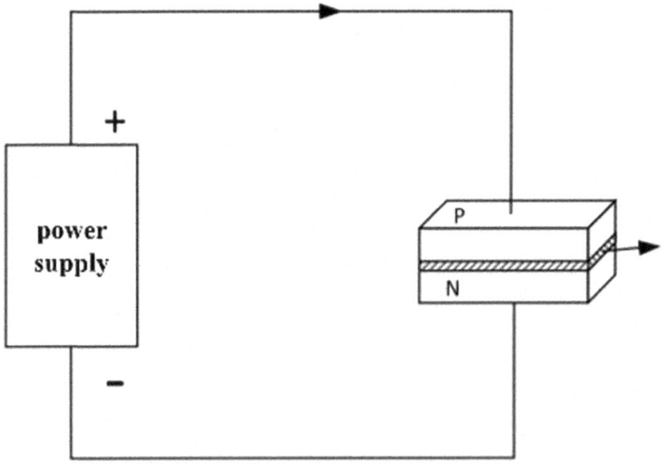

Fig. 3.1 Schematic diagram of the basic structure of a diode laser

and the valence band inside, and the bandgap between the conduction band and the valence band is called forbidden energy band, with the width, E_g. When electrons in the valence band are stimulated to transit to the conduction band, the valence band correspondingly generates holes (considered as positive charges). When a P-type semiconductor comes into contact with an N-type semiconductor, a PN junction is formed. When no voltage is applied to the ends of the PN junction, the further diffusion of carriers will be suppressed due to the built-in electric field of the PN junction.

The phenomenon of light emission in semiconductors is mainly caused by carrier recombination. When a positive voltage is applied to the PN junction of the semiconductor, the potential barrier function of the PN junction is reduced, allowing electrons to be injected into the N region and holes into the P region, resulting in recombination and the emission of photons of a specific wavelength. The wavelength of the photons emitted by recombination is determined by the bandgap width of the semiconductor E_g using the relation $\lambda = hc/E_g$.

Diode lasers, due to their small size, light weight, low cost, and long lifespan (up to tens of thousands of hours or more), and the use of a simple electrical pumping method, are easy to integrate with other components. These characteristics have led to their widespread use in fields such as optical communication, optical storage, and quantum precision measurement.

3.1.1 *Basic Principles*

As a coherent radiation light source, diode lasers require three fundamental conditions to generate coherent radiation:

(1) The population inversion condition corresponding to other lasers, for diode lasers, is that the number of electrons at the bottom of the high-energy conduction band is more than the number of holes at the top of the low-energy valence band, corresponding to a distribution of electron–hole pairs. This is achieved by applying a forward bias to the heterojunction, injecting the necessary carriers into the active layer, for which a pump source is needed.
(2) The presence of a suitable optical resonant cavity is essential to enable multiple reflections of the stimulated radiation within, leading to the establishment of a self-sustaining laser oscillation.
(3) To achieve stable oscillation, a sufficiently large gain must be provided such that the optical gain equals or exceeds the sum of all losses. In the case of current injection pumping, a certain current threshold condition must be met.

The following will discuss these issues separately.

1. The physical process of stimulated radiation

Considering a simple atomic system with two discrete energy levels E_1 and E_2 as an example, as shown in Fig. 3.2, where E_1 is the ground state energy level, and E_2 is the

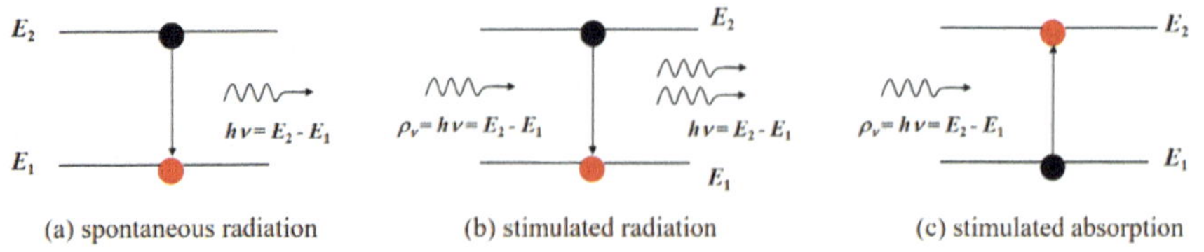

Fig. 3.2 Three basic atomic transition processes

excited state energy level. At normal temperature, most electrons are in the ground state. When atoms transit between E_1 and E_2, three basic processes of spontaneous radiation, stimulated radiation, and stimulated absorption will occur [2].

Figure 3.2a shows an atom at the high energy level E_2 transits spontaneously to a lower energy level E_1, and emits a photon with energy $hv = E_2 - E_1$. This process is called spontaneous emission. The characteristic of spontaneous emission is that each atomic transition occurs independently, thus emitting disorderly incoherent radiation. The lifetime of spontaneous emission, which is the average time when an atom is in an excited state, is generally between 10^{-9} and 10^{-3} s. For electrons in the gain semiconductor, the length of the lifetime depends on semiconductor parameters, such as the width of bandgap and the density of recombination centers.

Figure 3.2b shows that when a photon with energy hv acts on an atom at the excited energy level E_2, the atom is stimulated from the unstable excited energy level E_2 to return to the ground state, and emits a photon with the same frequency, phase, and direction as the incident radiation photon. This process is called stimulated emission.

Figure 3.2c shows the process of stimulated absorption where the atom receives radiation energy hv, and transitions from the ground state energy level E_1 to the excited energy level E_2. The atom in the excited state is unstable and tends to return to the ground state spontaneously, emitting a photon with energy hv.

In the laser working material, the external photon hv can cause stimulated emission of the excited state atom, and may also be absorbed by the ground state atom. These two processes coexist, and the probabilities of stimulated emission and absorption are the same. At room temperature, there are many more ground state atoms than excited state atoms, so absorption is greater than emission. To generate a laser, the total emission must be greater than the total absorption. Therefore, one of the necessary conditions for laser generation is that stimulated emission dominates.

2. Population (electron–hole) inversion distribution

In order to make stimulated emission dominate in the laser working material, it is necessary to inject energy into the working material from the outside, such as using optical excitation or forward PN junction injection. These methods can effectively increase the number of carriers in the excited state, significantly exceeding the number of carriers in the ground state, thereby achieving the inversion of carrier distribution. The population (electron–hole) inversion not only makes stimulated emission change from a secondary position to a dominant position, but is also a necessary condition for laser emission.

There are multiple approaches to achieve population inversion in laser materials. For solid-state lasers, this is often realized through intense light irradiation of the laser material at a specific spectral line. In contrast, gas lasers commonly employ gas ionization methods. In the context of semiconductors, electron energies are represented by energy bands consisting of a series of nearly continuous energy levels. To achieve population inversion in semiconductors, it is necessary to ensure that the number of electrons at the bottom of the conduction band, which represents the higher energy state, significantly exceeds the number of holes at the top of the valence band, which represents the lower energy state. This is typically achieved by applying a forward bias to the PN junction, injecting carriers into the active layer, causing atomic transition from ground-state atoms to excited states, and subsequently exciting electrons from the lower-energy valence band to the higher-energy conduction band. When a large number of electrons and holes in the population inversion state recombine, stimulated emission occurs.

3. Resonant cavity

The generation of laser light is inseparable from the core condition of population inversion distribution. While the gain of the active material is enhanced after achieving population inversion, this alone is insufficient for laser generation. To produce a coherent beam of light, it is further necessary to ensure the establishment of "oscillation" and the formation of resonance. This process relies on the function of an optical resonant cavity, which enables the output of coherent emission. The positive feedback mechanism of the resonant cavity contributes to the continuous accumulation of optical intensity, ultimately leading to the generation of laser light.

By placing two parallel mirrors on both sides of the laser working material, a "resonance" phenomenon of light is formed. This device is usually called a "Fabry–Pérot resonator" or "resonant cavity". In this resonant cavity, not all photons in all directions can resonate. Those photons whose direction is not parallel to the axis of the resonant cavity will be reflected out of the cavity. Only those photons whose direction is parallel to the axis can resonate in this cavity, get enhanced, and finally turn into stimulated radiation. Specifically, these photons parallel to the cavity axis reflect back and forth in the cavity, passing through the working material repeatedly. Each time they pass through, the photons are enhanced due to stimulated radiation, causing the number of photons to continue to grow. In the design of diode lasers, the resonant cavity and the working material are usually combined together, generally using the cleavage plane of the semiconductor crystal itself as the resonant cavity, simplifying the structure and improving efficiency.

When the photons of stimulated radiation oscillate in the resonant cavity, they will gradually be lost due to scattering, transmission, and absorption. If the number of new photons generated each time with each photonic oscillation in the cavity is far more than the number of losses, that is, the gain in the cavity is far greater than the loss, then lasing can be achieved. As the gain increases, the gain and loss eventually reach a dynamic balance, and this process can continue, producing stable laser resonance. In addition, when light reflects between the two reflective surfaces of the resonant cavity, two columns of light waves in opposite directions will be formed. These two columns of light waves need to superimpose to form standing waves, and this oscillation is stable. That is, these two columns of light waves need to reach a balance to produce a stable laser output. In order to produce stable oscillation, the length of the resonant cavity L must be exactly equal to an integer multiple of the half-wavelength of the radiation light. Because only when the cavity length and the wavelength of the light wave satisfy this relationship, the process of reflection and superposition of the light wave in the cavity can reach the best balance, thereby ensuring the stable output of the laser, given by

$$L = m\left(\frac{\lambda}{2n}\right) \tag{3.1}$$

In above, n is the refractive index of the medium, and m is a positive integer. Different values of m correspond to standing waves of different wavelengths.

The various possible standing waves formed in the resonant cavity along the direction of the cavity axis are usually referred to as the longitudinal modes of the resonant cavity. The resonant frequency of the resonant cavity, or the longitudinal mode frequency, can be derived from formula (3.1):

$$v = \frac{mc}{2nL} \tag{3.2}$$

The longitudinal mode frequency v is related to the cavity length L and the number of reflections in the medium material n.

There are various forms of optical resonant cavities that form light feedback, among which the simplest is the Fabry–Pérot cavity (F-P cavity). The characteristic of this resonant cavity is its simple structure, and it does not require the addition of extra mirrors, but directly uses the cleavage planes at both ends of the diode laser as mirrors. Cleavage planes are planes formed by cutting the semiconductor crystal in a specific way, so that they can have good reflectivity, and can reflect light back into the resonant cavity, thereby achieving light feedback. Figure 3.3 is a schematic diagram of the F-P cavity, with a reflectivity of:

$$R_m = \left(\frac{n' - 1}{n' + 1}\right)^2 \tag{3.3}$$

In Eq. 3.3, n′ is the refractive index of the gain medium, with a typical value of 3.5, so the reflectivity of the cleavage plane is about 30%.

In actual optical resonators, there are always various losses. If the optical gain is not sufficient, it will not be able to compensate for these losses, resulting in the laser's inability to stably output light. In order to achieve stable oscillation of the laser, the gain must reach or exceed the total loss. This critical point of gain is called the threshold gain. Therefore, a certain amount of current needs to be injected into the laser to reach the threshold gain, and this current is called the threshold current.

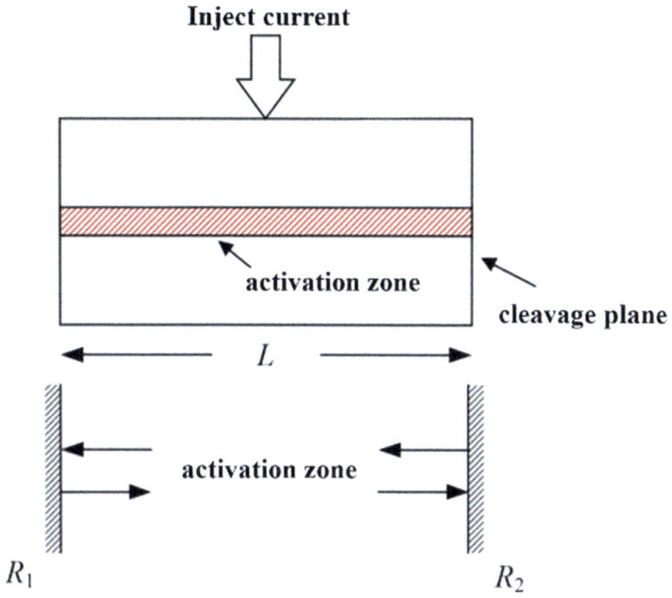

Fig. 3.3 Schematic diagram of Fabry–Pérot optical resonator. R_1, R_2 reflecting mirrors

Suppose a plane wave with an amplitude of E_0, a frequency of ω, wave number $k = n\omega/c$, after a round trip in a light cavity with a length of L and a power gain coefficient of g, its amplitude increases by $\exp[(g/2)(2L)]$ times, and the phase changes by $2kL$. Considering the various absorption and scattering losses in the laser and the end face transmission output, its amplitude change is $\sqrt{R_1R_2}\exp(-\alpha_{\text{int}} L)$, where R_1, R_2 are the end face reflectivity, α_{int} is the total loss rate inside the cavity. While working stably, the intensity of the light wave E_0 should remain unchanged after one round trip in the cavity. That is

$$E_0\exp(gL)\sqrt{R_1R_2}\exp(-\alpha_{\text{int}}L)\exp(i2kL) = E_0 \tag{3.4}$$

Let the amplitude and phase of both sides of the equation be equal, then we get

$$g = \alpha_{\text{int}} + \frac{1}{2L}\ln\left(\frac{1}{R_1R_2}\right) \tag{3.5}$$

$$2kL = 2m\pi, \quad \text{or } \nu = \nu_m = \frac{mc}{2nL}$$

where, $k = 2\pi n\nu/c$, m is an integer.

The stable operation of a laser requires satisfying both amplitude and phase conditions. The amplitude condition stipulates the minimum values for gain and current, meaning that the laser can only operate stably when the gain and current reach or

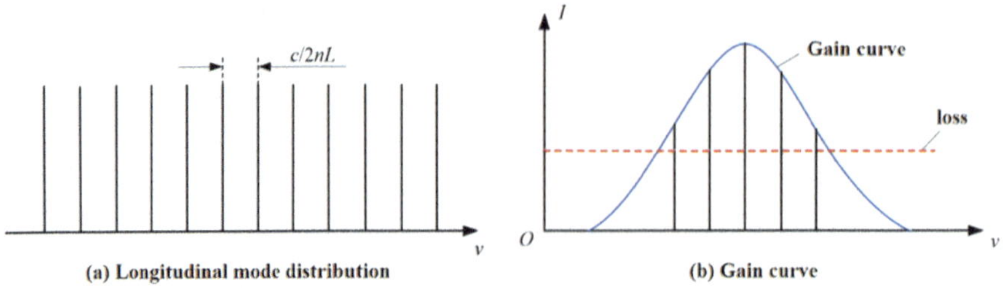

Fig. 3.4 Distribution of laser longitudinal modes and gain curve

exceed a certain value. The phase condition stipulates that the oscillation frequency of the laser must be one of the values in $v_m = mc/2nL$. A laser output can only be formed at a certain frequency when the gain of a particular longitudinal mode is greater than or equal to the loss. The distribution of laser longitudinal modes is shown in Fig. 3.4a, with the frequency interval between adjacent longitudinal modes being $c/2nL$. Figure 3.4b shows the gain curve, revealing the gain characteristics at different frequencies. For an F-P cavity, all photons at all frequencies have the same loss, so only those longitudinal modes with gains greater than the loss can produce laser radiation. Depending on the number of excited longitudinal modes, lasers can be divided into multi-longitudinal mode lasers and single-longitudinal mode lasers. Multi-longitudinal mode lasers mean that multiple longitudinal modes are excited at the same time, while single-longitudinal mode lasers only excite one longitudinal mode. To realize a single-longitudinal-mode laser, frequency-selective methods such as incorporating dispersive elements into the resonant cavity or employing external cavity feedback can be adopted.

3.1.2 Main Classification

Diode lasers can be divided into homojunction, single heterojunction, double heterojunction, and other types according to their structure. This section will focus on the introduction of heterojunction diode lasers.

1. PN junction diode injection laser structure and principle

The four different resonant cavity structures of junction lasers are shown in Fig. 3.5. Among them, the F-P cavity is the most commonly used structure, and the other three are less used in practical applications due to their difficult manufacturing and lower laser emission efficiency.

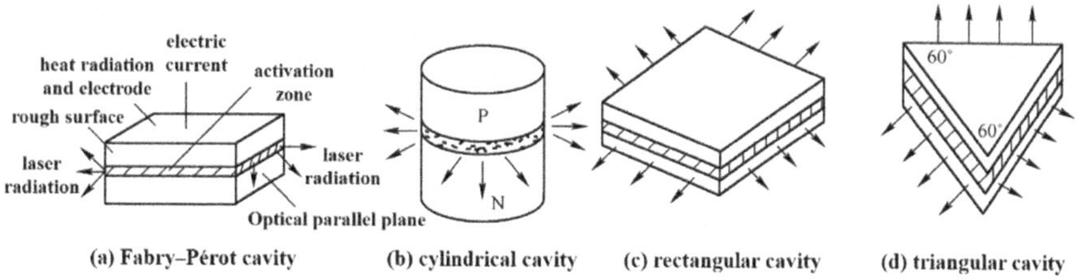

Fig. 3.5 Schematic diagram of junction laser structure [2]

GaAs is a semiconductor material with the capability of emitting laser light [2], and GaAs lasers are currently widespread applications. Group IV–VI compounds such as PbSe, PbTe, and PbS also possess the ability to emit laser light. The existence of these compounds further enriches the material options for lasers, enabling the selection of appropriate materials to fabricate diode lasers of different frequencies according to specific needs.

In order to make the number of carriers in the excited state far greater than the number of carriers in the ground state, high-concentration doped semiconductor materials need to be used on both sides of the PN junction, with doping concentrations usually reaching 10^{18}–10^{19} cm^{-3}. Such doping concentration allows the Fermi level to enter the conduction energy band and valence energy band, creating favorable conditions for the generation of lasers. In Fig. 3.5a, the left and right sides of the F-P cavity are the main laser output end faces, which are a pair of parallel cleavage or polished surfaces, and these two surfaces are perpendicular to the plane of the PN junction. These two end faces act like mirrors, forming the end reflection surfaces of the resonant cavity, allowing the laser to reflect back and forth in the cavity, enhancing the intensity of the laser. The left and right sides of the resonant cavity are designed to have rough surfaces, which can eliminate lasers in directions other than the main output direction, ensuring that the laser outputs from the front and back. After welding the lead wire on this resonant cavity, it provides the electrical energy required for the diode to work, which can drive the electrons and holes in the PN junction to move and generate lasers.

Junction lasers employ heavily doped semiconductor materials. In the absence of external excitation, the valence band top in the P-region is devoid of electrons, indicating that all electrons in this region have been filled to lower energy levels. Conversely, the conduction band bottom in the N-region contains a high concentration of electrons due to the doping in the N-region, which introduces more electrons into the conduction band, thereby increasing the electron concentration.

Figure 3.6 shows the energy band structure formed after applying a forward bias to the PN junction. In this structure, due to the effect of the external voltage U, the Fermi level of the N-region enters the conduction energy band, and the Fermi level of the P-region enters the valence energy band. This change reduces the potential barrier between the PN junctions. The reduction of the potential barrier allows a large number of electrons to easily cross the barrier from the N-region and recombine with

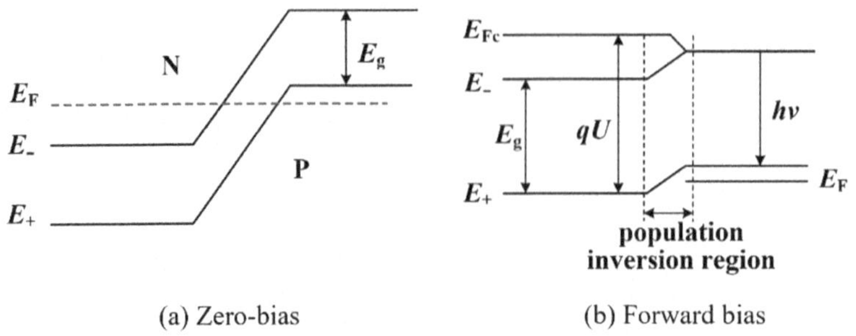

Fig. 3.6 Energy band structure of heavily doped semiconductor materials

the holes in the P-region. In this process, the energy of the electrons and holes is released in the form of photons, with energy *hv*. Simultaneously, the holes in the P-region can also flow to the N-region to recombine with the electrons, also emitting photons. When the external voltage is large enough, that is, qU is greater than or equal to the bandgap width Eg of the material, the energy of the photon *hv* will also be greater than or equal to E_g. In this case, a population inversion region will appear within the diffusion length range on both sides of the barrier region. This region is the working area for laser emission, where the distribution of electrons and holes has reversed, allowing more electrons and holes to recombine and emit photons. Finally, through the feedback mechanism of end-face reflection, these photons are repeatedly reflected and enhanced within the PN junction, eventually generating a laser. This structure and working mechanism enable the junction laser to efficiently generate lasers.

2. Heterojunction lasers

The PN junction (homojunction) laser has a high threshold current at room temperature, which prevents it from achieving continuous oscillation at normal temperature, severely limiting its practical application. To solve this problem, Kroemer proposed the use of a heterojunction structure in 1963 [3]. This method involves adding materials with a wider forbidden energy band on both sides of the material with a narrower forbidden energy band. The movement of carriers is effectively restricted, increasing the carrier concentration in the junction area. Due to the increase in carrier concentration, the efficiency of stimulated radiation has also been correspondingly improved. This technological innovation has shifted the development of diode lasers from homojunctions to heterojunctions.

(1) *Single heterojunction lasers*

Figure 3.7 shows the energy band structure of a single heterojunction laser. In this structure, a GaAs-$Al_xGa_{1-x}As$ heterojunction is formed on the PN junction of GaAs by molecular beam epitaxy or liquid phase epitaxy. The characteristic of this heterojunction lies in the difference in the bandgap widths between GaAs and $Al_xGa_{1-x}As$. This difference forms a higher potential barrier at the interface. When electrons are

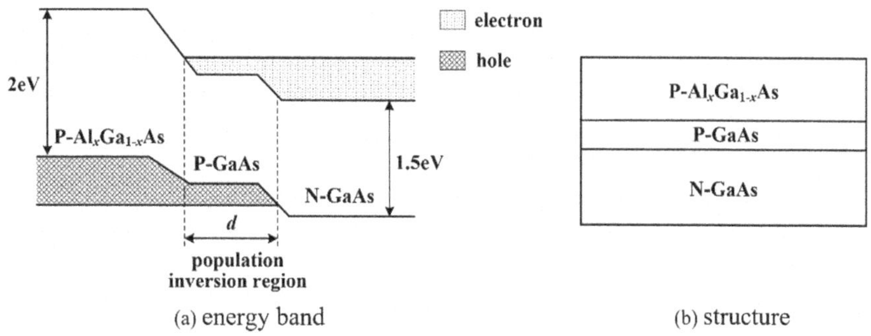

Fig. 3.7 Energy band structure of a single heterojunction laser

injected from N-GaAs into P-GaAs, this barrier hinders the electrons from further diffusing into P-$Al_xGa_{1-x}As$. Compared with the situation without this barrier, the electron concentration in the P-GaAs layer will increase, thereby increasing the probability of radiative recombination. This means that more electrons can combine with holes, producing more photons, thereby improving the luminous efficiency of the laser. In addition, P-type AlGaAs has a small light absorption coefficient from P-type GaAs, which means that the energy loss of light passing through P-type AlGaAs is less. Moreover, because the refractive index of AlGaAs is lower than that of GaAs, this restricts photons from entering the AlGaAs region, causing the light to be reflected and confined within the P-GaAs region. Since both electrons and photons are restricted at the heterojunction interface, the loss is reduced, thereby lowering the threshold.

At low temperatures, the threshold current density of a single heterojunction is almost the same as that of a homojunction. When the ambient temperature changes, the sensitivity of the threshold current density of the heterojunction laser to temperature is relatively low. For example, at room temperature, its threshold current density can be reduced to 8000 A/cm^2. Inspite of reduced threshold, this type of laser can only achieve pulse oscillation at room temperature and cannot achieve continuous oscillation. In order to further reduce the threshold current and achieve continuous oscillation at room temperature, researchers have developed double heterojunction lasers.

(2) *Double heterojunction lasers*

Figure 3.8a shows the energy band structure of a double heterojunction laser. In this structure, the two sides of GaAs are N-type $Al_xGa_{1-x}As$ and P-type $Al_xGa_{1-x}As$, respectively. The active region on both sides has symmetry, so the laser active region can be either in the N region or in the P region. Taking the P-type as an example, when a forward bias is applied, electrons will be injected into the active region and reach the interface of P-GaAs and P-$Al_xGa_{1-x}As$. At this interface, the electrons will encounter a potential barrier, which will prevent the electrons from entering the P-type $Al_xGa_{1-x}As$ layer. Due to this blocking effect, the electron concentration in

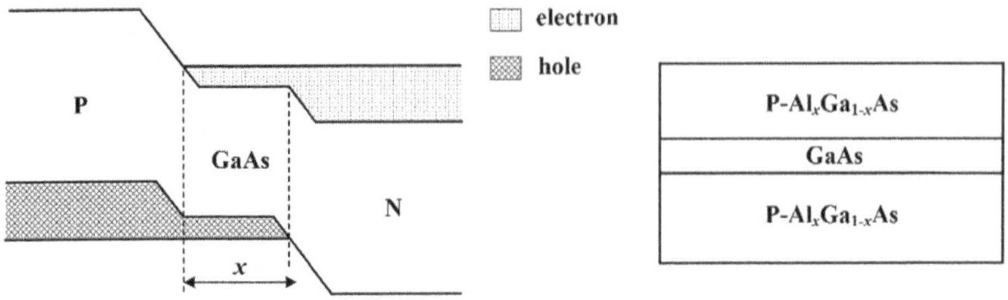

(a) energy band (b) structure

Fig. 3.8 Energy band structure of the double heterojunction laser

the P-type GaAs layer will increase, thereby increasing the gain of the laser. In addition, the potential barrier between N-type Al_xGa_{1-x} and P-type GaAs also avoids the common hole injection phenomenon in single heterojunction lasers. As a result, the concentrations of electrons and holes in the working area will increase, and the probability of electron–hole recombination will also increase. In addition, due to the significant changes in the refractive index at both interfaces, photons are more effectively confined within the active region, reducing energy loss. Therefore, the threshold current of the double heterojunction laser is further reduced to 1000 ~ 3000 A/cm^2, achieving continuous oscillation at room temperature. Experiments have shown that the threshold current also changes little with temperature.

Currently, the main development goals of heterojunction lasers are to further reduce their threshold current density and improve laser efficiency. In addition, expanding the spectral band of heterojunction lasers is also extremely important, which means that the laser can cover a wider range of light waves, meet the application needs of more fields, and provide a richer choice of light sources for scientific research and technical applications.

3.1.3 Performance Characteristics

1. Threshold current

In order to maintain the continuous oscillation of the laser, it is necessary to ensure that the photon generation rate is greater than its absorption and loss rate. This photon generation rate, which just compensates for absorption and loss, is called the threshold. In simple terms, the threshold is the minimum photon generation rate required for the laser to start generating lasers. Figure 3.9 shows the light output characteristic curve of the laser, where I_{th} represents the threshold current, which is the current density required to be injected when the laser reaches the threshold gain. When the injected current is low, spontaneous radiation is mainly generated, but as the current value gradually increases, the gain increases accordingly. When the current reaches the threshold current, the laser starts to generate.

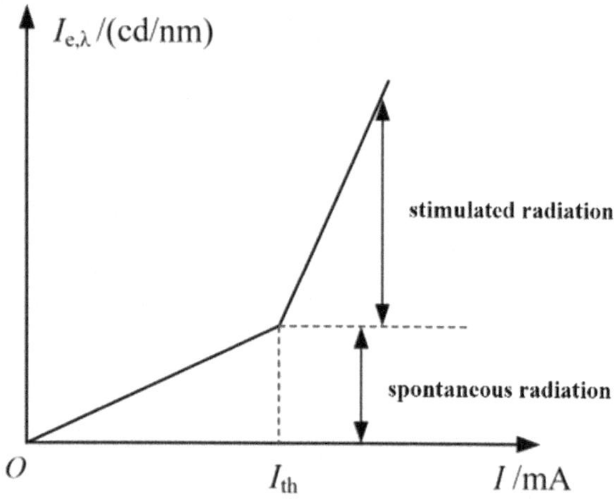

Fig. 3.9 Laser output characteristic curve

The main factors affecting the threshold are: (1) An increase in crystal doping concentration will lower the threshold. (2) The smaller the loss of the resonant cavity, the lower the threshold. For example, by increasing the reflectivity, the loss can be reduced, thereby lowering the threshold. (3) The threshold is also related to the type of semiconductor material. Generally speaking, the threshold current of a heterojunction is lower than that of a homojunction. The threshold current of a homojunction at room temperature exceeds 30,000 A/cm^2, a single heterojunction is about 8000 A/cm^2, and a double heterojunction is only about 1600 A/cm^2. (4) Temperature is also an important factor affecting the threshold current. At low temperatures, the threshold current will increase with the increase in doping concentration; whereas at high temperatures, the threshold current is proportional to the cube of the temperature. Hence, the higher the temperature, the higher the threshold. Therefore, in order to ensure the stable operation of the laser, it is best to use diode lasers at either low temperature or room temperature.

2. Spectral characteristics

The spectral distribution of a junction laser is closely related to the semiconductor material used. For example, lasers made from GaAs material emit a laser with a wavelength of 840 nm at a low temperature of 77 K, while InP material emits light of 910 nm at 30 K. The spectral distribution of a typical GaAs laser is shown in Fig. 3.10. When the current is below the threshold of the laser, the spectrum is mainly generated by spontaneous radiation, producing incoherent light, so the spectral line is wide, as shown in curve 1 of Fig. 3.10. However, as the current gradually increases, stimulated radiation begins to dominate. Once the current exceeds the threshold, the intensity of the stimulated radiation increases sharply, and the spectral line becomes narrow. At this moment, there is an exponential relationship between light intensity and gain coefficient, as shown in curves 2 and 3 in Fig. 3.10. At a certain frequency, the spectral line is enhanced most significantly. This special frequency is actually the standing wave frequency formed in the resonant cavity, which is also the laser frequency emitted by the laser.

3. Temperature characteristics

Diode lasers are very sensitive to temperature changes. As the temperature rises, the threshold current of the laser also increase accordingly, as shown in Fig. 3.11. In addition, as the temperature rises, the bandgap energy of the semiconductor gradually decreases, which causes the entire spectrum to shift towards the long-wave direction. Therefore, when using diode lasers, it is necessary to pay attention to the impact of temperature on its performance to ensure its stable operation.

4. Relative intensity noise

The relative intensity noise (RIN) of a laser is the slight fluctuation of the output light intensity caused by the random spontaneous radiation transition of the excited state particles when the laser light generated. The relative intensity noise of the laser is defined as the ratio of the statistical average of the square of the power fluctuation to the square of the average light power. This indicator can quantitatively measure the noise characteristics of the laser, which is of great significance for practical applications.

$$RIN = \frac{\delta P(t)^2}{P_0^2} \tag{3.6}$$

5. Spectral line width

The linewidth of a diode laser is mainly due to the unstable jitter of the output phase. This phase jitter mainly comes from two aspects: one is the random superposition of spontaneous radiation photons, which exists in all lasers and will make the coherence of the laser worse; the other is the change in carrier density. The gain and refractive index of the laser are directly related to the carrier density. Once the density jitters, the phase of the laser will be affected.

The spectral linewidth of a diode laser is a very critical parameter, which characterizes the spectral purity or coherence of the laser. Especially when the F-P cavity diode laser is in the direct modulation working state, spectral line broadening will occur. This broadening phenomenon will become more obvious with the increase of modulation rate, especially when the bias current is below the threshold, the broadening phenomenon will be more serious. This is because when the current is close to the threshold, the injected carrier concentration change rapidly, causing the phase of the spontaneous radiation light field to fluctuate, thereby widening the spectral line.

Although the oscillation mode of the laser and the formation mechanism of the spectral linewidth are not exactly the same, by taking appropriate effective measures, both single-mode oscillation and linewidth narrowing can be achieved. For this, a Faraday laser with an FADOF frequency selection mechanism is an example.

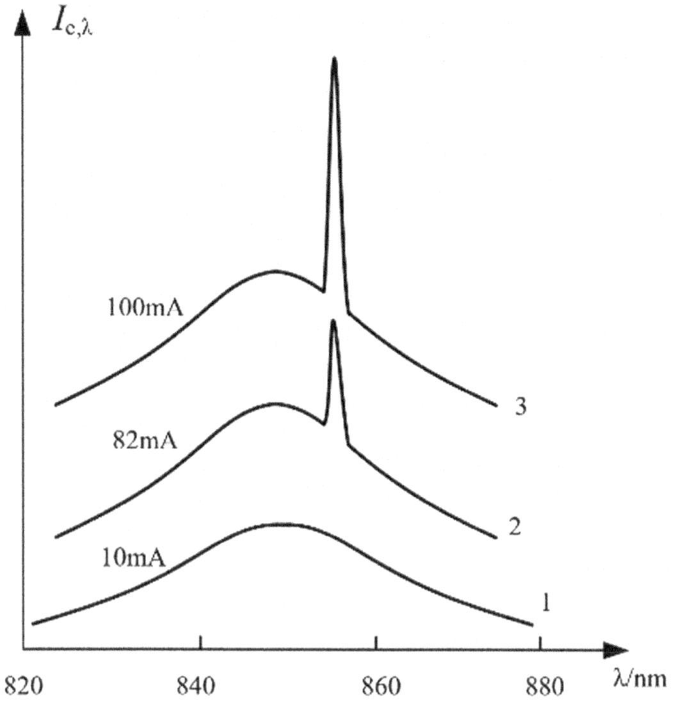

Fig. 3.10 Spectral distribution curve of GaAs laser

Fig. 3.11 Light power changes with temperature curve

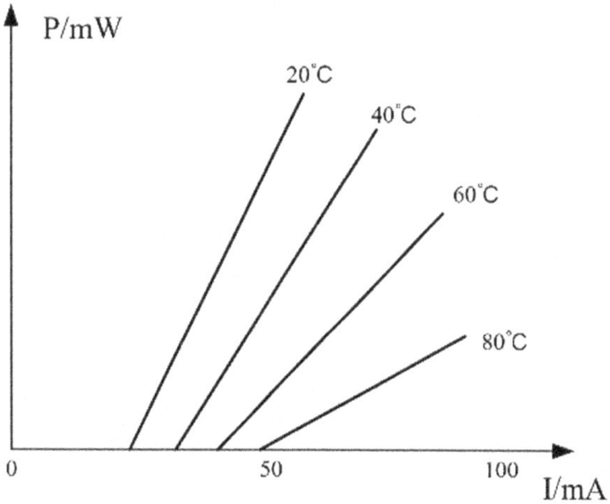

3.2 Special Laser Diode Manufacturing Process Technology for Faraday Lasers

3.2.1 Semiconductor Materials

The performance of diode lasers mainly depends on the properties of the selected materials. Only a very few semiconductor materials possess the characteristics required to manufacture high-quality lasers. When preparing double heterostructure lasers, at least two compatible materials need to be found, serving as the cladding and active region materials respectively. For more complex designs, three to four different bandgap materials may be needed, and these materials must have consistent crystal structures and similar lattice constants, because defects in the materials become the centers of non-radiative recombination, which consume the injected carriers that could have provided gain and light for the laser. Only materials with the same crystal structure and similar lattice constants can grow defect-free single-crystal thin films of another material on the surface of one material through epitaxial growth technology.

For diode laser materials, the ability to efficiently convert input electrical energy into light energy is crucial. This requires the injected electrons and holes to directly recombine and radiate, thereby generating sufficient light gain to reach or even exceed the threshold required for lasing. Therefore, semiconductor materials should be of the direct bandgap material type. Most III–V compounds, whether binary, ternary, or quaternary, belong to this type of semiconductor material. Therefore, III–V compounds are considered ideal materials for manufacturing diode lasers. Currently, researchers have extensively studied III–V, II–V, and V–VI compounds as light-emitting device materials. These compounds cover a very wide range of wavelengths, and Fig. 3.12 shows the emission wavelength ranges and application fields of several commonly used compounds.

3.2.2 Diode Laser Manufacturing Technology

In order to manufacture diode lasers, it is necessary to use epitaxial growth technology on an appropriate substrate, cultivate a lattice-matched single crystal layer, and strictly control its thickness to achieve the layered structure required for diode lasers. Below, a brief overview of several technical methods used to prepare the required thin layer materials is provided. Currently, the three most widely used core technologies are Liquid Phase Epitaxy (LPE), Molecular Beam Epitaxy (MBE), and Metal-organic Chemical Vapor Deposition (MOCVD). These three methods are based on material growth under liquid, vacuum, or gas flow conditions [4].

1. Liquid Phase Epitaxy (LPE)

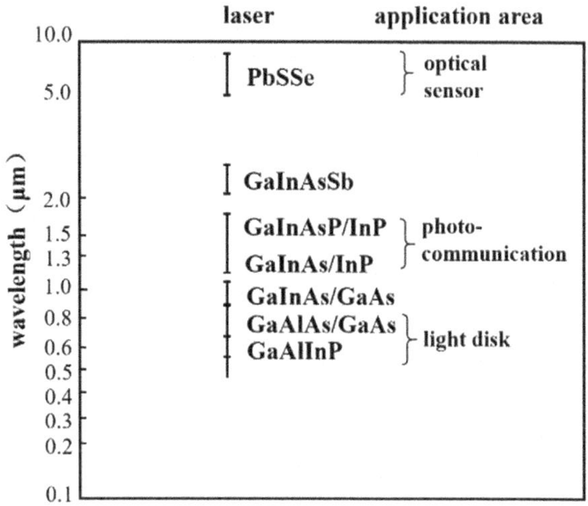

Fig. 3.12 Emission wavelength ranges and application fields of several commonly used compounds

In 1963, Nelson and others proposed the liquid phase epitaxy technology. This technology uses low-melting-point metals (such as Ga, In, etc.) as solvents, growth materials (such as Ga, As, Al, etc.) and dopants (such as Zn, Te, Sn, etc.) as solutes to form saturated or supersaturated solutions. By cooling, the solute precipitates, growing material with a similar crystal structure on the single crystal substrate, achieving epitaxial growth. LPE can grow single crystal layers of various semiconductor materials such as Si, GaAs, GaAlAs, GaP, etc., used to make optoelectronic devices and diode lasers, etc. Figure 3.13 shows a cross-section of the LPE system, with the substrate placed in the graphite slide groove, the solution filled in the small groove, and the material growing from the solution in the small groove. The entire structure is placed in a precisely controlled furnace, and the material growth is controlled by gradually cooling. Currently, in the manufacture of diode lasers, LPE has gradually been replaced by Metal-organic Chemical Vapor Deposition (MOCVD) technology, which has more advantages in material growth and quality control.

2. Metal-organic Chemical Vapor Deposition (MOCVD)

Metal-organic Chemical Vapor Deposition, or Organometallic Vapor Phase Epitaxy, also known as MOCVD or MOVPE. This technology uses the gas molecules of organic metal compounds, with the help of H_2 Gas is introduced into the reaction cell for thermal decomposition to produce compound semiconductors. By controlling the gas flow, the MOCVD method can flexibly adjust the composition of the compound and the doping concentration. Its equipment is simple, the growth speed is fast, the cycle is short, and it is suitable for mass production [5]. The MOCVD system mainly includes a reaction cell, a gas control and mixing system, a reaction source, and a waste gas treatment system [6]. The system obtains mixed gas through gas valves and gas pipelines and sends it to the growth reactor. The substrate is placed on a specially made base, and the base is usually heated by radio frequency induction or resistance heating technology. The MOCVD system can operate in low pressure or atmospheric pressure mode. Although the atmospheric pressure system efficiently uses the reactant gas, the film uniformity is poor and cleaning the reactor is time-consuming; the low-pressure system is suitable for manufacturing that requires highly pure interfaces, such as quantum well structures. Figure 3.14 is a schematic diagram of the Metal Organic Chemical Vapor Deposition (MOCVD) system.

3. Molecular Beam Epitaxy (MBE)

Molecular Beam Epitaxy is a technology that grows high-quality crystal thin films layer by layer on a single crystal substrate under ultra-high vacuum conditions by precisely controlling the deposition of molecular or atomic beams. Its advantages include low substrate temperature, slow growth rate, precise beam control, and flexible adjustment of film composition and doping concentration. MBE can prepare ultra-thin single crystal thin films and ultra-thin layer quantum microstructure materials [7]. Figure 3.15 shows the cross-section of the MBE growth cell using a solid source, where the ultra-high vacuum environment ensures the purity of the atomic beam, and the liquid nitrogen low-temperature shield maintains the system connectivity and growth environment purity. MBE uses stoichiometric control to precisely control the growth process by adjusting the substrate temperature and element flow. In the growth of the AlGaAs system, special attention must be paid to controlling the temperature to reduce the oxidation of aluminium elements, thereby ensuring the performance of the laser.

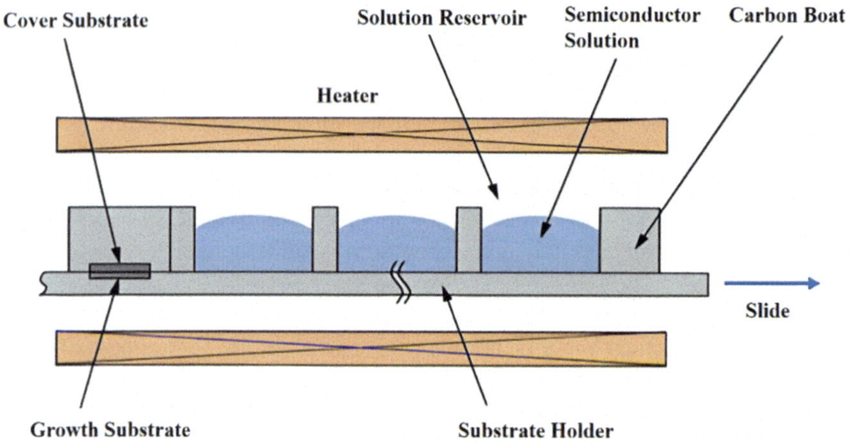

Fig. 3.13 Schematic diagram of the liquid phase epitaxy (LPE) system

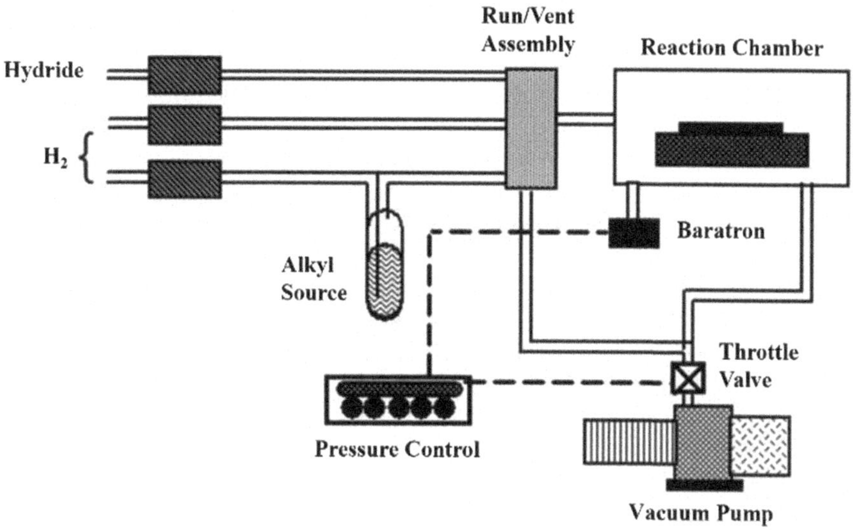

Fig. 3.14 Metal-organic Chemical Vapor Deposition (MOCVD) system schematic diagram

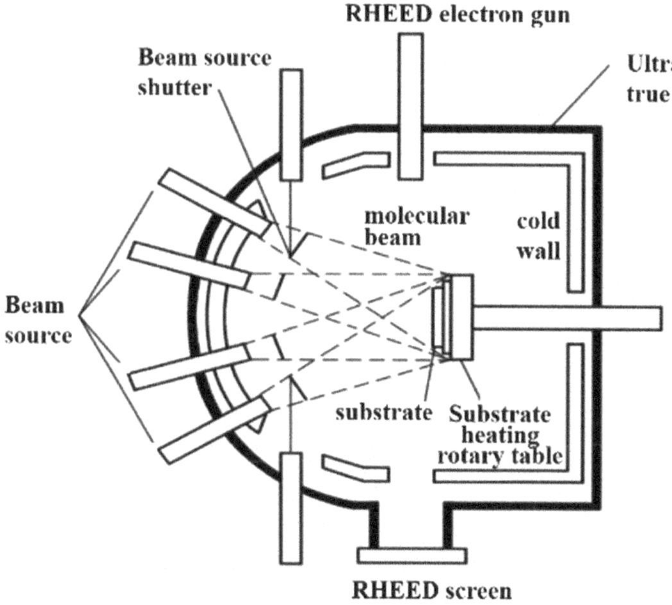

Fig. 3.15 Molecular beam epitaxy (MBE) system schematic diagram

3.2.3 Manufacturing Process Flow

Although the manufacturing process of diode lasers is similar in principle to the process of semiconductor electronic devices. The unique requirements of optoelectronic devices in terms of materials, structure, and performance require the use of a series of new processes and technologies during manufacturing. This is particularly prominent when manufacturing quantum wells and superlattice devices. Figure 3.16 shows the main process steps and manufacturing flow of laser diodes [8].

It can be clearly seen from the flow chart that the choice of substrate is the first step in device manufacturing. While choosing a substrate, several key factors need to be considered. First, the substrate must achieve lattice matching with the material forming the heterojunction, and a buffer layer needs to be added to improve the match. Secondly, it is necessary to determine the crystal orientation of the growth surface, or allow it to deviate by a certain angle, to ensure that the grown film has the required physical and chemical properties. In addition, the doping concentration of the substrate is also an important parameter, which directly affects the performance of the device. At the same time, it is necessary to ensure that the defect density on the surface and inside the substrate is as low as possibly, the surface is smooth, bright, and scratch-free. Finally, the substrate must have a certain thickness to ensure that the chip has sufficient mechanical strength during subsequent processing and use.

The epitaxial growth process is the core of diode laser manufacturing, playing a crucial role in determining the overall performance of the device and the final yield. During the manufacturing process, SiO_2 or Si_3N_4 thin films are deposited, the main purpose of which is to act as a mask during the diffusion or etching process after photolithography. Considering that some III–V group materials may undergo thermal corrosion at high temperatures, a low-temperature deposition process is preferred in actual operations to ensure the stability and reliability of the manufacturing process.

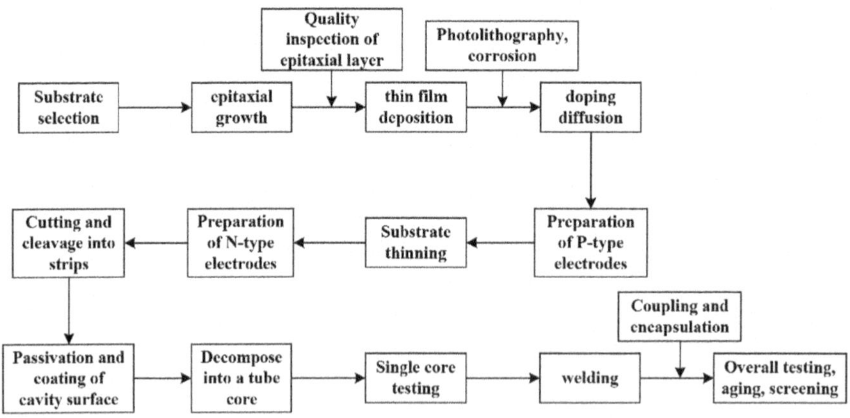

Fig. 3.16 Schematic diagram of diode laser manufacturing process

Etching is another key process in laser manufacturing. It precisely prepares various shapes according to the design structure and materials of the laser, usually implemented accurately based on the structural features of the laser and the selected materials. Etching techniques are mainly divided into dry etching and wet etching (or chemical etching). Dry etching is further subdivided into methods such as plasma etching, reactive ion etching, and magnetron sputtering etching, each with its specific application scenarios. In practical applications, except for particularly fine structures, chemical etching is usually used to prepare various shapes like V-shaped, trapezoidal, inverted trapezoidal grooves, or protruding ridge lines, steps, etc. This method is simple and effective, suitable for most scenarios. In addition, secondary epitaxial growth technology is used to manufacture refractive index-guided lasers, thereby achieving precise control of the side mode. Dry etching is more often used for fine etching work of small sizes, its high precision and efficiency make it significantly advantageous when dealing with complex, fine structures. Through the comprehensive and flexible use of these two types of etching techniques, the quality and performance of the laser manufacturing can be ensured, meeting the needs of practical applications.

Diffusion technology is an important process widely used in semiconductor device manufacturing. Through this process, the electrical and optical properties of semiconductor materials can be effectively changed, making it one of the key means to achieve device performance optimization and functional diversification.

Electrode fabrication, also known as ohmic contact preparation, is a seemingly simple but crucial process in diode laser manufacturing. The quality of ohmic contact directly affects the power conversion efficiency of the device and also affects the thermal state of the device during operation, which in turn has a profound impact on the reliability and lifespan of the device. Although the preparation process of ohmic contact for N-type and P-type materials such as GaAs, InP series is relatively mature, when it is necessary to develop lasers in new wavelength ranges, there is still the problem of high ohmic contact resistivity. Therefore, continuously optimizing and improving the preparation technology of ohmic contact is crucial for enhancing the performance and reliability of diode lasers.

Cleavage technology is a highly technical process, mainly used to finely decompose manufactured device chips into individual cores. Usually, a diamond knife or thin blade is used to apply moderate cutting pressure in the cleavage direction of the thinned substrate crystal, thereby cleaving the substrate crystal into bars. The cleavage surfaces on the bars form completely parallel F-P cavity surfaces. Generally, in order to obtain better performance and reliability, it is necessary to blunt and coat the F-P cavity surface, and the reflectivity of the cavity surface is designed as needed. After the cavity surface coating is completed, the designed individual cores are cut out in the direction perpendicular to the mirror surface. After the cutting is completed, the cores are subjected to strict testing and screening. Subsequently, the cores are welded to the heat sink of the tube shell to enhance its heat dissipation performance. The bonding of gold wire electrodes is completed by hot pressing welding or ultrasonic ball welding to ensure the stability and reliability of the electrode connection. Finally, after the coupling packaging process, a practical diode laser device is prepared. In the entire process, every step of operation needs to be finely controlled to ensure that the performance and quality of the final product reach the best state.

The post-process of diode lasers includes brazing, bonding, coupling, and packaging [9], which are briefly introduced one by one below.

1. Brazing

Brazing the qualified cores to the heat sink through soldering is an important process in the manufacture of diode lasers. When choosing heat sink materials, their thermal conductivity should be considered first to ensure that it matches the thermal expansion coefficient of the core, thereby reducing the generation of micro-cracks and preventing degradation. At the same time, when choosing heat sink materials, factors such as non-pollution and easy processing should also be considered. Generally speaking, oxygen-free copper coated with an Au layer is a commonly used heat sink material. In special cases, Si, diamond, AlN or BeO materials are also used to meet specific application needs. The choice of brazing method is also critical, mainly including vacuum brazing and forming gas protected brazing. The specific choice needs to be determined according to the nature of the solder and process requirements. The choice of solder is equally important, common ones include pure In, pure Sn, Au–Sn or Au–Ce fusible alloys, and sometimes Pb–Sn (containing Pb 40%) alloy is used. During the brazing process, it is necessary to ensure that the solder is firmly and evenly bonded to the core, while avoiding the use of too much solder and too high temperature to prevent the solder from overflowing the bottom surface, damaging the cleavage surface, and even contaminating the active area.

2. Bonding

Bonding is the process of connecting electrodes to the core using gold wire (about 50 μm) or gold foil tape (a few tens of μm wide) by ultrasonic welding or hot pressing welding, or both, to serve as the lead for current injection. During the welding process, the firmness of the welding point is crucial, but at the same time, it is necessary to avoid applying too much pressure to avoid micro-damage to the core. Especially when the pressure is too high, the root of the gold wire may be damaged, so the pressure during the welding process needs to be precisely controlled.

3. Coupling with optical fiber

In some application fields, the laser output of diode lasers needs to adopt the method of fiber output, but due to its large beam divergence angle, asymmetric direction, and elliptical cross-section, it is difficult to couple with the fiber. Generally, the fiber end can be made into a ball lens or a conical shape for coupling to improve the coupling efficiency. For fibers with a spherical lens end face, the coupling efficiency can be increased to 60%, and for a conical end, it can reach 80%. The ball lens is made by melting the fiber end, causing it to intrinsically shrink into a small ball, acting as a short focal length lens. The conical lens is formed by melting and thinning the front end of the fiber, increasing the numerical aperture of the fiber (increasing a_n/a_l times, where a_n is the radius of the fiber itself, and a_l is the radius of the front end of the fiber cone), increasing the numerical aperture of the fiber, and improving the focusing ability. These methods are simple, easy to manufacture, and have significant effects, and have been widely used in mass production.

4. Encapsulation

There are many types of laser diode package used for encapsulation, but there are mainly two types: dual in-line and coaxial. Structurally, there are mainly two forms: fiber-coupled output and optical window. To ensure the reliability of the encapsulation, full metallization and good sealing are crucial to ensure no air leakage. In addition, depending on specific needs, the laser diode package interior may also be equipped with temperature control, light control sensors, and semiconductor coolers, and sometimes even integrated drive circuits. The choice of these additional devices depends on actual application needs.

3.2.4 Laser Diodes with Ultra-low Reflectance Antireflection Coating

External cavity diode lasers (ECDL) constructed by using laser diodes offer advantages such as narrow linewidth, tunability, and high frequency stability. Besides applications in fundamental research areas like quantum frequency standards, laser cooling, and high-resolution spectroscopy, they are also widely utilized in optical storage, optical communications, laser sensing, lidar, and other fields. For such applications, laser frequency stability is of paramount importance. Conventionally, macroscopic frequency selection devices like gratings, Fabry–Pérot etalons, interference filters, and prisms have been employed. However, these macro-scale devices are sensitive to changes in temperature, current, and mechanical vibrations, rendering them inadequate for fulfilling the requirements of quantum precision measurements and other fields that demand high-precision frequency stability in laser sources.

In 2011, Miao and others from Peking University innovatively proposed the use of Faraday atomic filters (FADOF) as frequency selection components for Faraday diode lasers [9]. Based on the magneto-optical rotation effect, the output laser can directly correspond to atomic spectral lines, exhibiting excellent resistance to temperature and current disturbances, and providing effective long-term stability.

According to the basic principles of laser physics, the lasing frequency of an external cavity diode laser is the frequency of the mode with the maximum gain. Figure 3.17 shows the contribution of different components in the laser to the laser gain: ① Laser medium, which depends on the characteristics of the diode laser material. The gain of the medium is a wide spectrum, and its peak mainly depends on the temperature of the laser diode; ② In the internal cavity, the diode itself forms a small standard, and the frequency of the internal cavity gain curve changes with the temperature and current of the diode. As the temperature rises, the peaks of the medium gain and the internal cavity gain curve move towards longer wavelengths, but they do not move at the same rate, resulting in the laser mode jumping to different peaks of the internal cavity gain curve. Therefore, a diode laser without coating and feedback cannot be tuned to any wavelength; ③ Selective component, which is an atomic filter, where the transmission spectrum bandwidth Δv of the atomic filter is about 1 GHz, and the peak position corresponds to the atomic Doppler broadening line; ④ In the external cavity, the back of the laser diode and the feedback cavity mirror form the external cavity, and the mode interval can be adjusted by changing the distance between the feedback cavity mirror and the diode. The optical alignment of the external cavity laser needs to be precisely adjusted so that the gain of each component reaches its peak at v_0, in order to achieve optical feedback and obtain the required single-mode frequency v_0. If a laser diode without an anti-reflection coating is used, the laser diode itself contains an F-P cavity, and the cavity length of the diode is usually between 0.3 and 0.5 mm. Assuming a cavity length of 0.4 mm, the free spectral range of the internal cavity is FSR = c/2L = 3.75 GHz, while the external resonant cavity length of the Faraday laser is 40 cm, that is, the free spectral range FSR = c/2L = 375 MHz. Therefore, because the laser contains two resonant cavities, the output mode of the laser is affected by the two resonant cavities together. Although the gain curve width Δv of the selective component in the external cavity is about 1 GHz and remains unchanged, and there are only 2 or 3 external cavity modes competing, once the temperature or current of the diode changes over a large range, the internal cavity mode may jump and output alone, thereby affecting the output frequency. Therefore, in order to reduce the interference of the internal cavity mode, the Faraday laser innovatively uses a laser diode with an ultra-low reflectivity anti-reflection coating (ARLD), which was first proposed by Miao and others from Peking University [74]. When the end face of a laser diode is coated with an anti-reflective film, the end face reflectivity is significantly reduced, eliminating the internal cavity modes. Consequently, the laser loss increases, making it difficult to form a laser output, which instead occurs in the form of fluorescence.

The experimental setup of the Faraday laser based on the anti-reflection coating is shown in Fig. 3.18. Figure 3.19 shows the fluorescence spectrum of ARLD. ARLD shows an emission spectrum with a central wavelength of 760 nm. Due to the anti-reflection coating, the reflectivity of the front surface is 3×10^{-4}, and the ARLD, which only serves as a gain medium, has a fluorescence spectrum range as wide as 60 nm (see Fig. 3.19).

Figure 3.20a shows the FADOF transmission spectrum of Rb, with its highest transmission peak at 780.241 nm. The black dashed line is the natural spectrum of Rb filled only in the atomic vapor cell. As shown in Fig. 3.20, the transmission of the maximum transmission peak is 51%, so the feedback of the laser at the maximum transmission peak is 6.4%. The relationship between the output wavelength and the temperature change of the laser diode current is shown in Fig. 3.20b, c. Within a larger range of temperature and current, the laser wavelength is stable. The experimental results show that when the laser diode current is greater than the threshold, such as above 55 mA, to the current source limit of 142 mA, and within the controllable limit range of diode temperature from 15 to 35 °C, the laser wavelength is always stable within 2 pm. In addition, even if the fluorescence spectrum of the gain medium ARLD moves within a larger range with the current and temperature of ARLD,

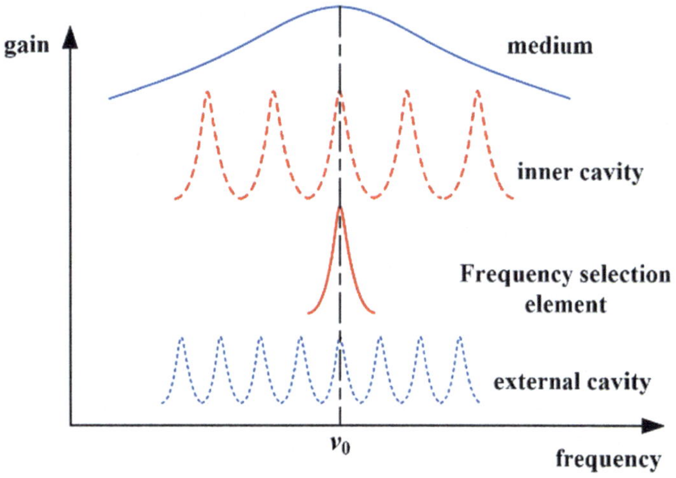

Fig. 3.17 Schematic diagram of the gain of various optical components of the external cavity diode laser

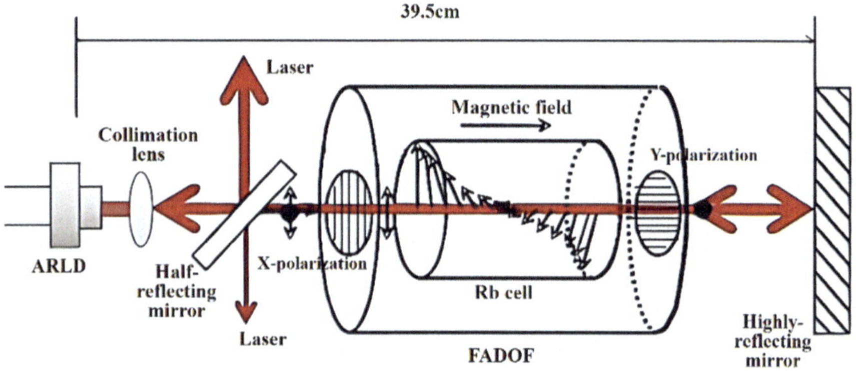

Fig. 3.18 Schematic diagram of ARLD external cavity laser

Fig. 3.19 ARLD fluorescence spectrum without external feedback (the temperature and current of ARLD are 21.24 °C and 78.93 mA respectively)

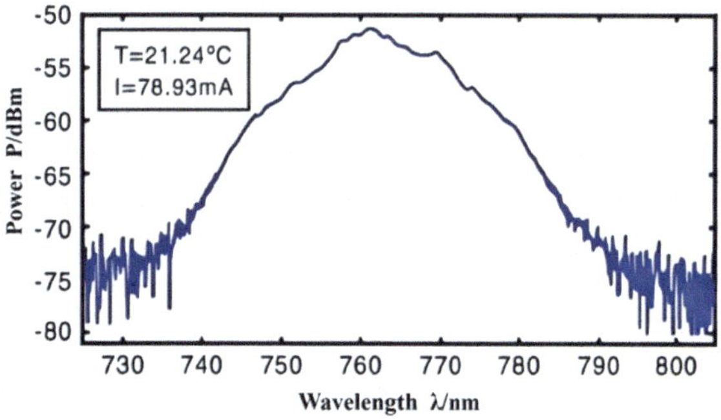

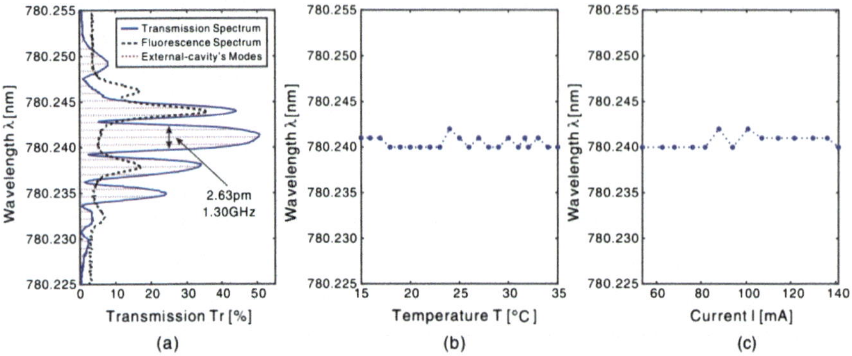

Fig. 3.20 **a** The transmission spectrum of Rb FADOF and the fluorescence spectrum of the Rb atomic vapor cell (the temperature and magnetic field of FADOF are 70 °C and 3.2×10^{-2} T. In the Rb atomic vapor cell, T = 24 °C and B = 0 T), as well as the external cavity longitudinal modes spaced at 379.7 MHz. The FWHM of the highest peak of the Rb FADOF transmission spectrum is 1.3 GHz. **b** The temperature-wavelength curve when the current is kept at 74.04 mA. **c** The current–wavelength curve when the temperature is kept at 21.31 °C. Comparing (**a**) with (**b**) and (**c**) horizontally, it shows that the wavelength of the laser is well maintained within the highest transmission peak

the transmission spectrum of FADOF can completely overlap with the fluorescence spectrum of ARLD, thus ensuring that the output laser wavelength of the Faraday laser is limited within the transmission peak of the Rb atom FADOF, effectively improving the stability of the output wavelength.

The red dashed line in Fig. 3.20a indicates that there are three longitudinal modes within the highest transmission peak of FADOF. Because the cavity length drifts slowly with external temperature and other factors without compensation, the output frequency of the Faraday laser jump between the three longitudinal modes and stabilize on the mode with the highest transmission. As shown in Fig. 3.20, the cavity length of the Faraday laser is 39.5 cm, so its free spectral range is 379.7 MHz. The resolution of the wavelength meter used for measurement is 1 pm, which is equivalent to 492.8 MHz at 780 nm. The longitudinal mode of the Faraday laser is measured by the Fabry–Pérot interferometer, and the measured side mode suppression ratio is 12.8 dB (see Fig. 3.21).

The above research results show that the Faraday laser, due to the use of ARLD and FADOF, exhibits robust immunity against current and temperature fluctuations of the laser diode. Regardless of how the current and temperature of the laser diode are adjusted and changed, the output wavelength of the Faraday laser is completely limited within the transmission peak of the Rb atom FADOF. The core of the Faraday laser is to use ARLD as the gain medium, and FADOF as the frequency selection element, keeping the laser frequency within a transmission peak of FADOF. Within the diode current range of 55–142 mA and the diode temperature range of 15–35 °C, the output wavelength of the above-mentioned Rb Faraday laser is stable within 2 pm (985.6 MHz at 780 nm), therefore, it has a wide range of application prospects and

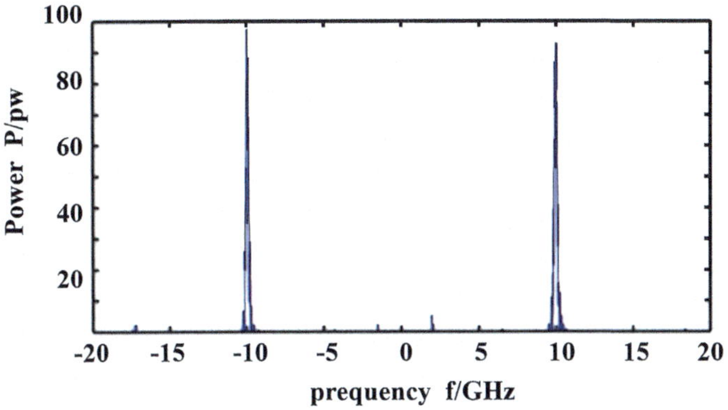

Fig. 3.21 Laser mode measured by Fabry–Pérot interferometer. The free spectral range of the Fabry–Pérot interferometer is 20 GHz

potential in fields of quantum precision measurement, atomic physics, and military defense equipment systems.

References

1. B. Tan, *Fundamentals of Optoelectronic Technology* (Electronics Industry Press, Beijing, 2014), p.54
2. Q. Wang, *Optoelectronic Technology*, 3rd edn. (Electronics Industry Press, Beijing, 2013), pp.130–136
3. H. Kroemer, A proposed class of hetero-junction injection lasers. Proc. IEEE **51**(12), 1782 (1963)
4. Coldren, *Diode Lasers and Integrated Optics*, vol. 6. Translated by H. Shi (Beijing University of Posts and Telecommunications Press, Beijing, 2006), pp. 9–11
5. C. Li, *Organometallic Compound Chemical Vapor Deposition Method* (Changchun Institute of Applied Chemistry, Chinese Academy of Sciences, 1985)
6. B. Xue, X. Zhang, *Applied Physics* (Xi'an University of Electronic Science and Technology Press, Xi'an, 2012)
7. M. Tian, *Thin Film Technology and Thin Film Materials* (Tsinghua University Press, Beijing, 2006), pp.369–374
8. J. Jiang, *Diode Lasers* (Electronics Industry Press, Beijing, 2000), pp. 320–322, 329–330
9. X. Miao, *External Cavity Diode laser Immune to Current Temperature Changes* (Peking University, 2011)

Chapter 4
Faraday Atomic Filter

In external cavity diode lasers, the filter serves as a frequency selector, playing a role in achieving stable single longitudinal mode output. The filter can filter out the broad-spectrum light emitted by the laser diode, select the light that resonates with the atomic transition frequency, and cause the longitudinal mode oscillation at a specific frequency, achieving the function of frequency selection, improving the monochromaticity and stability of the output laser. As described in Chap. 2, traditional frequency selection devices include gratings, interference filters, and F-P (Fabry–Pérot) cavities. Common grating-type lasers mainly include Littrow and Littman structures that serve as both frequency selection and feedback. The wavelength selection of the grating can be expressed by the diffraction formula, and its center wavelength is related to the incident angle; the output wavelength of the interference filter-type laser is related to the angle between the incident light and the interference filter; the F-P cavity is composed of two parallel reflecting mirrors, which can select narrow spectral lines, but its loss is large, the fine tuning requirements are high, and it is very sensitive to environmental influences such as mechanical vibration. It can be seen that traditional diode lasers use macroscopic devices for frequency selection, and the output laser frequency usually depends on the angle of the frequency selection device. They cannot intrinsically correspond to the atomic transition spectrum line. In recent years, with the rapid development and application of atomic filters using the interaction characteristics of light and atoms, it has provided a new idea for external cavity lasers to achieve ultra-narrow band and specific quantum transition frequency selection. At present, atomic filters suitable for laser frequency selection include Faraday anomalous dispersion atomic filters, Voigt atomic filters, and birefringence-induced atomic filters.

J. Chen et al., *Faraday Laser*, https://doi.org/10.1007/978-981-97-8023-5_4

4.1 Development History of Atomic Filters

At the beginning of the twentieth century, atomic physics and quantum mechanics developed rapidly. The predecessor of the atomic filter was a low-noise infrared quantum counter designed in 1959 [1] by Bloembergen, the Nobel laureate in nonlinear optics. This counter is based on Weber's quantum mechanics amplification theory, using crystals containing transition metal ions, resonantly absorbing infrared photons, and releasing visible light photons through spontaneous radiation, thereby achieving detection in the infrared wavelength region. This device is basically consistent with the principle of atomic resonance filters. In the 1970s, scientists found that replacing crystals with atomic vapor could improve the performance of quantum counters, so atomic vapor was widely used in atomic filters and related research. In 1978, Gelbwachs and others realized the infrared quantum counter with sodium atoms [2], with its principle and structure basically consistent with the subsequent atomic resonance filters. In 1979, Marling and others used atomic resonance transitions to achieve ultra-high Q isotropic optical filters [3], both of their works are considered pioneering achievements in the fields of atomic filters. In 1988, Gelbwachs applied the "wavelength conversion" effect of this quantum counter to filters. Based on cesium, Rb, and potassium atoms, they proposed atomic resonance filters with high Q ($10^5 \sim 10^6$), ultra-narrow bandwidth (0.001 nm), isotropy, and insensitivity of the center wavelength to environmental disturbances [4].

In the early development of quantum mechanics, scientists discovered that when a laser interacts with atoms in a magnetic field, the Faraday rotation effect is enhanced near the atomic resonance transition frequency. In 1956, Öhman first implemented a monochromator based on the principle of selective magnetic rotation in astronomical instrument research [5], consisting of two orthogonal polarizers on both sides, with an atomic medium interacting with light in the middle, laying the foundation for the basic structure of the Faraday atomic filter. In 1982, Yeh conducted related research on dispersion magnetic optical filters, sorting out the theoretical basis of Faraday atomic filters [6]. From 1991 to 1993, Shay and others coined the name "Faraday Anomalous Dispersion Optical Filter (FADOF)" [7–13], and conducted some experiments and explorations in this field. For example, in 1993, Yin and Shay proposed to connect a cell in front of the Faraday filter to absorb unnecessary transmission peaks [12]; Matthews and Shay proposed to apply the Faraday filter to Electronic Speckle Pattern Interferometry (ESPI) to enhance its resistance to environmental light [13]. In 1992, Menders and Korevaar applied for a related patent for the Voigt Anomalous Dispersion Atomic Filter [14], which changed the magnetic field structure on the basis of the Faraday filter, changing the direction of the magnetic field from axial to perpendicular to the direction of light propagation, and forming a 45° angle with the polarization direction of the incident light. In the design of some specific filter structures, this makes its volume more compact and simple, suitable for the design of miniaturized atomic filters.

In addition, in the development history of atomic filters, there have been many filters based on different physical principles, such as birefringence-induced atomic filters, spectral lamp-based atomic filters, calcium atomic beam-based atomic filters, etc., which will be briefly introduced to readers later. Although the term Faraday Anomalous Dispersion Optical Filter (FADOF) is still used today, not all FADOFs have applied the anomalous dispersion effect. Generally speaking, anomalous dispersion occurs in the transition spectral line area, and currently, the transmission peaks of most wing-working mode FADOFs are located within the frequency range of normal atomic dispersion. In 2018, Gerhardt discussed this phenomenon and its related expressions [15].

4.2 Basic Principles and Main Classifications

The phenomena of the interaction between light and atoms are diverse, and various optical filters can be made using these phenomena. For example, a Doppler atomic filter can be made using the resonant absorption of atoms to light, an atomic resonance filter can be made by utilizing the absorption and reemission of light from atoms. A Faraday anomalous dispersion atomic filter can be made using the Faraday rotation effect of atoms in a magnetic field, and a birefringence-induced atomic filter can be made using circularly polarized light polarization pumping instead of the magnetic field to disrupt the distribution symmetry of atoms on the magnetic sublevels. In the Faraday anomalous dispersion atomic filter, the magnetic field structure can be modified to make a Voigt-type atomic filter; improvements to the structure of the atomic cell have led to the birth of atomic filters based on spectral lamps, hollow cathode lamps, and calcium atomic beams. This section will provide a brief introduction to these atomic filters. In addition, there are single-peak atomic filters, Faraday molecular filters, and Faraday atomic filters indirectly pumped by lasers. However, due to their lack of characteristic features, this section will not introduce these contents individually.

4.2.1 Doppler Atomic Filter

The Doppler atomic filter consists of an atomic cell with adjustable temperature, as shown in Fig. 4.1a. Since atoms only interact with light that resonates with their transition frequency, only light near the transition frequency will be absorbed by the atomic cell, which acts as a band-stop filter.

Temperature has a significant influence on the performance of the Doppler atomic filter, as shown in Fig. 4.1b. On the one hand, an increase in temperature will increase the number of atoms interacting with light per unit volume, thereby enhancing the absorption of light. On the other hand, in the Doppler atomic filter, the main contribution to the absorption line shape comes from Doppler broadening due to the motion of the atomic vapor. According to the Maxwell speed distribution, higher temperatures will increase the speed of the atoms, thereby increasing the Doppler broadening of

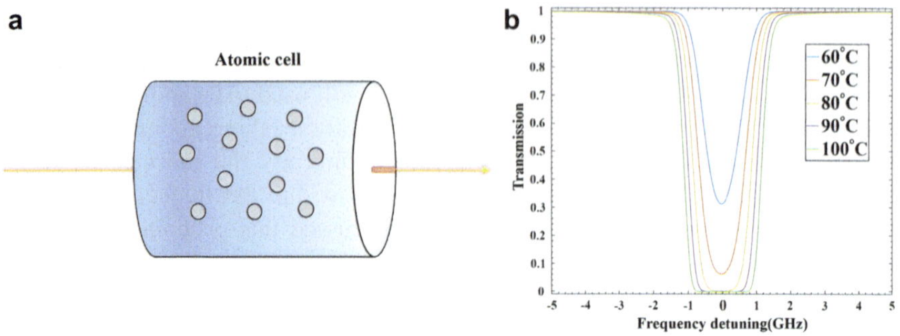

Fig. 4.1 **a** Schematic diagram of the Doppler atomic filter. **b** Transmission spectrum of the ^{39}K Doppler atomic filter

the atoms. Therefore, by adjusting the temperature, the in-band suppression rate and suppression bandwidth of the Doppler filter can be adjusted.

4.2.2 Atomic Resonance Filter

In 1958, Bloembergen proposed a solid-state infrared quantum counter, which is the prototype of the atomic resonance filter and is considered to be the pioneering work in the field of atomic filters. The atomic resonance filter is the earliest filter that uses the principle of interaction between light and atoms to achieve ultra-narrow band filtering. It ingeniously utilizes the resonant absorption characteristics of atoms to achieve photon frequency conversion, converting broadband incident photons into spontaneously emitted photons of another frequency with better monochromaticity, thereby realizing narrowband optical filtering. Its basic structure and atomic energy levels are shown in Fig. 4.2.

In the atomic resonance filter, an atomic cell is sandwiched between two traditional band-pass filters, with the wavelengths, λ_i and λ_o, respectively, corresponding to the

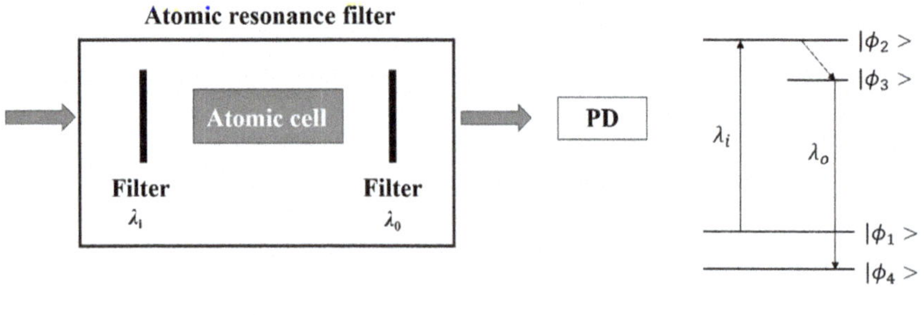

Fig. 4.2 Passive atomic resonance filter schematic

two transition wavelengths of the atoms in the atomic cell, with non-overlapping passbands. Broadband incident light enters the filter from the left, first passing through the first filter piece, being coarsely filtered into light with a central wavelength of λ_i, and enters the atomic cell. Among these coarsely filtered photons, those with frequencies closer to λ_i will be absorbed by atomic resonance, causing an atomic transition from the $|\phi_1\rangle$ state to the excited state $|\phi_2\rangle$, and then, to the metastable state $|\phi_3\rangle$. Atoms in the $|\phi_3\rangle$ state will produce spontaneous radiation, emitting photons with a wavelength of λ_0 and fall onto $|\phi_4\rangle$. In this process, selective frequency conversion is achieved. The role of the second filter is to allow the photons of light λ_o generated by spontaneous emission to pass through while blocking the photons of light λ_i that were not resonantly absorbed due to excessive detuning in the previous stage. After passing through the filter, the wide-band incident light is converted into narrow-band, low-noise light with a central wavelength of λ_o. The internal photon conversion efficiency η_ϕ of the atomic resonance filter can be expressed as follows:

$$\eta_\phi = \frac{v_T}{v_T + A_i} \tag{4.1}$$

where, v_T is the collision transfer rate, which is directly proportional to the vapor pressure of the atomic vapor and buffer gas in the cell, A_i is the Einstein coefficient of the corresponding transition line. In 1992, Gelbwachs and Chan designed a passive atomic resonance filter based on the 5S and 5P energy levels of strontium atoms, with $\lambda_i = 460.7\,\text{nm}$, $\lambda_o = 689.3\,\text{nm}$ and studied the effects of different types and pressures of buffer gases on photon conversion efficiency and filter bandwidth. Their experiments showed that increasing the vapor pressure of the buffer gas improves the photon conversion efficiency, but increases the passband bandwidth, and among various rare gases, the effect of argon is the most significant [16].

Atomic resonance filters exist in two types: active and passive. In passive atomic resonance filters, $|\phi_1\rangle$ state is the ground state, which does not require the action of an external pump source, has a simpler structure, lower cost, but the applicable atomic types and central wavelengths are more limited; however, in the active atomic filters, it is necessary to use a pump light with a wavelength of λ_p to pump the atoms from ground state $|\phi_0\rangle$ to the excited state $|\phi_1\rangle$ (according to practical applications, electric pumping, chemical pumping, etc. can also be used), as shown in Fig. 4.3. For some alkaline earth metals (such as calcium, magnesium, etc.), a double pumping method can be used, using the second beam of pump laser and the collision of buffer gas to pump the atoms that have absorbed the incident photons to a higher energy level, making the selection of atomic media more flexible. Moreover, the buffer gas can also improve the conversion efficiency to a certain extent.

In the twenty-first century, with the rapid development of high-speed optical communication, high-resolution imaging, and high-stability light sources, the limitations of atomic resonance filters have gradually emerged. Among many applications, the importance of atomic filters in laser communication is particularly prominent, because the monochromaticity of the light source is one of the core parameters of laser communication applications. However, the maximum information rate that an

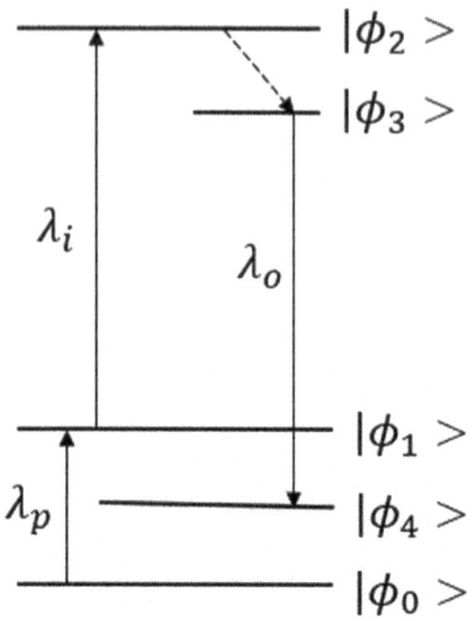

Fig. 4.3 Energy level diagram of active atomic resonance filter

atomic resonance filter can provide is constrained by its response time, which is mainly affected by the relaxation rate of the atomic system's corresponding output wavelength energy level. Usually, the relaxation time caused by the spontaneous radiation process of gaseous atomic media is relatively long, about tens of ns, and the corresponding frequency is only tens of MHz, which will severely limit the optical information processing rate. To overcome this limitation, a common practice is to convert the gaseous atomic medium into a semiconductor quantum structure, but this loses the low noise and narrow bandwidth, which are advantages of the gaseous atomic system. In contrast, Faraday atomic filters using the magneto-optical effect can effectively solve such problems and are more suitable for high-speed laser communication systems. At the same time, the direction consistency of the output light path of the atomic resonance filter is very poor, because the direction of the photons of spontaneous radiation is random, which has a serious impact on its application in imaging systems. In order to achieve the purpose of imaging, the imaging system needs to overcome this impact by greatly reducing the length of the cell, but this method greatly sacrifices the wavelength conversion efficiency, which is also an another important factor restricting the application of atomic resonance filters.

4.2.3 Faraday Anomalous Dispersion Atomic Filter

FADOF uses the Faraday anomalous dispersion phenomenon to filter light. By placing an atomic cell between two polarizers with orthogonal polarization directions, adding an axial magnetic field, and controlling the temperature of the atomic cell, a basic FADOF structure is formed, as shown in Fig. 4.4.

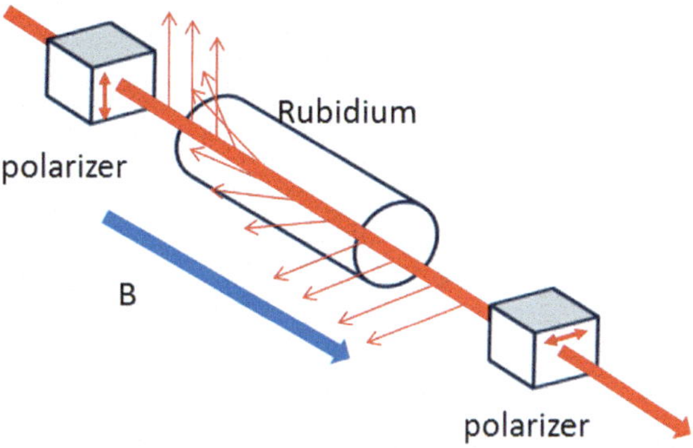

Fig. 4.4 Schematic diagram of the basic structure of FADOF based on Rb atoms

Atoms undergo Zeeman splitting in a magnetic field, causing the splitting of the absorption spectra of the atoms for the left and right polarizations as well as the dispersion spectra. The phase changes of the left and right circularly polarized components near the resonance frequency are different, causing the rotation of the polarization direction of the light signal, and thus transmitting from the rear polarizer. Light signals with non-resonant frequencies will be blocked by orthogonal polarizing elements, thereby achieving the filtering effect. Figure 4.5 shows the principle of FADOF rotation.

Resonance transitions can occur between the ground state and the excited state, or between the two excited states. Based on the location where the transition occurs, FADOF can be divided into passive and active categories. The filtering process of passive FADOF is established between the ground state and the excited state of the atom, while active FADOF (also known as excited state Faraday anomalous dispersion filter ES-FADOF) relies on pumping to establish the filtering process between the first excited state of the atom and higher excited states. Figure 4.6 shows the structure of active FADOF.

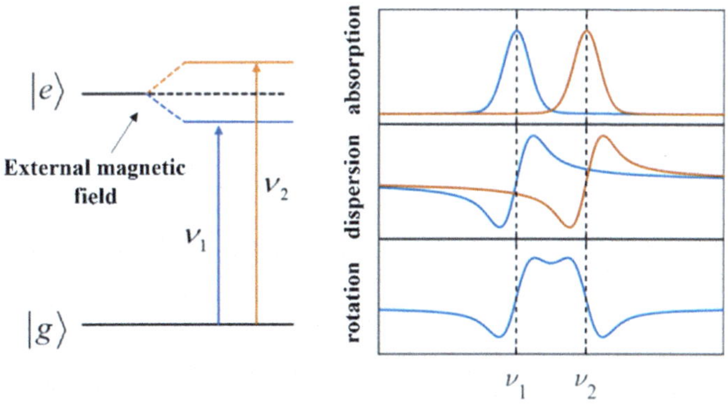

Fig. 4.5 FADOF rotation principle

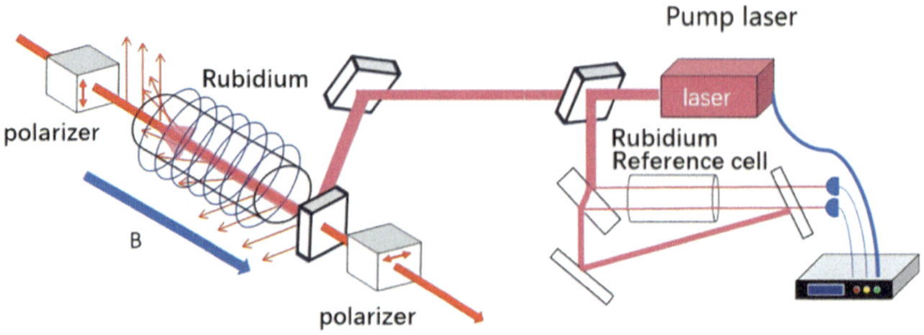

Fig. 4.6 Schematic diagram of the structure of active FADOF

The transition frequency between excited states is much richer than that of the ground state, allowing for a wider range of filter wavelength selection. The transition from the first excited state to higher excited states of alkali metals exists in the blue-green band, such as the transition of potassium atom $4P_{1/2}$-$8S_{1/2}$ corresponding to 532 nm, and the transition of Rb atom $5P_{3/2}$-$8D_{5/2}$ corresponding to 543 nm, making ES-FADOF suitable for underwater optical communication (400–550 nm).

4.2.4 Voigt Anomalous Dispersion Atomic Filter

The Voigt Anomalous Dispersion Atomic Filter (VADOF) is a variant of the Faraday Anomalous Dispersion Atomic Filter (FADOF), also based on the optical rotation effect produced by the Zeeman splitting of atomic energy levels in a magnetic field. The only difference is the direction of the working magnetic field of the two atomic filters. The working magnetic field direction of FADOF is along the propagation direction of the incident signal light, for a cylindrical atomic cell, that is, along the axial direction of the atomic cell. The working magnetic field direction of VADOF is perpendicular to the propagation direction of the signal light, as shown in the typical VADOF model structure in Fig. 4.7.

Fig. 4.7 VADOF model structure diagram

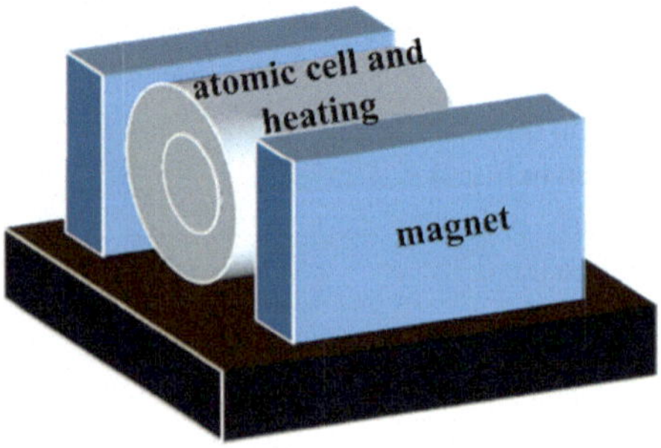

This type of atomic filter with a transverse working magnetic field configuration is easier to manufacture than FADOF. Because the working magnetic field direction of FADOF is parallel to the propagation direction of the signal light, holes must be left in the magnetic field design, and the holes often need to be drilled in the permanent magnets or iron cores at both ends of the atomic cell. For VADOF, as shown in Fig. 4.7, because the magnetic field direction is perpendicular to the optical path, the magnet used does not block the propagation of the light, and the designed whole magnet can be placed directly on both sides of the atomic cell. In addition to the convenience of manufacturing in structure, this magnetic field configuration is more conducive to form a high magnetic field value in the atomic cell to meet the working magnetic field requirements. The length of the atomic cell is usually greater than its width. In FADOF, because the magnet is often placed at both ends of the cell, due to the technical limitations of actual magnet manufacturing, as their distance increases, the magnetic field value generated in the cell will rapidly decrease, and the uniformity of the magnetic field also decreases which, to some extent changes the performance of this filter. If a coil is used to generate a magnetic field in actual operation, it may bring inconvenience to the application, and a large current is required to reach the required working magnetic field. For the transverse magnetic field configuration of VADOF, the width of a common atomic cell is about 1 cm. Regardless of how the length of the atomic cell changes, the size of the working magnetic field generated in the cell is not affected, and the uniformity of the magnetic field is better than the former, and the requirements for the magnetic field source are also reduced.

Although VADOF has its conveniences in magnetic field design and production compared to FADOF, it also has certain limitations in practical use. This limitation comes from the difference between the way VADOF and FADOF produce optical rotation effects. Under the FADOF configuration, because the direction of the magnetic field is parallel to the direction of signal light propagation, the polarization direction of the incident linearly polarized signal light is always perpendicular to the magnetic field. Its optical rotation effect comes from the Zeeman splitting brought by the working magnetic field, which causes the difference in the refractive index of the left and right circularly polarized components of the linearly polarized signal light. However, VADOF is different. Because the direction of the magnetic field is perpendicular to the direction of signal light propagation, if the polarization direction of the incident linearly polarized signal light is parallel to the magnetic field, there is no optical rotation effect, nor is a transmission curve. It is necessary to manually control the polarization direction of the incident signal light to make its angle with the direction of the working magnetic field 45°. The relationship between the polarization direction of the VADOF signal light and the direction of the magnetic field is shown in Fig. 4.8. Under this condition, the incident signal light decomposes into two linearly polarized components with polarization directions parallel and perpendicular to the direction of the magnetic field, and the optical rotation effect is produced by the phase difference between them.

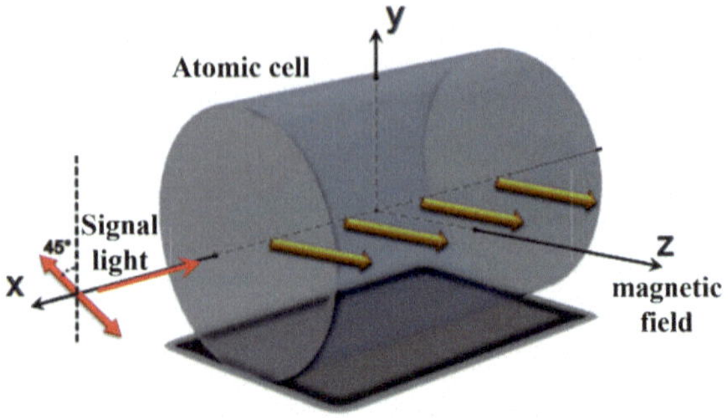

Fig. 4.8 Relationship between the polarization direction of VADOF signal light and the direction of the magnetic field

4.2.5 Induced Dichroism Atomic Filter

Dichroism refers to the dependence of a material's absorption coefficient on the polarization state of the incident light. When a beam of light is incident on a specific medium, it is decomposed into two polarized lights with perpendicular vibration directions, different propagation speeds, and unequal refractive indices. The medium absorbs these two polarized lights very differently, resulting in dichroism. Among the several types of atomic filters introduced earlier, except for the Doppler broadened Faraday atomic filter and the atomic resonance filter that do not need to apply a magnetic field, both FADOF and VADOF filters need to use a magnetic field to cause the asymmetry of the resulted polarization rate to achieve filtering. Therefore, the realization of these two types of atomic filters cannot be separated from the magnetic field, which is one of the indispensable factors for both. Although under the action of the magnetic field, the two filters can provide excellent filtering functions, but for some special application scenarios, the presence of the working magnetic field may become a factor of influence. This is because the magnetic field usually affects other experimental equipment, especially for some filters that can only rotate the polarization plane 90° under a super large magnetic field. Therefore, the presence of the magnetic field limits the application of the filter to some extent.

In 1995, Gayen and others designed a new type of atomic filter that does not need to apply a magnetic field, called the induced-dichroism atomic line filter [17]. This filter is also based on atomic energy levels, but it uses the phenomenon of photo-induced birefringence to achieve the filtering function. Taking the ground state induced-dichroism atomic line filter as an example, as shown in Fig. 4.9, a circularly polarized pump laser is first used to polarize the hyperfine energy level structure of the atom, thereby disrupting the uniform population distribution on the atomic magnetic sublevels, forming polarization, so that the left and right circularly polarized light of the incident signal is subjected to different refractive index effects when passed through such an atomic cell, thereby rotating the polarization plane of the signal light, and achieving the filtering function without a magnetic field. Using the above polarization pumping method, the rotation of the polarization plane no longer depends on the strength of the magnetic field, but depends on the uneven population distribution prepared by the circularly polarized pump laser, so it can successfully avoid the use of magnetic field. Under appropriate working conditions, it even has the potential to replace a super-large magnetic field. Since the proposal of induced-dichroism atomic line filter, the relevant theoretical calculation model has been successfully established. In 2005, Zhang and others from Harbin Institute of Technology used semi-classical theory to establish the theoretical calculation model of induced-dichroism atomic line filter, and their calculation results were in good agreement with the experimental results [18].

In addition, induced-dichroism atomic line filter can actually work between excited state energy levels, as shown in Fig. 4.10. In the excited state induced-dichroism atomic line filter, after selecting the first excited sublevel of polarization pumping, a beam of circularly polarized pump light is used to achieve polarization pumping, so that only some sublevels on the first excited state have a population distribution. When the detection signal light passes through, only the right-handed circularly polarized light part can interact with the atom and thus be subjected to different refractive index effects, while the left-handed circularly polarized light part "transparently" passes through the atoms. Therefore, at the exit stage, due to the two polarized light parts experiencing different refractive indices, the direction of the synthesized light polarization is rotated, and then the filtering function can be completed through the polarizer.

Since there is no magnetic field in the induced-dichroism atomic line filter structure, the Zeeman effect does not occur, and the atomic energy levels involved do not split further, so the transmission spectrum bandwidth obtained by induced-dichroism atomic line filter under appropriate working parameters is narrower than FADOF and VADOF. In 2002, Turner and others used a narrow linewidth pump laser to achieve

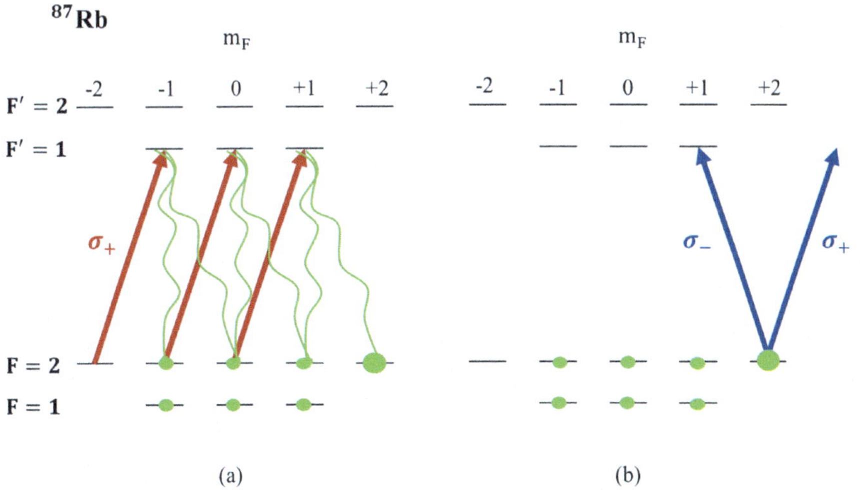

Fig. 4.9 Schematic diagram of the ground state induced-dichroism atomic line filter. **a** Pumping process; **b** detection process

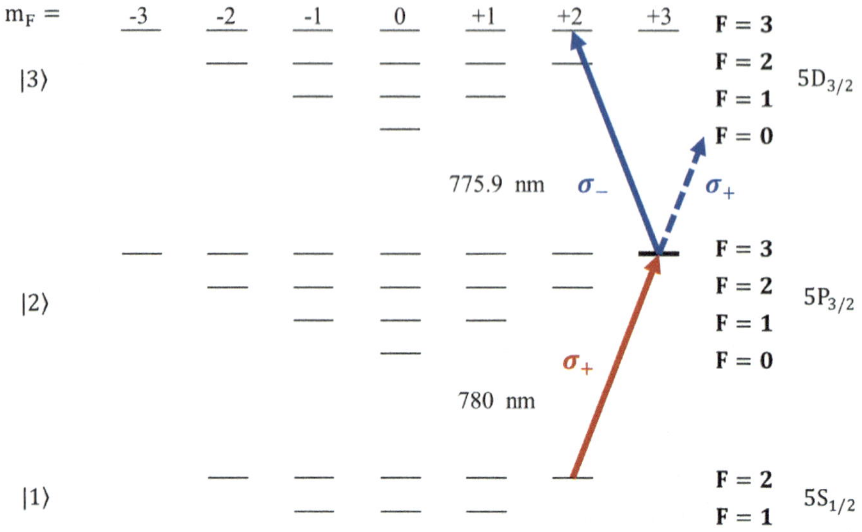

Fig. 4.10 Schematic diagram of the excited state induced-dichroism atomic line filter

velocity-selective pumping, and built a potassium atom induced-dichroism atomic line filter with a transmission spectrum of sub-Doppler linewidth, and the working bandwidth of this filter was only 170 MHz [19]. In 2007, He and others built a Rb atom induced-dichroism atomic line filter, and the working bandwidth of this filter was only 398 MHz [20]. In 2009, Cere also used velocity-selective pumping to achieve a Rb atom induced-dichroism atomic line filter, and the working bandwidth of this filter was only 80 MHz [21].

In 2012, Wang and others from Peking University obtained an ultra-narrow linewidth atomic filter close to the natural linewidth of atomic transition at 455 nm [22]. Corresponding to $6S_{1/2}, F = 4 \rightarrow 7P_{3/2}, F' = 5$ peak transmission of this transition can reach 9.7%, and the linewidth is 6.2 MHz; corresponding to $6S_{1/2}, F = 3 \rightarrow 7P_{3/2}, F' = 2, 3$ transition, the peak transmission can reach 6.1%, and the linewidth is 3.9 MHz. The experimental schematic of the filter is shown in Fig. 4.11. The output beam of the 455 nm external cavity diode laser is divided into strong and weak parts by beam splitter 1. The strong light is used as pump light, and the weak light is used as probe light. G1 and G2 are Glan–Taylor prisms, with an extinction ratio of up to 10^5:1. Half-wave plate 1 changes the polarization direction of the pump light to reduce the impact on the detection signal. Half-wave plate 2 changes the polarization direction of the probe light relative to the polarizing beam splitter, so that one of the two signals of the polarizing beam splitter is parallel to the polarization direction of the probe light, and the other is perpendicular to the polarization direction of the probe light. The signal whose polarization direction is parallel to the polarization direction of the probe light is the saturated absorption spectrum signal, which is detected by photodetector 1 and is used as a reference signal. The linewidth of the filter can be obtained from the saturated absorption spectrum. The signal whose polarization direction is perpendicular to the polarization direction of the probe light is the filter signal, which is detected by photodetector 2. In order to eliminate the influence of stray light on the experimental results, a pinhole aperture is added to the experimental system. Two lenses are added to focus the beam, both of which are convex lenses. The focal length of lens 1 is 25 cm, which is used to enhance the light intensity of the pump light. The focal length of lens 2 is 3.5 cm, which is used to increase the intensity of the probe light. The adjustable optical attenuator 1 is used to change pump light intensity, whereas the adjustable optical attenuator 2 is used to change probe light intensity. The cesium cell has length of 5 cm and is wrapped in a heating sheet to control its temperature. The temperature of the cell can be continuously adjusted between room temperature and 150 °C. Two permanent magnets are placed on both sides of the cell. The size of the magnetic field is adjusted by changing the distance between the magnets. Since the required magnetic field is very weak, the magnet used is small, and the magnetic field experienced by the cell can be approximated as a uniform magnetic field. In order to eliminate the influence of the external magnetic field, a magnetic shielding device is added to the filter system, as shown by the dashed line in Fig. 4.11.

Figure 4.12 shows the filter signal of the cesium atom $6S_{1/2}$, $F = 4 \rightarrow 7P_{3/2}$, $F' = 3, 4, 5$ transition. The horizontal axis represents frequency, and the vertical axis represents transmission. The topmost curve is the saturated absorption spectrum used as a reference signal received by photodetector 1. Using the saturated absorption spectrum as a reference, the linewidth of the filter can be accurately obtained. The three data curves from the bottom to the top represent the filter signals received by photodetector 2 under the conditions of no magnetic field and no pump light, magnetic field and no pump light, and magnetic field and pump light, respectively. The magnetic field strength is 4 G, the temperature is 140 °C, and the pump light intensity is 30 mW/cm^2. $6S_{1/2} \rightarrow 7P_{3/2}$ transition has a saturation intensity of 1.6 mW/

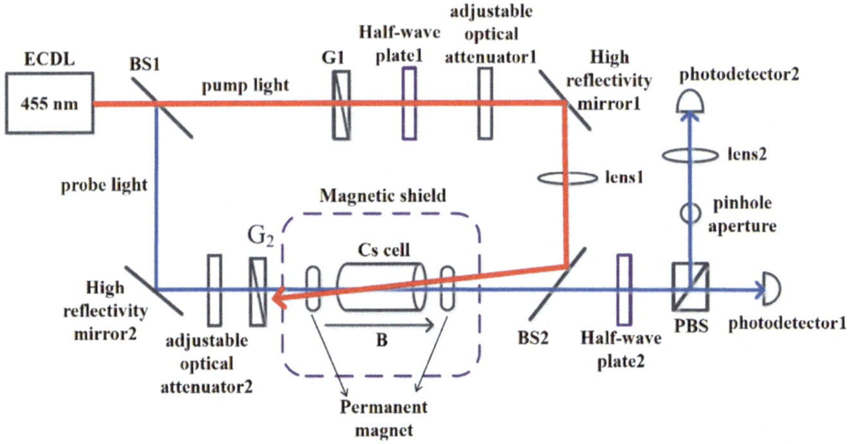

Fig. 4.11 Schematic diagram of the 455 nm ultra-narrow linewidth atomic filter experiment

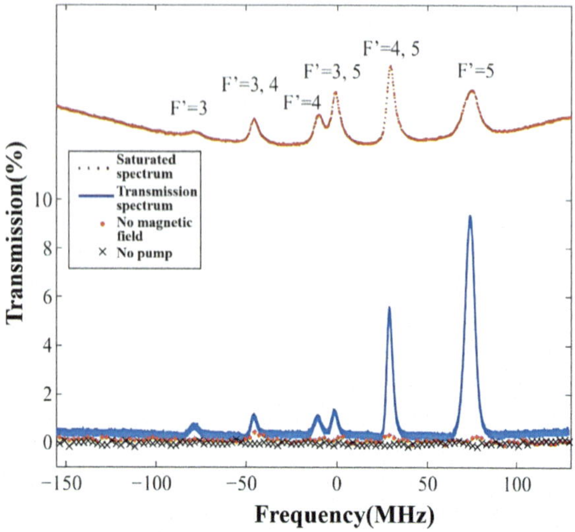

Fig. 4.12 Filter signal of $6S_{1/2}$, $F = 4 \to 7P_{3/2}$, $F' = 3, 4, 5$ transitions

cm^2. The probe light intensity is 0.3 mW/cm^2, which is much less than the saturation intensity. For the $F = 4 \to F' = 5$ transition, the maximum transmission of the filter can reach 9.7%, leading the linewidth of 6.2 MHz. In addition, for the $F = 4 \to F' = 4, 5$ transitions, the linewidth reaches only 3.7 MHz, when the transmission reaches 5.8%.

Similar to Fig. 4.12, Fig. 4.13 shows the filter signal of the cesium atom $6S_{1/2}$, $F = 3 \to 7P_{3/2}$, $F' = 2, 3, 4$ transitions. The topmost image is the saturated absorption spectrum used as a reference signal received by photodetector 1. The three data curves from the bottom to the top represent the filter signals received by photodetector 2 under the conditions of zero magnetic field without pump light, magnetic field without pump light, and magnetic field with pump light. The magnetic field strength is 5 G, the temperature is 110 °C, and the pump light intensity is 32 mW/cm^2. For the $F = 3 \to F' = 2, 3$ transitions, the maximum transmission of the filter can reach 6.1%, and the linewidth is 3.9 MHz. From Figs. 4.12 and 4.13, it can be seen that a good filter signal only appears when both the magnetic field and the pump light are present, and the transmission becomes very small if any one is missing. Once the magnetic field and pump light are added, the transmission will increase rapidly.

Although induced-dichroism atomic line filter has eliminated the dependence on the magnetic field, simplifying its structure to a certain extent and no longer needing to consider the adjustment of the magnetic field, the use of additional circularly polarized pump light complicates the structure of induced-dichroism atomic line filter. Normally, different states of circularly polarized pump light have varying degrees of impact on the filtering performance. Therefore, the adjustment of circularly polarized pump light becomes a necessary condition for the implementation of induced-dichroism atomic line filter, which means that the realization and application of induced-dichroism atomic line filter have not been simplified. So far, there

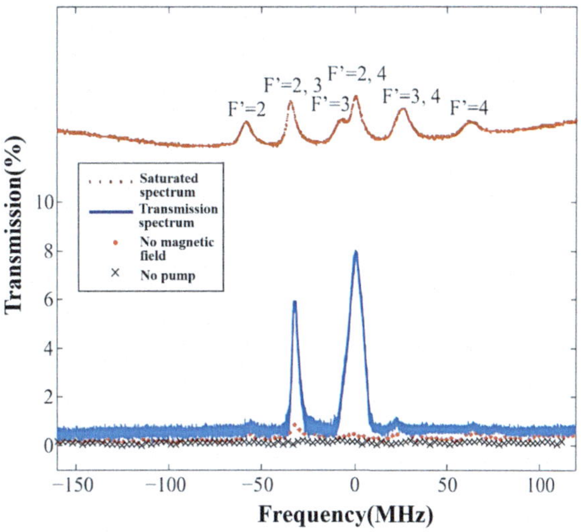

Fig. 4.13 Filter signal of $6S_{1/2}, F = 3 \rightarrow 7P_{3/2}, F' = 2, 3, 4$ transitions

is not much research related to induced-dichroism atomic line filter, and researchers have only explored potassium and Rb elements to varying degrees.

4.2.6 Atomic Filters Based on Spectral Lamps

The pump laser is an indispensable device in the excited state Faraday anomalous dispersion optical filter (ES-FADOF) system. For the required filtering frequency, it is necessary to find a pump laser of the corresponding frequency. Moreover, the stability of the pump laser has a significant impact on the filtering effect, requiring a high stability for the pump laser. This leads to the problem of complex equipment and high cost in ES-FADOF. To solve these problems, alternative means of exciting atoms need to be found.

Taking Rb atoms as an example, the Rb spectral lamp (hereinafter referred to as the Rb lamp) is an important pump light source for microwave Rb atomic clocks. Its function is to achieve atomic pumping between the hyperfine energy levels of the Rb atomic ground state, to obtain a stable microwave frequency signal for resonance transition. The structure of the Rb lamp mainly includes a Rb atomic cell filled with buffer gas, an RF coil and circuit, and a temperature control device. When the atomic cell temperature reaches about 100 °C and a radio frequency signal of 70–200 MHz is supplied, the Rb lamp will be lit. During the lighting process of the Rb lamp, under the excitation of the radio frequency signal, ions and electrons are generated in the Rb cell, which collide with the buffer gas, exciting more ions and electrons. High-energy electrons and ions excite the buffer gas to a high energy level, and then release photons through spontaneous radiation, thus, the spectrum of the buffer gas appears; when these buffer gases collide with Rb atoms, they excite Rb atoms to a high energy state. Due to spontaneous radiation, Rb atoms release photons and fall to the ground state, producing the spectrum of Rb, and the Rb lamp presents a brilliant purple-red color. The Rb lamp can emit a rich spectrum, which indicates that Rb atoms are populated on multiple excited states through electrical excitation. Using electrical excitation, the pump laser system can be replaced in ES-FADOF, thus creating a small, low-cost filter system based on spectral lamps.

As shown in Fig. 4.14, the atomic cell in FADOF is replaced with an atomic discharge lamp forms an lamp-based excited-state Faraday anomalous dispersion optical filter (LESFADOF). When the lamp is off, the filter can be equivalent to a common ground state filter; when the lamp is lit, the frequency of the filter can operate in the excited state.

Based on the power of the radio frequency signal, the Rb lamp can display three spectral modes: ring, red, and weak modes, as shown in Fig. 4.15. When the temperature of the lamp is below 100 °C and the radio frequency power is relatively high, the Rb lamp operates in the ring mode, with the center of the glass bulb turning white and a circle around it appearing purplish-red; when the temperature is above 100 °C and the radio frequency power is relatively low, the Rb lamp operates in the red mode, with the entire glass bulb appearing purplish-red; when the radio frequency power is very low, the Rb lamp operates in the weak mode, with the light being very dim, the center no longer glowing, and only a circle around it appearing dark purplish-red.

In the application of FADOF, it is hoped to gather the most atoms on the excited state energy level. In the experiment, it is found that when the Rb lamp operates in red mode and is at the transition point between red mode and weak mode, the Rb lamp has the greatest absorption of laser at the working wavelength of the filter, and the filter works best. According to the spectral line intensity observed in the experiment, the formula

$$P_\lambda = n_j A_{ji} \hbar v \tag{4.2}$$

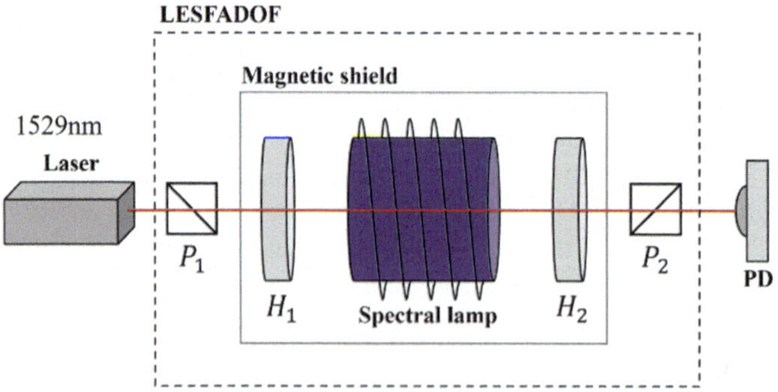

Fig. 4.14 The basic structure of LESFADOF, H_1 and H_2 are permanent magnets, P_1 and P_2 are orthogonal polarizers, and PD is a photodetector

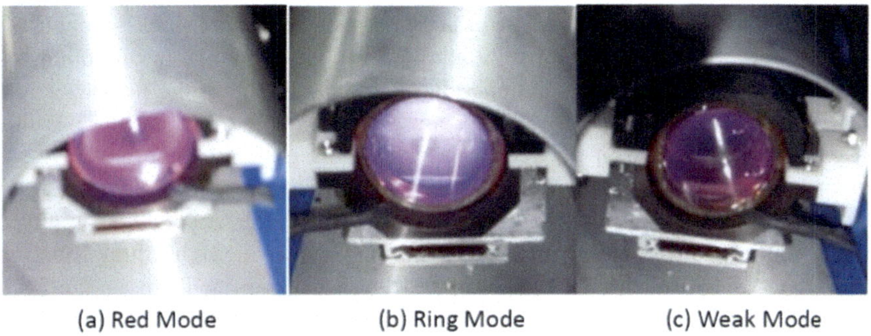

Fig. 4.15 The three operating modes of the Rb lamp

can be used to estimate the atomic density on a certain energy level. Here, λ represents the wavelength, n_j represents the atomic density on the jth energy level, A_{ji} represents the spontaneous radiation probability between energy levels i and j, $\hbar$ is the Planck constant, v is the transition frequency.

Figure 4.16 shows the experimental results of the filtering effect of 775.9 nm (corresponding to the Rb atom $5P_{3/2} \rightarrow 5D_{5/2}$ transition). Compared with 780 nm, the transmittance of the filter for light near 776 nm is much lower. Under three spectral modes, the background signals of the transmission spectrum are different, and almost, all these background signals come from the lamp light. The filtering signal of 775.9 nm presents a single peak shape. As can be seen from Fig. 4.16, in the weak mode, the background signal is not strong, but the transmission signal is too weak; while in the ring mode, the lamp light background is too strong. The red mode is the most suitable spectral mode for the 775.9 nm filter, at this time the signal-to-noise ratio is the highest, and the background signal is not as strong as in the ring mode, and the transmission bandwidth is about 650 MHz. At the same time, the electromagnetic leakage in the red mode is relatively smaller.

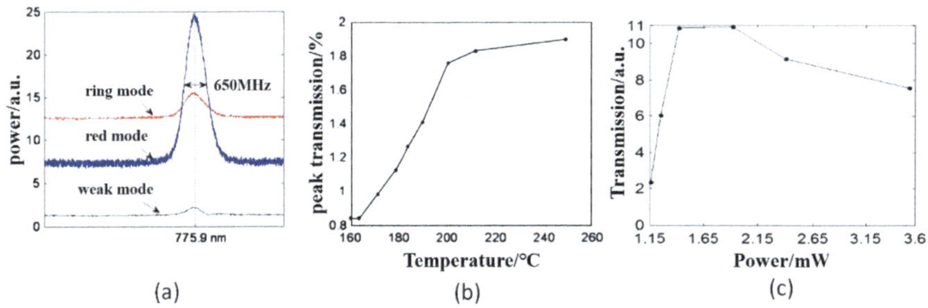

Fig. 4.16 The filtering effect of the Rb atom 775.9 nm $5P_{3/2} \rightarrow 5D_{5/2}$ spectral line. **a** The relationship between the transmission spectrum of LESFADOF and the spectral lamp mode, when the temperature is 242 °C. **b** The relationship between the transmittance of LESFADOF and temperature. **c** The relationship between the transmittance of LESFADOF and the incident power

The wavelengths corresponding to the two transition lines of the Rb atom $5P_{3/2} \rightarrow 4D_{5/2}$ and $5P_{3/2} \rightarrow 4D_{3/2}$ are 1529 nm, which belongs to the optical communication band and has high application value. Figure 4.17 shows the filtering effect of LESFADOF at the Rb atom 1529.37 nm $5P_{3/2} \rightarrow 4D_{5/2}$ transition line. At this wavelength, the relationship between the spectral mode of the Rb lamp and the transmittance is similar to that at 776 nm. In the red mode, the transmittance of the filter is the highest, the transmittance in the ring mode is very poor, and the transmittance in the weak mode is between the two. In the weak mode, the light of the Rb lamp is very weak, close to the extinguished state, and there are few atoms in the $5P_{3/2}$ state, which is the reason for the lower transmittance in the weak mode. Under the three spectral modes, the line shapes of the transmission spectrum are different: in the weak and ring modes, the transmission spectrum has three transmission peaks, the middle peak corresponds to the absorption frequency of 1529 nm; while in the red mode, there are only two peaks. The reason is that in the red mode when the spectral lamp is turned on, the number of atoms gathered in the first excited state is the highest, and the absorption intensity is the greatest, resulting in a decrease in transmittance, and the three peaks in the transmission spectrum become two peaks.

Overall, in the red mode, it is best to reduce the RF power as much as possible and work near the spectral mode transition point, which is most beneficial for the operation of the filter. In the red mode, as the RF signal power decreases, the transmittance of the two small peaks continuously strengthens, and the transmittance at the center frequency continuously decreases. This is due to the increase the number of atoms on the $5P_{3/2}$ state and the enhancement of the absorption at the center frequency, while the atoms on the $5P_{3/2}$ state have a high absorption of the 1529 nm laser, close to 100%. Similarly, the temperature of the Rb spectral lamp affects the transmission spectrum to some extent, and the variation pattern is close to that of the 776 nm filter, with the transmittance gradually increasing with temperature and then approaching saturation.

4.2.7 Atomic Filters Based on Hollow Cathode Lamps

In most Faraday atomic filters, low-melting-point metals such as mercury, sodium, potassium, Rb, and cesium are often used as media. For high-melting-point elements such as alkaline earth metals calcium and strontium, the difficulty of implementing Faraday atomic filters significantly increases because it becomes difficult to prepare high-density atomic gas. In traditional systems, the atomic number density is usually determined by thermal distribution. Therefore, for high-melting-point metals, the atomic sample needs to be heated to a high temperature to achieve a sufficient atomic number density. So far, only the calcium atom Faraday atomic filter has been reported. Such filters often have complex experimental structures, such as calcium atom beam tubes [23], or stainless steel atomic cells with sapphire windows installed at both ends [24]. To avoid the difficulty of achieving high-temperature atoms and thus expand the types of elements that can be used for Faraday atomic filters, it is imperative to

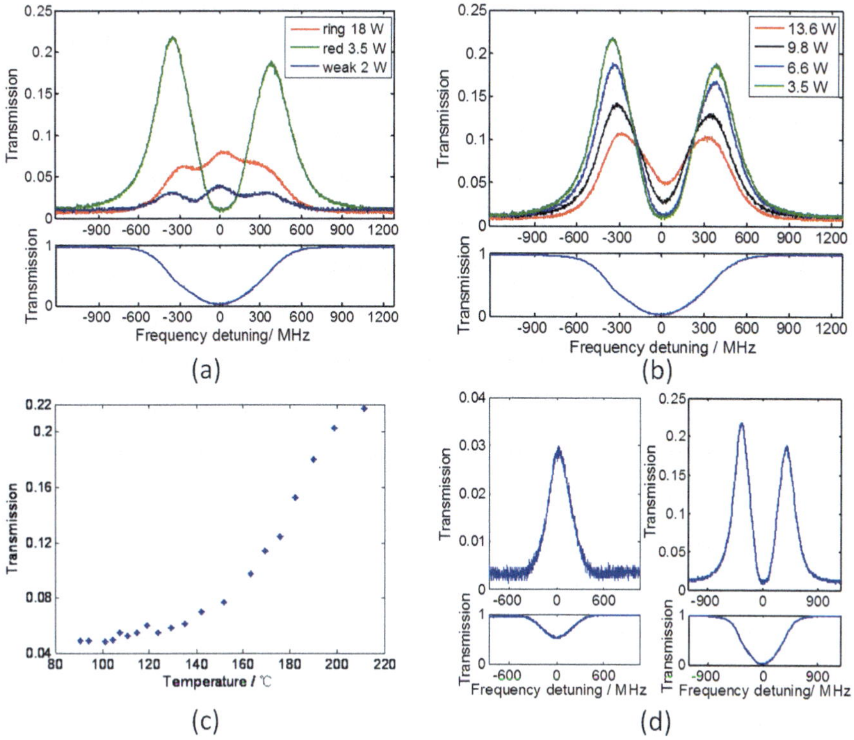

Fig. 4.17 The filtering effect of Rb atoms at 1529 nm. **a** At the $5P_{3/2}$-$4D_{5/2}$ spectral line, the relationship between the LESFADOF transmission spectrum and the spectral lamp mode. **b** When the spectral lamp is in red mode, the relationship between the LESFADOF transmission spectrum at the $5P_{3/2}$-$4D_{5/2}$ spectral line and the RF power. **c** When the spectral lamp is in red mode, the relationship between the LESFADOF transmission spectrum at the $5P_{3/2}$-$4D_{5/2}$ spectral line and temperature. **d** Comparison of the LESFADOF transmission spectra at the $5P_{3/2}$-$4D_{3/2}$ and $5P_{3/2}$-$4D_{5/2}$ spectral lines when the spectral lamp is in the red mode; the spectra in the lower parts of figures **a**, **b**, and **d** are the reference transmission spectra in the red mode with the orthogonal polarizers removed

increase the atomic number using an additive mechanism rather than mere thermal equilibrium. The hollow cathode lamp is a device that obtains high-density atomic gas by electric shock, and the current technology is quite mature, with more than seventy types of atoms. It can excite natural atoms or ions to different energy levels, thus providing a large number of usable atomic transition lines. In 2014, for the first time, Pan and others of Peking University realized a Faraday atom filter based on the 461 nm transition line of ^{88}Sr atoms using a hollow cathode lamp, and measured its transmittance as a function of magnetic field and hollow cathode lamp discharge current, with a maximum transmittance of 62.5% [25]. This realization of the atomic filter greatly expands the application range of Faraday atomic filters, making them applicable to more than seventy refractory metals and their numerous transition lines, beyond their previous limitation to low melting point metal gases. In traditional

Faraday atomic filters, the required atomic number density is determined by the saturated vapor pressure under thermal equilibrium, so the atomic system needs to be heated to a high temperature. Taking strontium atoms as an example, its saturated vapor pressure expression is [26]:

$$\lg P(Pa) = 10.750 - 8427/T \tag{4.3}$$

Combining with the ideal gas equation, the atomic number density at different temperatures is expressed as:

$$\rho = \frac{1}{kT} \times 10^{10.750 - 8427/T} \tag{4.4}$$

On the other hand, the atomic number density can also be represented by optical density (OD), where the expression defining optical density is:

$$OD = -\ln(P/P_0) \tag{4.5}$$

P_0 and P represent the power of the probe light before and after being absorbed by the atomic system, respectively [27, 28]. Using this, the effective atomic density (i.e., the Doppler shift is within the natural linewidth) can be calculated by the following formula:

$$P/P_0 = e^{-\sigma \rho l} \tag{4.6}$$

Combining the above two formulas and considering Doppler broadening, if the optical density in the hollow cathode lamp reaches 2.6 or above, the atomic density should be higher than $2.1 \times 10^{16}/m^3$, and the heating temperature required by thermal equilibrium method needs to be above 300 °C. The atomic cell needs to replace the common glass material with steel and sapphire materials. At the same time, optical coating at this temperature is more challenging. However, in the experiment conducted by the research group at Peking University, by increasing the discharge current of the hollow cathode lamp to 16 mA, the same optical density can be easily obtained at room temperature. Therefore, this method can greatly simplify the system complexity of the Faraday atomic filter, thus providing a new feasible scheme for applying high melting point metals to atomic filters.

The experimental setup of the Faraday atomic filter based on the hollow cathode lamp is shown in Fig. 4.18. The core device is a Faraday atom filter using an ^{88}Sr hollow cathode lamp (HCL), H1 and H2 are a pair of permanent magnets, generating a magnetic field up to 1748 G at the center of the hollow cathode lamp. G1 and G2 are a pair of Glan–Taylor prisms perpendicular to each other, with an extinction ratio of 1×10^5. After being placed in the hollow cathode lamp, due to the birefringence effect introduced by the windows at both ends of the hollow cathode lamp, the extinction ratio of the Glan–Taylor prism drops to 2×10^2. A 461 nm external cavity diode laser is used as the probe light, and the spectrum is measured by a photodetector (PD). The

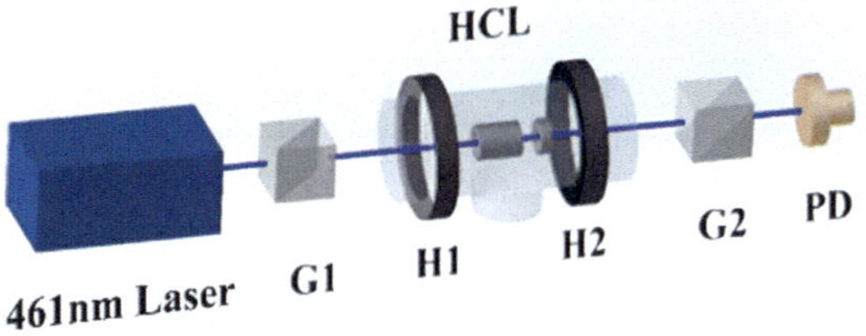

Fig. 4.18 Experimental setup of the Faraday atomic filter based on the hollow cathode lamp: G1, G2, Glan–Taylor prisms; H1, H2, permanent magnets; HCL, hollow cathode lamp; PD, photodetector

model of the strontium atom hollow cathode lamp is HAMAMATSUL2783, which is composed of a ring anode and a cylindrical cathode inside. The cathode is 20 mm long and has an inner diameter of 3 mm. By adjusting the discharge current of the hollow cathode lamp, the internal atomic gas pressure, i.e., the atomic number density, can be adjusted. The optical density spatial distribution inside the hollow cathode lamp is not uniform, and it is distributed more near the wall of the cathode cylinder. In the experiment, the position with the optimal atomic beam density is selected, and the optical density can reach 2.6 under a discharge current of 16 mA. During the measurement of optical density, the waist radius of the probe light is 0.23 mm, and the power of the probe light is 50 μW. Therefore, the calculated light intensity of the probe light is 31 mW/cm^2, which is smaller than the saturation light intensity of the strontium atom 461 nm transition of 42.75 mW/cm^2, which is measured under weak light conditions.

When G1 and G2 are perpendicular to each other and the frequency of the probe light is in resonance with the atomic system, a magnetic field is applied to the hollow cathode lamp, and the transmitted light power P1 is measured by the photodetector. When G1 and G2 are parallel to each other and the frequency of the probe light is far off-resonance with the atomic system, no magnetic field is applied to the hollow cathode lamp, and the transmitted light power is measured by the photodetector. Since the laser power varies linearly with the current within the measurement range, the power value measured in the off-resonance state can be converted to the resonance state power, P2, through linear fitting. The fluorescence intensity of the hollow cathode lamp is detected by the photodetector, denoted as P′. When the detection distance is 5.8 cm, the discharge current of the hollow cathode lamp is at 7 mA, 10 mA, 13 mA, 16 mA, 19 mA, 22 mA, the detected fluorescence intensities are 1.3 μW, 2.4 μW, 3.5 μW, 7.3 μW, 8.7 μW and 10.0 μW, respectively. The transmission T of the Faraday atomic filter is defined as $T = (P1 - P')/(P2 - P')$. In this definition, since the optical loss of the system appears in both the numerator and the denominator, the impact of system optical loss is eliminated.

Under a 1748 G magnetic field and a cathode lamp discharge current of 16 mA, the peak transmittance of the transmission peak was measured to be 62.5%, and the transmission bandwidth was found to be 1.19 GHz. The transmission spectrum is shown as the blue solid line in Fig. 4.19, with a measurement uncertainty of 2.1%, including 2.1% from the extinction ratio of the Glan–Taylor prism, 0.1% from laser power measurement, 0.02% from magnetic field fluctuations, and 0.03% from fluctuations in the hollow cathode lamp discharge current. The transmittance of the hollow cathode lamp Faraday laser varies with the applied magnetic field and cathode lamp discharge current. The transmission spectra under different magnetic fields are shown in Fig. 4.19b, and both the maximum transmittance and transmission bandwidth increase with the strength of the magnetic field. The transmission spectra under different discharge currents are shown in Fig. 4.19c. When the amplification current increases, the maximum transmittance initially increases and then decreases, while the transmission bandwidth always increases with the discharge current. The peak transmittance of the Faraday atomic filter under different magnetic fields and discharge currents is shown in Fig. 4.19d, and the maximum transmittance of 62.5% is reached when the magnetic field strength is 1748 G and the hollow cathode lamp discharge current is 16 mA.

Under different magnetic field strengths, the transmittance changes with the hollow cathode lamp discharge current as shown in Fig. 4.20a. When the magnetic field strength is below 1043 G, the maximum transmittance gradually increases with the discharge current. When the magnetic field strength is above 1143 G, when the discharge current is small, the atomic number density is the main limiting factor for the transmittance. The rotation angle increases with the increase of the atomic number density, and the central transmission intensity also rapidly increases. However, as the discharge current continues to increase, the higher atomic number density and the exacerbation of Doppler broadening enhance the absorption of the atoms to the

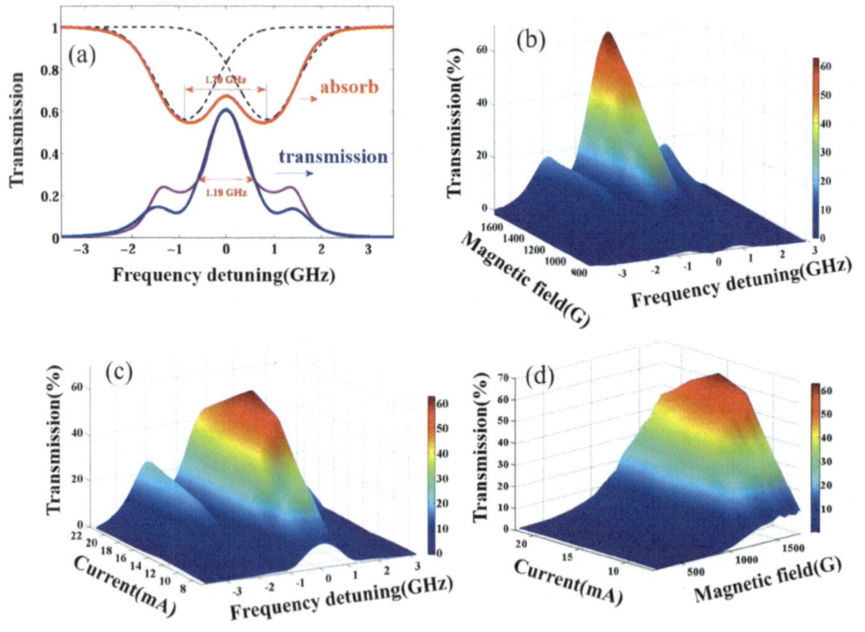

Fig. 4.19 **a** Faraday atomic filter transmission spectrum line. **b** Faraday atomic filter transmission spectrum line under different magnetic fields. **c** Faraday atomic filter transmission spectrum line under different discharge currents. **d** Peak transmittance under different magnetic fields and currents

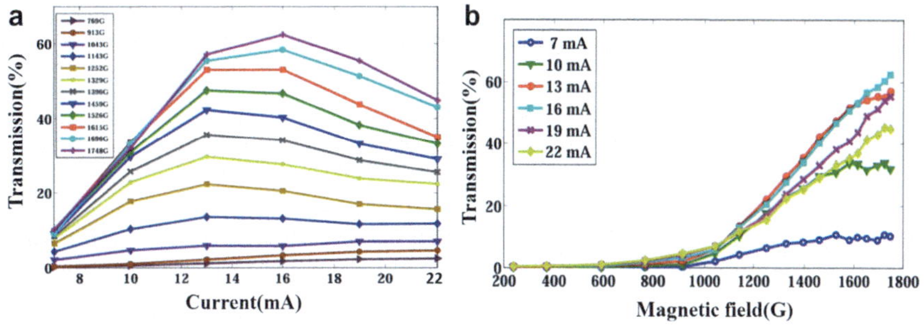

Fig. 4.20 **a** Peak transmittance changes with discharge current. **b** Peak transmittance changes with magnetic field strength

probe light, and the final transmission intensity decreases. Under different discharge currents, the transmittance changes with the strength of the magnetic field as shown in Fig. 4.20b. When the discharge current is below 13 mA, the transmittance increases with the increase of the magnetic field strength and gradually saturates. When the discharge current increases to 16 mA, the transmittance increases with the magnetic field strength throughout the measurement area.

The current problems and future development directions of the hollow cathode lamp Faraday atomic filter are analyzed as follows:

1. There is a fluorescence background. The hollow cathode lamp has a considerable fluorescence background. In the experiment, it was found that when the discharge current was 7, 10, 13, 16, 19, and 22 mA, the fluorescence background was 1.3, 2.4, 3.5, 7.3, 8.7, and 10.0 μW at a distance of 5.8 cm from the lamp body. Under such constraints, the Faraday atomic filter based on the hollow cathode lamp is not suitable for weak light detection, but its rich atomic transition spectrum provides potential applications in frequency stabilization systems.
2. There is a magnetic shielding effect. It is worth mentioning that the magnetic field strength measured in the above experiment is not accurate. The experimental measurement is the magnetic field strength obtained at the corresponding center position in the absence of a hollow cathode lamp. However, since the cathode base of the hollow cathode lamp is made of iron material, and the cathode is made of strontium, nickel, and tin alloy, both the cathode seat and the cathode material will produce a magnetic shielding effect on the magnetic field at the center, and the actual magnetic field in the hollow cathode lamp is far less than 1748 G. Due to the limitation of measurement accuracy, the precise magnetic field strength at the center position can only be estimated by the separation degree of the Zeeman spectrum. After measurement, the Zeeman sublevel splitting is 1.70 GHz, corresponding to a magnetic field strength of 607 G. Under this magnetic field, the transmission spectrum is theoretically fitted, as shown by the purple solid line in Fig. 4.19a, which matches the experimental results well, only there is some difference in the height of the two wings, which may be due to the uneven atomic number density in the hollow cathode lamp. It can be

speculated that the magnetic shielding coefficient of the strontium atom hollow cathode lamp used is 2.9. Due to the existence of the magnetic shielding effect, a higher actual magnetic field needs to be added in the experiment to achieve the magnetic field strength required by the Faraday atomic filter. This problem can be solved by using non-magnetic materials as the electrode materials of the hollow cathode lamp.
3. There is optical loss. The experiment measured that the optical loss from the two ends of the cathode lamp is 33%, which can be solved by improving the flatness of the window and using an optical window with an anti-reflection coating.

As a possible future development direction, the Faraday atomic filter based on the hollow cathode lamp has application potential in optical frequency atomic clocks. This atomic filter can be used as a frequency selector, reflecting the transmitted light back to the laser diode, replacing the frequency selection function of the grating with the atomic filter, and realizing a small Faraday laser system with a definite frequency value. In this system, after the laser is turned on, it will work accurately on the atomic transition spectrum line, and its frequency is immune to fluctuations in the injection current and laser diode temperature. Taking the strontium atomic clock as an example, the laser cooling and probe light are usually provided by a 461 nm diode laser, which are extremely sensitive to external vibrations and require precise temperature and current control to ensure an accurate output frequency. Therefore, Faraday lasers operating at the ^{88}Sr $(5s^2)^1S_0$–$(5s5p)^1P_1$ resonance line (461 nm) are undoubtedly a better choice, and they can be further locked to either the saturation spectrum or the modulation transfer spectrum to provide long-term stability.

4.2.8 Atomic Filter Based on Calcium Atomic Beam

The atomic filter was originally designed to isolate weak laser signals from sunlight and other broadband background interference, achieving narrow-band optical filtering. High sensitivity and ultra-narrow spectral bandwidth make it a core device applied in laser frequency stabilization, metrology, laser communication, laser radar, and other fields. In addition, if the atomic filter matches a certain number of dark feature lines in the solar spectrum, namely Fraunhofer lines [29], the signal after filtering can significantly reduce the background noise of the solar spectrum. Among them, the calcium atom 423 nm $^1S_0$–$^1P_1$ transition spectral line has been widely used in satellite-submarine communication, and this frequency corresponds to the Fraunhofer line, and the external cavity diode laser currently working in the 423 nm band has been realized.

At present, the calcium atom filter is mainly reported as an atomic resonance filter, in which a double-peak structure was observed in the experiment of the magneto-optical atomic filter [30], each transmission peak has a bandwidth of 1.5 GHz, and the actual bandwidth of the entire magneto-optical atomic filter is 3 GHz. However, the working temperature of the calcium atom glass cell has reached 450 °C, and the corresponding Doppler bandwidth has reached 2.2 GHz, which limits the further reduction of the bandwidth of this atomic filter. Xu and others of Peking University have realized a Faraday atomic filter based on a calcium atom beam with a single-peak transmission spectrum and a transmission bandwidth smaller than the Doppler broadening [29].

The working wavelength of the atomic filter based on the calcium atom beam is 423 nm, and the corresponding atomic energy levels are shown in Fig. 4.21. Unlike the general Faraday atomic filter, using a calcium atom beam instead of a calcium atom glass cell as the absorption medium of the atomic filter. There is no collision between the atoms of the absorption medium, which can effectively reduce the passband bandwidth of the calcium atom filter to below the Doppler linewidth. In order to improve the transmission of the calcium atom beam filter, a 9-path light scheme is adopted, that is, a total reflection mirror is used to reflect the 423 nm incident probe light back and forth 9 times on both sides of the calcium atom beam, as shown in Fig. 4.22. In addition, corner-cube retroreflector can be used to implement a scheme for multiple reflections of probe light, as shown in Fig. 4.23. Using a staggered corner-cube retroreflector, the probe light is reflected multiple times through the atomic beam. In comparison to the plane mirrors, it has higher mechanical robustness and stability.

The experimental results of the narrow linewidth atomic filter of the Ca atom beam can be seen in Figs. 4.24 and 4.25. Figure 4.24 corresponds to the transmission spectrum of the single-path light scheme and the 9-path light scheme, where the external magnetic field is 120 G. The transmission peak corresponds to the atomic transition frequency without an external magnetic field. The dashed line represents the theoretical values, and the solid line represents the experimental measurements. Both the transmission and the transmission bandwidth of the Ca atom filter increase with the increase in atomic density in the Ca atom beam. This paper measures that the remaining Doppler linewidth in the direction perpendicular to the atomic beam is about 320 MHz, which is about 1/7 of the 2.2 GHz Doppler broadening. In the

Fig. 4.21 Related energy levels of calcium atoms [29]

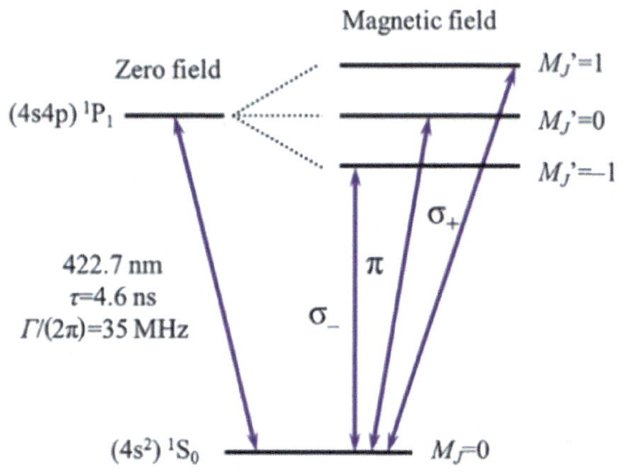

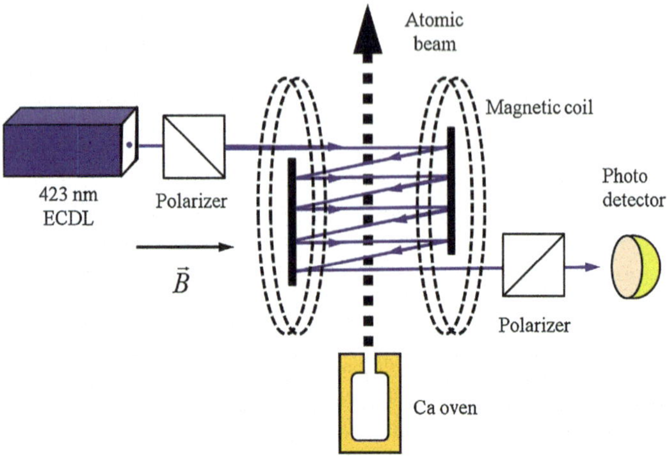

Fig. 4.22 Experimental principle diagram of the magneto-optical atomic filter based on the calcium atom beam (9-path light scheme) [29]

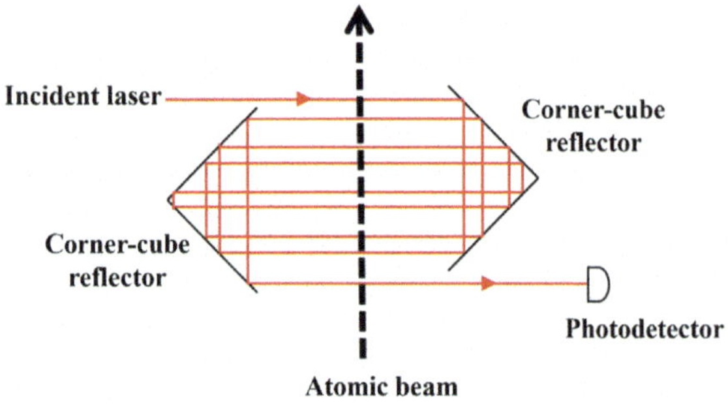

Fig. 4.23 Scheme for multiple reflections of detection light based on an angular cone mirror corner-cube retroreflector

Ca atom beam filter of the single-path light scheme, the 423 nm probe laser passes vertically through the Ca atom beam. At this time, the maximum transmission of the filtered light is measured to be 0.1%, and the bandwidth of the transmission spectrum is 350 MHz. In this case, the highest transmission of the filter transmission spectrum is measured to be 7.3%, and the bandwidth of the transmission spectrum is found to be 590 MHz, which is much smaller than the Doppler linewidth of 2.2 GHz. It is also found that when the 423 nm probe light is far from resonance with the corresponding 423 nm transition frequency, a non-zero transmission background is observed, which increases with the increase of the furnace temperature and the interaction length of the atomic beam. This is because the atomic beam ejected from the Ca oven collides with the wall of the collimating tube and enters the vacuum cavity as well. These atoms have a larger transverse velocity component perpendicular to the direction of the atomic beam, corresponding to a wider velocity distribution. However, these atoms contribute to the non-zero transmission background, and the increase in furnace temperature and interaction length of the atomic beam increases this contribution.

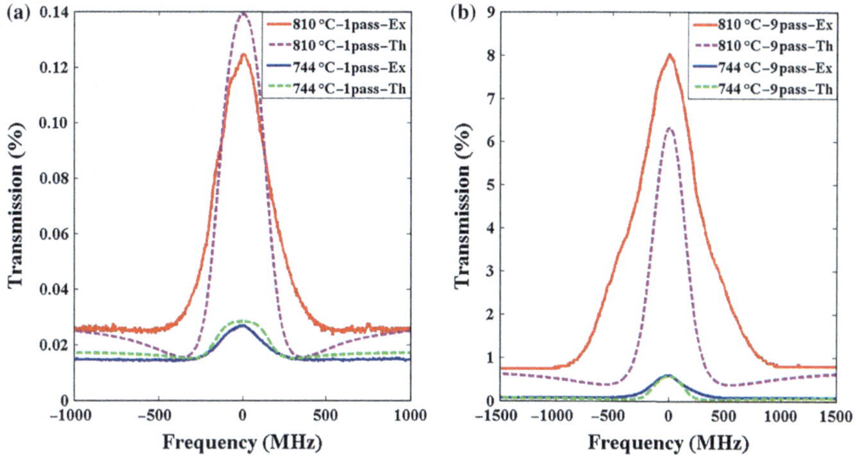

Fig. 4.24 Transmission spectrum of the magneto-optical atomic filter based on Ca atomic beam **a** single light scheme; **b** 9-light scheme [29]

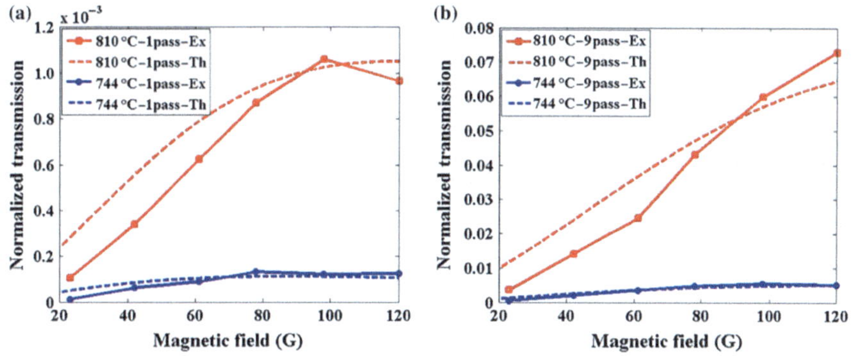

Fig. 4.25 Variation curves of transmission with applied magnetic field strength for Ca-beam-based atom filter at 744 and 810 °C oven temperatures [29] **a** single light scheme; **b** 9-light scheme

Figure 4.25 shows the transmission of the single light path and 9-light path atomic filters under different Ca oven temperatures and external magnetic field strengths. As shown in Fig. 4.25, the theoretical and experimental values are very consistent at zero detuning, and the peak transmission of the atomic filter based on the Ca atomic beam is determined by the strength of the external magnetic field. When the frequency shift of σ_+ and the transition equals half of the transverse Doppler linewidth of the Ca atomic beam, the highest transmission can be obtained. The transverse Doppler linewidth of the collimated Ca atomic beam increases with the increase of the oven temperature. When the oven temperature is 744 °C and 810 °C, the Doppler linewidths in the single light scheme are 240 MHz and 320 MHz, respectively. In the nine-light scheme, because the probe light beam is not completely perpendicular to the direction of the atomic beam, the measured values of the transverse Doppler width increase to 350 MHz and 480 MHz, respectively.

Internationally, this is the first time that a single-peak, sub-Doppler transmission bandwidth Ca atomic magneto-optical atomic filter based on an atomic beam has been realized. By further improving the collimation efficiency and beam intensity of the Ca atomic beam, the performance indicators of the magneto-optical atomic filter based on the Ca atomic beam can be further improved.

4.3 Principle, Process and Performance of Faraday Atomic Filter

4.3.1 Basic Principle

The Faraday atomic filter uses the resonant Faraday magneto-optical effect to rotate the light polarization near the atomic transition frequency by a certain angle, and then uses two orthogonal polarizers (or a combination of a half-wave plate and PBS) as the polarizer and analyzer to achieve the filtering effect. Its basic structure mainly includes the polarizer P_1, atomic cell, axial magnetic field, and polarizer P_2, as shown in Fig. 4.26 [31].

P_1 and P_2 have orthogonal polarization axes. Assuming that the polarization axis of P_1 is along the x direction, and P_2 is along the y direction, the electric field intensity vector of the linearly polarized light entering the atomic cell through P_1 can be written as

$$\boldsymbol{E} = \boldsymbol{e}_x E_0 \cos(\omega t) \tag{4.7}$$

Any beam of linearly polarized light can always be decomposed into a combination of left-handed circularly polarized light and right-handed circularly polarized light. Further, we can write E in exponential form and decompose it into two components as

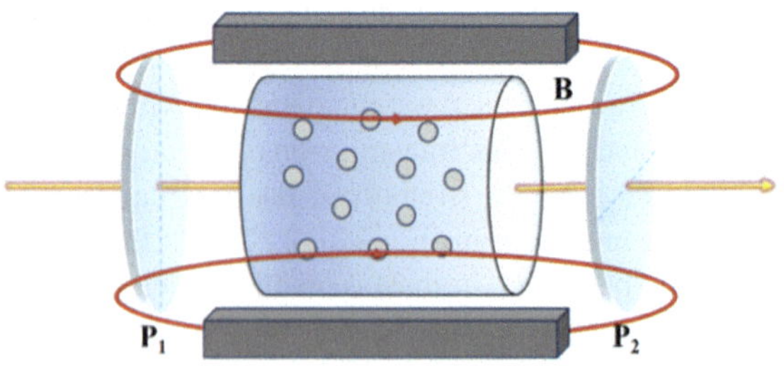

Fig. 4.26 Schematic diagram of the Faraday atomic filter

$$\boldsymbol{E} = \frac{\sqrt{2}}{4} E_0 \boldsymbol{e}_+ \left(e^{-i\omega t} + e^{i\omega t}\right) + \frac{\sqrt{2}}{4} E_0 \boldsymbol{e}_- \left(e^{-i\omega t} + e^{i\omega t}\right) \tag{4.8}$$

where $\mathbf{e}_+, \mathbf{e}_-$ represents the unit vectors of two circularly polarized rotations. Assuming l to be the length of the atomic cell, once the laser passes through the atomic cell, its electric field strength becomes

$$\begin{aligned}\mathbf{E} = & \frac{\sqrt{2}}{4} E_0 \mathbf{e}_+ \left(e^{i(k_+ l - \omega t)} + e^{-i(k_+ l - \omega t)}\right) \\ & + \frac{\sqrt{2}}{4} E_0 \mathbf{e}_- \left(e^{i(k_- l - \omega t)} + e^{-i(k_- l - \omega t)}\right)\end{aligned} \tag{4.9}$$

where, k_+ and k_- are the wave numbers of the left and right circularly polarized components, respectively. Due to the Zeeman effect of the magnetic field in the cell, the atomic vapor has different polarizabilities for the left and right circularly polarized components, which will cause $k_+ \neq k_-$ (for a detailed theoretical derivation, see Sect. 4.4). This will cause a phase difference $\Delta\phi$ between the circularly polarized components of the output laser, which, after superposition, causes the polarization direction of the linearly polarized light entering the atomic cell to be rotated by an angle. At the same time, for different incident wavelengths, atomic types, temperatures, magnetic fields, atomic cell lengths, and other parameters, $\Delta\phi$ is different. Moreover, the change in $\Delta\phi$ is particularly significant near the atomic resonance frequency, which is caused by the resonant Faraday magneto-optical effect. Since P_1 and P_2 are orthogonal, therefore, in the absence of optical rotation effect, theoretically, no light will pass through the filter. However, in the presence of optical rotation effect, as the rotation angle is different at different wavelengths/frequencies, the light intensity passing through the polarizer P_2 is also different, forming a filter. The change in transmittance with respect to the frequency of the incident light is called the transmission spectrum of the filter. In practical applications, the filter parameters are usually set so that during operation, the polarization direction of the light at the atomic resonance frequency is just rotated odd multiples of $\frac{\pi}{2}$, to allow it to pass through the polarizer with as large as possible. Therefore, the filter has the maximum transmission at a certain atomic transition frequency. Using a Faraday atomic filter to select the longitudinal mode of the laser, the laser output frequency can intrinsically correspond to the atomic transition line.

4.3.2 Preparation Process

The preparation of the Faraday atomic filter is a complex process that requires precise design and meticulous operation to ensure high-performance filtering effects. To ensure that the filter can efficiently and accurately filter out light of a specific wavelength, the design and preparation of its main components are particularly critical, directly affecting the working performance of the filter.

1. Atomic cell

Depending on the required wavelength of the output laser, select the center wavelength of the filter, which is determined by atom. The commonly used alkali metal transition wavelengths are shown in Table 4.1.

Different types of alkali metal atoms (such as sodium, potassium, Rb, cesium, etc.) have different absorption lines, which can undergo a resonant Faraday rotation effect at their respective resonance frequencies, thereby filtering light of a specific wavelength. Alkali metal atom vapor is filled into a sealed vacuum quartz cell, and inert buffer gas can be selectively filled into the cell as needed. In order to improve light transmission performance, reduce loss and beam deformation, the end face of the cell often adopts a bonding process, and an anti-reflection mirror of the target wavelength is coated. Also, the diameter of the end face should be designed with reference to the beam area to ensure that all beams enter the cell. At the same time, the length of the cell represents the interaction length between light and atoms, which will affect the rotation angle after passing through cell, and has a significant impact on the transmission spectrum of the atomic filter.

2. Design of the magnetic field

In the Faraday atomic filter, the commonly used magnetic field configurations are axial or transverse. The axial magnetic field refers to the direction of the magnetic field in the cell being the same as the direction of light propagation. The axial magnetic field can be generated by several bar-shaped or ring-shaped permanent magnets wrapped around the cell (depending upon the applications, coils can also be used to generate a magnetic field), and the poles of the magnets are directed towards the two ends of the cell, as shown in Fig. 4.27. In contrast, the transverse magnetic field refers to the direction of the magnetic field being perpendicular to the direction of light propagation. In the transverse magnetic field, if the magnetic field forms a 45° angle with the polarization direction of the incident light, this configuration is also called the Voigt filter. The strength of the magnetic field has a significant influence on the transmission spectrum, so it is necessary to choose an appropriate magnetic field strength according to the working parameters of other components in actual design and manufacturing. At the same time, the shape and arrangement of the magnets will affect the uniformity of the magnetic field in the cell. Since the magnetic field is mostly uneven at both ends of the magnet, extending the length of the magnet can improve the uniformity of the magnetic field in the cell and thus improving the quality of its transmission spectrum.

Table 4.1 Center wavelengths of common alkali metal transition lines

Atomic species	D_1 line center wavelength (nm)	D_2 line center wavelength (nm)
Sodium Na	589.756	589.158
Potassium K	770.108	766.701
Rb Rb	794.979	780.241
Cesium Cs	894.593	852.347

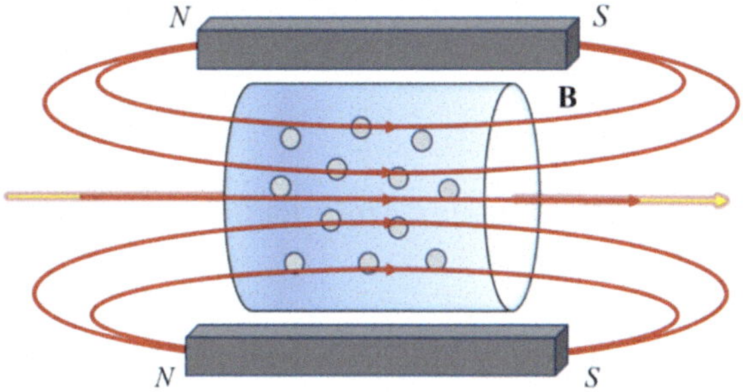

Fig. 4.27 Magnetic field in the Faraday atomic filter

3. System packaging and temperature control

For alkali metal atom cell, due to the high melting point of alkali metals, temperature affects the atomic density. When the temperature rises by 10 °C, the atomic density is approximately twice that of the original, and when the temperature rises by 30 °C, the atomic density increases by about an order of magnitude. At the same time, the Doppler broadening of atomic spectral lines also undergoes significant changes with temperature fluctuations. These factors cause the atomic absorption characteristics to change with temperature, ultimately affecting the transmission spectrum of the filter. Therefore, in Faraday atomic filters, it is crucial to precisely control the temperature of the atomic optical rotation medium to form a stable transmission spectrum, so as to ensure the stability of the output laser. In the design, since the working temperature of atomic vapor is generally higher than room temperature, flexible heating devices such as polyimide film can be used to wrap the outer surface of the atomic cell, and thermistors and PID temperature controllers can be used for precise temperature control of the atomic cell. Polytetrafluoroethylene is easy to process and has good insulation properties, making it commonly used for the packaging of atomic cell and heating devices. At the same time, since air is a poor conductor of heat, an air gap can be appropriately introduced between the atomic cell and the filter casing to achieve good insulation, provided that the installation is secure.

4. Matching and tuning of working parameters

At the beginning of the filter design, parameters and working conditions (such as cell length, working temperature, magnetic field strength, etc.) for each component need to be set based on actual volume, material requirements, and target wavelength. Theoretical simulations are then conducted to calculate the ideal target transmission spectrum. After processing and assembly according to the drawings, the actual transmission spectrum is measured by a scanning optical path or spectrometer. Due to the individual errors and non-uniformity of the magnetic field, temperature, and atomic vapor pressure, there is always a certain difference between the actual transmission spectrum and the theory, as well as between the transmission spectra of each filter. Therefore, after preparation, it is necessary to fine-tune the working parameters of the filter to compensate for these differences as much as possible, thereby achieving the best performance closest to the theoretical value and improving the quality of the filter.

4.3.3 Alkali Metal Atom Cell

Atomic vapor cell is a small atomic vapor container. This container is an important experimental device in the fields of atomic physics, quantum optics, etc., helping scientists study the basic behavior of atoms and the interaction between atoms and external environments such as light and magnetic field. The atomic cell is the core component of FADOF, and it is the medium for light and atoms to interact and produce the resonant Faraday rotation effect. In active optical clock technology, the stimulated radiation optical frequency standard signal is also generated by the atomic cell; in atomic spectral laser frequency stabilization technology, the atomic cell is used as an external quantum frequency reference for the laser, providing a standard for active correction of the output laser frequency.

The atomic cell is made through high-precision glass fusion or meticulous micro-fabrication processes, forming a transparent and sealed space. This space is filled with alkali metal atoms, and according to application requirements, a specific proportion of inert gas can also be filled in. Technologies such as glass precision fusion or microfabrication processes are keys to prepare high-precision atomic cells. In addition, the quality of the glass shell of the atomic cell, the accurate shape production, the precision of the internal wall surface, and the precise ratio of the gas also have a significant impact on its overall performance. The main core process technologies in the preparation of atomic cells are as follows:

1. Precision machining technology of the vapor cell shell

The shell material and its structure are key factors affecting the performance of the atomic cell. When the atomic cell is used in detection scenarios, spatial light usually only passes through the atomic cell once, and there is no high requirement for the flatness of the end faces of the cell. However, in the application scenario of the atomic filter for Faraday laser, the atomic cell is placed in the resonant cavity, and the laser will pass through the front and back end faces of the atomic cell multiple times in the resonant cavity. Therefore, to improve the beam quality of the output laser, wavefront aberration and feedback loss must be reduced and the flatness of the end face of the cell is particularly ensured. Usually, an optically standardized end glass piece is attached to the sidewall of the cell using bonding process. At the same time, the optical transmittance of the atomic vapor cell can be improved, and the spatial uniformity of the window refractive index can be improved through surface coating and material processing technology. In addition, attention should also be paid to the gas tightness of the atomic system to reduce the leakage of alkali metal atom vapor and buffer gas.

2. Precise filling technology of atomic cell

In the atomic cell, it is crucial to precisely manipulate the number of alkali metal atoms. If the filled atoms' number is too small, it weakens the interaction between light and atoms, affecting the performance of the atomic cell; if the filled atoms' number is too large, it makes the alkali metal vapor easy to condense, reducing the transmittance of the cell. To make the transmission spectrum of FADOF show a single transmission peak, it is also necessary to fill in the buffer gas in a specific ratio. As the volume of the cell decreases and the scale of production expands, ensuring the uniformity of alkali metal atom filling and the pressure and ratio of the gas becomes more challenging.

In order to improve the performance of the atomic cell, it is necessary to systematically analyze the complex interactions between alkali metal atoms, working gas, and buffer gas under specific light field and magnetic field conditions. In addition, studying the influence of the filling material, cavity structure, and internal wall surface state in the cell on atomic spin polarization and relaxation can provide a scientific basis for designing more efficient atomic cell structures and adjusting gas component ratios.

4.3.4 Faraday Anomalous Dispersion Optical Filter (FADOF)

The core preparation process of FADOF has been introduced above, this section mainly introduces the assembly process of FADOF and research testing methods.

1. FADOF assembly process

Let us consider a 30 × 10 mm atomic cell as an example. First, according to the size of the atomic cell, cut a suitable size of tin foil and wrap a layer of tin foil tightly along the cell starting from the cold finger position. Make sure that the shiny side of the tin foil is facing inward and fix it with heat-conducting tape. The two ends of the atomic cell should have an extra 2 mm or so to ensure that the normal light diameter of the cell is greater than 6 mm. The cold finger of the cell can be wrapped with a 3 ~ 10 cm heat conduction pipe made of tin foil, as shown in Fig. 4.28.

Take a pair of heat-conducting copper seats, choose a hole diameter on the copper block that matches the size of the thermistor. Use a syringe or toothpick, fill as much as possible with thermally conductive silicone grease to ensure good thermal contact, and plug in the measured thermistor (resistance value of about 10 k, two thermistors of similar resistance value). Clean the edges of the hole of the heat-conducting silicone grease, apply the adjusted AB glue and fix the thermistor. Place it for about one or two hours, the longer the better, so that the glue is completely solidified. Put the heat-conducting copper seat on both ends of the atomic cell, as shown in Fig. 4.29.

Take a matching size heating sheet, with an initial resistance of 9.6 Ω. Starting from the cold finger position, precisely fit the heat-conducting copper seat around one circle, and fix it with insulating tape. Wrap another layer of polytetrafluoroethylene

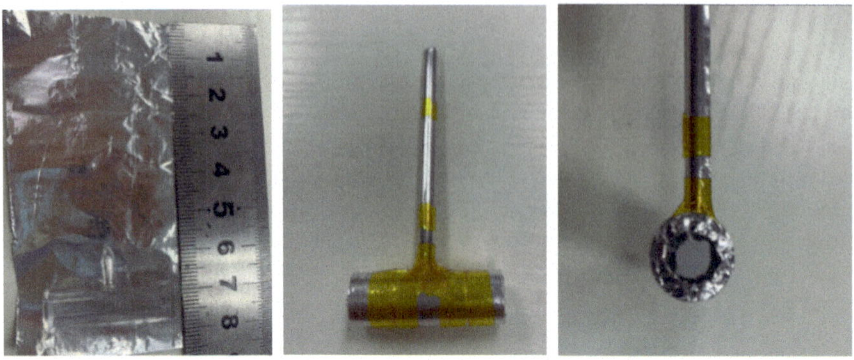

Fig. 4.28 Cut tin foil to the proper size (left); wrap the atomic cell with tin foil (middle, right)

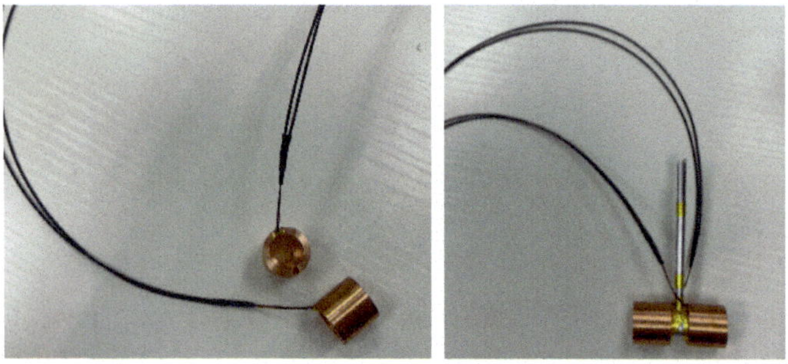

Fig. 4.29 Insert the heat-conducting copper seat with the thermistor (left); install the atomic cell into the heat-conducting copper seat (right)

(PTFE) film outside the heating sheet, as shown in Fig. 4.30, so that it serves the function of fixation and has a better thermal insulation effect.

Before putting the wrapped atomic cell into the polytetrafluoroethylene insulation case, a layer of insulation cotton needs to be added to the two ends of the atomic cell in contact with the copper block and the insulation case. The insulation cotton is trimmed into a ring according to the end face of the copper block, for the insulation without affecting the light transmission. In addition, due to the relatively distant location of the heating element wires from the cold finger and the presence of protrusions, the area where the insulation shell comes into contact with the heating element wires should be slightly thinned using a file to prevent excessive tightness, which could potentially damage the wires. The insulation case is fixed with insulating tape, as shown in Fig. 4.31.

The wire of the wrapped atomic cell is first stuffed into the small hole in the center of the cylindrical magnet, and then the entire atomic cell is stuffed into the cylindrical magnet. Because the outer diameter of the PTFE insulation tube in Fig. 4.31 is smaller than the inner diameter of the cylindrical magnet, the cell may shake inside the magnet, affecting the long-term stability of the atomic filter. It is necessary to

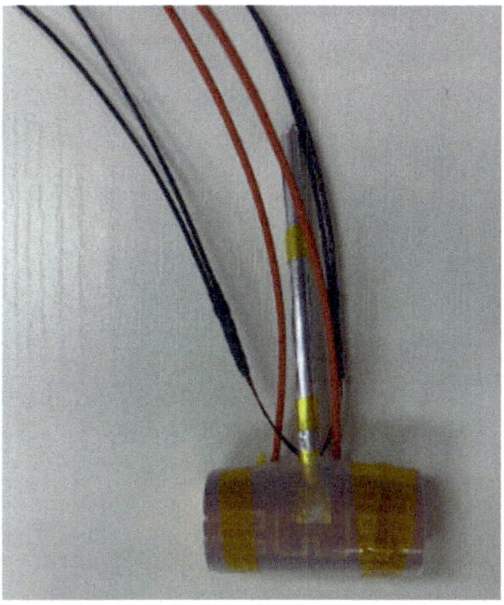

Fig. 4.30 The heating sheet is fitted and wrapped

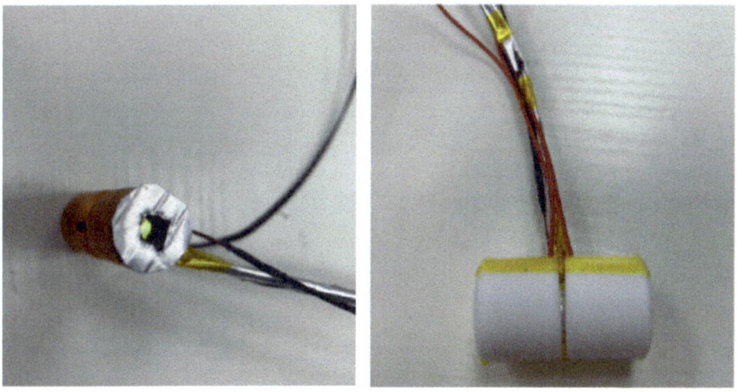

Fig. 4.31 Ring-shaped insulation cotton is pasted at both ends of the atomic cell (left); the atomic cell is put into the polytetrafluoroethylene insulation tube (right)

insert a PTFE sheet to fix the atomic cell, so that the cell remains stable inside the magnet. In addition, besides using a cylindrical magnet, magnetic strips can also be symmetrically placed around the atomic cell to generate a magnetic field, with the polarity of each magnetic strip placed needs to be the same, as shown in Fig. 4.32.

Place the entire magnet into a PTFE cylinder of corresponding size, with the outer diameter of the magnet being close to the inner diameter of the PTFE cylinder. Then assemble the PTFE cylinder and its corresponding PTFE structure, and place the whole assembly into an aluminum shell by tightening the screws. The steps for using a magnetic strip structure are the same, as shown in Figs. 4.33 and 4.34.

Connect the thermistor and heater to the temperature control circuit board or TC300. Insert two polarizing beam splitters into the slot structures near the two light

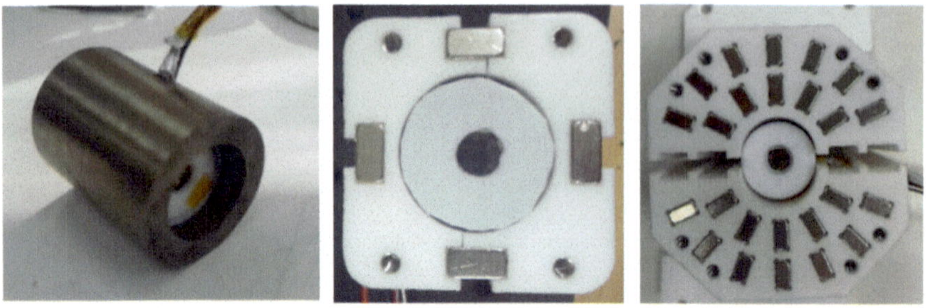

Fig. 4.32 Put the wrapped atomic cell into the cylindrical magnet (left); the structure of the atomic filter with four magnetic strips (middle) the structure of the atomic filter with twenty-eight magnetic strips (right)

Fig. 4.33 Assembly of atomic filter using cylindrical magnet

Fig. 4.34 Assembly of atomic filter using magnetic strip

holes of the aluminum shell, with the polarization directions of the two prisms being orthogonal, as shown in Fig. 4.35.

2. FADOF test research

The performance indicators of the atomic cell mainly include the transmittance, surface flatness, and anti-condensation ability.

The transmittance of the atomic cell directly affects the transmittance of FADOF. The test light path is shown in Fig. 4.36. The specific method is: ① The laser output

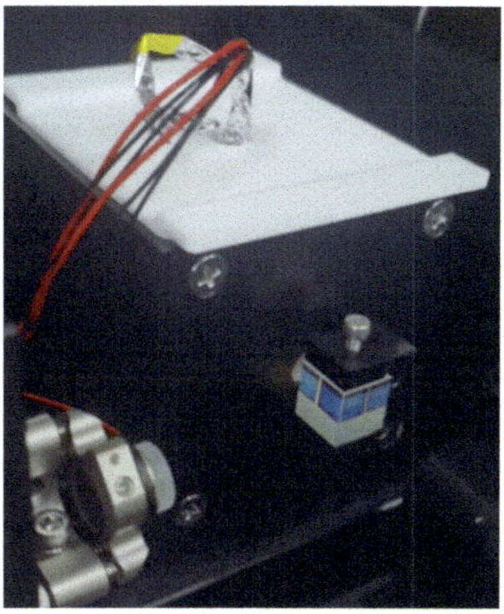

Fig. 4.35 Atomic filter with installed polarizing beam splitter

from the 780/852 nm diode laser light is divided into two beams by the partial reflection mirror M1, the reflected light enters the saturated absorption spectroscopy (SAS) light path with a reference atomic cell, and the transmitted light enters the atomic cell to be tested; ② Adjust the current and PZT parameters of the laser to ensure that the output light frequency resonates with the Rb/Cs atoms in the SAS light path, ensuring that the SAS signal can be observed; ③ Maintain the current and PZT parameters of the laser, and use a power meter to detect the power P1 before and the power P2 after passing through the atomic cell, leading to the transmittance T = P2/P1; ④ Repeat the experiment multiple times and take the average value as final result.

The surface flatness of the atomic cell affects the divergence of the laser. The test light path is shown in Fig. 4.37. The specific method is: ① Collimate the laser output from the diode laser; ② The collimated laser beam enters the atomic cell; ③ Use a photosensitive card to observe the spot size of the laser exiting the atomic cell,

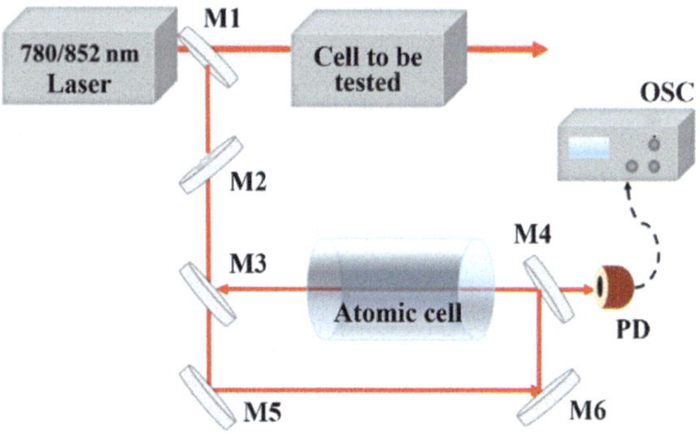

Fig. 4.36 Test light path of atomic cell transmittance. OSC, oscilloscope; PD, photodetector; M: reflector

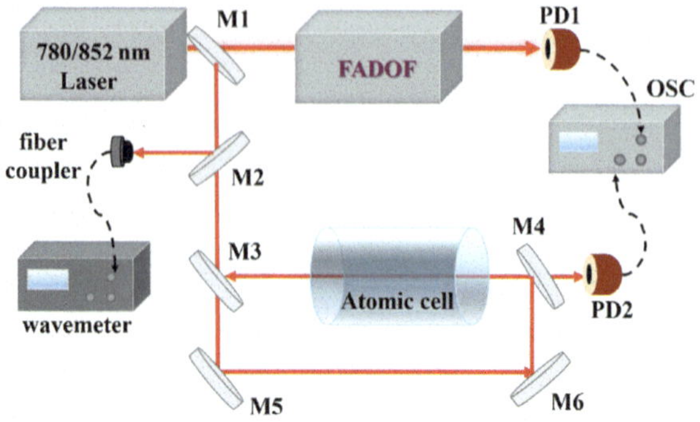

Fig. 4.37 FADOF transmission spectrum test optical path

change the distance between the photosensitive card and the atomic cell, and observe whether the spot diverges.

Due to the heating plate being located in the cell cylinder, the end face usually has the lowest temperature, and sudden cooling can easily cause atoms to condense on the end face. Once the atoms in the cell condense, the transmittance of FADOF rapidly decreases, therefore it is necessary to ensure the anti-condensation ability of the atomic cell. The test method for the anti-condensation ability of the atomic cell is as follows: ① Heat the atomic cell to 100 °C; perform a transmittance test of the atomic cell, and record the cell transmittance T1. After the test is over, stop heating; ② Heat the atomic cell to 100 °C again after an overnight interval; ③ Perform the transmittance test again, and record the transmittance T2; ④ Compare the transmittance T2 with T1. If T2 is significantly less than T1, condensation exists; if the values of T2 and T1 are close, there is basically no condensation.

The performance indicators of FADOF mainly include the transmission spectrum, and the shape of the transmission spectrum line determines the application value of FADOF. Among them, the most critical indicators in the transmission spectrum are the peak transmittance and transmission bandwidth.

The test method for the FADOF transmission spectrum is as follows: ① The laser output from the external cavity diode laser light is divided into two beams by the partial reflection mirror M1. The transmitted light is used to detect the transmission spectrum of the atomic filter at the photodetector (PD1), and the reflected light enters another partial reflection mirror M2; ② Before the reflected light enters the SAS optical path, it is divided into two beams by M2. The reflected light is connected to the wavelength meter through the fiber coupler FC to measure the laser wavelength, thereby adjusting the current and PZT parameters of the laser to ensure that the output light of the laser resonates with the Rb/Cs atoms in the SAS optical path, ensuring that the SAS signal can be observed; ③ The transmitted light after M2 is used to detect the SAS signal at the photodetector (PD2) and serve as a reference; the photodetector PD1 converts the FADOF transmission spectrum light signal, and the photodetector PD2 converts the SAS light signal, into electrical signals, and then inputs these signals into the oscilloscope to obtain the transmission spectrum of the atomic filter and the reference SAS signal.

The test method for the absolute peak transmittance of FADOF is as follows: ① Use the 780/852 nm laser as the probe light, input it into FADOF for frequency selection. After frequency selection, only light within the FADOF transmission spectrum bandwidth can pass through, and the transmitted light is input into the photodetector PD1 to convert the light signal into an electrical signal, then input this signal into the oscilloscope to observe the FADOF transmission spectrum; ② Make the polarization directions of the two Glan–Taylor prisms or PBSs in FADOF parallel and record the maximum value V1 of the FADOF transmission spectrum signal through the oscilloscope. Then make the polarization directions of the two Glan–Taylor prisms in FADOF perpendicular and record the maximum value V2 of the FADOF transmission spectrum signal through the oscilloscope. Calculate T1 = V2/V1, which is the relative transmittance of FADOF; ③ Measure the transmittance of GT1, GT2, and the atomic cell, the three components in FADOF for 780/852 nm light. Multiply them to get T2, multiply T2 with the relative transmittance T1 to get the absolute transmittance of FADOF, T = T1 × T2.

The test method for the peak transmission bandwidth of FADOF is as follows: ① The laser output from the external cavity diode laser light is divided into two parts by the partial reflection mirror M1. The transmitted light is used to detect the transmission spectrum of the atomic filter using the photodetector PD1, and the reflected light enters another partial reflection mirror M2; ② Before the reflected light enters the SAS optical path, it is divided into two parts by M2. The reflected light is connected to the wavelength meter through the fiber coupler FC to measure laser wavelength, thereby adjusting the current and PZT parameters of the laser to ensure that the output light resonates with the Rb/Cs atoms, ensuring that the SAS signal can be observed; ③ The transmitted light after M2 is used to detect the SAS signal through PD2 and serves as a reference; ④ PD1 converts the FADOF transmission spectrum light signal, and PD2 converts the SAS light signal, into electrical signals, which are then input into the oscilloscope. The SAS signal is used as a reference to calibrate the peak transmission bandwidth of the FADOF transmission spectrum.

Taking the cesium atom 852 nm FADOF as an example, it consists of a 30 mm long Cs atomic cell, a cylindrical magnet, a temperature control system, and two polarizing beam splitters. The atomic cell with anti-reflection coating has a transmission of 95% for 852 nm lasers. In the Cs atomic cell, in addition to Cs atoms, it also contains inert gases Ar or Xe at different pressures, ranging from 0 to 50 torr. The temperature of the atomic cell is controlled by an electric heating element, which keeps the cell temperature within a range of 0.1 °C. The axial magnetic field can vary from 200 to 3000 G, as generated by cylindrical magnets of different sizes. Also the maximum working temperature of the magnet does not exceed 150 °C. Two polarizing beam splitters are placed on both sides of the atomic cell, where the polarization direction of the rear prism can be rotated, allowing the two prisms to be orthogonal or parallel.

Figure 4.38 shows the variation of the transmission spectrum of the atom filter with light intensity and temperature at different magnetic fields when the atomic cell is filled with pure cesium gas. At 100 G, there are four transmission peaks with higher transmission in the transmission spectrum, but the transmission peaks do not coincide with the points of 0 detuning and 9.192 GHz detuning. The wavelength of

Faraday laser is challenging when used for MTS frequency locking. When the light intensity changes, the transmission spectrum changes slightly due to the internal light intensity. As the temperature rises, the side peaks on both sides move away, and the transmission at the center of the spectrum continues to rise. However, the atomic filter with a magnetic field of 300 G has a transmission peak near the two points of 0 detuning and 9.192 GHz detuning, with the 0 detuning transmission peak being higher. This atomic filter used in the Faraday laser, can produce a laser with a wavelength corresponding to the atomic transition line, which is convenient for laser frequency locking. However, because the bandwidth of the transmission peak is narrow, the performance of the laser after frequency locking cannot reach its best. If an atomic filter with a magnetic field strength of 1000 G is used, its transmission peak corresponds to the atomic transition line, and the bandwidth is close to 3 GHz. This indicates that when atomic filter with these parameters is used in the Faraday laser, the laser wavelength is easy to correspond to the atomic transition line. Then the MTS locking technology can be used to achieve frequency locking, and the laser can work stably for a long time after locking.

Figure 4.39 shows the changes in the atomic filter transmission spectrum with light intensity and temperature after filling with inert gas. Compared with the transmission spectrum of the pure cesium atomic filter at 1000 G in Fig. 4.38, the biggest difference is that after filling with inert gas, as the light intensity changes, the line shape of the transmission spectrum of the atom filter also changes. This is unfavorable for predicting the transmission spectrum of the internal atomic filter during normal operation. In addition, after adding inert gas, the overall transmittance of the transmission spectrum is suppressed, which increases the threshold current of the corresponding laser. While bringing these disadvantages, the inert gas broadens the transmission peak in the transmission spectrum to some extent, further highlighting the advantages of the Faraday laser being ready to use and lock immediately upon turning on.

4.3.5 Voigt Atomic Filter

The assembly process of VADOF is consistent with that of FADOF, the difference lies only in their structures. According to the structure of VADOF, adjusting the assembly details appropriately can complete the assembly of VADOF. Figure 4.40 shows the physical picture of the Voigt atomic filter.

The VADOF transmission spectrum test method is consistent with FADOF, just replace the FADOF in the test optical path with VADOF.

Taking the Rb atom 780 nm VADOF as an example, it consists of a 30 mm long ^{87}Rb atomic cell, a bar magnet, a temperature control system, and two polarizing beam splitters. The atomic cell with anti-reflection coating has a transmission of 95% for 780 nm laser. The temperature of the ^{87}Rb atomic cell is controlled by an electric heating element, which keeps the cell temperature within a range of 0.1 °C. The radial magnetic field range can vary from 200 to 3500 G, generated by moving two

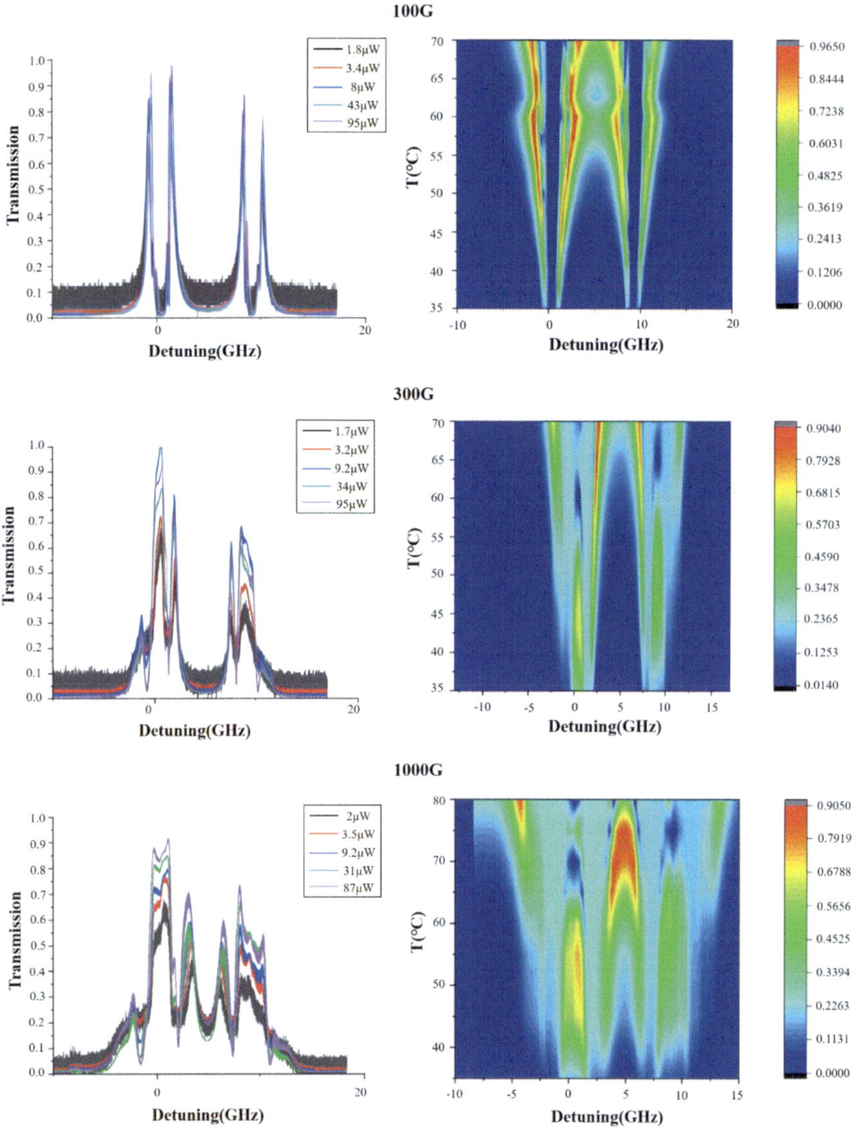

Fig. 4.38 Pure cesium atomic cell; 100 G, transmission spectra as a function of light intensity and temperature (above); 300 G, transmission spectra as a function of light intensity and temperature (middle); 1000 G, transmission spectra as a function of light intensity and temperature (bottom)

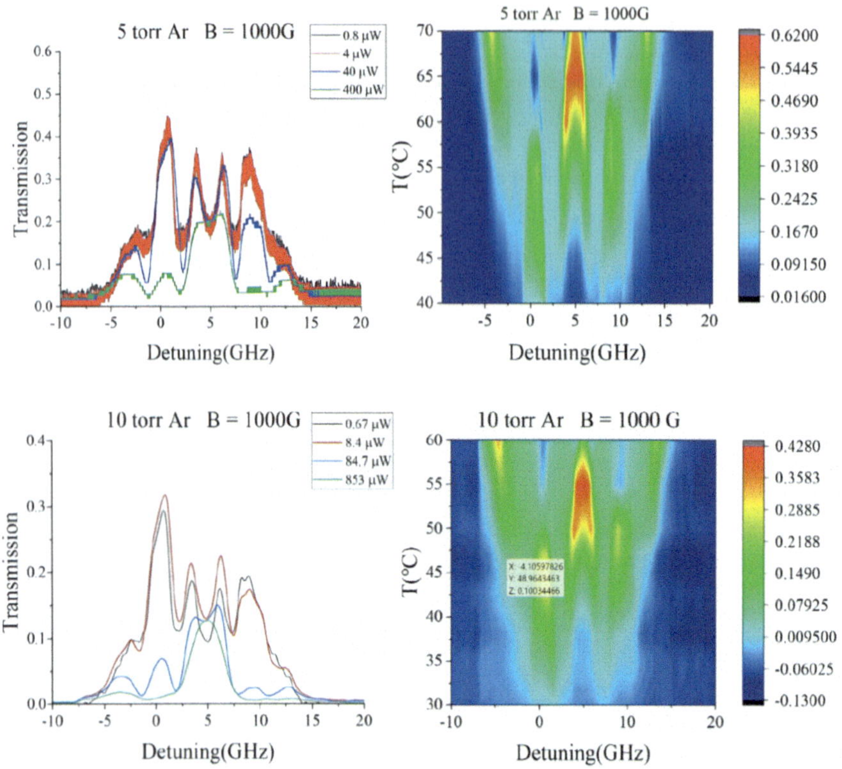

Fig. 4.39 Cesium atomic cell filled with inert gas. The Ar pressure is 5 torr, transmission spectra as a function of light intensity and temperature (top); transmission spectra as a function of light intensity and temperature (bottom)

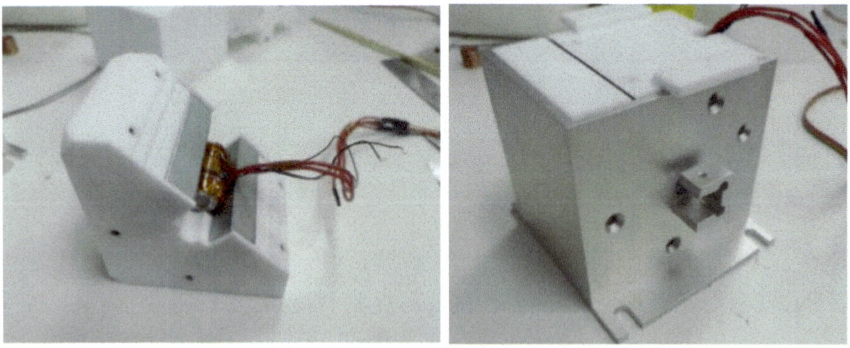

Fig. 4.40 Physical picture of Voigt atomic filter

15 mm thick permanent magnets. To obtain a better Voigt effect, there is a 45° angle between the direction of the magnetic field and the direction of laser polarization. Two polarizing beam splitters are placed on both sides of the atomic cell, where the polarization direction of the rear prism can be rotated so that the two prisms can be orthogonal or parallel. A self-made 780 nm interference filter laser is used to detect the transmission spectrum of VADOF, and the beam diameter of this laser is about 2 mm. The detection laser light is divided into two beams by a polarizing beam splitter, one of which is used to detect the saturated absorption spectroscopy of ^{87}Rb as a frequency reference, and the other is shot into a photodetector through VADOF to measure the transmission spectrum.

Figure 4.41 shows the transmission spectra of VADOF using a magnetic field of 3500 G at different temperatures. The test results show that the highest transmission peak of 3500 G Rb VADOF corresponds to a wavelength of 780.255 nm, which cannot correspond to the atomic transition line. After the actual laser is set up, the output laser wavelength meets the expectation, which is 780.255 nm, but it cannot correspond to the atomic spectrum, and additional frequency shifting optics are needed for modulation transfer spectrum locking. To solve this problem, the parameters of VADOF need to be adjusted, the magnetic field is replaced with 3700 G, and the newly replaced VADOF is tested for transmission spectrum. Figure 4.42 shows the transmission spectrum of VADOF with a magnetic field of 3700 G at different temperatures, and the topmost saturated absorption spectrum is used as a frequency reference.

When the temperature is below 75 °C, the transmission of VADOF is too low, and the laser is not enough to oscillate. When the temperature reaches 80 °C, the highest transmission peak can correspond to the atomic transition line, with the highest transmission meeting the requirements. When the temperature exceeds 85 °C, the spectrum lines are messy, and the zero point of detuning gradually moves to the boundary of the transmission peak, which does not meet the requirement that the transition spectrum line would correspond to the transmission peak. Therefore, the initial working temperature of this VADOF should be set to 80 °C corresponding to wavelength of 780.243 nm.

4.3.6 MEMS Type Atomic Filter

The FADOF is a quantum selective filter with high transmittance, narrow bandwidth, and high noise suppression ratio. The output frequency of the diode laser using the FADOF as the selective device can intrinsically resonate with the atomic spectrum, and has absolute advantages in suppressing the temperature sensitivity and current sensitivity of laser diode. However, the existing Faraday lasers are large in size, which introduce problems such as poor magnetic field uniformity in the atomic cell, easy mode hopping of the output laser, low integration, and high cost. The Micro-Electro-Mechanical Systems (MEMS) process to prepare MEMS type Faraday atomic filters that can effectively solve the above problems, and to

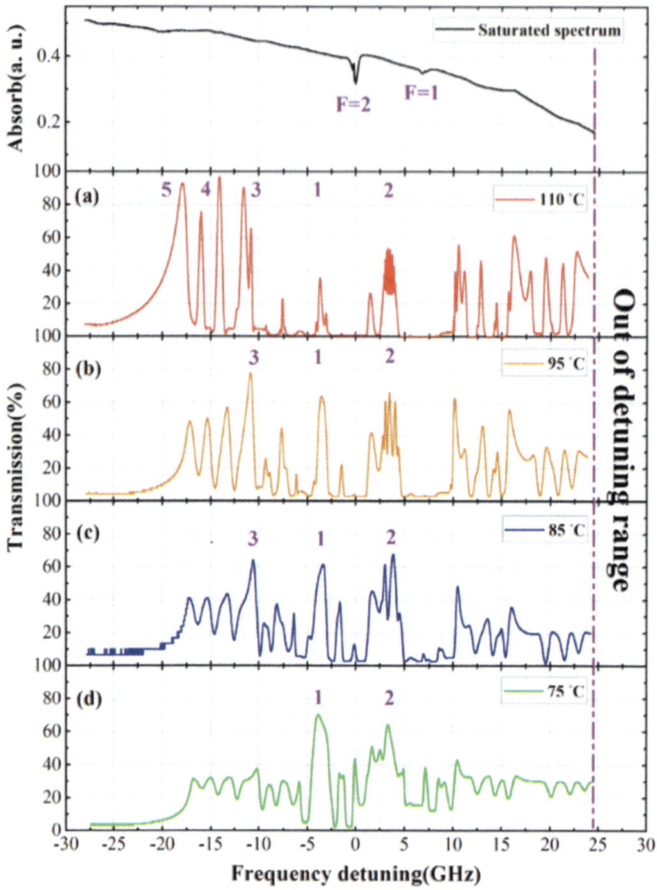

Fig. 4.41 Transmission spectrum of 780 nm VADOF at different temperatures (magnetic field 3500 G)

achieve mass production and application, reduced production costs, with prospective applications in mobile, portable time–frequency equipment, serving chip-level quantum metrology related aspects, as well as laser radar, laser sonar, free space optical communication, star-to-earth optical communication link, laser frequency stabilization and other fields.

To solve the above problems, in 2024, Peking University proposed an implementation plan for preparing atomic filters based on MEMS technology, as the core selective device for micro Faraday lasers. Using the MEMS process to prepare high airtightness micro atomic cells, compressing the volume of the magnetic control temperature module, adopting a more compact packaging, realizing the MEMS cell micro Faraday atomic filter, as shown in Fig. 4.43. Based on this filter, the volume of the Faraday laser is greatly reduced, making it suitable for compact application scenarios. At the

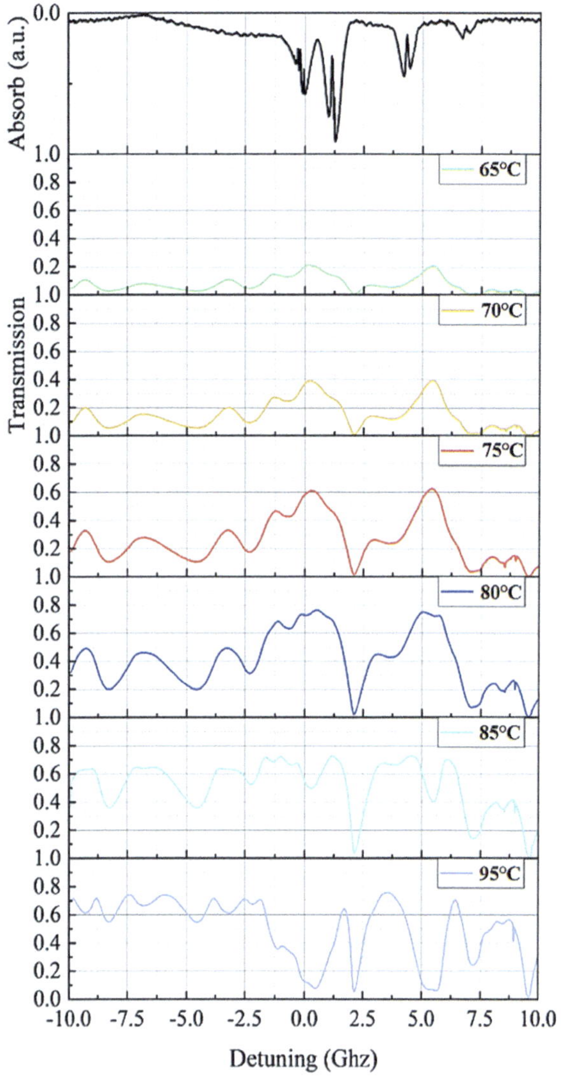

Fig. 4.42 Transmission spectrum of 780 nm VADOF at different temperatures (magnetic field 3700 G)

same time, due to the shortening of the resonant cavity length, the Faraday laser has a larger free spectral range, further improving the mechanical stability and frequency stability. The following will introduce the two key components of the micro atomic filter: MEMS atomic cell and compact permanent magnet.

1. Resonant Faraday rotation effect based on mems atomic cell

The alkali metal gas sealed in the atomic cell can change the polarization direction of the signal light, which is key to realizing Faraday magneto-optical rotation. Therefore, high requirements are placed on the atomic cell. Compared with traditional

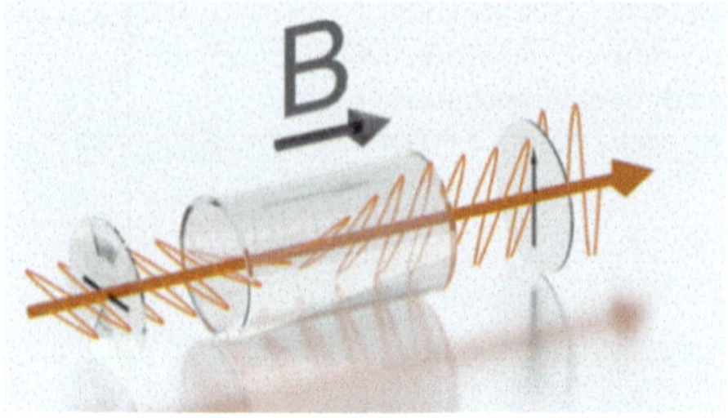

Fig. 4.43 Schematic diagram of the working principle of the MEMS cell micro Faraday atomic filter

glass blowing technology, MEMS technology is easy to manufacture atomic cells of millimeter or even sub-millimeter size, and the whole process is easy to scale up, enabling mass production.

Using MEMS technology to make small, airtight atomic cells is key to realizing miniature atomic filters. The preparation process of MEMS atomic cells can be divided into 5 types: ① In-situ chemical reaction method, which reacts alkali metal compounds with a certain solution, dries to obtain powder, and generates the required alkali metal atoms through vacuum heating. This method is simple and easy to integrate, but it can also easily introduce non-metal impurities in the process of generating alkali metal atoms, thereby affecting the transmittance of the cell and causing frequency shift; ② Direct filling method of alkali metal, this method makes a silicon hole on the silicon wafer, then bonds it with a glass piece through an anode bonding process, directly drips liquid Cs or Rb into the silicon hole in a low-temperature vacuum anaerobic environment, simultaneously fills in buffer gas, and then seals the atomic cell through secondary anode bonding. Although this method is quite perfect, it requires special equipment and is costly, hence, cannot be widely used; ③ Photo-decomposition method, first deposit a layer of alkali metal atom thin film inside the cell, such as cesium nitride, then after the cell is sealed by anode bonding, use ultraviolet light to irradiate the cesium nitride thin film in the cell to decompose and generate pure cesium and nitrogen gas. This method is simple to operate and easy to mass produce, but it requires special equipment; ④ Electrochemical decomposition method, which prepares the atomic cell by depositing alkali metal element from solid electrolyte into the sealed cavity. Although this method has good airtightness, it is extremely complex to operate and is not conducive to mass production; ⑤ Paraffin wrapping method, by depositing a layer of SiN thin film on the surface of the silicon hole, then sticking the wax package wrapped with alkali metal element under the SiN thin film. After the cell is sealed, using laser ablation to decompose the SiN and wax package, so that the alkali metal element enters the silicon hole. This method obtains a high-purity cell, but it is also not conducive to mass production.

The above methods each have their advantages and disadvantages, but considering from the perspective of practicality and industrialization, the in-situ chemical reaction method combined with the photo-decomposition method to make atomic cells is a preferred scheme. The physical objects of traditional glass blown atomic cells and MEMS cells are shown in Fig. 4.44.

Fig. 4.44 Small size glass blown cell (left) and MEMS atomic cell (right)

Heating and insulation devices are applied to the MEMS cell, and a saturated absorption spectroscopy path is set up. The saturated absorption spectral line is observed, and the signal-to-noise ratio of the spectral line is comparable to the results achieved using traditional glass blowing technology. The test object and saturated absorption spectrum are shown in Fig. 4.45.

2. Uniform magnetic field based on compact permanent magnet

By applying a uniform axial magnetic field to the external cell, the atomic magnetic sublevels undergo Zeeman splitting, causing the dispersion curves of the left and right circularly polarized light near the atomic resonance absorption line to split, thereby producing the Faraday rotation effect. When the polarization angle of the signal light in the atomic cell rotates about 90°, the light can pass through the atomic filter at the highest transmission. Under the premise of reducing the size of the cell, in order to meet the above conditions, combined with the absorption effect of the atomic cell

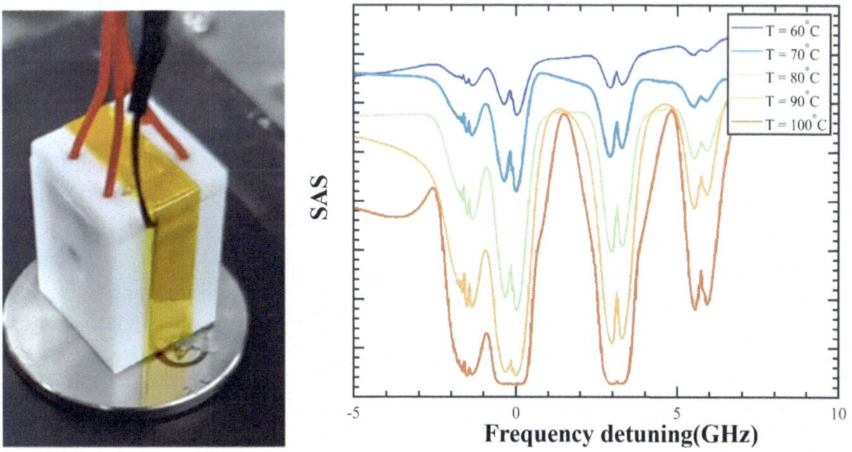

Fig. 4.45 Physical diagram of MEMS atomic cell heating and insulation structure (left) and saturated absorption spectrum at different temperatures (right)

on the signal light, the magnet structure needs to be optimized. After magnetic field simulation, for an atomic cell of millimeter-scale length, the required magnetic field strength is as high as 3000 G. However, at the same time, the compression of the atomic cell structure also makes the magnetic field strength applied in the direction of light transmission in the cell sufficiently uniform. The results of the magnetic field simulation are shown in Fig. 4.46. The volume of the atomic filter is compressed using a circular magnet structure, as shown in Fig. 4.47.

3. Assembly of atomic filter

Based on the design and processing of the above key components, and combined with the polarizer and analyzer, the MEMS atomic cell Faraday atomic filter is realized. Figure 4.48 shows the physical picture of the atomic filter and the transmission spectrum test results. The transmission spectrum of the miniaturized Faraday atomic filter corresponds to the Doppler broadening of atomic transition line, and the output Faraday laser realized by this atomic filter is restricted near the atomic transition frequency. The laser frequency intrinsically corresponds to the quantum transition wavelength, which has significant advantages in resisting temperature and current fluctuations of the laser diode, and is a laser source urgently needed in the field of quantum precision measurement.

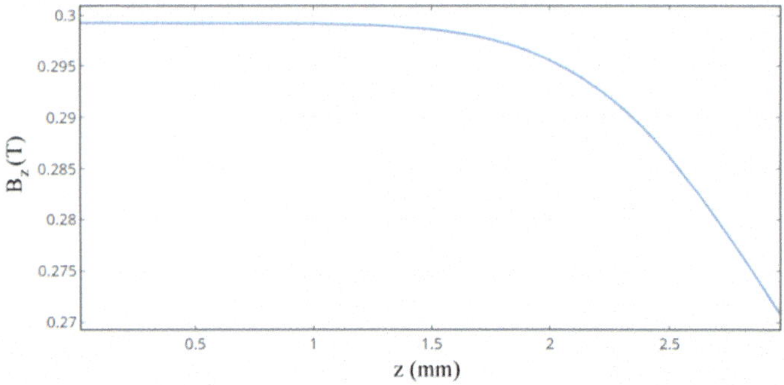

Fig. 4.46 Simulation results of axial magnetic field strength in MEMS atomic cell

Fig. 4.47 Cylindrical magnet design

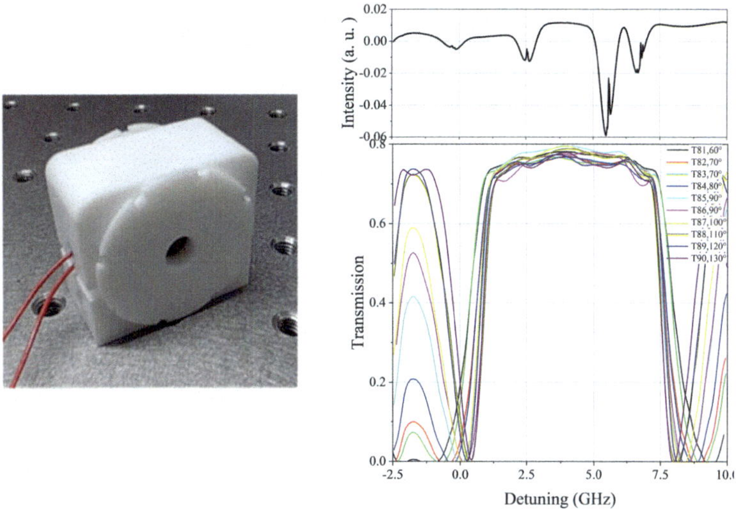

Fig. 4.48 Physical picture (left) and transmission spectrum (right) of the miniaturized Faraday atomic filter

It should be pointed out that the MEMS atomic filter realized based on the MEMS atomic cell, in addition to the FADOF, can also be extended to the VADOF and the birefringence-induced atomic filter, meeting the requirements of the quantum micro-nano field for miniaturization, narrow bandwidth, high transmission, and high out-of-band suppression ratio of the frequency selection device.

4.4 Theory and Simulation of Faraday Atomic Filter

As mentioned earlier, the working principle of the atomic filter uses the Faraday rotation effect produced when light passes through an atomic medium, but the previous analysis was only qualitative. In this section, the transmission spectrum of the atomic filter will be calculated quantitatively, which essentially involves the interaction between light and atoms. The specific calculation method is as follows [6, 32–34]: Firstly, write out the Hamiltonian equation that the atom satisfies in a given magnetic field, and solve the eigenvalue problem to obtain the eigen energy of the atom and the corresponding transition frequency. Secondly treat the atom as an isolated two-level system, and use the density operator method to obtain the polarization response of the atom. Finally, from the polarization response of the atom, obtain the expression for the light propagation in atomic vapor, and thus obtain the transmission spectrum of the atomic filter.

The above method is in the order of calculation. If the principle of the atomic filter is introduced directly in this order, readers will inevitably have the question of "why should we do this?". To facilitate readers' understanding, this book will embark on an in-depth exploration, starting from Maxwell's equations and unraveling the complexities layer by layer.

4.4.1 Electromagnetic Waves in Medium

Now suppose a plane light beam enters from a vacuum into a medium filled with atomic vapor, then the electromagnetic field components of the light field satisfies the following Maxwell's equations:

$$\begin{cases} \nabla \cdot \boldsymbol{D} = \rho_f \\ \nabla \cdot \boldsymbol{B} = 0 \\ \nabla \times \boldsymbol{E} = -\frac{\partial \boldsymbol{B}}{\partial t} \\ \nabla \times \boldsymbol{H} = \boldsymbol{J}_f + \frac{\partial \boldsymbol{D}}{\partial t} \end{cases} \tag{4.10}$$

Because the atomic vapor satisfies the following conditions:

There are no free charges in the medium, i.e. $\rho_f = 0$.

There are no free currents in the medium, i.e. $\boldsymbol{J}_f = 0$.

The medium is non-ferromagnetic, i.e. $\boldsymbol{B} = \mu_0 \boldsymbol{H}$.

According to the constitutive relationship $\boldsymbol{D} = \varepsilon_0 \boldsymbol{E} + \boldsymbol{P}$, the electric field component of the light field in the medium satisfies the following wave equation:

$$\left(\nabla^2 - \frac{1}{c^2}\frac{\partial^2}{\partial t^2}\right)\boldsymbol{E} = \frac{1}{\varepsilon_0 c^2}\frac{\partial^2}{\partial t^2}\boldsymbol{P} \tag{4.11}$$

where c is the speed of light in vacuum, ε_0 is the dielectric constant in vacuum, $\boldsymbol{P}$ is the polarization intensity vector, also known as the macroscopic polarization response of atoms to the externally applied driving electric field.

Assume that the driving electric field incident into the medium propagates along the z-axis and can be written in the following harmonic form:

$$\boldsymbol{E} = \boldsymbol{E}_0 \cos(kz - \omega t) = \frac{1}{2}\boldsymbol{E}_0 e^{-i(\omega t - kz)} + c.c. \tag{4.12}$$

This assumption is based on the fact that the atomic scale is much smaller than the light wavelength, so the light field can be regarded as a uniform plane wave.

Also assume that the atom itself is a linear optical system, so its polarization response should satisfy the following linear relationship with the driving electric field:

$$\boldsymbol{P} = \frac{1}{2}\varepsilon_0 \chi(\omega)\boldsymbol{E}_0 e^{-i(\omega t - kz)} + c.c. \tag{4.13}$$

where $\chi(\omega)$ is the tensor of the dielectric polarization rate, and its real part and imaginary part satisfy:

$$\begin{cases} Re(\chi(\omega)) = \frac{PV}{\pi}\int\limits_{-\infty}^{+\infty}\frac{Im(\chi(\omega'))}{\omega'-\omega}d\omega' \\ Im(\chi(\omega)) = -\frac{PV}{\pi}\int\limits_{-\infty}^{+\infty}\frac{Re(\chi(\omega'))}{\omega'-\omega}d\omega' \end{cases} \tag{4.14}$$

This set of equations is known as the Kramers–Krönig relations, which are derived from the causality of physical systems, describing the mutual constraints between absorption and dispersion in linear optical systems.

Substituting Eqs. (4.12) and (4.13) into Eq. (4.11), the relationship between the electric field wave number k and the polarization rate is:

$$k = \frac{\omega}{c}\sqrt{1+\chi(\omega)} \tag{4.15}$$

From Eq. (4.15), it can be seen that the propagation of the electric field in the medium is closely related to the polarization rate of the medium, and the polarization rate depends on the polarization response of the medium. Therefore, the key to the problem of light-atom interaction is to solve the polarization response of atoms to the external light field.

4.4.2 Atomic Polarization Response

It is very difficult to directly study the evolution of an atomic system under the action of an electromagnetic field. Therefore, some reasonable approximations need to be made. Common approximation methods include:

(1) Electric field approximation: Since the electric field component in the light field is much larger than the magnetic field component, only the interaction between the electric field and the atom is considered.
(2) Electric dipole approximation: The atomic nucleus and inner electrons are regarded as an ion entity, which forms a pair of electric dipoles with the outer electrons. The interaction between different dipoles is ignored, and only the interaction between the dipole and the electric field is considered.

For a system composed of electric dipoles, its macroscopic polarization intensity vector is the sum of the microscopic electric dipole moments per unit volume, that is:

$$\boldsymbol{P} = N_a\boldsymbol{d} = -N_a e\boldsymbol{r} \tag{4.16}$$

where, N_a is the atomic number per unit volume, $\boldsymbol{d} = -e\boldsymbol{r}$ is the electric dipole moment of a single dipole.

Since the atomic system is a mixed state, quantum mechanically, it is generally described by density operators $\hat{\rho}$, for any mechanical quantity $\hat{O}$, its ensemble average is given by:

$$\langle \hat{O} \rangle = Tr(\hat{\rho}\hat{O}) \tag{4.17}$$

Similarly, according to Eq. (4.17), if the density operator $\hat{\rho}$ of the system is known, then the average value of the system's polarization intensity is given by:

$$\langle \hat{\boldsymbol{P}} \rangle = -N_a Tr(\hat{\rho} e \hat{\boldsymbol{r}}) \tag{4.18}$$

For alkali metal atoms, it is difficult to directly write out their density operators. Taking the D_2 transition 852 nm line of Cs atom as an example, it involves 16 ground state magnetic sublevels and 32 excited state magnetic sublevels. The density operator of the system will be a 48-dimensional matrix, and it is extremely difficult to solve such a problem. Therefore, further approximation of the atomic energy levels is needed here.

The simplest approximation is the two-level approximation. The so-called two-level approximation refers to the approximation when light of a certain frequency is incident, only the transition between the pair of energy levels that resonate with its frequency is considered, and the transition between other energy levels is not considered, that is, the atom is treated as a two-level system.

For a two-level system, the symbol 'g' represents the ground state, and the symbol 'e' represents the excited state, then the density operator of the considered atomic system can be written as:

$$\hat{\rho} = \begin{bmatrix} \rho_{gg} & \rho_{ge} \\ \rho_{eg} & \rho_{ee} \end{bmatrix} \tag{4.19}$$

Quantum mechanically, the time evolution of a closed system can be described by the Liouville equation. Since the atomic system is an open system, there is always dissipative interaction between the system and the environment. Such a system with dissipation can be described by the Lindblad master equation. Its general form is as follows:

$$\frac{d\hat{\rho}}{dt} = \frac{1}{i\hbar}\left[\hat{H}, \hat{\rho}\right] + L(\hat{\rho}) \tag{4.20}$$

where, $\hat{H} = \hat{H}_0 + \hat{H}_I$ is the Hamiltonian of the atom, including the free Hamiltonian of the atom $\hat{H}_0$ and the interaction Hamiltonian $\hat{H}_I$; $L(\hat{\rho}) = \hat{L}^{\dagger}\hat{\rho}\hat{L} - \left(\hat{L}^{\dagger}\hat{L}\hat{\rho} + \rho\hat{L}^{\dagger}\hat{L}\right)/2$ is the dissipation term, $\hat{L}$ is the Lindblad operator, describing the dissipative interaction between the system and the environment.

Here we use the energy representation, with the eigenstates of the atom in the free state as the basis vectors to expand each term. Then we have:

$$\hat{H}_0 = \hbar \sum_i \omega_i |i\rangle\langle i| = \hbar \begin{bmatrix} \omega_g & 0 \\ 0 & \omega_e \end{bmatrix} \tag{4.21}$$

According to the electric dipole approximation, the interaction Hamiltonian $\hat{H}_I$ due to interaction between the atom and the electric field is expressed as:

$$\begin{aligned} \hat{H}_I &= -\hat{\boldsymbol{d}} \cdot \boldsymbol{E} \\ &= e\hat{\boldsymbol{r}} \cdot \boldsymbol{E}_0 \cos \omega t = (er_0 E_0 - er_+ E_- - er_- E_+) \cos \omega t \end{aligned} \tag{4.22}$$

where, $\hat{r}_+, \hat{r}_-, \hat{r}_z$ and E_+, E_-, E_z are respectively the three components of $\hat{\boldsymbol{r}}$ and $\boldsymbol{E}_0$ under the spherical basis vectors. The vectors using above stated spherical basis vectors can be converted into the Cartesian:

$$\boldsymbol{e}_\pm = \mp \frac{1}{\sqrt{2}} \left(\boldsymbol{e}_x \pm i \boldsymbol{e}_y \right), \ \boldsymbol{e}_0 = \boldsymbol{e}_z \tag{4.23}$$

The reason for using spherical basis vectors is that the dipole transition operator $e\hat{\boldsymbol{r}}$ is easier to obtain under the spherical basis vectors, and the result directly corresponds to the transition selection rules (which will be detailed in Sect. 4.4.3). Further, Eq. (4.22) can be written in matrix form for simplification as:

$$\hat{H}_I = \boldsymbol{E}_0 \cdot \langle g | e\hat{\boldsymbol{r}} | e \rangle \cos \omega t \begin{bmatrix} 0 & 1 \\ 1 & 0 \end{bmatrix} \tag{4.24}$$

Substituting (4.19), (4.21), and (4.24) into (4.20), the first term on the right side of the main equation can be obtained as:

$$\begin{aligned} \frac{1}{i\hbar} \left[\hat{H}, \hat{\rho} \right] = &- i\omega_0 \begin{bmatrix} 0 & \rho_{ge} \\ \rho_{eg} & 0 \end{bmatrix} \\ &- i\Omega \cos \omega t \begin{bmatrix} \rho_{eg} - \rho_{ge} & \rho_{ee} - \rho_{gg} \\ \rho_{gg} - \rho_{ee} & \rho_{ge} - \rho_{eg} \end{bmatrix} \end{aligned} \tag{4.25}$$

where, $\omega_0 = \omega_e - \omega_g$, is the difference in transition frequencies between the ground state and the excited state; $\Omega = \boldsymbol{E}_0 \cdot \langle g | e\hat{\boldsymbol{r}} | e \rangle / \hbar$, is known as the Rabi frequency.

The next problem to solve is the second term on the right side of the main equation. The specific form of the Lindblad operator can be phenomenologically written as:

$$\hat{L} = \sqrt{\Gamma} |g\rangle\langle e| = \begin{bmatrix} 0 & \sqrt{\Gamma} \\ 0 & 0 \end{bmatrix} \tag{4.26}$$

The reason why it is phenomenological is that the evolution of the quantum system it describes is not derived from the microscopic mechanism of the interaction between the system and the environment. The Lindblad operator here is also called the transition operator. This operator, when applied to the excited state, causes an atomic transition from the excited state to the ground state. The main cause of the dissipation effect is spontaneous radiation. The dissipation rate is related to the constant Γ. The larger the constant Γ, the faster the rate at which the atom transits from the excited state to the ground state due to the dissipation effect, the shorter the lifetime of the atom in the excited state. Therefore, Γ is represented by the natural linewidth of the transition spectrum, and its relationship with the lifetime of the atom in the excited state τ is:

$$\Gamma = \frac{1}{\tau} \tag{4.27}$$

From Eq. (4.26), the dissipation term can be obtained as

$$L(\hat{\rho}) = \hat{L}^{\dagger}\hat{\rho}\hat{L} - \frac{1}{2}(\hat{L}^{\dagger}\hat{L}\hat{\rho} + \rho\hat{L}^{\dagger}\hat{L}) = \frac{1}{2}\Gamma\begin{bmatrix} -2\rho_{ee} & \rho_{ge} \\ \rho_{eg} & 2\rho_{ee} \end{bmatrix} \tag{4.28}$$

By substituting Eqs. (4.19), (4.25), and (4.28) into Eq. (4.20), a set of ordinary differential equations can be obtained:

$$\begin{cases} \dot{\rho}_{gg} = -i\Omega\cos\omega t\left(\rho_{eg} - \rho_{eg}\right) + \Gamma\rho_{ee} \\ \dot{\rho}_{ee} = -i\Omega\cos\omega t\left(\rho_{ge} - \rho_{eg}\right) - \Gamma\rho_{ee} \\ \dot{\rho}_{ge} = \left(i\omega_0 - \frac{\Gamma}{2}\right)\rho_{ge} - i\Omega\cos\omega t\left(\rho_{ee} - \rho_{gg}\right) \\ \dot{\rho}_{eg} = \left(-i\omega_0 - \frac{\Gamma}{2}\right)\rho_{eg} - i\Omega\cos\omega t\left(\rho_{gg} - \rho_{ee}\right) \end{cases} \tag{4.29}$$

Observing this set of differential equations, it is not difficult to find that the two diagonal elements and the off-diagonal elements have a certain relationship:

$$\begin{cases} \rho_{eg} = \rho_{eg}^{*} \\ \dot{\rho}_{ee} + \dot{\rho}_{gg} = 0 \Rightarrow \rho_{ee} + \rho_{gg} = \left(\rho_{ee} + \rho_{gg}\right)|_{t=0} \end{cases} \tag{4.30}$$

Therefore, the system of differential equations can be simplified as:

$$\dot{\rho}_{eg} = \left(-i\omega_0 - \frac{\Gamma}{2}\right)\rho_{eg} + i\Omega\cos\omega t\left(\rho_{ee} - \rho_{gg}\right) \tag{4.31}$$

$$\begin{aligned} \dot{\rho}_{ee} - \dot{\rho}_{gg} = {} & 2i\Omega\cos\omega t\left(\rho_{eg} - \rho_{eg}^{*}\right) \\ & - \Gamma\left[\rho_{ee} - \rho_{gg} + \left(\rho_{ee} + \rho_{gg}\right)|_{t=0}\right] \end{aligned} \tag{4.32}$$

This system of differential equations is difficult to solve analytically, but we can try to find its steady-state solution. The so-called steady-state solution refers to ignoring the rapid changes in the system response to the driving field applied to the atoms in a short time, and only caring about the solution when the system reaches a stable state after a long time of evolution. Of course, this "long time" is relative to the atomic transition rate. For people to observe, it is just a moment, that is, the steady-state solution is sufficient to describe the behavior of atoms changing with time in reality.

When introducing Eq. (4.13), we assumed that the atomic system's response to the external driving electric field is a linear response, that is, the polarization response is proportional to $\cos \omega t$, so we might as well assume that the form of the steady-state solution is:

$$\rho_{eg} = \sigma e^{-i\omega t} \tag{4.33}$$

Substituting this steady-state solution into Eqs. (4.31) and (4.32), and setting the left side of Eq. (4.32) equal to 0, we can get:

$$\left(\frac{\Gamma}{2} - i\Delta\right)\sigma = \frac{i\Omega}{2}\left(\rho_{ee} - \rho_{gg}\right) \tag{4.34}$$

$$i\Omega\left(\sigma - \sigma^*\right) = \Gamma\left[\rho_{ee} - \rho_{gg} + \left(\rho_{ee} + \rho_{gg}\right)|_{t=0}\right] \tag{4.35}$$

where $\Delta = \omega - \omega_0$, is the detuning of the driving light frequency and the atomic transition frequency. The derivation here uses the rotating wave approximation, that is, ignoring the high-frequency oscillations that appear in the derivation process (including $e^{2i\omega t}$ and $e^{-2i\omega t}$ terms), because in comparison to the driving frequency, these high-frequency terms are averaged out over time and cannot be observed in experiments. Because the form of Eqs. (4.34) and (4.35) is very similar to the Bloch equation in magnetic resonance, it is also called the optical Bloch equation.

Combining Eqs. (4.34) and (4.35), we get:

$$\rho_{ee} - \rho_{gg} = -\frac{\Delta^2 + \Gamma^2/4}{\Delta^2 + \Gamma^2/4 + \Omega^2/2}\left(\rho_{ee} + \rho_{gg}\right)|_{t=0} \tag{4.36}$$

$$\sigma = \frac{-\Omega/2}{\Delta + i\Gamma/2}\left(\rho_{ee} - \rho_{gg}\right) \tag{4.37}$$

By substituting Eq. (4.37) into Eq. (4.33), we can obtain the steady-state solution of the density matrix element. Once the steady-state solution of the density matrix element is obtained, the ensemble average of the polarization response can be solved by Eq. (4.18):

$$\begin{aligned}\langle \boldsymbol{P} \rangle &= -N_a \langle g|e\hat{\boldsymbol{r}}|e\rangle \left(\rho_{eg} + \rho_{ge}\right) \\ &= -N_a \langle g|e\hat{\boldsymbol{r}}|e\rangle \left(\sigma e^{-i\omega t} + \sigma^* e^{i\omega t}\right)\end{aligned} \tag{4.38}$$

Now let's discuss the relationship between the atomic polarization response and the polarizability. Under the spherical basis vectors, we can get from Eq. (4.13):

$$\boldsymbol{P} = \frac{1}{2}\varepsilon_0(\chi_+ E_+ \boldsymbol{e}_+ + \chi_- E_- \boldsymbol{e}_- + \chi_0 E_0 \boldsymbol{e}_0)e^{-i\omega t} + c.c. \tag{4.39}$$

By comparing Eq. (4.38) with Eq. (4.39), we can obtain the expression of atomic polarizability:

$$\begin{aligned}\chi_q(\omega) &= -\frac{2N_a\langle g|e\hat{r}_q|e\rangle}{\varepsilon_0 E_q}\sigma \\ &= \frac{N_a\left|\langle g|e\hat{r}_q|e\rangle\right|^2}{\varepsilon_0\hbar(\Delta + i\Gamma/2)}\left(\rho_{ee} - \rho_{gg}\right)\end{aligned} \tag{4.40}$$

where, $q = 0, \pm 1$ is the three components under the base vector of the quantized axial sphere. It should be noted that the formula (4.40) only applies to the three components under the two-level atoms. The ground state and excited state of alkali metal atoms obviously have multiple levels, thus there exist multiple transition lines from the ground state to the excited state. When we previously dealt with alkali metal atoms, we made a two-level approximation, assuming that the incidence of light at a certain frequency will only cause transitions between the pair of levels that resonate with its frequency, so it is reasonable to assume that these transition lines are independent of each other, and their contributions to the polarizability satisfy additivity.

At the same time, since the diagonal elements of the density matrix represent the probability of the system being in the corresponding pure state, and the initial state of the system should follow the thermal equilibrium state Maxwell–Boltzmann statistical distribution, we have:

$$\chi_q(\omega) = \frac{N_a}{\varepsilon_0\hbar}\sum_{i=1}^{n}\sum_{j=1}^{m}\frac{\left|\langle g_j|e\hat{r}_q|e_i\rangle\right|^2}{\Delta_{ij} + i\Gamma_{ij}/2}\left(\rho_{ii} - \rho_{jj}\right) \tag{4.41}$$

where, m and n are the number of ground state and excited state energy levels respectively. The initial state of the system should obey the Maxwell–Boltzmann statistical distribution of the equilibrium state. Since the energy of the excited state is much larger than that of the ground state, and the energy difference between each ground state is very small, it can be considered that the atoms initially show a uniform distribution in the ground state, so there are:

$$\rho_{ii} - \rho_{jj} = -\frac{\Delta_{ij}^2 + \Gamma_{ij}^2/4}{\Delta_{ij}^2 + \Gamma_{ij}^2/4 + \Omega_{ij}^2/2}\cdot\frac{1}{m} \tag{4.42}$$

Γ_{ij}, Δ_{ij} and Ω_{ij} are the spontaneous emission rate, laser angular frequency mismatch and Rabi frequency of the transition between excited state i and ground state j, respectively.

Considering the Doppler effect, the light field frequency perceived by atoms with a velocity of should be $\omega - kv$, and the atomic velocity v follows the Maxwell speed distribution:

$$g(\upsilon) = \frac{1}{\sqrt{\pi} u} e^{-\frac{\upsilon^2}{u^2}} \tag{4.43}$$

where $u = \sqrt{2k_b T/m}$ is the most probable speed of the atom.

Therefore, considering the Doppler broadening, the polarizability of the atom should be:

$$\chi^D_{\pm,0}(\omega) = \int_{-\infty}^{\infty} \chi_{\pm,0}(\omega - k\upsilon) g(\upsilon) d\upsilon \tag{4.44}$$

4.4.3 Microstructure of Alkali Metal Atoms and Zeeman Effect

As can be seen from Eq. (4.42), to calculate the polarizability of the atom, we first need to know the electric dipole operator $\hat{\boldsymbol{r}}$, the transition frequency of the atom ω_{ij} and the natural linewidth of the corresponding transition line Γ_{ij}. These physical quantities are determined by the microstructure of the atom and the external environment. The microstructure of the atom includes the fine structure and hyperfine structure. At the same time, under the action of an external static magnetic field, the energy levels of the atom shift and split, exhibiting Zeeman effect. In this section, we provide the Hamiltonian of these microstructures and obtain the energy level structure by solving the eigenvalue equation.

Due to the complexity of atomic energy levels, we only provide a brief summary of atomic energy level structure in this section. First, only considering the Coulomb interaction between the atomic nucleus and the outer electrons, at this time the atom only has orbital angular momentum $\hat{\boldsymbol{L}}$. Its energy level is determined by the principal quantum number n, orbital quantum number L and magnetic quantum number m_L. Considering that the electron spin also has an angular momentum $\hat{\mathbf{S}}$, the splitting of the atomic energy level produced after its coupling with the orbital angular momentum is called the fine structure. At this time, the atomic orbital angular momentum $\hat{\boldsymbol{L}}$ and the spin angular momentum $\hat{\boldsymbol{S}}$ couple into a new angular momentum $\hat{\boldsymbol{J}} = \hat{\boldsymbol{L}} + \hat{\mathbf{S}}$. This type of coupling is called $L - S$ coupling. At this time, the atomic energy level introduces an additional quantum number S, m_S, J, m_J to describe. Further, considering that the atomic nucleus spin also has an angular momentum $\hat{\boldsymbol{I}}$, it couples with the angular momentum $\hat{\boldsymbol{J}}$ to get a new angular momentum $\hat{\boldsymbol{F}} = \hat{\boldsymbol{J}} + \hat{\boldsymbol{I}}$, the corresponding quantum number is I, m_I, F, m_F.

Here, relativistic dynamics corrections, Darwin terms, and Lamb shifts are not considered because these terms have little effect on the calculation of the polarizability. Due to space limitations, the detailed derivation process of the Hamiltonian of various microscopic structures of atoms is not given here. Please refer to professional books on atomic physics for detailed processes. To write the Hamiltonian of the atom, we first need to choose a suitable representation. Here, choose $\hat{\boldsymbol{L}}^2, \hat{\boldsymbol{S}}^2, \hat{\boldsymbol{J}}^2, \hat{\boldsymbol{I}}^2, \hat{\boldsymbol{F}}^2, \hat{F}_z$'s common eigenstate $|n, L, S, J, I, F, m_F\rangle$ as the base vector, abbreviated as $|F, m_F\rangle$.

In the $|F, m_F\rangle$ representation, the Hamiltonian of the Coulomb interaction can be expressed as:

$$\hat{H}_c = -\frac{m_e c^2}{2}\left(\frac{Z\alpha}{n}\right)^2 \delta_n \tag{4.45}$$

Generally, the energy level transition involved in the atomic filter does not cause the main quantum number n to change, so this term is a constant times the unit matrix, which has no effect on the transition frequency (the frequency difference between the ground state and the excited state), so this term can be ignored. The Hamiltonian of the fine structure can be expressed as:

$$\hat{H}_f = \frac{\gamma_f}{\hbar^2}(\hat{\boldsymbol{L}} \cdot \hat{\boldsymbol{S}}) \tag{4.46}$$

In the formula, γ_f is the spin–orbit interaction constant. For a specific transition, its value can be obtained by looking up the table. The Hamiltonian of the hyperfine structure can be expressed as:

$$\hat{H}_{hf} = \hat{H}_{md} + \hat{H}_{eg} \tag{4.47}$$

$$\hat{H}_{md} = \frac{A_{hf}}{\hbar^2}(\hat{\boldsymbol{I}} \cdot \hat{\boldsymbol{J}}) \tag{4.48}$$

$$\hat{H}_{eq} = \frac{B_{hf}}{\hbar^4} \cdot \frac{3(\hat{\boldsymbol{I}} \cdot \hat{\boldsymbol{J}})^2 + \frac{3}{2}(\hat{\boldsymbol{I}} \cdot \hat{\boldsymbol{J}})\hbar^2 - \hat{I}^2 \cdot \hat{J}^2}{2I(2I-1)2J(2J-1)} \tag{4.49}$$

where $\hat{H}_{md}$ is the Hamiltonian of the magnetic dipole interaction, $\hat{H}_{eq}$ is the Hamiltonian of the electric quadrupole interaction, A_{hf} and B_{hf} are the constants of the magnetic dipole interaction and the electric quadrupole interaction, respectively, both of which can be obtained by looking up in tables.

Under the action of an external static magnetic field, the atomic energy level structure also undergoes Zeeman splitting. Assuming that the direction of the magnetic field is z direction, then, the Hamiltonian of Zeeman splitting can be expressed as:

$$\hat{H}_B = -\hat{\boldsymbol{\mu}} \cdot \hat{\boldsymbol{B}} = \mu_B B_Z\left(g_J \hat{J}_Z + g_I \hat{I}_Z\right) \tag{4.50}$$

where g_J and g_I are the Landé factors of the total electronic angular momentum $\hat{\boldsymbol{J}}$ and the nuclear spin angular momentum $\hat{\boldsymbol{I}}$, the value of g_I is given by experiment and can be obtained by looking up in tables, the Landé factors g_J can be calculated by the following formula:

$$
\begin{aligned}
g_J &= g_L \frac{J(J+1) - S(S+1) + L(L+1)}{2J(J+1)} \\
&\quad + g_s \frac{J(J+1) - L(L+1) + S(S+1)}{2J(J+1)} \\
&\approx 1 + \frac{J(J+1) - L(L+1) + S(S+1)}{2J(J+1)}
\end{aligned}
\tag{4.51}
$$

Next, we will calculate each of the above Hamiltonians in detail:

Below, we proceed with the detailed calculation of the aforementioned Hamiltonians:

Firstly, in the fine and hyperfine structure Hamiltonians,

$$
\hat{\boldsymbol{L}} \cdot \hat{\boldsymbol{S}} = \frac{1}{2}(\hat{J}^2 - \hat{L}^2 - \hat{S}^2) \tag{4.52}
$$

$$
\hat{\boldsymbol{I}} \cdot \hat{\boldsymbol{J}} = \frac{1}{2}(\hat{F}^2 - \hat{J}^2 - \hat{I}^2) \tag{4.53}
$$

Simultaneously, for hyperfine state $|F, m_F\rangle$, $\hat{\boldsymbol{L}}^2, \hat{\boldsymbol{S}}^2, \hat{\boldsymbol{J}}^2, \hat{\boldsymbol{I}}^2, \hat{\boldsymbol{F}}^2, \hat{F}_z$ are all diagonal matrices, and we have:

$$
\begin{cases}
\langle F, m_F | \hat{\boldsymbol{L}}^2 | F', m_F' \rangle = L(L+1)\hbar^2 \delta_{L,L'} \\
\langle F, m_F | \hat{\boldsymbol{S}}^2 | F', m_F' \rangle = S(S+1)\hbar^2 \delta_{S,S'} \\
\langle F, m_F | \hat{\boldsymbol{J}}^2 | F', m_F' \rangle = J(J+1)\hbar^2 \delta_{J,J'} \\
\langle F, m_F | \hat{\boldsymbol{I}}^2 | F', m_F' \rangle = I(I+1)\hbar^2 \delta_{I,I'} \\
\langle F, m_F | \hat{\boldsymbol{F}}^2 | F', m_F' \rangle = F(F+1)\hbar^2 \delta_{F,F'} \\
\langle F, m_F | \hat{F}_z | F', m_F' \rangle = m_F \hbar \delta_{m_F, m_F'}
\end{cases}
\tag{4.54}
$$

Therefore, $\hat{H}_f$ and $\hat{H}_{hf}$ are also diagonal matrices, which can be directly written out from formulas (4.53) and (4.54).

However, since $\hat{\boldsymbol{J}}$ and $\hat{\boldsymbol{I}}$ do not commute with any component of $\hat{F}^2$, that is, $|F, m_F\rangle$ is not the eigenstate of $\hat{J}_z$ and $\hat{I}_z$, the Hamiltonian representing Zeeman splitting cannot be directly written out. It must be obtained through the transformation of representation to get its matrix elements under the representation of $|F, m_F\rangle$.

We know that $|J, m_J, I, m_I\rangle$ is the common eigenstate of $\hat{J}_z$ and $\hat{I}_z$, and we have:

$$
\hat{J}_z |J, m_J, I, m_I\rangle = m_J \hbar |J, m_J, I, m_I\rangle \tag{4.55}
$$

$$
\hat{I}_z |J, m_J, I, m_I\rangle = m_I \hbar |J, m_J, I, m_I\rangle \tag{4.56}
$$

Expanding $|F, m_F\rangle$ with the orthogonal normalized complete set formed by $|J, m_J, I, m_I\rangle$, we can get:

$$|F', m_F'\rangle = \sum_{m_J, m_I} |J, m_J, I, m_I\rangle\langle J, m_J, I, m_I|F', m_F'\rangle \tag{4.57}$$

Therefore, the matrix elements of $\hat{J}_z$ in the terms of $|F, m_F\rangle$ are:

$$\begin{aligned}
&\langle F, m_F|\hat{J}_z|F', m_F'\rangle \\
&= \sum_{m_J, m_I} \langle F, m_F|\hat{J}_z|J, m_J, I, m_I\rangle\langle J, m_J, I, m_I|F', m_F'\rangle \\
&= \sum_{m_J, m_I} \langle F, m_F|J, m_J, I, m_I\rangle m_J \hbar\langle J, m_J, I, m_I|F', m_F'\rangle
\end{aligned} \tag{4.58}$$

In the formula, $\langle F, m_F|J, m_J, I, m_I\rangle$ represents the Clebsch–Gordan coefficient for the transformation between the coupled representation $|F, m_F\rangle$ and the uncoupled representation $|J, m_J, I, m_I\rangle$, which can be calculated by the following formula:

$$\langle F, m_F|J, m_J, I, m_I\rangle = (-1)^{-J+I-m_F}\sqrt{2F+1}\begin{pmatrix} J & I & F \\ m_J & m_I & -m_F \end{pmatrix} \tag{4.59}$$

Writing formula (4.58) in matrix form, we can get a more concise expression:

$$\hat{J}_z = \mathbf{U}\hat{J}_z'\mathbf{U}^\dagger \tag{4.60}$$

where $\hat{J}_z'$ is the matrix representation of $\hat{J}_z$ in $|J, m_J, I, m_I\rangle$, $\boldsymbol{U}$ is a unitary transformation matrix, its matrix elements are determined by formula (4.59).

Similarly, if $\hat{I}_z'$ is the matrix representation of $\hat{I}_z$ in $|J, m_J, I, m_I\rangle$, then $\hat{I}_z$ can be written in the terms of $|F, m_F\rangle$ as follows:

$$\hat{I}_z = \boldsymbol{U}\hat{I}_z'\boldsymbol{U}^\dagger \tag{4.61}$$

Thus, we have obtained the operators $\hat{J}_z$ and $\hat{I}_z$ in the terms of $|F, m_F\rangle$. By substituting them into formula (4.50), we can calculate the Hamiltonian representing the Zeeman splitting caused by the magnetic field. By summing $\hat{H}_f$, $\hat{H}_{hf}$, and $\hat{H}_B$ obtained from the aforementioned calculations, the total Hamiltonian of the atom can be derived. The next step is to write out the atomic Hamiltonian, establish the following Hamiltonian equation, and solve for the eigenvalues and eigenstates of the Hamiltonian:

$$\hat{H}|k\rangle = E_k|k\rangle \tag{4.62}$$

The eigenvalues of the Hamiltonian are the energy levels of the atomic splitting. By subtracting the ground state energy from the excited state energy, we can find the transition energy, and thus obtain the corresponding transition frequency:

$$\omega_{ij} = \frac{E_j - E_i}{\hbar} \tag{4.63}$$

Next, let's discuss the atomic electric dipole transition operator $e\hat{\boldsymbol{r}}$ and the natural linewidth of the transition line Γ_{ij}. The natural linewidth is mainly caused by the spontaneous radiation of the atom, which is caused by the zero-point field in the vacuum. To strictly explain the principle of spontaneous radiation, the radiation field needs to be quantized. Further, relationship between the natural linewidth and the dipole transition matrix element is given by:

$$\Gamma_{ij} = A_{ij} = \frac{\omega_{ij}^3}{3\pi\varepsilon_0\hbar c^3}\left|\langle g_i|e\hat{\boldsymbol{r}}|e_j\rangle\right|^2 \tag{4.64}$$

where A_{ij} is known as the Einstein spontaneous radiation coefficient, which describes the spontaneous radiation rate between the excited state and the ground state, ω_{ij} is the transition frequency obtained through Eq. (4.63), $\langle g_i|e\hat{\boldsymbol{r}}|e_j\rangle$ is the matrix element under the energy representation of the dipole transition operator.

Next, let's calculate the electric dipole transition matrix element. According to the Wigner–Eckart theorem, in terms of $|F, m_F\rangle$, the matrix elements of each component of the dipole transition operator can be given by:

$$\begin{aligned}\langle F, m_F|e\hat{r}_q|F', m_F'\rangle &= \langle F\|e\hat{\boldsymbol{r}}\|F'\rangle\langle F, m_F|F', m_F', 1, q\rangle \\ &= (-1)^{F'-1+m_F}\sqrt{2F'+1}\begin{pmatrix} F' & 1 & F \\ m_F' & q & -m_F \end{pmatrix}\langle F\|e\hat{r}\|F'\rangle\end{aligned} \tag{4.65}$$

where $e\hat{r}_q$ is the q-component of the electric dipole transition operator $e\hat{\boldsymbol{r}}$. The values of q are 0 and ± 1, which correspond to $\boldsymbol{e}_z$ and $\boldsymbol{e}_\pm$ in the spherical basis vectors. The non-zero condition of the Wigner-3j coefficient is known:

$$\begin{cases} F' = |1 - F|, \ldots, 1 + F \\ \Delta m_F = m_F' - m_F = q = 0, \pm 1 \end{cases} \tag{4.66}$$

This is the selection rule for electric dipole transitions. Only when the difference in total magnetic quantum numbers before and after the transition is 0 or ± 1, the transition matrix element is not zero, and the electric dipole transition can occur. $\Delta m_F = 0, +1$ and -1 correspond to π, σ_+, and σ_- transition respectively, it is called transition; when, it is called transition; when, it is called transition.

Further, by decoupling $\boldsymbol{F} = \boldsymbol{J} + \boldsymbol{I}$, and applying the Wigner–Eckart theorem again, we can get:

$$\begin{aligned}\left\langle F\|e\hat{\boldsymbol{r}}\|F'\right\rangle &= \left\langle J, I, F\|e\hat{\boldsymbol{r}}\|J', I', F'\right\rangle \\ &= (-1)^{F'+J+1+I}\sqrt{(2F'+1)(2J+1)}\begin{pmatrix} 1 & J & J' \\ I & F & F' \end{pmatrix}\left\langle J\|e\hat{\boldsymbol{r}}\|J'\right\rangle\end{aligned} \tag{4.67}$$

where $\left\langle J\|e\hat{\boldsymbol{r}}\|J'\right\rangle$ is the reduced matrix element, its value is only related to the total angular momentum quantum number, but it is still difficult to calculate. However, from Eq. (4.64), it is known that it has the following relationship with the spontaneous radiation rate:

$$\Gamma_0 = \frac{\omega_0^3}{3\pi\varepsilon_0\hbar c^3}\frac{2J+1}{2J'+1}\left|\left\langle J\|e\hat{\boldsymbol{r}}\|J'\right\rangle\right|^2 \tag{4.68}$$

Here, Γ_0 is the natural linewidth of the $J' \to J$ transition, and the coefficient $(2J+1)/(2J'+1)$ is caused by the degeneracy of the energy level. Γ_0 is generally determined by experiments, it can be obtained by looking up the table, and the reduced matrix element $\left\langle J\|e\hat{\boldsymbol{r}}\|J'\right\rangle$ can be obtained by substituting into Eq. (4.68), then the components of the dipole transition operator can be determined by Eqs. (4.65) and (4.67) in the representation of $|F, m_F\rangle$.

Next, we will transform the electric dipole transition operator from the representation of $|F, m_F\rangle$ to the energy representation, following the previous treatment of the operator $\hat{J}_z$ in terms of $|F, m_F\rangle$, we can get in the energy representation:

$$\left\langle k_i|e\hat{r}_q|k_j\right\rangle = \sum_{F,m_F}\sum_{F',m'_F}\left\langle k_i|F', m'_F\right\rangle\left\langle F, m_F|e\hat{r}_q|F', m'_F\right\rangle\left\langle F, m_F|k_j\right\rangle \tag{4.69}$$

In matrix form, it can be written as:

$$e\hat{r}_q = \boldsymbol{U}\left(e\hat{r}_q\right)'\boldsymbol{U}^\dagger \tag{4.70}$$

where, $\left(e\hat{r}_q\right)'$ is the electric dipole transition operator in the representation of $|F, m_F\rangle$, its matrix element is obtained by Eq. (4.65). For convenience of description, let $|\alpha\rangle = |F, m_F\rangle$, then the form of the unitary transformation matrix $\boldsymbol{U}$ is:

$$\begin{aligned}\boldsymbol{U} &= \begin{bmatrix} \langle k_1|\alpha_1\rangle & \langle k_1|\alpha_2\rangle & \cdots & \langle k_1|\alpha_n\rangle \\ \langle k_2|\alpha_1\rangle & \langle k_2|\alpha_2\rangle & \cdots & \vdots \\ \vdots & \vdots & \ddots & \vdots \\ \langle k_n|\alpha_1\rangle & \cdots & \cdots & \langle k_n|\alpha_n\rangle \end{bmatrix} \\ &= \begin{bmatrix} \langle k_1| \\ \langle k_2| \\ \vdots \\ \langle k_n| \end{bmatrix}\left[|\alpha_1\rangle, |\alpha_2\rangle, \cdots |\alpha_n\rangle\right]\end{aligned} \tag{4.71}$$

where $|k_i\rangle$ is the eigenvector obtained from the Hamiltonian equation (4.62). Since the initially chosen $|F, m_F\rangle$ is the base vector, so $|\alpha_i\rangle$ is a unit column vector, $[|\alpha_1\rangle, |\alpha_2\rangle, \ldots |\alpha_n\rangle]$ forms a unit matrix, so we have:

$$\boldsymbol{U} = \begin{bmatrix} \langle k_1| \\ \langle k_2| \\ \vdots \\ \langle k_n| \end{bmatrix} \begin{bmatrix} 1 & 0 & \cdots & 0 \\ 0 & 1 & \cdots & 0 \\ \vdots & \vdots & \ddots & \vdots \\ 0 & 0 & \cdots & 1 \end{bmatrix} = \begin{bmatrix} \langle k_1| \\ \langle k_2| \\ \vdots \\ \langle k_n| \end{bmatrix} \tag{4.72}$$

Substituting Eq. (4.72) into Eq. (4.70), we can obtain the electric dipole transition operator in the energy terms as:

$$e\hat{r}_q = \begin{bmatrix} \langle k_1| \\ \langle k_2| \\ \vdots \\ \langle k_n| \end{bmatrix} \begin{bmatrix} \left\langle \alpha_1 | e\hat{r}_q | \alpha_1 \right\rangle & \left\langle \alpha_1 | e\hat{r}_q | \alpha_2 \right\rangle & \cdots & \left\langle \alpha_1 | e\hat{r}_q | \alpha_n \right\rangle \\ \left\langle \alpha_2 | e\hat{r}_q | \alpha_1 \right\rangle & \left\langle \alpha_2 | e\hat{r}_q | \alpha_2 \right\rangle & \cdots & \left\langle \alpha_2 | e\hat{r}_q | \alpha_n \right\rangle \\ \vdots & \vdots & \ddots & \vdots \\ \left\langle \alpha_n | e\hat{r}_q | \alpha_1 \right\rangle & \left\langle \alpha_n | e\hat{r}_q | \alpha_2 \right\rangle & \cdots & \left\langle \alpha_n | e\hat{r}_q | \alpha_n \right\rangle \end{bmatrix} [|k_1\rangle, |k_2\rangle, \ldots, |k_n\rangle] \tag{4.73}$$

By distinguishing between the ground state and the excited state, we can obtain the dipole transition matrix element $\left\langle g_i | e\hat{\boldsymbol{r}} | e_j \right\rangle$ between the ground state and the excited state, and then substitute it into (4.64) to get the natural linewidth Γ_{ij} of the transition between any two energy levels. At this point, all the physical quantities needed to calculate the atomic polarizability have been obtained, and the atomic polarizability can be calculated by substituting them into formula (4.42).

4.4.4 Faraday Rotation Effect

In Sect. 4.3, we have already introduced how the atomic filter causes the polarization direction of the incident light to rotate: this is caused by the phase difference caused by the different propagation speeds of left-handed and right-handed circularly polarized lights. However, the difference in propagation speeds of left-handed and right-handed circularly polarized lights has not been explained. In Sect. 4.4.1, we derived that the wave number k of light in the medium is closely related to the polarizability, but as can be seen from the above process of solving the polarizability, the atomic polarizability shows anisotropy under the spherical basis vectors with the quantization axis (magnetic field direction) as the main axis, so the atomic polarizability should be represented by a tensor. In addition, for the Faraday atomic filter, the direction of the magnetic field is parallel to the direction of light propagation, but in actual applications, the direction of the magnetic field is not necessarily the optical axis direction. In general, this section will discuss the working principle of the atomic filter in any magnetic field direction.

As shown in Fig. 4.49, the direction of light propagation is along the z-direction, the angle between the direction of the magnetic field and the z-axis is, and the azimuth angle is φ. For convenience, it is stipulated here that all vectors are described by column vectors. For example, the base vectors of this Cartesian coordinate system can be written as:

$$\boldsymbol{e}_x = \begin{bmatrix} 1 \\ 0 \\ 0 \end{bmatrix},\ \boldsymbol{e}_y = \begin{bmatrix} 0 \\ 1 \\ 0 \end{bmatrix},\ \boldsymbol{e}_z = \begin{bmatrix} 0 \\ 0 \\ 1 \end{bmatrix} \tag{4.74}$$

Since the quantization axis is the direction of the magnetic field, to study the rotation effect under different magnetic field directions, it is necessary to establish the connection between the spherical base vectors under the quantization axis and the orthogonal base vectors under the light propagation axis. This connection can be derived using a simple vector decomposition method.

First, the unit vector in the x axis direction can definitely be decomposed into the components along the magnetic field direction and the component perpendicular to the magnetic field direction. Here, it is assumed that the direction of the component perpendicular to the magnetic field direction is the x' axis under the quantization axis, so we have:

$$\boldsymbol{e}_x = \boldsymbol{e}_0 \cos\theta_x + \boldsymbol{e}_{x'} \sin\theta_x = \boldsymbol{e}_0 \cos\theta_x - \frac{\boldsymbol{e}_+ - \boldsymbol{e}_-}{\sqrt{2}} \sin\theta_x \tag{4.75}$$

Similarly, the unit vectors in the y direction and z direction can also be decomposed as follows:

$$\begin{aligned} \boldsymbol{e}_y &= \boldsymbol{e}_0 \cos\theta_y + \frac{\boldsymbol{e}_+ e^{i\alpha} - \boldsymbol{e}_- e^{-i\alpha}}{\sqrt{2}} \sin\theta_y \\ \boldsymbol{e}_z &= \boldsymbol{e}_0 \cos\theta + \frac{\boldsymbol{e}_+ e^{i\beta} - \boldsymbol{e}_- e^{-i\beta}}{\sqrt{2}} \sin\theta \end{aligned} \tag{4.76}$$

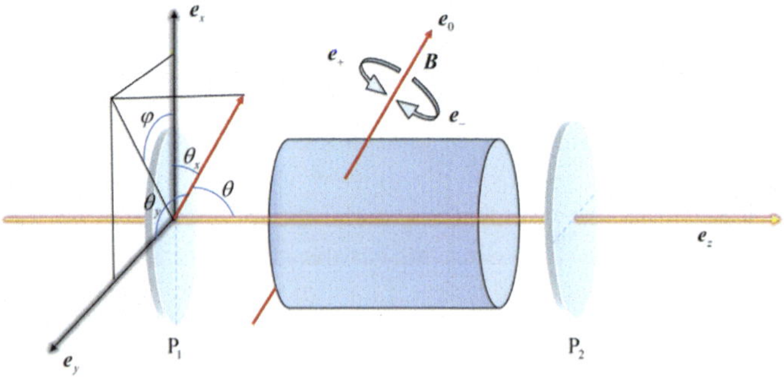

Fig. 4.49 Arbitrary magnetic field direction

where θ_x and θ_y satisfy simple geometric relationships:

$$\begin{aligned} \cos\theta_x &= \sin\theta\cos\varphi \\ \cos\theta_y &= \sin\theta\sin\varphi \end{aligned} \tag{4.77}$$

Then, based on the orthogonality between the base vectors, we can get:

$$\begin{aligned} \boldsymbol{e}_x^{\dagger}\boldsymbol{e}_y &= \cos\theta_x\cos\theta_y - \sin\theta_x\sin\theta_y\cos\alpha = 0 \\ &\Rightarrow \alpha = \arccos\left(\cot\theta_x\cot\theta_y\right) \\ \boldsymbol{e}_x^{\dagger}\boldsymbol{e}_z &= \cos\theta_x\cos\theta - \sin\theta_x\sin\theta\cos\beta = 0 \\ &\Rightarrow \beta = \arccos(\cot\theta_x\cot\theta) \end{aligned} \tag{4.78}$$

Therefore, as long as we know the elevation angle θ and the azimuth angle φ of the magnetic field, we can find the conversion relationship between the orthogonal base vector and the quantum axis spherical base vector:

$$\begin{aligned} \boldsymbol{I} &= \left[\boldsymbol{e}_x\ \boldsymbol{e}_y\ \boldsymbol{e}_z\right] \\ &= \left[\boldsymbol{e}_0\ \boldsymbol{e}_+\ \boldsymbol{e}_-\right]\begin{bmatrix} \cos\theta_x & \cos\theta_y & \cos\theta \\ \frac{-1}{\sqrt{2}}\sin\theta_x & \frac{e^{i\alpha}}{\sqrt{2}}\sin\theta_y & \frac{e^{i\beta}}{\sqrt{2}}\sin\theta \\ \frac{1}{\sqrt{2}}\sin\theta_x & \frac{-e^{-i\alpha}}{\sqrt{2}}\sin\theta_y & \frac{-e^{-i\beta}}{\sqrt{2}}\sin\theta \end{bmatrix} \end{aligned} \tag{4.79}$$

$$\boldsymbol{I} = \left[\boldsymbol{e}_0\ \boldsymbol{e}_+\ \boldsymbol{e}_-\right]\ \boldsymbol{U} = \left[\boldsymbol{e}_0\ \boldsymbol{e}_+\ \boldsymbol{e}_-\right]\begin{bmatrix} \boldsymbol{e}_0^{\dagger} \\ \boldsymbol{e}_+^{\dagger} \\ \boldsymbol{e}_-^{\dagger} \end{bmatrix} \tag{4.80}$$

where $\boldsymbol{U}$ represents the transformation matrix between the two base vectors, it is not difficult to see that $\boldsymbol{U}$ is a unitary matrix, which satisfies:

$$\boldsymbol{U}^{\dagger} = \boldsymbol{U}^{-1} = \left[\boldsymbol{e}_0\ \boldsymbol{e}_+\ \boldsymbol{e}_-\right] \tag{4.81}$$

The conversion relationship of the electric field strength under different base vectors can be obtained from the transformation matrix:

$$\boldsymbol{E} = \left[\boldsymbol{e}_x\ \boldsymbol{e}_y\ \boldsymbol{e}_z\right]\begin{bmatrix} E_x \\ E_y \\ E_z \end{bmatrix} = \left[\boldsymbol{e}_0\ \boldsymbol{e}_+\ \boldsymbol{e}_-\right]\boldsymbol{U}\begin{bmatrix} E_x \\ E_y \\ E_z \end{bmatrix} \tag{4.82}$$

$$\begin{bmatrix} E_0 \\ E_+ \\ E_- \end{bmatrix} = \boldsymbol{U}\begin{bmatrix} E_x \\ E_y \\ E_z \end{bmatrix} \tag{4.83}$$

Here, the exponential evolution and complex conjugate terms are omitted. Similarly, under the spherical base vectors of the quantization axis, the atomic polarizability has three independent components, which are χ_0, χ_+, χ_-, the polarization intensity vector can be expressed by the polarizability tensor:

$$\begin{aligned} \boldsymbol{P} &= \varepsilon_0 \begin{bmatrix} \boldsymbol{e}_0 & \boldsymbol{e}_+ & \boldsymbol{e}_- \end{bmatrix} \begin{bmatrix} \chi_0 & 0 & 0 \\ 0 & \chi_+ & 0 \\ 0 & 0 & \chi_- \end{bmatrix} \begin{bmatrix} E_0 \\ E_+ \\ E_- \end{bmatrix} \\ &= \varepsilon_0 \begin{bmatrix} \boldsymbol{e}_x & \boldsymbol{e}_y & \boldsymbol{e}_z \end{bmatrix} \boldsymbol{U}^\dagger \begin{bmatrix} \chi_0 & 0 & 0 \\ 0 & \chi_+ & 0 \\ 0 & 0 & \chi_- \end{bmatrix} \boldsymbol{U} \begin{bmatrix} E_x \\ E_y \\ E_z \end{bmatrix} \end{aligned} \tag{4.84}$$

From this, the polarizability tensor of the atom under the orthogonal base vectors can be obtained:

$$\boldsymbol{\chi}_{xyz} = \boldsymbol{U}^\dagger \begin{bmatrix} \chi_0 & 0 & 0 \\ 0 & \chi_+ & 0 \\ 0 & 0 & \chi_- \end{bmatrix} \boldsymbol{U} \tag{4.85}$$

Assuming that the propagation of the electric field in the atomic vapor still satisfies the harmonic form:

$$\boldsymbol{E} = \boldsymbol{E}_0 e^{i(\boldsymbol{k}\cdot\boldsymbol{r} - \omega t)} \tag{4.86}$$

Substituting this expression into Maxwell's equations (4.10) yields:

$$\boldsymbol{k} \times (\boldsymbol{k} \times \boldsymbol{E}) + \frac{\omega^2}{c^2}(\boldsymbol{I} + \boldsymbol{\chi})\boldsymbol{E} = 0 \tag{4.87}$$

For orthogonal base vectors, we have:

$$\frac{\omega^2}{c^2}(\boldsymbol{I} + \boldsymbol{\chi}_{xyz})\boldsymbol{E} = \begin{bmatrix} k_y^2 + k_z^2 & -k_x k_y & -k_x k_z \\ -k_y k_x & k_z^2 + k_x^2 & -k_y k_z \\ -k_z k_x & -k_z k_y & k_x^2 + k_y^2 \end{bmatrix} \boldsymbol{E} \tag{4.88}$$

For orthogonal base vectors, the wave vector $\boldsymbol{k}$ only has z component, that is:

$$\boldsymbol{k} = \begin{bmatrix} 0 \\ 0 \\ k \end{bmatrix} \tag{4.89}$$

Substituting (4.89) into Eq. (4.87), the equation satisfied by the electric field under the orthogonal basis can be obtained:

$$\frac{\omega^2}{\varepsilon_0 c^2}\boldsymbol{\varepsilon}_{xyz}\begin{bmatrix}E_x\\E_y\\E_z\end{bmatrix}=\frac{\omega^2}{\varepsilon_0 c^2}\begin{bmatrix}\varepsilon_{xx}\ \varepsilon_{xy}\ \varepsilon_{xz}\\ \varepsilon_{yx}\ \varepsilon_{yy}\ \varepsilon_{yz}\\ \varepsilon_{zx}\ \varepsilon_{zy}\ \varepsilon_{zz}\end{bmatrix}\begin{bmatrix}E_x\\E_y\\E_z\end{bmatrix}$$

$$=k^2\begin{bmatrix}1\ 0\ 0\\0\ 1\ 0\\0\ 0\ 0\end{bmatrix}\begin{bmatrix}E_x\\E_y\\E_z\end{bmatrix} \tag{4.90}$$

where $\boldsymbol{\varepsilon}_{xyz}=\varepsilon_0(\boldsymbol{I}+\boldsymbol{\chi}_{xyz})$ is the complex dielectric constant tensor of the medium under the orthogonal basis, $\boldsymbol{I}$ is the unit matrix, and from the third component of this equation:

$$\varepsilon_{zx}E_x+\varepsilon_{zy}E_y+\varepsilon_{zz}E_z=0 \tag{4.91}$$

By expressing E_z with E_x and E_y, Eq. (4.90) can be simplified into a second-order characteristic equation:

$$E_z=-\frac{\varepsilon_{zx}E_x+\varepsilon_{zy}E_y}{\varepsilon_{zz}} \tag{4.92}$$

$$\frac{\omega^2}{\varepsilon_0 c^2}\begin{bmatrix}\varepsilon_{xx}-\frac{\varepsilon_{xz}\varepsilon_{zx}}{\varepsilon_{zz}} & \varepsilon_{xy}-\frac{\varepsilon_{xz}\varepsilon_{zy}}{\varepsilon_{zz}}\\ \varepsilon_{yx}-\frac{\varepsilon_{yz}\varepsilon_{zx}}{\varepsilon_{zz}} & \varepsilon_{yy}-\frac{\varepsilon_{yz}\varepsilon_{zy}}{\varepsilon_{zz}}\end{bmatrix}\begin{bmatrix}E_x\\E_y\end{bmatrix}=k^2\begin{bmatrix}E_x\\E_y\end{bmatrix} \tag{4.93}$$

Solving this characteristic equation can obtain the eigenvalues and corresponding eigenvectors of k^2. Substituting the eigenvectors into Eq. (4.92), we can obtain the z component of the eigenvectors. However, it should be noted that only the x and y components of the eigenvectors are obtained from the solution of the characteristic equation. These two components are independent of each other, while the z component is obtained by the linear combination of the first two components, which means that the two eigenvectors correspond to two independent polarization directions. It should be noted that the two eigenvectors are not necessarily orthogonal to each other, because all three components of the polarization tensor are complex numbers, which makes the diagonal elements of the coefficient matrix in the above characteristic equation complex, making the coefficient matrix not a Hermitian matrix. In some special cases, the eigenvectors of non-Hermitian matrices can also be orthogonal (such as the Faraday-type and Voigt-type atomic filters introduced later), but readers should not generalize this special case to all situations.

In isotropic media, the direction of the wave vector $\boldsymbol{k}$ is perpendicular to the direction of the electric field $\boldsymbol{E}$, it can be considered that the direction of $\boldsymbol{k}$ is the same as the direction of light propagation. However, in anisotropic media, the direction of $\boldsymbol{k}$ is generally not perpendicular to the direction of the electric field $\boldsymbol{E}$, only perpendicular to the direction of the displacement vector $\boldsymbol{D}$. But the electric field $\boldsymbol{E}$ still has only two degrees of freedom, as its z-direction component is constrained by Eq. (4.92). Therefore, when we calculate the transmission spectrum later, we only consider the transverse component of the electric field, and its longitudinal component can naturally be represented by the linear combination of the transverse components.

Only considering the x and y components, the eigenvectors can be normalized to obtain the eigen basis vectors. Suppose the two eigenvalues and corresponding eigen basis vectors obtained are k_1, k_2 and $\boldsymbol{e}_1, \boldsymbol{e}_2$, their meaning is: only when the polarization direction of the transverse electric field is any eigen basis vector $\boldsymbol{e}_i$, the wave number of the polarized light propagating in space is the corresponding eigenvalue k_i. For other polarized light whose polarization direction is not on the eigen basis vector, its wave number cannot be determined, but it can be regarded as the linear superposition of two characteristic polarization directions. Therefore, to analyze the propagation of light in anisotropic media, its polarization direction must be decomposed onto two eigen basis vectors.

The matrix composed of eigen basis vector columns is denoted as $\boldsymbol{V}$, the diagonal matrix composed of wave number eigenvalues is denoted as $\boldsymbol{K}$, the propagation form of the electric field in space in terms of the eigen basis can be written as:

$$\begin{aligned}\boldsymbol{E}(z) &= \begin{bmatrix}\boldsymbol{e}_1 & \boldsymbol{e}_2\end{bmatrix}\begin{bmatrix}E_1 e^{ik_1 z}\\ E_2 e^{ik_2 z}\end{bmatrix}\\ &= \begin{bmatrix}\boldsymbol{e}_1 & \boldsymbol{e}_2\end{bmatrix}\begin{bmatrix}e^{ik_1 z} & 0\\ 0 & e^{ik_2 z}\end{bmatrix}\begin{bmatrix}E_1\\ E_2\end{bmatrix}\\ &= \boldsymbol{V}e^{i\boldsymbol{K}z}\begin{bmatrix}E_1\\ E_2\end{bmatrix}\end{aligned} \tag{4.94}$$

For polarized light under the orthogonal basis, it needs to be decomposed into the superposition of each characteristic component:

$$\begin{bmatrix}\boldsymbol{e}_x & \boldsymbol{e}_y\end{bmatrix}\begin{bmatrix}E_x\\ E_y\end{bmatrix} = \boldsymbol{V}\boldsymbol{V}^{-1}\begin{bmatrix}E_x\\ E_y\end{bmatrix} = \begin{bmatrix}\boldsymbol{e}_1 & \boldsymbol{e}_2\end{bmatrix}\begin{bmatrix}E_1\\ E_2\end{bmatrix} \tag{4.95}$$

Therefore, the transformation relationship between the characteristic base vector and the orthogonal base vector under polarization components is:

$$\begin{bmatrix}E_1\\ E_2\end{bmatrix} = \boldsymbol{V}^{-1}\begin{bmatrix}E_x\\ E_y\end{bmatrix} \tag{4.96}$$

Substituting into Eq. (4.94), the propagation form of light under the orthogonal base vector can be obtained:

$$\boldsymbol{E}(z) = \boldsymbol{V}e^{i\boldsymbol{K}z}\boldsymbol{V}^{-1}\begin{bmatrix}E_x\\ E_y\end{bmatrix} \tag{4.97}$$

Assume that the polarization directions of the polarizer P_1 and the analyzer P_2 are $\boldsymbol{p}_1$ and $\boldsymbol{p}_2$ respectively. The incident light becomes linearly polarized light along the

$\boldsymbol{p}_1$ direction after passing through the polarizer P_1. Since the transverse component of the electric field is continuous at the interface of the medium, when entering the atomic filter, the transverse component of the electric field is still the linearly polarized light in the direction. Assuming that the amplitude of the incident light has been normalized, then after passing through an atomic filter of length l, its electric field strength becomes:

$$\boldsymbol{E}(l) = \boldsymbol{V}e^{i\boldsymbol{K}l}\boldsymbol{V}^{-1}\boldsymbol{p}_1 \tag{4.98}$$

The analyzer P_2 only allows components polarized in the $\boldsymbol{p}_2$ direction to pass through, so the outgoing electric field is the projection of the electric field incident on P_2 in the $\boldsymbol{p}_2$ direction:

$$E_{out} = \boldsymbol{p}_2^T\boldsymbol{E}(l) = \boldsymbol{p}_2^T\boldsymbol{V}e^{i\boldsymbol{K}l}\boldsymbol{V}^{-1}\boldsymbol{p}_1 \tag{4.99}$$

Not only is this expression concise, but its physical meaning is also very clear: $\boldsymbol{p}_1$ represents the action of the polarizer, $\boldsymbol{V}^{-1}$ represents the projection of the light's polarization direction onto the characteristic polarization direction, $e^{i\boldsymbol{K}l}$ represents the transmission of light in the characteristic polarization direction in the atomic cell, $\boldsymbol{V}$ represents the projection of light from the characteristic polarization direction back into the orthogonal base, the entire $\boldsymbol{V}e^{i\boldsymbol{K}l}\boldsymbol{V}^{-1}$ is the action of the atomic cell on the light, $\boldsymbol{p}_2^T$ represents the action of the analyzer, E_{out} is the field amplitude of the light after passing through the polarizer, atomic cell, and analyzer.

Since the incident light amplitude is assumed to be normalized, the transmittance of the atomic filter is:

$$Trs = |E_{out}|^2 = E_{out}E_{out}^* \tag{4.100}$$

The above is the entire process of calculating the transmission spectrum of the atomic filter. Specifically, there are two special configurations of atomic filters in practical applications, one is called the Faraday atomic filter, and the other is called the Voigt atomic filter. We have discussed these two configurations in detail below.

1. Faraday Anomalous Dispersion Optical Filter (FADOF)

The magnetic field of the Faraday atomic filter is parallel to the direction of light propagation, and its quantization axis coincides with the z axis of the orthogonal base. The transformation matrix of the two coordinate systems can be known from formula (4.79):

$$\boldsymbol{U} = \begin{bmatrix} -\frac{\sqrt{2}}{2} & \frac{\sqrt{2}}{2}i & 0 \\ \frac{\sqrt{2}}{2} & -\frac{\sqrt{2}}{2}i & 0 \\ 0 & 0 & 1 \end{bmatrix} \tag{4.101}$$

The polarizability tensor of the atomic vapor medium in the orthogonal coordinate system can be obtained from formula (4.85):

$$\boldsymbol{\chi}_{xyz} = \boldsymbol{U}^{\dagger}\boldsymbol{\chi}_{0\pm}\boldsymbol{U} = \begin{bmatrix} \frac{\chi_+ + \chi_-}{2} & -i\frac{\chi_+ - \chi_-}{2} & 0 \\ i\frac{\chi_+ - \chi_-}{2} & \frac{\chi_+ + \chi_-}{2} & 0 \\ 0 & 0 & \chi_0 \end{bmatrix} \tag{4.102}$$

Substituting this equation into formula (4.93), the characteristic equation satisfied by the light propagating in the Faraday atomic filter can be obtained:

$$\begin{bmatrix} 1 + \frac{\chi_+ + \chi_-}{2} & -i\frac{\chi_+ - \chi_-}{2} \\ i\frac{\chi_+ - \chi_-}{2} & 1 + \frac{\chi_+ + \chi_-}{2} \end{bmatrix} \begin{bmatrix} E_x \\ E_y \end{bmatrix} = \frac{c^2 k^2}{\omega^2} \begin{bmatrix} E_x \\ E_y \end{bmatrix} \tag{4.103}$$

Solving this characteristic equation, we can obtain its eigenvalues and their corresponding eigenvectors:

$$\boldsymbol{K} = \begin{bmatrix} k_1 & 0 \\ 0 & k_2 \end{bmatrix} = \frac{\omega}{c} \begin{bmatrix} \sqrt{1+\chi_+} & 0 \\ 0 & \sqrt{1+\chi_-} \end{bmatrix} \tag{4.104}$$

$$\boldsymbol{V} = \begin{bmatrix} \boldsymbol{e}_1 & \boldsymbol{e}_2 \end{bmatrix} = \frac{1}{\sqrt{2}} \begin{bmatrix} \begin{pmatrix} -i \\ 1 \end{pmatrix} & \begin{pmatrix} i \\ 1 \end{pmatrix} \end{bmatrix} \tag{4.105}$$

It can be seen that for the Faraday atomic filter, its characteristic eigenvectors of polarization direction are the left-handed and right-handed polarization components. This is why, we decompose the incident linearly polarized light into a superposition of a left-handed circularly polarized light and a right-handed circularly polarized light in Sect. 4.3 when we first introduce the principle of the Faraday atomic filter, rather than decomposing it into other directions. Because only when the polarization direction is these two characteristic polarization directions, the wave number of its propagation in space is the eigenvalue determined by (4.104), and the wave number of any other polarization direction cannot be determined.

2. Voigt Anomalous Dispersion Optical Filter (VADOF)

The magnetic field direction of the Voigt atomic filter is perpendicular to the direction of light propagation and forms a 45° angle with the direction of incident light polarization. Assuming the magnetic field direction is along the x axis, the transformation matrix of the quantization axis and the orthogonal coordinate system is:

$$\boldsymbol{U} = \begin{bmatrix} 1 & 0 & 0 \\ 0 & \frac{\sqrt{2}}{2} & \frac{\sqrt{2}}{2}i \\ 0 & -\frac{\sqrt{2}}{2} & -\frac{\sqrt{2}}{2}i \end{bmatrix} \tag{4.106}$$

Similarly, the expression of the polarization tensor of the Voigt atomic filter in the orthogonal coordinate system can be obtained:

$$\boldsymbol{\chi}_{xyz} = \boldsymbol{U}^{\dagger}\boldsymbol{\chi}_{0\pm}\boldsymbol{U} = \begin{bmatrix} \chi_0 & 0 & 0 \\ 0 & \frac{\chi_+ + \chi_-}{2} & i\frac{\chi_+ - \chi_-}{2} \\ 0 & -i\frac{\chi_+ - \chi_-}{2} & \frac{\chi_+ + \chi_-}{2} \end{bmatrix} \tag{4.107}$$

Similarly, the characteristic equation satisfied by the light propagating in the Voigt atomic filter is:

$$\begin{bmatrix} 1+\chi_0 & 0 \\ 0 & \frac{2(1+\chi_+)(1+\chi_-)}{2+\chi_+ + \chi_-} \end{bmatrix}\begin{bmatrix} E_x \\ E_y \end{bmatrix} = \frac{c^2k^2}{\omega^2}\begin{bmatrix} E_x \\ E_y \end{bmatrix} \tag{4.108}$$

The coefficient matrix of this characteristic equation has been diagonalized, so its eigenvalues and corresponding eigenvectors can be directly written as:

$$\boldsymbol{K} = \begin{bmatrix} k_1 & 0 \\ 0 & k_2 \end{bmatrix} = \frac{\omega}{c}\begin{bmatrix} \sqrt{1+\chi_0} & 0 \\ 0 & \sqrt{\frac{2(1+\chi_+)(1+\chi_-)}{2+\chi_+ + \chi_-}} \end{bmatrix} \tag{4.109}$$

$$\boldsymbol{V} = \begin{bmatrix} \boldsymbol{e}_1 & \boldsymbol{e}_2 \end{bmatrix} = \begin{bmatrix} \begin{pmatrix} 1 \\ 0 \end{pmatrix} & \begin{pmatrix} 0 \\ 1 \end{pmatrix} \end{bmatrix} \tag{4.110}$$

It can be seen that for the Voigt atomic filter, one of the characteristic polarization directions is parallel to the magnetic field direction, and the other characteristic polarization direction is perpendicular to the magnetic field direction. Therefore, when studying the Voigt atomic filter, it is necessary to decompose the incident linearly polarized light into components along the magnetic field direction and perpendicular to the magnetic field direction. At this time, the wave numbers k_1 and k_2 of light propagating in space are given in (4.109). From the above derivation process, it can be seen that whether it is Faraday or Voigt atomic filters, their working principles are essentially the same, which is nothing more than the transmission of light in anisotropic media, but the direction of the magnetic field is different. The atomic cell is like a wave plate, but its most prominent feature is that its working band is limited to the frequency range corresponding to the atomic transition spectrum. This characteristic makes the atomic cell to have unique advantages in optical devices and a widely used device in laser technology, optical communication, quantum imaging and other fields.

According to all the content discussed in this section, the transmission spectrum of alkali metal atomic filters can be calculated. Here, consider the ^{87}Rb atomic filter as an example, under a 300 G axial magnetic field, the D_2 transition line transmission spectrum at different temperatures is shown in Fig. 4.50.

The light intensity in the experiment is 0.2 mW/mm^2, which has exceeded the saturation light intensity of the ^{87}Rb atomic D_2 transition line. It can be seen that the theoretical calculation still matches the experimental data quite well, but does this mean that this set of calculation theories can fully cope with the FADOF transmission spectrum under any light intensity? Actually not, we have made quite a lot of approximations in the derivation process, including the plane wave approximation

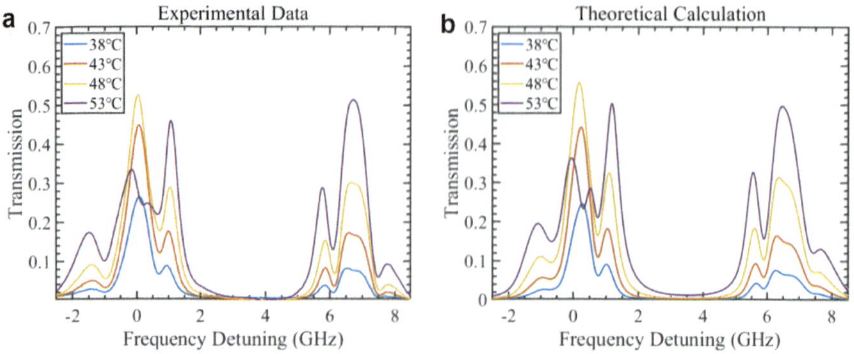

Fig. 4.50 ^{87}Rb atomic FADOF transmission spectrum **a** experimental data; **b** theoretical calculation

of the electric field, the electric dipole approximation, the two-level approximation, etc., and these approximations have their conditions of establishment. Among them, when the light intensity far exceeds the saturation light intensity, the two-level approximation is no longer valid.

Perhaps some readers may be puzzled, doesn't the two-level model itself include items related to light intensity (Rabi frequency)? Does this mean that the model has already considered the factor of light intensity? There is no problem with this idea. The two-level model has taken light intensity into account since its inception, and the process of solving its steady-state solution only uses the rotating wave approximation. The effect of this approximation is very good and hardly has any impact on the results. It can be said that there is no problem with the solution of the two-level model itself. The key is that it is only applicable to two-level systems. Why can alkali metal atoms be regarded as a two-level system? This is due to the assumption that light only induces transitions between the pair of energy levels that resonate with its frequency, and that transitions between each pair of energy levels are independent of each other, with their contributions to the atomic polarizability being additive. Where does this assumption come from? In some quantum optics textbooks, the method of perturbation theory is first introduced when dealing with the interaction between light and atoms. Perturbation theory can handle multi-level atoms, but its condition is that the intensity of light is small enough to ensure that the population of atoms in the initial state remains unchanged. The result derived from this condition indicates that only when the light resonates with the transition frequency between a pair of energy levels, the atomic transition may occur. The hyperfine energy level spacing of alkali metal atoms is much larger than the natural linewidth of the transition, so each energy level transition is independent. If only the pair of energy levels that play a leading role are considered, then the atom can be regarded as a two-level system. But if the light intensity is large enough to make the saturation broadening of the atom much larger than the natural linewidth, are the transitions of each energy level of the atom still independent? Is there still only one pair of energy levels that play a leading role in the transition? Can atoms still be treated as a two-level system? This

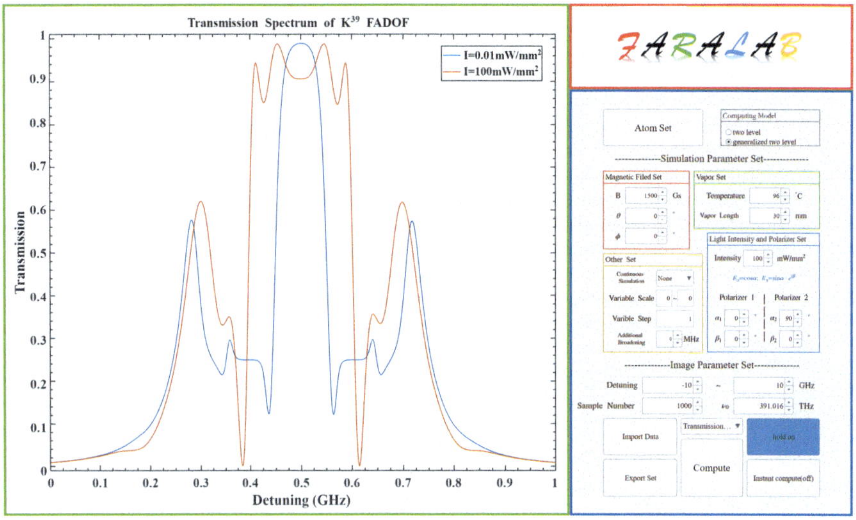

Fig. 4.51 Schematic diagram of FARALAB

is the challenge of studying atomic filters under strong light. To solve this problem, a basic model of strong field interaction with atoms must be established. This model must be able to consider the complex energy level structure of atoms and the effect of strong light at the same time. This is an important topic for future research on Faraday atomic filters. Once this model is established, the calculation problem of the transmission spectrum of Faraday atomic filters under strong light can be solved. Currently, Peking University has achieved a significant breakthrough in the core theory related to this challenging problem. For the first time, Peking University has overcome the generalized two-level atomic model. Additionally, they have developed a software named FARALAB, which is capable of calculating the interaction process between light and atoms under high intensities beam, as well as the transmission spectrum of atomic filters in such conditions. In terms of the range/upper limit of light intensity, the variety of atomic transition energy levels, and the computational ability for multiple parameters at the same interface, FARALAB surpasses the performance of ElecSus software developed by Durham University in the United Kingdom. The application for software copyright has currently been accepted, and the software interface is shown in Fig. 4.51. The software is capable of calculating the transmission spectrum of FADOF under the intense light exceeding the saturation intensity by at least six orders of magnitude, as well as various parameters such as magnetic field, temperature, and vapor cell length. The results obtained are in good agreement with the experimental data. Additionally, the software supports data import and export, as well as continuous simulation of parameters, providing reliable theoretical support for the design of FADOF under intense light conditions and the development of high-power Faraday lasers.

References

1. N. Bloembergen, Solid state infrared quantum counters. Phys. Rev. Lett. **2**(3), 84 (1959)
2. J. Gelbwachs, C. Klein, J. Wessel, Infrared detection by an atomic vapor quantum counter. IEEE J. Quantum Electron. **14**(2), 77–79 (1978)
3. J.B. Marling, J. Nilsen, L.C. West et al., An ultrahigh-Q isotropically sensitive optical filter employing atomic resonance transitions. J. Appl. Phys. **50**(2), 610–614 (1979)
4. J.A. Gelbwachs, Atomic resonance filters. IEEE J. Quantum Electron. **24**(7), 1266–1277 (1988)
5. Y. Öhman, *On Some New Auxiliary Instruments in Astrophysical Research* (Almqvist & Wiksell, 1956)
6. P. Yeh, Dispersive magnetooptic filters. Appl. Opt. **21**(11), 2069–2075 (1982)
7. B. Yin, T.M. Shay, Theoretical model for a Faraday anomalous dispersion optical filter. Opt. Lett. **16**(20), 1617–1619 (1991)
8. B. Yin, T.M. Shay, Faraday anomalous dispersion optical filter for the Cs 455 nm transition. IEEE Photonics Technol. Lett. **4**(5), 488–490 (1992)
9. B. Yin, T.M. Shay, Faraday anomalous dispersion optical filters, in *Lasers 1991; Proceedings of the 14th International Conference on Lasers and Applications* (1992), pp. 641–648
10. B. Yin, T.M. Shay, A potassium Faraday anomalous dispersion optical filter. Opt. Commun. **94**(1–3), 30–32 (1992)
11. T.M. Shay, B. Yin, L.S. Alvarez, Faraday anomalous dispersion optical filters. NASA STI/Recon Tech. Rep. A **95**, 829–836 (1993)
12. B. Yin, Q.C. Liu, L.S. Alvarez et al., Single transmission band Faraday anomalous dispersion optical filter, in *Conference on Lasers and Electro-Optics* (Optica Publishing Group, 1993), p. 1993CTuN52
13. L.K. Matthews, T.M. Shay, G.V. Garcia, Incorporation of a FADOF to an ESPI system, in *Industrial Applications of Optical Inspection, Metrology, and Sensing*, vol. 1821 (SPIE, 1993), pp. 226–231
14. J.H. Menders, E.J. Korevaar, Voigt filter, US005731585A.1998.5.24, 1998
15. I. Gerhardt, How anomalous is my Faraday filter? Opt. Lett. **43**(21), 5295–5298 (2018)
16. J.A. Gelbwachs, Y.C. Chan, Passive Fraunhofer-wavelength atomic filter at 460.7 nm. IEEE J. Quantum Electron. **28**(11), 2577–2581 (1992)
17. S.K. Gayen, R.I. Billmers, V.M. Contarino et al., Induced-dichroism-excited atomic line filter at 532 nm. Opt. Lett. **20**(12), 1427–1429 (1995)
18. Y. Zhang, X. Sun, Z. He et al., Theoretical investigation of induced-dichroism-excited atomic line filter, in *Information Optics and Photonics Technology*, vol. 5642 (SPIE, 2005), pp. 391–396
19. L.D. Turner, V. Karaganov, P. Teubner et al., Sub-Doppler bandwidth atomic optical filter. Opt. Lett. **27**(7), 500–502 (2002)
20. Z. He, Y. Zhang, S. Liu et al., Transmission characteristics of an excited-state induced dispersion optical filter of Rb at 775.9 nm. Chin. Opt. Lett. **5**(5), 252–254 (2007)
21. A. Cere, V. Parigi, M. Abad et al., Narrowband tunable filter based on velocity-selective optical pumping in an atomic vapor. Opt. Lett. **34**(7), 1012–1014 (2009)
22. Y. Wang, S. Zhang, D. Wang et al., Nonlinear optical filter with ultra-narrow bandwidth approaching the natural linewidth. Opt. Lett. **37**, 4059–4061 (2012)
23. Z. Xu, X. Xue, X. Zhang et al., Narrower atomic filter at 422.7 nm based on thermal ca beam. Chin. Sci. Bull. (28), 3543 (2014)
24. Y. Chan, J. Gelbwachs, A Fraunhofer-wavelength magnetooptic atomic filter at 422.7 nm. IEEE J. Quantum Electron. **29**, 2379 (1993)
25. D. Pan, X. Xue, H. Shang et al., Hollow cathode lamp based Faraday anomalous dispersion optical filter. Sci. Rep. **6**(1), 1–6 (2016)
26. M. Asano, K. Kubo, Vapor pressure of strontium below 660°K. J. Nucl. Sci. Technol. **15**, 765 (2008)
27. U. Dammalapati, I. Norris, E. Riis, Saturated absorption spectroscopy of calcium in a hollow-cathode lamp. J. Phys. B At. Mol. Opt. Phys. **42**, 165001 (2009)

28. Y. Shimada, Y. Chida, N. Ohtsubo et al., A simplified 461-nm laser system using blue laser diodes and a hollow cathode lamp for laser cooling of Sr. Rev. Sci. Instrum. **84**, 063101 (2013)
29. Z. Xu, X. Xue, D Pan et al., Narrower atomic filter at 422.7 nm based on thermal Ca beam. Chin. Sci. Bull. **59**, 3543–3548 (2014)
30. Y.C. Chan, J.A. Gelbwachs, A Fraunhofer-wavelength magnetooptic atomic filter at 422.7 nm. IEEE J. Quantum Electron. **29**, 2379–2384 (1993)
31. L. Zhang, Study on the working mechanism of active Faraday anomalous dispersion optical filter (FADOF) and the application of passive FADOF. Doctoral dissertation of Peking University, 1996
32. Y. Wang, *Laser Cooling and Trapping of Atoms* (Peking University Press, 2007)
33. B. Luo, *Principles and Technology of Atomic Filters* (Beijing University of Posts and Telecommunications Press, Beijing, 2018)
34. M.D. Rotondaro, B.V. Zhdanov, R.J. Knize, Generalized treatment of magneto-optical transmission filters. J. Opt. Soc. Am. B **32**(12), 2507–2513 (2015)

Chapter 5
Faraday Laser Scheme and Technology

Faraday laser uses a laser diode coated with an antireflection film as the gain medium, and a Faraday atomic filter as the frequency selection device. The laser frequency intrinsically corresponds to the atomic transition lines. Due to its significant quantum frequency selection technology, the Faraday laser has become increasingly important in the field of frequency-stabilized lasers. It has become a frequency-stabilized diode laser technology scheme with significant advantages, and has been widely used in atomic clocks, interferometers, gravimeters and other quantum precision measurement fields. It has significantly improved the technical level of related system equipment, and can further be extended to precision spectroscopy, atomic physics, atomic radar, laser weapons and other fields in the future. In this chapter, we introduce the overall technical scheme, optical cavity, circuit system, and main performance indicators of the Faraday laser.

5.1 Overall Technical Scheme

The Faraday laser is a special type of atomic frequency selection diode laser. Its main optical structure includes: laser diode, integrated atomic filter, collimator, feedback cavity mirror and related components, as shown in Fig. 5.1.

During operation, the higher the working current of the laser diode, the greater the heat power of the device, which causes the diode temperature to rise, affecting the performance of the diode. Therefore, it is necessary to perform heat sink analysis on the laser diode, use a thermistor as a temperature detector, and a thermoelectric cooling chip (TEC) as a temperature control device to feedback control the temperature of the laser diode to ensure efficient and stable operation of the diode. The fluorescence emitted from the laser diode has a divergence angle usually between 15 and 30°. It needs to be collimated with an aspherical lens of appropriate focal length

J. Chen et al., *Faraday Laser*, https://doi.org/10.1007/978-981-97-8023-5_5

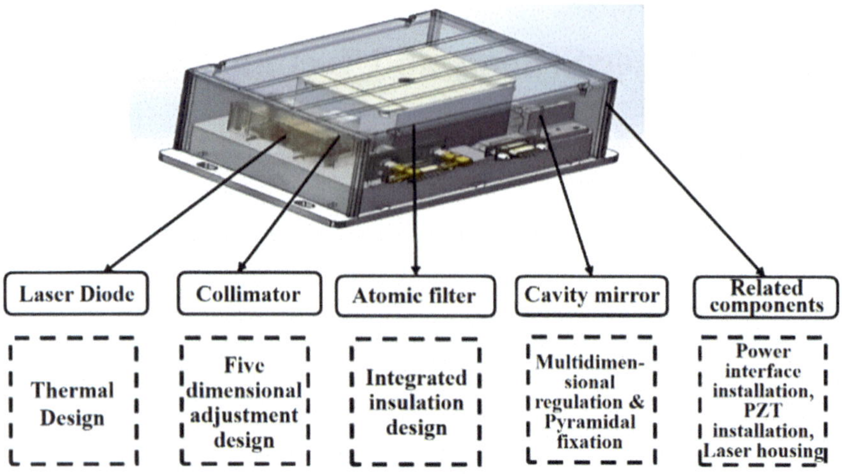

Fig. 5.1 Mechanical design of the Faraday laser

so that its light spot does not diverge within 2 ~ 3 m. The size of the light spot is determined by the focal length of the aspherical lens used. The larger the focal length, the larger the light spot. The most critical device of the Faraday laser is the atomic filter. Its principle has been described in detail in the previous chapter. In practical applications for frequency selection in Faraday lasers, once the kinds of atoms, buffer gas type, gas pressure magnitude, and magnetic field intensity are determined, the temperature of the atomic filter emerges as the most crucial factor influencing its performance. Consequently, a high-precision temperature control system is required. The same thermistor is used as a temperature detector, but because the working temperatures of both the Rb and Cs atomic filters are above room temperature, only heating is required. A smaller volume polyimide heating film is used as a temperature control device to ensure that the working temperature fluctuation of the atomic filter is less than 0.1 °C. After the collimated laser beam passes through the atomic filter, it is reflected by the feedback cavity mirror, and then reflected back to the laser diode. The better the light path coincides before and after the cavity mirror reflection, the better the laser oscillation effect, the greater the output laser power, and the better the frequency stability. Therefore, the mechanical stability of the feedback cavity mirror is extremely important, and can be optimized by appropriate mechanical design or the use of a cube corner retroreflector. After considering the size of the laser circuit board, the size of each interface, and the volume requirements, the overall size of the laser has been realized $(85 \times 50 \times 56)\ mm^3$, as shown in Fig. 5.2.

Depending on the type of the atom in the atomic filter, it is mainly divided into two categories, namely, Cs and Rb atom Faraday lasers, with corresponding laser wavelengths of 780 and 852 nm. Its operational parameters include atomic kinds, buffer gas composition and pressure, operating temperature, magnetic field magnitude and uniformity, as well as the length of the gas chamber, among others. The selection of

Fig. 5.2 Appearance of the laser

optimal operational parameters requires comprehensive consideration. Its working parameters include the type of atom, the composition and pressure of the buffer gas, the working temperature, the size and uniformity of the magnetic field, the length of the atomic cell, etc., and the best parameters need to be considered comprehensively. For the 852 nm cesium atom Faraday laser, the currently suitable magnetic field strengths are 1000 G, 1500 G, and 1750 G, respectively, corresponding to different atomic transition lines and tunable ranges, and the specific choice is determined by actual needs, as shown in Fig. 5.3. Correspondingly, for different magnetic field strengths, the working temperature of the atomic filter also needs to be adjusted appropriately, and there will be a large difference, usually setting between 55 and 75 °C.

Fig. 5.3 Internal structure of the laser

5.1.1 Overview of Atomic Frequency Selection Laser

Since the advent of diode lasers in 1962 [1–3], due to its high efficiency, small size, low cost, and tunability, it is widely used in fields such as laser cooling, atomic clocks, quantum optics, optical sensing, laser communication, etc. In some special application scenarios, diode lasers with narrower linewidth and larger tunable range are needed. Traditional diode lasers, such as Distributed Bragg Reflector Lasers (DBR) and Distributed Feedback Lasers (DFB), are difficult to further reduce the linewidth due to the influence of large optical cavity loss and spontaneous radiation of gain medium. At the same time, the tuning methods based on carrier dispersion effect or thermal optical effect have a tuning range of several nanometers, which can meet most application needs. With the urgent demand for narrow linewidth, high frequency stability, and large tunable range diode lasers, in 1978, Velichansky and others found that by using an external reflection mirror to provide feedback, i.e., constructing an external resonant cavity, the linewidth of the diode laser can be effectively narrowed, and the working principle and method of the external cavity diode laser were proposed [4].

The frequency selection of traditional external cavity diode lasers mainly includes three methods: one is the grating external cavity diode laser, which introduces a grating as the optical feedback and frequency selection component of the diode laser, increases the quality factor (Q value) of the resonant cavity, and also has the effect of narrowing the linewidth of the laser. Grating external cavity diode laser can be divided into diffraction grating external cavity diode lasers, fiber grating external cavity diode lasers, volume holographic grating external cavity diode lasers, and volume Bragg grating external cavity diode lasers, etc., according to the type of grating. Another is the filter-type external cavity diode laser, which uses devices with optical filtering functions as the frequency selection components of the external cavity diode laser, such as Fabry–Pérot etalons, interference filters, acousto-optic and electro-optic tunable filters, birefringent filter, etc. There is also a waveguide-type external cavity diode laser, generally composed of an amplifier and an optical waveguide coupling, the optical waveguide as an external cavity can achieve frequency selection. The optical waveguide is generally designed as two micro-ring structures with slightly different radii. The double micro-ring structure has a difference in the free spectral range (FSR) due to the different radii, and the output spectrum overlapping one other, and then only one laser mode survives and outputs after mode competition.

In recent years, narrow linewidth diode lasers with frequencies corresponding to atomic transition lines have been widely used in the fields of atomic physics and quantum precision measurement. For the above-mentioned traditional external cavity diode lasers, the output frequency of the laser depends on the parameters of the frequency selection device, such as the angle between the interference filter and the incident light, the angle between the grating and the incident light, the cavity length of the Fabry–Pérot etalon, etc., and precise control of the parameters of the frequency selection device is needed to make the output frequency of the laser correspond to the target atomic transition line. In addition, external vibrations and temperature fluctuations affect the long-term continuous operation of the laser, and manual adjustment and control of the parameters of the frequency selection device are needed to restore the initial working frequency, which affects the actual use efficiency.

To meet the demand for automatic corresponding atomic transition lines in fields such as quantum precision measurement, atomic filters can be used as frequency selection devices to realize Faraday lasers. These lasers have the advantage of directly corresponding to the atomic transition line, and are ready-to-use narrow linewidth diode lasers. Using the Faraday laser as a laser resource, turn-key atomic devices such as atomic clocks and atomic gravimeters can be built. In this case, the development cost and usage difficulty of these high-end scientific instruments related to atoms can be reduced.

Due to the characteristics of atomic filters, as described in Chap. 4 of this book, only wavelengths corresponding to atomic transition lines can pass through the filter. The output wavelength of this laser can always be limited within the transmission bandwidth of the atomic filter, directly corresponding to atomic transition lines, ensuring the long-term continuous operation of the laser. Moreover, when the parameters of the laser change, the output wavelength of the laser still remains within the transmission bandwidth of the atomic filter, making it immune to fluctuations in the driving current and operating temperature of the laser diode.

Generally, the direction of the magnetic field in the atomic filter is parallel to the direction of light propagation. This structure of atomic filter is called a Faraday anomalous dispersion atomic filter (FADOF), and the diode laser using it as a frequency selection device is called a "Faraday laser". In addition, there is another type of atomic filter, where the direction of the magnetic field is perpendicular to the direction of light propagation. This structure of atomic filter is called a Voigt anomalous dispersion atomic filter (VADOF), and the diode laser using it as a frequency selection device is called a "Voigt laser". Since the Faraday laser was realized by Miao of Peking University in 2011 [5], it has received widespread attention due to its excellent performance, and related work progress and breakthroughs have followed one after another. The research on Voigt lasers was not proposed and realized by Liu of Peking University until 2023 [6]. Therefore, when introducing this type of laser, the main content is about the Faraday laser, while the Voigt laser will be introduced separately in Sect. 5.1.4. Further, the development process of the Faraday laser can be roughly divided into three stages: exploration and preparation, establishment and naming, and maturity and application. The main work of each stage is shown in Table 5.1.

Table 5.1 Stages of Faraday laser development

Stage	Time	Milestone work
Exploration and preparation	1845 ~ 2011	Faraday rotation effect, proposal and realization of Faraday anomalous dispersion atomic filter, use of atomic filter for frequency stabilization of dye lasers
Establishment and naming	2012 ~ 2018	Realization of Faraday laser and naming of "Faraday laser"
Maturity and application	2019 ~ present	Implemented Faraday lasers corresponding to different alkali metal atom transition lines, used in atomic clocks, gravimeters, and other equipment. Developed brand new Voigt lasers and dual-frequency Faraday lasers

5.1.2 Magneto-Optical Faraday Effect Lasers

1. Exploration and preparation stage of Faraday lasers (1845 ~ 2011)

The exploration and preparation stage of Faraday lasers began in 1845 when Faraday first discovered the Faraday effect, an experimental discovery that linearly polarized light undergoes polarization rotation in a medium under the influence of a magnetic field [7]. In 1956, Öhman, based on the Faraday effect, first realized and applied the Faraday anomalous dispersion atomic filter [8]. In September 1969, Sorokin and others at the IBM Research Center in the United States placed a Faraday filter in the resonant cavity of a dye laser, controlled the heating temperature of the Na atomic cell at 170 ~ 220 °C, the atomic density was between 5×10^{11} and 3×10^{12} cm^{-3}, by applying a magnetic field, the direction of light polarization can rotate an integer multiple of 90° in a 25 cm long cell, successfully stabilizing the frequency of the dye laser on the Na atom resonance absorption line. This work provided the possibility for dye lasers to be locked to the transition lines of other atoms [9].

The research on the combination of Faraday filters and diode lasers began with the use of optical loops to lock the output frequency of diode lasers to atomic transition lines. From the emergence of diode lasers in 1962 to the late 1980s and early 1990s, nearly three decades, with the widespread application of diode lasers in remote sensing, communication, and other fields, the requirements for the performance of diode lasers, such as output mode, laser linewidth, frequency stability, etc., are getting higher and higher. In this context, technologies such as using gratings, F-P cavities, etc., to construct external cavity narrowing laser linewidth, using saturated absorption spectroscopy to construct electrical feedback locking loops with atomic transition lines as frequency references, have developed rapidly, thereby further improving the frequency stability and other performances of external cavity diode lasers. At the same time, due to the excellent background noise filtering effect, Faraday filters are playing an increasingly important role in applications such as communication reception. They usually only have a GHz-level passband, so they also have higher requirements for the linewidth of the emitted laser, so the electrical feedback locking loop technology is indispensable. Although the electrical feedback loop can improve

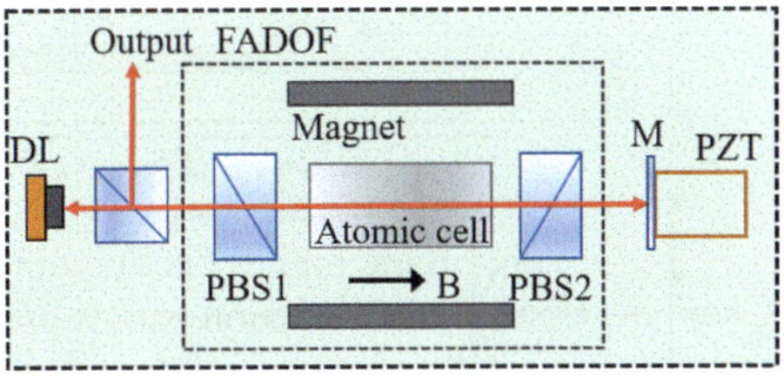

Fig. 5.4 Principle diagram of FADOF optical loop lock construction. *DL* laser diode with anti-reflection coating, *PBS1* polarizing beam splitter 1, *PBS2* polarizing beam splitter 2, *M* cavity mirror, *PZT* piezoelectric ceramic

the performance of the external cavity diode laser, the electrical feedback locking loop system is relatively complex and costly, so its application scenario is relatively limited. In this context, thanks to the excellent performance of the Faraday atomic filter and its increasingly important role in communication, the technology of constructing diode laser optical loop locking with the Faraday filter as an important component is born. Its principle diagram is shown in Fig. 5.4. By reflecting the light emitted by the diode laser and passing through the FADOF with a very low reflectance mirror, an optical loop is formed to achieve locking.

In December 1990, scientists Lee and Campbell from the University of Texas at Austin in the United States used a Rb Faraday atomic filter to construct a narrow-band optical feedback loop, achieving frequency locking of the $Al_xGa_{1-x}As/GaAs$ material 780 nm diode laser [10]. The research team used FADOF and a low reflectivity (~ 10^{-5}) mirror to construct a feedback loop, allowing the output light to return to the laser only when it is close to the atomic resonance frequency. Through the construction of the optical loop, not only is the laser linewidth narrowed, but the laser frequency is also locked to the sub-Doppler feature of the Rb atomic spectrum. The team used an optical heterodyne detection method to evaluate the laser linewidth, and found that through the construction of the optical loop feedback, the laser linewidth decreased from 30 MHz to 35 kHz. In this work, the light intensity (BS1 transmission) used to construct the feedback loop is 50 mW/cm^2, accounting for about 90% of the diode laser output light. Although there was a large energy loss, the diode laser frequency was locked based on the optical loop feedback. Subsequently, based on the above technology, the team narrowed the laser linewidth to 500 kHz while stabilizing the frequency. In the experiment, they used heterodyne interference technology, that is, using two systems for beat frequency, and completed the first measurement of the laser output frequency stability (Allan deviation) using this method. When the average time is 200 s, an Allan deviation of 4.9×10^{-12} can be obtained.

2. Faraday laser establishment and naming phase (2012 ~ 2018)

In August 2011, Miao and others from Peking University used Rb atom FADOF as a frequency selection component, analyzed the advantages and disadvantages in previous research, and for the first time used a laser diode with anti-reflection coating

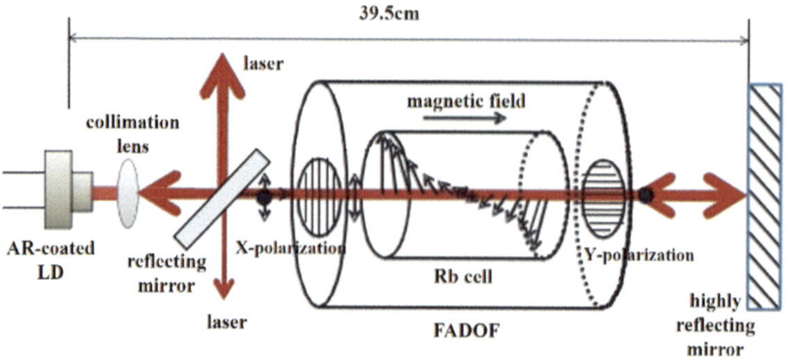

Fig. 5.5 Principle diagram of Faraday laser with anti-reflection coated laser diode

to build a Faraday laser. Under the condition of 6.4% feedback intensity, they achieved a diode laser immune to fluctuations in the diode current and temperature, with a laser linewidth of 69 kHz [5].

The experimental device is shown in Fig. 5.5. In order to eliminate the influence of the internal cavity mode of the laser diode and stabilize the laser frequency corresponding to the atomic transition line, the team creatively used a laser diode with anti-reflection coating (Anti-reflection coated laser diode, ARLD) as a gain medium. Without feedback, the emission spectrum of this laser diode is a 60 nm wide gain fluorescence spectrum, avoiding the need to heat or cool the laser diode to tune its emission spectrum characteristics.

The light emitted by the laser diode with an anti-reflective coating is collimated by a lens, passes through a beam splitter prism, then through an Rb atomic Faraday atomic filter, and is returned to the laser diode by a cavity mirror with a reflectivity of up to 98.3%. The diameter of the Rb atomic cell in the Faraday atomic filter is 20 mm, and the length is 30 mm. The atomic cell contains naturally abundant Rb atoms, with ^{85}Rb accounting for 72.2% and ^{87}Rb accounting for 27.8%. The temperature of the Rb atomic cell is controlled at 70 °C with a temperature control accuracy of 0.1 °C by a circuit, and an axial magnetic field of 320 G is applied at the same time, causing the polarization direction of the incident light to rotate. Only light that satisfies the rotation of the polarization direction to $n\pi + \pi/2$ can pass through the Faraday atomic filter, achieving atomic frequency selection. Under these conditions, the peak transmission of the Faraday atomic filter is 51%. In this work, a diode laser is built by combining the Faraday atomic filter with the laser diode with an anti-reflective coating, resulting in a Faraday laser that is immune to fluctuations in diode current and temperature. When the laser diode current varies within 55 ~ 142 mA and the diode temperature varies within the range of 15 ~ 35 °C, the laser output wavelength remains stable within 2 pm, as shown in Fig. 5.6. The fully functional Faraday laser reported in this work has broaden application prospects in many fields such as quantum precision measurement, military metrology, remote sensing, and optical communication.

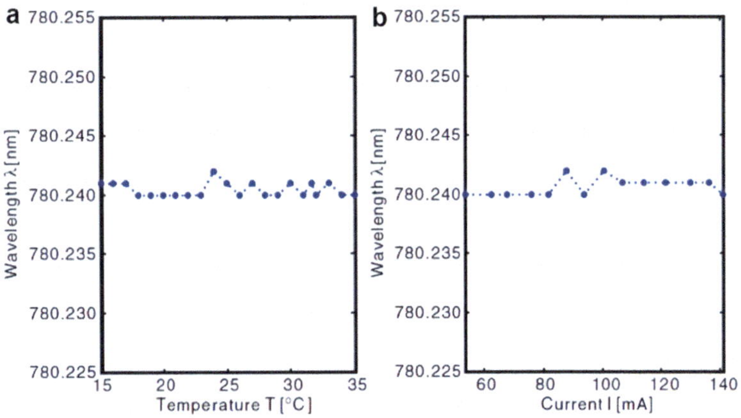

Fig. 5.6 Faraday laser performance: **a** Faraday laser output wavelength changes with diode temperature control; **b** Faraday laser output wavelength changes with diode drive current

In 2015, Tao and others from Peking University first named the Faraday laser [11] internationally. The team used an Rb atomic FADOF as an atomic frequency selection device and a laser diode with an anti-reflective coating as a gain medium to achieve a laser that is immune to fluctuations in laser diode temperature and current. Considering the problems that exist when using natural Rb, namely that the presence of different isotopes makes it difficult to limit the laser frequency to precisely resonate with a specific atomic transition, they chose to use ^{87}Rb atomic isotopes with a concentration of 96.5%, a cell length of 5 cm, and a polarization extinction ratio of 1×10^{-5} for the two Glan–Taylor prisms. In the experiment, by measuring the transmission spectrum of the FADOF at a constant temperature under different magnetic fields, and measuring the transmission spectrum at different temperatures under a constant magnetic field, they analyzed the effects of the magnetic field and temperature on the highest transmission of the transmission spectrum, thereby determining the optimal operating parameters of FADOF to obtain the best laser performance. This Faraday laser can achieve a change in the drive current of the laser diode from 55 to 140 mA, a temperature change from 15 to 35 °C, and the laser output wavelength always near 780.2456 nm, with a wavelength fluctuation standard deviation less than or equal to 0.27 pm, which can correspond to the atomic transition line.

In 2016, Tao and others from Peking University used 1.2 km optical fiber serving as an extended cavity for the diode laser, using the Rb atomic FADOF as the atomic frequency selection component, always limiting the laser frequency near the cross peak for $5S_{1/2}F = 4 \rightarrow 5P_{3/2}F' = 1, 3$, realizing the ultra-long cavity Faraday laser [12]. In this work, the laser output wavelength is first tuned to the 780 nm wavelength corresponding to the natural Rb atom $5S_{1/2} \rightarrow 5P_{3/2}$ transition line, and then the 5 mW output from the laser is divided into two parts by a spectroscope, one part is used for saturated absorption spectroscopy to provide frequency reference, the other part is divided into two parts by a polarizing beam splitter, where the probe light is used to form an extended cavity and provide optical feedback, the pump light becomes circularly polarized after passing through a quarter-wave plate, and then interacts with the probe light in the Rb atomic cell. By extending the cavity length to 1.2 km through the fiber, the relative frequency stability of the Faraday laser reaches 10^{-10} within a sampling time of 0.06 ~ 1 s, and the long-term relative frequency stability reaches 10^{-11} without controlling the fiber length.

In 2016, Keaveney and others from Durham University in the UK, using the Rb atomic FADOF as the atomic frequency selection component, achieved a Faraday laser with a linewidth better than 400 kHz [13]. The team used a laser diode with an anti-reflection coating, the back end of the laser diode and a reflection cavity mirror with a 90% reflectivity form a resonant cavity, the cavity length is 7 cm, and the free spectral range is 2.1 GHz. The atomic cell in the Faraday anomalous dispersion atomic filter is also coated with an anti-reflection film, with a length of 5 mm, a magnetic field size of 250 G, and a heating temperature of 90 °C. After forming the Faraday laser, by monitoring the output power of the Faraday laser, i.e., whether the laser cavity mode always coincides with the highest point of the transmission spectrum, it serves as a PID feedback signal for electrical feedback control of the Faraday laser. In the experiment, the team also measured the linewidth of the Faraday laser by beating with a commercial external cavity diode laser locked to the $5S_{1/2}F = 3 \rightarrow 5P_{3/2}F' = 4$ transition line.

In 2017, Chang and others from Peking University used a discharge lamp to realize the excited state Faraday laser for the first time, which can directly output the Faraday laser corresponding to the Rb atom $5P_{3/2} \rightarrow 4D_{5/2}$ transition at 1529 nm [14]. Unlike traditional excited-state FADOF that require additional frequency-locked lasers to excite atoms from the 5S state to the 5P state, this work is based on an excited-state FADOF implemented with a discharge lamp for atomic frequency selection. Combined with a laser diode coated with an anti-reflection film, the laser diode's driving current can stably operate within a range of 85–171 mA, and the temperature of the laser diode can vary from 11 to 32 °C. The long-term frequency fluctuation of the laser within 24 h is maintained within 600 MHz, as shown in Fig. 5.7. This work also measured the laser linewidth and relative intensity noise (RIN). The back end of the laser diode and an 80% reflective cavity mirror form a resonant cavity with a length of 60 cm and a free spectral range (FSR) of 250 MHz. In the excited-state FADOF based on the discharge lamp, the length of the Rb atomic cell is 15 cm and the diameter is 2 cm. The cell is filled with natural Rb and buffer gas, the buffer gas is 2 torr Xe, and the magnetic field used is 500 G. When the cell is heated to 135 °C, the maximum transmission reaches 46%, forming an oscillating output laser. The Faraday laser, by beating with a fiber Bragg grating laser with a nominal linewidth of 1 kHz, measures the linewidth of the Faraday laser to be 15.5 kHz, and the laser intensity noise is better than – 110 dBc/Hz at 10 kHz. The emergence of this Faraday laser provides a good solution for some Faraday lasers at excited-state wavelengths to be realized, making the system more compact and reliable, and able to provide lasers corresponding to atomic excited-state transition lines, with great application potential.

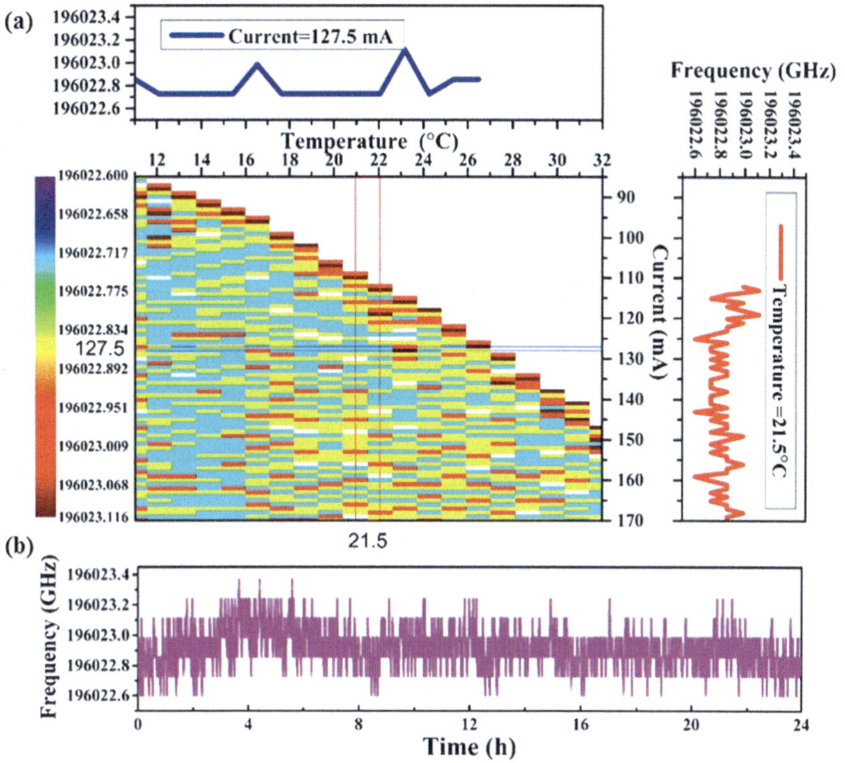

Fig. 5.7 Performance of the excited-state Faraday laser: **a** the output frequency of the Faraday laser changes with the diode drive current and temperature control; **b** when the diode drive current is 141.5 mA, the temperature control is 21.5 °C, and the FADOF temperature control is 135 °C, the output frequency of the Faraday laser changes

In 2018, Rotondaro and others from the United States Air Force Academy used the Cs atom FADOF's $6^2S_{1/2} \rightarrow 6^2P_{3/2}$ transition to achieve frequency selection and output based on a 600 W anti-reflective coated diode laser. As an atomic frequency selection component, the FADOF achieved a laser output power of 518 W, narrowing the spectrum of the laser diode from 3 THz to 10 GHz [15]. The FADOF was placed in a resonant cavity formed by the back end of the laser diode array and a 20% reflective output cavity mirror. The 6-in. Cs atomic cell was evacuated to 10^{-7} torr and heated to 96 °C, corresponding to an atomic density in the cell of 10^{13} cm^{-3}, and a 700 G magnetic field was used for optimal performance. When the FADOF was placed in the resonant cavity, the spectral width was narrowed from 3 THz to 10 GHz. Due to the extremely high laser intensity in this work, there is significant optical pumping of the ground state atoms, leading to a decrease in alkali metal density, and therefore the transmission spectrum does not quite match the theoretical calculations. The transmission spectrum of FADOF under these conditions is dual-frequency, and the decrease in alkali metal density caused by optical pumping causes the dual transmission spectrum to shift towards the center. This provides a reference for the shape of the transmission spectrum of the FADOF under extremely high laser intensity.

3. Faraday laser maturity and application stage (2019 ~ present)

In 2019, Chang and others from Peking University used Cs atom FADOF as an atomic frequency selection device, realized a Faraday laser immune to temperature and current fluctuations of the laser diode based on Cs atom 852 nm [16]. The 40 cm long resonator is composed of the back end of the laser diode and a high reflectivity plane resonator mirror. The atomic cell in the FADOF is a cylinder with a diameter of 15 mm and a length of 30 mm. By measuring the transmission of the transmission spectrum corresponding to atomic transitions at different temperatures and magnetic fields, the optimal working conditions are determined to be a temperature of 41 °C and a magnetic field of 330 G. The temperature control accuracy is 0.1 °C, and the inhomogeneity of the magnetic field can be ignored at 7%. Finally, a transmission spectrum with a bandwidth of 800 MHz is obtained. When the operating temperature of the laser diode changes within the range of 14 ~ 35 °C, and the drive current changes within the range of 74 ~ 130 mA, the output wavelength of the laser always remains near the atomic transition line. Regarding wavelength stability, the wavelength change is kept within ± 2 pm within 48 h. By beating with another external cavity diode laser, the linewidth of the Faraday laser is measured and calculated to be 17 kHz. This work provides a practical possibility for some application scenarios that require the laser wavelength to correspond to the atomic transition spectrum and have good long-term stability, and has important practical application value.

In 2020, Shi and others from Peking University used the Cs atom FADOF as a frequency selection component, and for the first time realized a free-running Faraday laser with dual-frequency output, with each frequency linewidth is less than 33 kHz, and the most probable beat frequency linewidth of the dual-frequency is 902.95 Hz. The dual-frequency laser has potential applications in precision measurement, dual-frequency interference, and other fields [17]. The 69 cm long resonator is composed of a laser diode back-end face with an anti-reflection coating and a reflection cavity mirror. The FADOF is placed in the resonator, and its transmission spectrum profile is determined by the magnetic field size, atomic cell length, and atomic cell temperature. By adjusting the parameters, the transmission peak corresponds to the atomic spectrum line and has a high transmission, thus forming a dual-frequency laser output during the frequency selection process. In this work, the magnetic field size is 330 G, the atomic cell is 30 mm long, and is filled with pure Cs atom. The atomic cell temperature varies within the range of 36 ~ 55 °C, and the frequency difference between the dual frequencies is adjustable within the range of 8.8 ~ 7.4 GHz. The intensity noise of the dual-frequency Faraday laser is also measured in this work, which is found to be − 134 dBc/Hz at 10 kHz. The realization of the dual-frequency Faraday laser shows excellent application potential, such as by precisely designing the resonator length, making the frequency interval between the dual frequencies exactly an integer multiple of FSR, as a coherent light source for coherent population trapping atomic clocks. The dual-frequency Faraday laser can also be analogized with the optical frequency comb, connecting optical frequency and microwave frequency. The emergence of the dual-frequency Faraday laser expands a new research field for Faraday lasers and has an important value.

In 2021, Tang and others from the National University of Defense Technology used a FADOF based on ^{85}Rb as an atomic frequency selection component, and a laser diode array with an output power of 22 W composed of 9 steps as the light source, achieving a Faraday laser output of 18 W under a drive current of 4.9 A. At the same time, the laser output spectrum is narrowed from 4 to 0.002 nm (1.2 GHz) [18]. To avoid mode competition, the output surface of the laser diode module is coated with an anti-reflection film with a reflectivity of less than 0.5%. In the experiment, the resonant cavity is composed of the back end of the laser diode and a total reflection cavity mirror, and the FADOF is placed in it. After getting the laser output, it is first collimated by a pair of lenses with focal lengths of 5 and 15 cm, and then controlled by a half-wave plate and a polarizing beam splitter to control the light intensity (transmission) and output light intensity (reflection) ratio to find the best working conditions. In this work, the FADOF uses a 124 G magnetic field, the atomic cell is a cylinder with a length of 10 mm and a diameter of 25.4 mm, filled with pure ^{85}Rb atom. To improve the transmission, the outer end faces of the two light windows are coated with an anti-reflection film. A resistance wire with a temperature control accuracy of 1 °C is used to heat the atomic cell, and the transmission spectrum of the FADOF at different temperatures is tested to determine that 139.8 °C is the best working temperature. While determining the best power for feedback, since it is a strong light condition, the transmission spectrum of the FADOF under the same temperature and magnetic field conditions is significantly different from that under weak light due to the presence of optical pumping. The team placed a 1/2 plate within the resonant cavity and determined the optimal operating conditions by rotating the 1/2 plate to find a balance between laser output power and the suppression effect of other modes formed within the FADOF. Finally, it is concluded that the best working condition was when the transmission and reflection energy ratio of the pre-set polarizing beam splitter inside the FADOF is 1:4, effectively avoiding the thermal deposition effect. Although it is a dual-frequency output, 90% of the energy is concentrated in one mode, and the other mode only has 10% of the energy. Finally, the output Faraday laser is limited to the Rb atom D_2 line, achieving a laser output wavelength always near the Rb atom transition line at 780.244 nm with the laser diode module temperature varying between 16 and 30 °C and the drive current varying between 1 and 5 A. This high-power Faraday laser is applicable in alkali metal laser pumping, metastable rare gas laser pumping, spin-exchange optical pumping, and quantum optics.

In 2022, Chang and others from Peking University locked the output frequency of the Cs atom 852 nm Faraday laser to the atomic transition spectrum line through modulation transfer spectroscopy. In this work, the Faraday laser can be adjusted to single-frequency or dual-frequency output. When working in single-frequency, the remaining error is evaluated after locking, and the short-term stability reaches $3 \times 10^{-14}/\sqrt{\tau}$; when working in dual-frequency, the beat frequency linewidth of the dual-frequency laser can be narrowed to 85 Hz through locking [19]. The experimental setup is shown in Fig. 5.8. The resonant cavity is formed by the back end of a laser diode with an anti-reflective coating and a reflective cavity mirror. A FADOF is placed in the resonant cavity for frequency selection. The magnetic field of the FADOF used in this work is 330 G, with a inhomogeneity of 7%; the atomic cell is a cylinder with a length of 30 mm and a diameter of 15 mm, filled with pure Cs atom; the cell is heated to 38 °C, which can produce transmission spectra with transmittance of about 50% and 25% at $F = 4$ and $F = 3$, respectively, allowing the laser to produce dual-frequency output. By precisely adjusting the working conditions, such as the driving current of the laser diode and the heating temperature of the cell, it can also operate in a single-frequency state. After the Faraday laser output, it passes through an isolator to avoid external interference with the resonant cavity. It is then divided into transmission (probe light) and reflection (pump light) parts by a polarizing beam splitter, with powers of 0.5 mW and 1.5 mW, respectively, for frequency locking using modulation transfer spectroscopy. The atomic cell used in the modulation transfer spectroscopy optical path has a cylindrical shape with a diameter of 2 cm and a length of 10 cm, and its temperature is controlled at around room temperature of 25 °C. In the single-frequency state, the laser frequency is locked to the cesium atom $6^2S_{1/2}F = 4 \rightarrow 6^2P_{3/2}F' = 5$ cycling transition line through modulation transfer spectroscopy, greatly improving the frequency stability of the Faraday laser; in the dual-frequency state, one of the laser frequencies is locked to the cesium atom $6^2S_{1/2}F = 3 \rightarrow 6^2P_{3/2}F' = 2$ transition line, effectively narrowing the beat signal linewidth between the dual-frequency lasers, reducing the noise of the microwave signal, and opening up a new way for optically generated microwaves. This work is of great significance for scientific research.

In 2022, Tang and others from the National University of Defense Technology used a FADOF based on ^{87}Rb as an atomic frequency selection component, and a laser diode module with an output power of 47 W (driving current of 5 A) composed of 18 units as the light source, achieving a Faraday laser output of 38.3 W under a driving current of 5 A, while narrowing the laser output spectrum from 4 to 0.005 nm (2.6 GHz) [20]. To avoid mode competition under free running conditions, the back end of the laser diode module is coated with an anti-reflection film, which has a reflectivity of less than 0.5%. The experimental setup is shown in Fig. 5.9. In order to improve the frequency selection related characteristics of the Faraday anomalous dispersion optical filter (FADOF), the team adopted the ring resonator structure shown in Fig. 5.9 to increase the external cavity efficiency and ensure high power output. The FADOF magnetic field in this work is 111 G, provided by two cylindrical magnets with an inner diameter of 55 mm, an outer diameter of 80 mm, and a length of 10 mm; the atomic cell is cylindrical with a length of 20 mm and a diameter of 25.4 mm, filled with pure ^{87}Rb atom. To improve the transmission, the outer end of the two light windows is coated with an anti-reflection film. The atomic cell is heated by a resistance wire with a temperature control accuracy of 1 °C. By obtaining the transmission spectrum of FADOF at different temperatures and comparing them, 80 °C is determined to be the optimal working temperature. Under the above working conditions, stable output of high-power Faraday laser was achieved. With the temperature of the laser diode module varying between 28 and 38 °C and the driving current

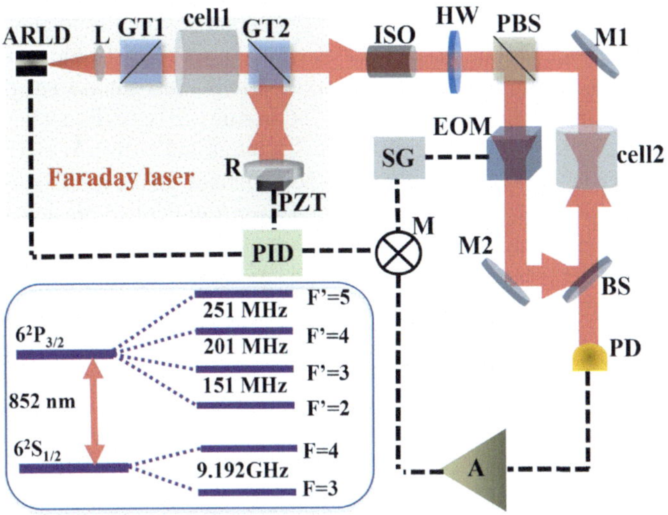

Fig. 5.8 Experimental setup diagram. *ARLD* anti-reflective coated laser diode, *L* lens, *GT1* Glan–Taylor prism 1, *GT2* Glan–Taylor prism 2, *R* cavity mirror, *PZT* piezoelectric ceramic, *PID* proportional integral derivative feedback circuit, *PBS* polarizing beam splitter, *EOM* electro-optic modulator, *SG* signal generator, *A* amplifier

varying between 1.5 and 5 A, the laser output wavelength is always around 780.23 nm (Rb atomic spectral line). Although the laser output is multi-frequency under this optimal condition, the laser energy is well concentrated after the output spectrum is narrowed, which has important potential application value for diode-pumped alkali metal lasers, quantum optics, etc.

In 2022, Shi and others from Peking University achieved Faraday laser based on Cs atomic Faraday anomalous dispersion optical filter, and locked its output frequency using modulation transfer spectroscopy, obtaining a residual error evaluation of the output laser frequency short-term stability of $5.8 \times 10^{-15}/\sqrt{\tau}$, and 2.0×10^{-15} [21]

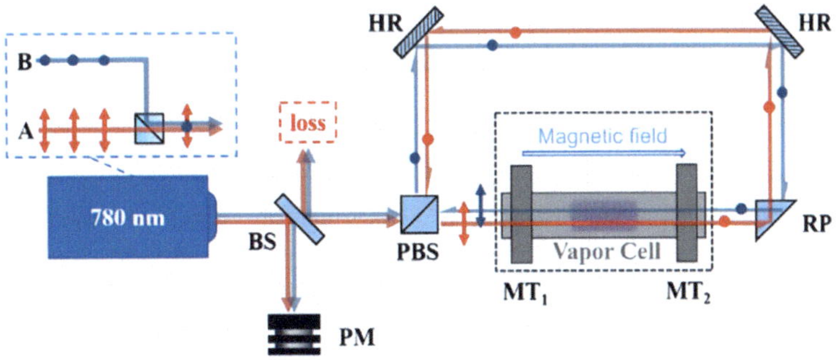

Fig. 5.9 Experimental setup

at 50 s. The experimental setup is similar to Fig. 5.8. The 155 mm long resonator is composed of the back end of the laser diode coated with an anti-reflection film and a reflection cavity mirror, with the FADOF placed in the resonator. In this work, the FADOF uses a magnetic field size of 900 G. The atomic cell is a cylinder with a length of 30 mm and a diameter of 15 mm, filled with Cs atoms and 10 torr Ar as a buffer gas. The cell is heated to 72 °C to ensure optimal performance. In the modulation transfer spectroscopy section, the reference atomic cell has a length of 11 cm and a diameter of 2.6 cm cylindrical body filled with pure Cs adopts a double-layer structure, with a vacuum between the two layers of glass, which can effectively suppress temperature fluctuations in the reference atomic cell, thereby reducing the impact on frequency stability. When locking the frequency using modulation transfer spectroscopy, the effect of different pump powers on frequency stabilization is measured by fixing the detection power, in order to find the optimal pump power. Then, with the optimal pump power fixed, the effect of different detection powers on frequency stabilization is measured to determine the optimal values of pump power and detection power for the best frequency stabilization effect. This work improves the stability of the Faraday laser output frequency by filling the cell of FADOF with buffer gas and by using a double-layer atomic cell as the reference signal source in the modulation transfer spectroscopy. It has a wide range of applications in quantum optics, quantum precision measurement, and other fields.

Currently, Faraday lasers are rapidly developing due to their excellent performance and huge application potential, and have made a series of important progress. Although there are still some research difficulties, such as how to accurately obtain the transmission spectrum of FADOF when forming a resonant cavity inside the laser, how to avoid the influence of atomic depolarization inside FADOF on the performance of Faraday lasers, and how to achieve high power, narrow linewidth, single longitudinal mode Faraday lasers, which can inevitably be solved with the continuation of research. At the same time, in the future development of Faraday lasers, cross-integration play an important role. Due to the performance characteristics of Faraday lasers, research combining them with precision spectroscopy and cold atom physics becomes an important research direction for future Faraday lasers.

5.1.3 Induced Dichroism Faraday Laser

The induced dichroism Faraday laser uses an induced-dichroism atomic line filter which has been discussed in Sect. 4.2.5 as a frequency selection component, a laser diode with an anti-reflection coating as a gain medium, combined with resonant cavity feedback, to achieve laser output. The induced-dichroism atomic line filter, as a core device, is different from the FADOF and the VADOF based on the principle of magneto-optic birefringence. It does not need to apply a magnetic field and specifically uses circularly polarized light to polarize the pump to generate induced birefringence for light transmission. Therefore, for some special application scenarios sensitive to magnetic fields, the induced birefringence Faraday laser is more suitable.

The basic structure of the induced dichroism Faraday laser is shown in Fig. 5.10. Two polarizers G1 and G2 are orthogonal to each other, with the atomic cell located between them. The linearly polarized light emitted by the laser diode becomes σ^+ or σ^- circularly polarized pump laser after passing through the quarter-wave plate, and passes through the atomic cell in reverse. Taking the σ^+ circularly polarized pump laser as an example, the working principle of the induced birefringence laser is specifically explained in conjunction with Fig. 5.11.

Under the action of σ^+ circularly polarized pump lasers, ground state atoms gradually populate the highest magnetic quantum number from a uniform distribution in a thermal equilibrium state, causing the atoms to accumulate at the highest magnetic level. This asymmetric atomic distribution allows the right-handed circularly polarized component of the linearly polarized light emitted by the laser diode to exist on the excitable upper energy level, thereby resonating with the atoms and changing the refractive index. However, the left-handed circularly polarized component cannot resonate with the atoms due to the absence of an excitable upper energy level, so its refractive index remains unchanged. This is the process of induced dichroism (photo-induced birefringence). Similar to the FADOF and VADOF, based on the birefringence effect, the phase difference brought by the right-handed and left-handed circularly polarized components after propagating a certain distance in the atomic cell causes the polarization plane of the superimposed linearly polarized output light

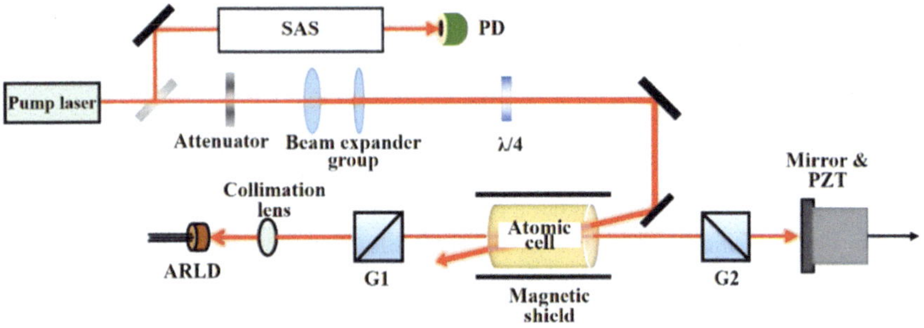

Fig. 5.10 Basic structure of the induced dichroism Faraday laser

Fig. 5.11 Energy level structure of the ground state induced dichroism Faraday laser

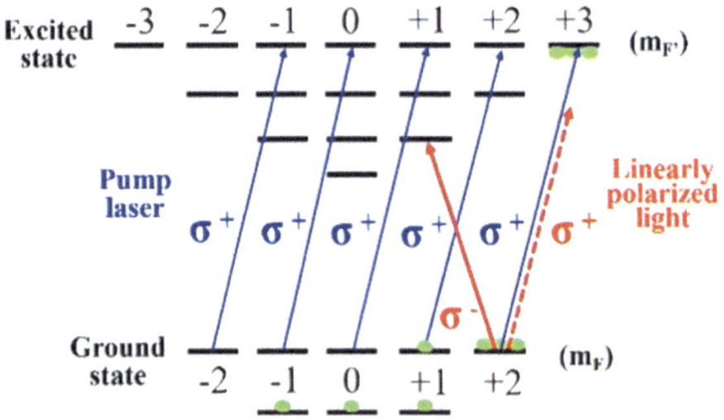

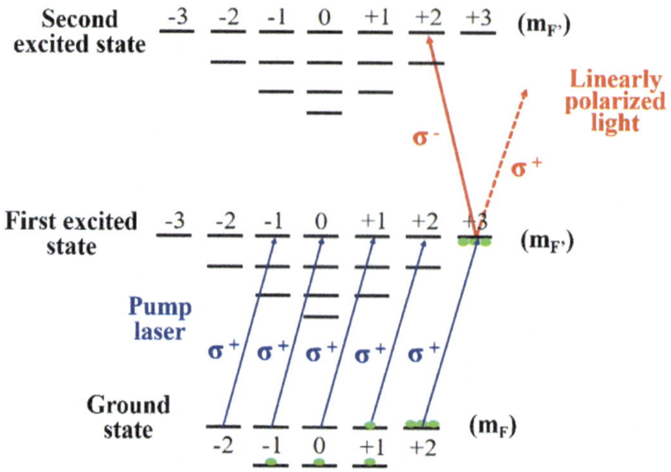

Fig. 5.12 Energy level structure of the excited state induced dichroic Faraday laser

to rotate, and then forms a transmission spectrum through the second polarizer G2. Afterwards, under the feedback resonance of the back surface of the laser diode and the coupling cavity reflector, the laser output of the induced dichroic Faraday laser is formed. Here, it should be noted: the induced dichroic Faraday laser only works when the maximum magnetic quantum number of the upper transition level is not greater than the maximum magnetic quantum number of the lower transition level, only then can induced dichroism be produced.

In addition, the induced dichroic Faraday laser can also work between excited states. As shown in Fig. 5.12, the σ+ circularly polarized pump laser continuously pumps to make the atoms only have a certain population distribution at the highest magnetic level of the first excited state. Next, similar to the ground state induced dichroic laser, the linearly polarized light emitted by the laser diode produces induced dichroism between the first and second excited states, and finally achieves oscillation output with the feedback of the resonant cavity. In comparison to the Faraday laser based on the magneto-optical effect, the induced dichroic Faraday laser does not need to apply a magnetic field, which simplifies the structure of the laser to a certain extent. The disadvantage is that the induced dichroic Faraday laser requires an additional pump laser, which also brings some limitations to its application.

5.1.4 Voigt Laser Based on Magneto-Optical Effect

As mentioned in Sect. 5.1.1, the performance of the Voigt laser is similar to the latest reported Faraday laser, with its frequency limited within the 1 GHz transmission bandwidth of the atomic filter. Therefore, regardless of how the current and temperature of the laser diode fluctuate, the Voigt laser can always stay near the atomic transition line. Within the GHz range, immunity to fluctuations in laser diode parameters is achieved. As shown in Fig. 5.13, there are two types of laser magnetic field generators, the Voigt laser and the Faraday laser. The biggest difference between the Voigt laser and the Faraday laser lies in their structural differences. In the Faraday laser, magnets are often placed in the axial direction of the atomic cell, while in the Voigt laser, magnets are placed in the radial direction of the atomic cell. Generally, considering the issue of atomic interaction length, the axial length of the atomic cell used in both cases is much greater than the radial length. Therefore, the Voigt laser is more likely to generate a sufficiently strong magnetic field, which can explore the performance of the laser under a larger magnetic field range. In fact, the maximum magnetic field currently used by the Voigt laser is 3500 G, while the maximum magnetic field currently used by the Faraday laser is 1750 G. In addition, the placement of magnets on both sides of the atomic cell in the Faraday laser inevitably increases the length of the Faraday laser's resonant cavity, which is not conducive to further miniaturization. The structure of the Voigt laser is relatively simpler, without the need to increase the length of the laser cavity, which is more conducive to the portability and miniaturization of subsequent lasers.

In 2023, Liu and others from Peking University proposed and implemented the Voigt laser, studied the main influencing factors of ^{87}Rb VADOF, and explored and verified the performance parameters of VADOF [6]. They used ^{87}Rb VADOF as a frequency selection component in the laser resonant cavity, successfully realizing the Voigt laser. The output frequency of this Voigt laser is maintained within the transmission bandwidth of 500 MHz of VADOF, showing stability under different laser diode currents (73 ~ 150 mA) and temperature changes (12 ~ 30 °C). This ideal laser source can be used to design compact and plug-and-play atomic devices.

Figure 5.14 is a schematic diagram of the experimental setup for studying the performance parameters of VADOF. Considering the scarcity of research on VADOF, a detailed investigation into the selection of magnetic field parameters is

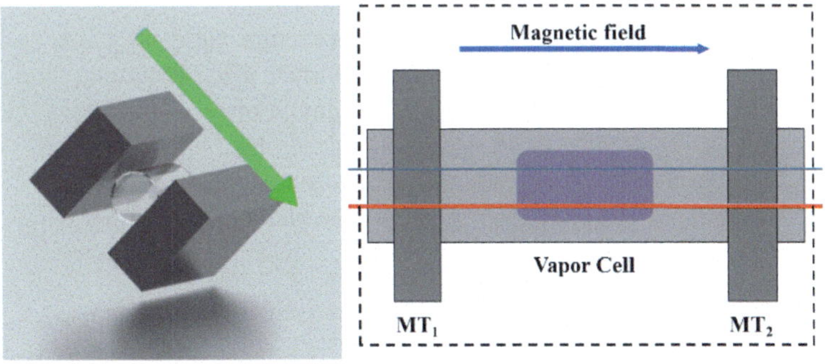

Fig. 5.13 Magnetic field generators of the Voigt laser and the Faraday laser, MT_1 magnet 1, MT_2 magnet 2

required when applying it to Voigt lasers. Therefore, a detailed simulation design of the magnetic field generator in VADOF is carried out, making its magnetic field adjustable from 50 to 3500 G. In addition, the intensity of the input laser and the temperature of VADOF will have a huge impact on the transmission performance of VADOF. After detailed research on the performance of VADOF under various parameters, the optimal working parameters of VADOF in the Voigt laser are obtained.

The principle of the Voigt laser is shown in Fig. 5.15. The experimental device consists of a ^{87}Rb VADOF, a laser diode with an anti-reflective coating (ARLD), an aspherical lens, a high reflector (HR), and a piezoelectric ceramic tube (PZT), with the optical cavity length of the Voigt laser being 20 cm. After collimation by the

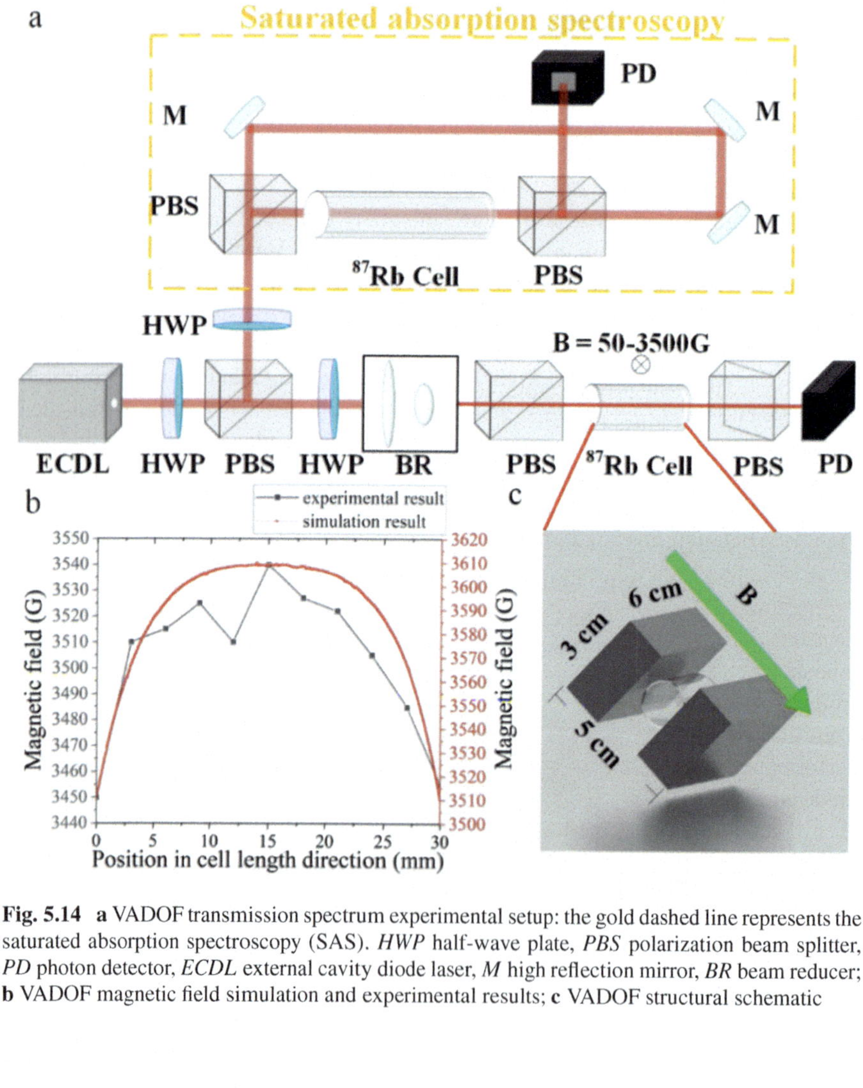

Fig. 5.14 **a** VADOF transmission spectrum experimental setup: the gold dashed line represents the saturated absorption spectroscopy (SAS). *HWP* half-wave plate, *PBS* polarization beam splitter, *PD* photon detector, *ECDL* external cavity diode laser, *M* high reflection mirror, *BR* beam reducer; **b** VADOF magnetic field simulation and experimental results; **c** VADOF structural schematic

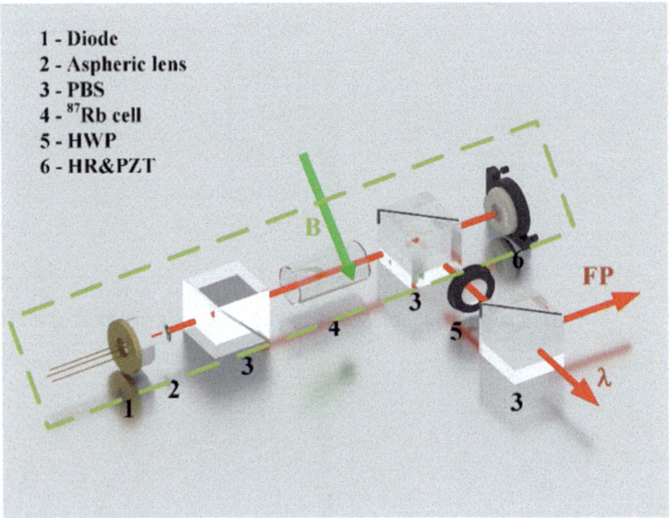

Fig. 5.15 Experimental setup of the Voigt laser. *HWP* half-wave plate, *PBS* polarizing beam splitter, *HR* high reflector, *PZT* piezoelectric ceramic tube, *FP* scanning Fabry–Pérot interferometer, λ optical wavelength meter

aspherical lens, the beam diameter emitted by the ARLD expands to 2 mm. Then, the collimated light is fed back to the ARLD through the HR and passes through the VADOF twice. The HR, which has a reflectivity of 99.9% for the 780 nm laser, is fixed on the PZT to tune the cavity length. The second PBS outputs the laser beam and monitors its wavelength and scans the Fabry–Pérot spectrum to evaluate the performance of the Voigt laser.

Based on the research results of the factors affecting VADOF, the magnetic field B = 3500 G and the cell temperature T = 75 °C were chosen in the parameter settings. The Voigt laser outputs a single longitudinal mode frequency of 780.255 nm, and remains stable as the laser diode current changes from 73 to 150 mA and the diode temperature changes from 12 to 30 °C, with the output wavelength fluctuation less than 0.5 pm (see Fig. 5.16a, b). This indicates that the Voigt laser has excellent robustness to changes in laser diode parameters. As the laser diode current increases, an increase in output power jitter is detected (as shown in Fig. 5.16c, which comes from the influence of light intensity on the VADOF transmission). In addition, the performance of the Voigt laser running freely for 48 h is monitored, with the laser diode current set to 150 mA and the temperature set to 22 °C. As shown in Fig. 5.16d, even if mode hopping occasionally occurs, the output wavelength always remains within 0.5 pm, proving that the Voigt laser has good long-term repeatability. Furthermore, after more than half a year of repeated experiments, the output wavelength of the Voigt laser remains unchanged within the range of 0.5 pm without manual adjustment, demonstrating its good reproducibility. Therefore, the Voigt laser is particularly

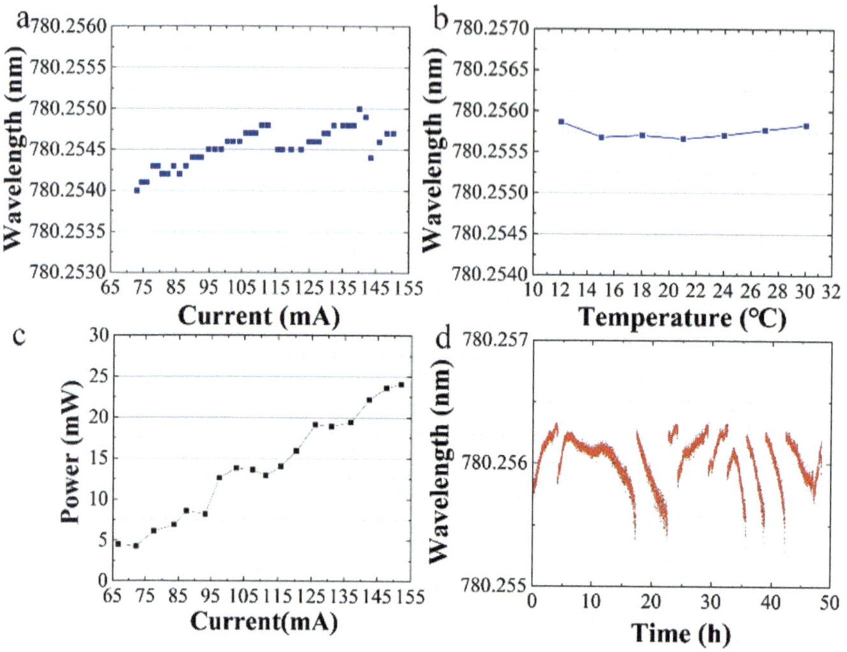

Fig. 5.16 Characteristics of the Voigt laser: **a** relationship between output wavelength and diode current; **b** relationship between output wavelength and diode temperature; **c** relationship between output power and diode current; **d** fluctuation of output wavelength during 48 h of free running

suitable for the design and application of equipment related to atomic physics and other fields that require high long-term repeatability.

Currently, the Voigt laser only has a corresponding 780 nm Rb Voigt laser. Using other types such as potassium, cesium and other alkali metal atoms, a Voigt laser can also be realized. However, it should be noted that in order to achieve stable output, the required magnetic field strength of the Voigt laser is generally greater than that of the Faraday laser, which needs special attention while designing experiments. Secondly, the angle between the magnetic field direction and the laser polarization direction of the Voigt laser also affect the output wavelength of the Voigt laser. It is necessary to theoretically calculate the transmission spectrum of the atomic filter at the corresponding angle in advance and select the required parameters.

5.2 Optical Resonator

The optical resonator is the foundation for laser oscillation and frequency selection in Faraday lasers, and the optical components contained in the resonator and the external cavity feedback optical path determine the frequency selection effect and linewidth narrowing effect of the Faraday laser. To date, Faraday lasers have achieved significant results at the 780 nm wavelength corresponding to Rb atoms and the 852 nm wavelength corresponding to Cs atoms, and have developed Faraday lasers using different feedback structures and types of optical components. The following subsections discuss them separately.

5.2.1 Resonator Optical Path Structure

The resonator optical path of the Faraday laser can mainly be divided into straight cavity feedback and L-cavity feedback. The main difference between the two is: in the Faraday laser with straight cavity feedback, the fluorescence emitted by the laser diode does not reflect off other optical components before reaching the feedback cavity mirror, and its optical path diagram is a straight line, hence it is called straight cavity feedback; while in the Faraday laser with L-cavity feedback, the fluorescence emitted by the laser diode is reflected by a polarizing beam splitter or high reflection mirror before reaching the feedback cavity mirror, and the optical path diagram presents an L shape or a more complex shape. In addition to classification based on the feedback optical path structure, Faraday lasers can also be categorized into plane mirror-feedback Faraday lasers and corner cube-feedback Faraday lasers, depending on whether the feedback cavity mirror used is a plane mirror or a corner cube mirror.

As shown in Fig. 5.17, it is a straight cavity Faraday laser with plane mirror feedback. The fluorescence emits from the laser diode, passes through a PBS, an atomic filter before reaching the plane mirror (reflection cavity mirror), and then returns along the same path to form feedback. The entire feedback optical path is a straight line. Obviously, the structure of the straight cavity feedback laser is the simplest, and is also simple to assemble the laser and adjust the feedback. At the same time, since most optical components basically do not affect the transmission optical path, when environmental fluctuations cause internal optical components to shift, the resistance of the straight cavity feedback Faraday laser is stronger. However, if a longer cavity length laser needs to be built, the straight cavity laser can only achieve a longer cavity length by continuously stretching the distance between the feedback cavity mirror and the laser diode, which causes the size of the laser in the optical axis direction to become unreasonable, affecting the aesthetics and limiting the portability of the laser, which is not conducive to the subsequent integration of the laser into other complex systems.

As shown in Fig. 5.18, it is an L-cavity Faraday laser with plane mirror feedback. After the fluorescence is emitted from the laser diode, it passes through the atomic filter. After being reflected by the second PBS and a high reflection mirror in the atomic filter, it reaches the plane mirror (feedback cavity mirror) and returns along the original path to form feedback. In this laser device, the feedback light path reaches the feedback cavity mirror after multiple reflections. Although its optical path diagram is more complex than the L-type, it can also be called an L-cavity feedback laser. In fact, if the high reflection mirror is cancelled and the feedback

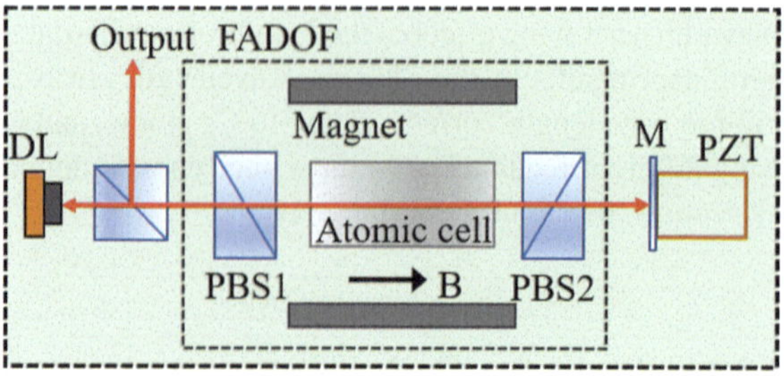

Fig. 5.17 Straight cavity Faraday laser. *DL* laser diode with antireflection coating, *PBS1* polarizing beam splitter 1, *PBS2* polarizing beam splitter 2, *M* cavity mirror, *PZT* piezoelectric ceramic

cavity mirror is moved to the position of the high reflection mirror, the optical path diagram at this time is L-shaped. Since the reflection path is greatly affected by the position and angle of the reflection element, when environmental fluctuations cause the internal optical elements to shift, especially when the PBS and high reflection mirror are affected, it poses a negative impact on the feedback of the laser. However, as compared to the straight cavity feedback, L-cavity feedback has a great advantage in building a laser with a large cavity length. Comparing this L-cavity Faraday laser with the aforementioned straight cavity Faraday laser, under the same cavity length, the volume of the L-cavity Faraday laser is likely to be reduced by half. Therefore, the L-cavity Faraday laser can play a greater role in miniaturization and engineering.

Whether it is for straight cavity lasers or L-cavity lasers, external environmental interference (such as mechanical vibrations and temperature fluctuations) is a major obstacle in further improvement of the long-term performance of lasers. The laser diode and the feedback cavity mirror, as the first and last elements of the feedback light path, enhancing their resistance to environmental interference is the most effective way to improve the long-term performance of the laser. The anti-interference ability of the laser diode can be achieved by designing a corresponding anti-vibration support system and a high-precision temperature control system. For the feedback cavity mirror, replacing the plane mirror with a corner-cube retroreflector can greatly enhance the resistance to external interference. Because compared to plane mirrors, corner-cube retroreflector has extremely low requirements for the angle and position of incident fluorescence, basically achieving the function of oscillation as long as the

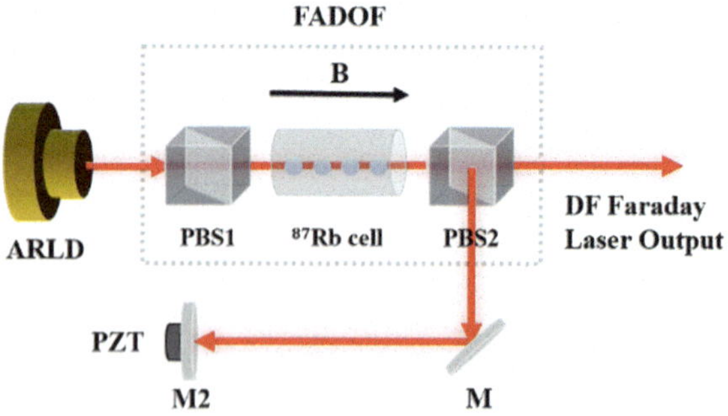

Fig. 5.18 L-cavity Faraday laser. *ARLD* laser diode with anti-reflection coating, *PBS1* polarizing beam splitter 1, *PBS2* polarizing beam splitter 2, *M* high reflection mirror, *M2* cavity mirror, *PZT* piezoelectric ceramic

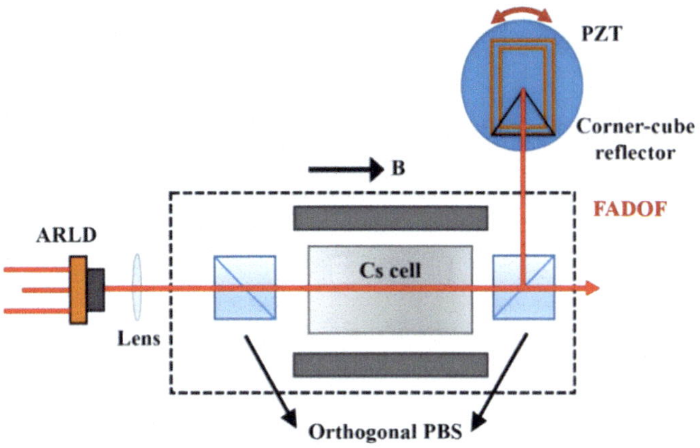

Fig. 5.19 Cone feedback L-cavity Faraday laser

laser is shot into the cone, without further adjustment. Using a corner-cube retroreflector can greatly improve assembly efficiency, and can enhance the anti-interference ability of the laser thereby, improving long-term performance. This kind of design is shown in Fig. 5.19, and is known as an L-cavity Faraday laser with cone feedback.

5.2.2 Linewidth Compression Technology

Lasers have good monochromaticity and coherence, and are widely used in the fields of precision measurement, optical communication, atomic and molecular spectroscopy, and atomic clock. Since 1970s, with the improvement and perfection of the diode laser medium manufacturing process, the performance of diode lasers has been continuously improved, which is one of the important ways to obtain narrow linewidth laser sources. The linewidth of a general F-P type diode laser is at the level of several hundred MHz due to spontaneous radiation. Based on the thermodynamic model of phase diffusion, its radiation spectrum is Lorentzian, and the full width at half maximum can be represented by the modified Schawlow–Townes linewidth.

$$\Delta\nu_{\mathrm{LD}} = \frac{2\pi h\nu_0\left(\Delta\nu_{1/2}\right)^2\mu}{P}\left(1+\alpha^2\right) \tag{5.1}$$

where, P is the output power, $\mu = N_2/(N_2 - N_1)$, is the parameter describing population inversion, $\Delta\nu_{1/2}$ is the linewidth of the resonant cavity mode, α is the Henry coupling parameter, which depends on the diode laser material.

As can be seen from formula (5.1), the linewidth can be narrowed by reducing losses, increasing power, and extending photon lifetime. The main methods can be divided into external cavity extension, self-injection, optimizing semiconductor

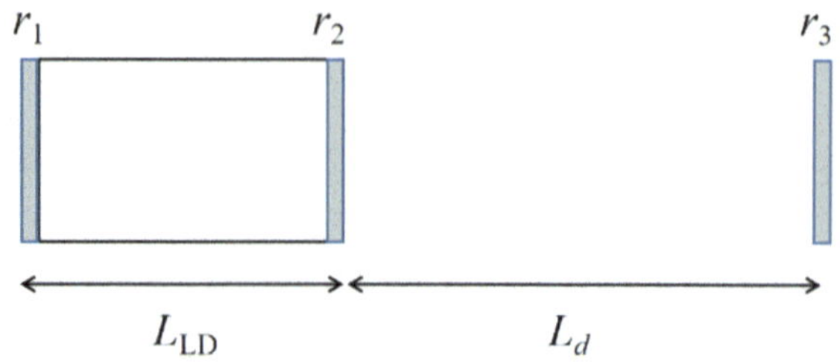

Fig. 5.20 Narrowing linewidth by external cavity optical feedback

processes, mutual injection locking, and electrical negative feedback to narrow the linewidth. First, let's introduce the method of narrowing the linewidth by the external cavity.

1. Extented cavity method for narrowing linewidth

In 1978, Velichansky and others proposed that the method of using external cavity feedback can effectively suppress spontaneous radiation noise, thereby narrowing the laser linewidth, and called it an external cavity diode laser. The schematic diagram is shown in Fig. 5.20. After the cavity extension, the laser linewidth becomes

$$\Delta v = \frac{\Delta v_{\mathrm{LD}}}{[1 + (L_d/nL_{\mathrm{LD}})]^2} \tag{5.2}$$

n is the refractive index of the laser diode, L_{d} and L_{LD} are the lengths of the external cavity and the laser diode, respectively. The formula shows that by using the external cavity method, the linewidth of the diode laser can be effectively narrowed to the level of hundreds of kHz. The external cavity can be as long as 30 cm, so the interval between adjacent longitudinal modes is about 500 MHz. This longitudinal mode interval is very small compared to the diode gain bandwidth, making it difficult to achieve stable single longitudinal mode output. Therefore, gratings, prisms, and interference filters are generally used as frequency selection devices for mode selection.

For grating frequency selection, the wavelength selection can be calculated by the diffraction formula,

$$m\lambda = a(\sin\theta_i + \sin\theta_d) \tag{5.3}$$

The output wavelength depends on the incident angle θ_{i}, the diffraction angle θ_{d}, the diffraction constant a, and the diffraction order m. Common grating diode lasers include Littrow and Littman structures. The external cavity diode laser of the Littrow structure uses a reflective grating as the output coupling mirror of the external cavity. By adjusting the diffraction angle, making $\theta_{\mathrm{i}} = \theta_{\mathrm{d}}$, the first-order diffracted light coincides with the incident light and feeds back to the laser diode, and the zero-order diffracted light is used as the output light. The output wavelength of the diode laser is adjusted by rotating the grating angle. The Littman structure uses a folded laser cavity. Compared with the Littrow structure, the incident light beam and the diffracted light beam are not collinear. The diffracted light is reflected back from the mirror, and after the second diffraction through the grating, it directly feeds back to the laser diode. The wavelength adjustment is achieved by rotating the mirror. The advantage of the Littman structure is that tuning does not change the direction of the output beam, and the wavelength is independent of the incident angle. Therefore, a larger incident angle can be chosen to improve the resolution. However, the second diffraction will increase the loss in the cavity, so a high reflectivity grating is needed.

Both of the above external cavity diode lasers use a grating as a frequency selector placed in the external cavity structure, and use grating diffraction to select specific frequency laser for optical feedback, to reduce the impact of spontaneous emission noise on the laser frequency. Because the grating is used for both frequency selection and optical feedback, external environmental vibrations affect the mode and power of the output laser. The laser is sensitive to external vibrations and temperature fluctuations, and the laser linewidth is generally at the level of hundreds of kHz. If a narrower linewidth laser is needed, a narrow-band interference filter can be used as a frequency selector placed in the external cavity structure. This structure laser has more independent mode selection and optical feedback, better stability, and can achieve a narrow linewidth tunable external cavity diode laser.

2. Self-injection locking narrow linewidth

The external cavity self-injection locking of a laser involves adding an external resonant cavity to the laser to form an external cavity self-injection locking system. The laser operates in two modes: the free oscillation mode and the locking mode triggered by stimulated radiation. The laser produced by the self-injection locking technology is generated with spontaneous radiation. Due to the simultaneous existence of multiple modes of laser in the resonant cavity, the resonant cavity set outside the diode laser plays a role in selecting the frequency of the laser generated by spontaneous radiation, and realizes the selection and filtering of the output light of the laser. This allows the light of one mode of the initial output to return to the diode laser, also known as feedback light or seed light. When the seed light reaches the diode laser, it gives this mode of light stronger competitive ability and suppresses the growth of other modes. When the strength of the feedback light exceeds the minimum limit required for the laser to lock, the stimulated light field of the laser is locked to the injected light. At this time, when the ratio of the light power of the injected laser to the light field power of the laser itself reaches a certain value, the output frequency of the laser becomes the same as the injected light. In the working mode of spontaneous radiation, the center wavelength of the laser still remains unchanged. The light enters the external resonant cavity from the laser, and after its feedback, it returns to the laser to complete the stimulated radiation. In this process, the carriers in the cavity change, leading to an increase in the gain of the feedback mode and a significant increase in intensity; while the gain of other modes outside the feedback mode further decreases, and the intensity also decreases. The gain of the laser changes with the change of the carriers in the cavity, which in turn causes the refractive index of the gain medium to change, and the working frequency of the laser changes accordingly.

The self-injection locking systems currently studied include: grating external cavity injection locking, F-P cavity injection locking, master–slave injection locking, and echo wall injection locking.

(1) *Grating external cavity injection locking*

As shown in Fig. 5.21, the grating external cavity is used to select the frequency of the laser, couple and feedback its output light, and narrow the linewidth of the laser. It can effectively avoid the diffraction and scattering loss of the light wave of the integrated grating in the cavity, and make the output linewidth narrower. However, there are problems such as low reliability, sensitivity to noise, and high coupling.

(2) *F-P cavity injection locking*

Due to the high Q value and narrow full width at half maximum (FWHM) the F-P cavity, the working frequency of the laser can be locked to the resonant point of the F-P cavity to narrow the linewidth of the laser, as shown in Fig. 5.22. The stability of the resonant frequency of the F-P cavity and the laser frequency are closely related. However, F-P cavity injection locking is significantly affected by external temperature and mechanical noise, and precise control of the system is required. The initial light frequency needs to be consistent with the resonant frequency of the F-P cavity.

(3) *Master–slave injection locking*

This method injects the output light of the master laser into the slave laser, making the working frequency of the slave laser consistent with that of the master laser, as shown in Fig. 5.23. Because the master and slave lasers in this locking method are two independent laser systems, it increases the volume of the experimental device, and include devices and a complex structure, which is very limited in practical applications.

(4) *Whispering gallery mode injection locking*

The whispering gallery mode is a resonant mode formed by total internal reflection of light in the resonant cavity. By injecting the whispering gallery mode produced by the laser in the microcavity into the back laser, linewidth narrowing is achieved, as shown in Fig. 5.24. Although the whispering gallery mode cavity has advantages such as small size and high Q value, and has mode selection frequency characteristics, it still has disadvantages such as high preparation cost, challenging manufacturing difficulty, and difficult coupling.

Self-injection locking has a wide range of applications. Simply put, as long as the gain spectrum of the laser includes the resonance spectrum of the external cavity, self-injection locking can be theoretically achieved. The difference from the first method is that the external cavity is part of the resonant cavity in the first method, while in the self-injection locking method, it is not. At the same time, self-injection locking does not depend on other electrical devices, so it can provide better miniaturization capabilities and higher integration for the self-injection system. In addition, it also makes significant contributions to thermal stability optimization and noise suppression. At the same time, due to the frequency selection effect of the external

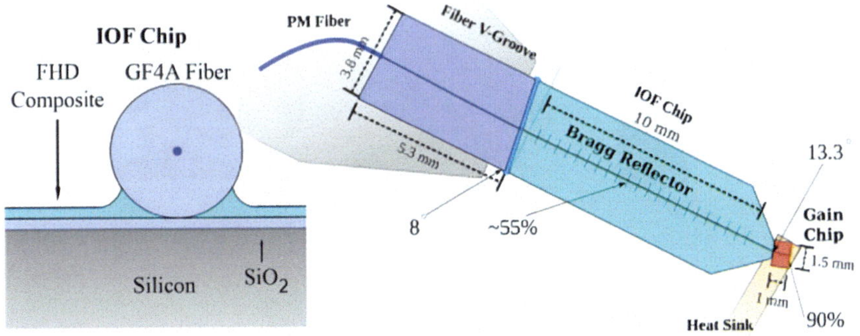

Fig. 5.21 Grating external cavity injection locking

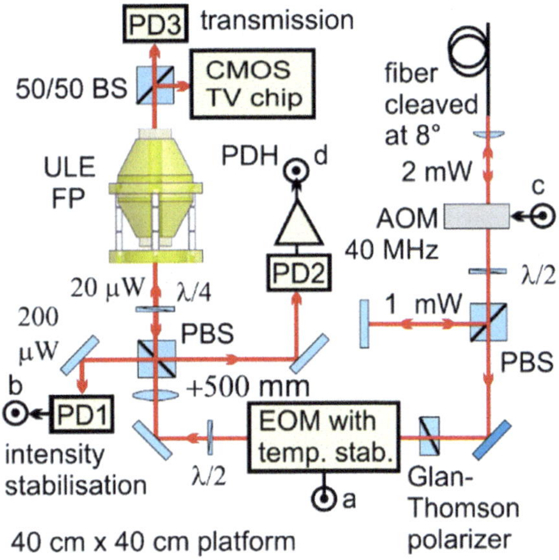

Fig. 5.22 F-P cavity injection locking. *PD* photodetector, *CMOS* photosensitive chip, *ULE-FP* Fabry–Pérot cavity, *PBS* polarizing beam splitter, *EOM* electro-optic modulator, *AOM* acousto-optic modulator, *GT* Glan–Taylor prism, *λ/2* half-wave plate, *λ/4* quarter-wave plate

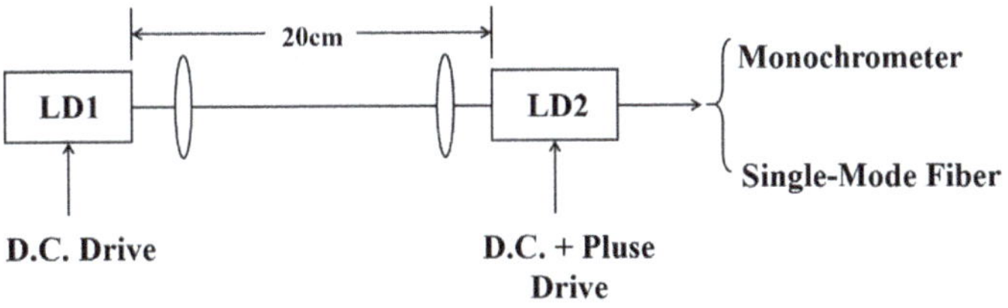

Fig. 5.23 Master–slave injection locking

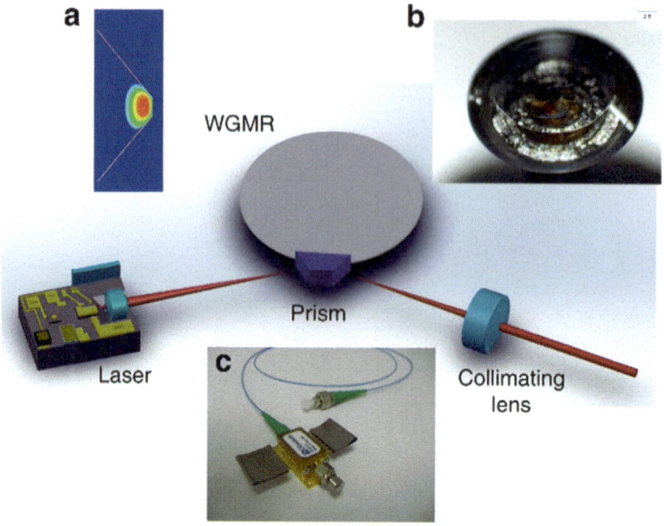

Fig. 5.24 Whispering gallery mode injection locking, *WGMR* whispering gallery microcavity

cavity on the original light of the diode laser, the linewidth of the output light after stimulated radiation by self-injection locking is sharply narrowed compared to the original linewidth, achieving linewidth narrowing.

3. Electrical negative feedback linewidth narrowing

The linewidth of diode lasers is easily affected by temperature and drive current. Therefore, through the temperature compensation circuit and current control unit, the center frequency of the laser can be stabilized while narrowing the linewidth of the laser. The temperature control part and the drive current control part are designed based on the principle of negative feedback, thereby achieving automatic control of the injection current power and temperature of the laser. The principle diagram of narrowing the linewidth of the laser through electrical negative feedback is shown in Fig. 5.25.

The electrical feedback method mainly detects the frequency fluctuation signal of the laser through a frequency discriminator (such as an F-P interferometer), then converts this signal into an electrical signal, and controls the temperature and driving current of the laser through a servo control loop. The laser in Fig. 5.25 is fixed on a heat sink composed of a copper block and a Peltier element. The thermistor senses the temperature fluctuation of the heat sink, so the temperature control circuit can adjust the working temperature of the diode laser in real time. The isolator ensures that the reflection signal of external optical devices does not enter the laser. The two ends of the F-P interferometer are coated with a high-reflection multilayer dielectric film. The light incident on the interferometer identifies the frequency modulation noise of the laser through the slope of the resonance curve, and then enters the phase compensation current to form a correction current after the broadband operational amplifier. The

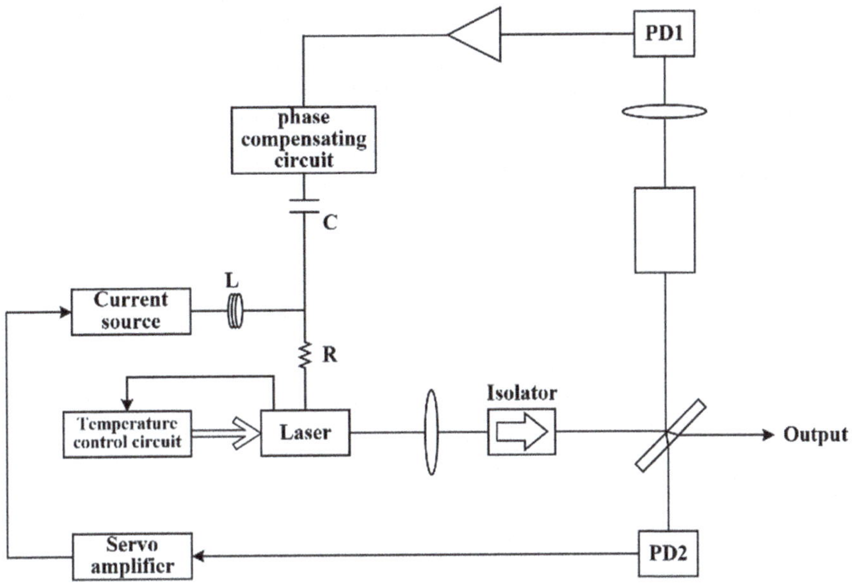

Fig. 5.25 Electrical negative feedback linewidth narrowing method

electrical feedback method can effectively narrow the linewidth of the laser without adjusting the cavity structure of the laser. However, the noise reduction bandwidth of electrical feedback is limited, and it is not easy to achieve the suppression effect on high-frequency noise. Therefore, diode lasers based on electrical feedback can be used for low bit rate coherent optical communication, narrowband sensing systems: such as fiber optic gyroscopes and optically pumped atomic clocks, etc.

4. Mutual injection locking narrow linewidth

The weak coupling mutual injection locking is used to reduce the linewidth of the diode laser. The main mechanism is to suppress their phase noise by locking the two lasers, and at the same time, to extend the photon life by coupling the cavity, as shown in Fig. 5.26. Short delay and long delay can be achieved by integrated devices and fiber link systems. For short-delay integrated lasers, sub-MHz linewidths have been achieved, and for fiber link coupled long-delay lasers, the intrinsic linewidth has been reduced to below 100 Hz.

Among the above various methods, the Faraday laser uses the increase in cavity length and atomic selection to narrow the laser linewidth. Since the transmission spectrum bandwidth of FADOF is about 1 GHz, which is much smaller than the

Fig. 5.26 Mutual injection locking narrow linewidth method

several hundred GHz bandwidth of the grating or interference piece, the suppression effect on laser mode competition is better. Moreover, the transmission spectrum of FADOF is not sensitive to mechanical vibrations, so the laser linewidth of the Faraday laser also becomes smaller than that of the external cavity diode laser of the grating or interference piece. The Faraday laser may have better anti-mechanical vibration capabilities.

5.2.3 Main Optical Devices

1. Laser diode

(1) *Laser diode fluorescence spectrum and self-excitation phenomenon*

The F-P type laser diode, as a type of diode laser, its basic principle is: under the action of an external electric excitation source, an appropriate voltage is applied to the two ends of the PN junction of the semiconductor crystal, causing the carriers to form a reverse distribution, that is, there are electrons in the conduction band, and the corresponding valence band is left with holes. The electrons in the conduction band transit down to the lower energy valence band, and the recombination of electrons and holes occurs. During the transition, photons are emitted. Due to the feedback effect of the resonant cavity (the reflectivity of the rear cavity is > 99%, and the reflectivity of the front cavity is 14 ~ 70%), it produces a laser.

Early external cavity diode lasers use the aforementioned laser diodes. By controlling the external cavity length and feedback intensity, linewidth compression and single-frequency output are achieved. However, due to the disturbance of the intracavity mode in the diode, the laser can only work at certain bias current points, and the laser is prone to mode hopping and low-frequency fluctuations. By reducing the reflection on the coupling end face of the chip (the front cavity reflection surface), that is, by coating an anti-reflective film, the longitudinal mode of the internal cavity can be suppressed and the role of external cavity feedback can be enhanced. The Faraday laser, as a type of atomic selection diode laser, uses ARLD to eliminate the internal cavity mode. ARLD, as a gain medium, has a spectral (half-width) width greater than 10 nm, which can easily cover the required atomic spectral lines. Thus, even if its gain center wavelength deviates from the expected atomic spectral line center wavelength by 5 nm, or even more, it will not affect the output wavelength of the Faraday laser.

When the LD without an anti-reflective film increases the working current, the LD output window and the rear end face forms a resonant cavity, and the laser diode spontaneously oscillates, producing an internal cavity mode laser output. In the Faraday laser, the internal cavity mode competes with the mode after FADOF selection, causing the Faraday laser output to be unstable and unable to maintain single longitudinal mode output for a long time. Therefore, an anti-reflective film should be coated on the output window of the laser diode to prevent the return light from producing an internal cavity mode, ensuring that there is no internal cavity mode laser output in the laser diode working in the current range of 0 ~ 200 mA, and its spectrum should always be a fluorescence spectrum.

(2) *Laser diode beam quality*

The beam quality of a laser diode includes its divergence angle and transverse mode distribution. For the Faraday laser, the light emitted by the laser diode is collimated and then feedback is provided by the resonant cavity mirror. The proportion of light fed back to the diode directly affects the output efficiency of the external cavity laser. When the diode itself outputs a single transverse mode, the output spot of the external cavity laser may have a fundamental Gaussian distribution.

Due to the asymmetry of its own waveguide structure ($1 \times 3\ \mu m^2$), the light emitted by the LD has a typical divergence angle of about 10° in the direction parallel to the p–n junction (slow axis), and in the direction perpendicular to the p–n junction (fast axis) the typical divergence angle is about 20 ~ 30°, and the light spot diverges in an elliptical shape. The LD FWHM divergence angle $\theta_{x,y}$ and the far-field divergence half-angle $\alpha_{x,y}$ relationship is:

$$\alpha_{x,y} = \frac{\theta_{x,y}}{\sqrt{2\ln 2}} \tag{5.4}$$

The 780 and 852 nm laser diodes used in Faraday lasers adopt an asymmetric large optical cavity design for the external structure. The waveguide layer and the limiting layer use a mature AlGaAs material system, and the quantum well uses a GaAsP or InGaAsP aluminum-free material system, which has extremely high reliability and consistency. The diode uses a ridge waveguide structure with a ridge width of 4 μm. To accurately control the transverse mode of the laser, an etching barrier layer is designed on the epitaxial layer to accurately control the width and thickness of the ridge waveguide and avoid mode hopping. The diode's fast-axis FWHM divergence angle is 20°, and the slow-axis FWHM divergence angle is 10°. To suppress the impact of the intracavity mode on the Faraday laser, the front cavity reflectivity is less than three ten-thousandths, thereby controlling the self-excitation current of the laser diode above the operating current. The diode design has a cavity length of 1500 μm, a diode width of 500 μm, a thickness of 130 μm, uses a TO9 packaging form, and the packaging solder uses highly reliable gold-tin solder packaging, fully meeting the high power output and high reliability requirements of laser diodes.

The output beam of the edge-emitting LD has astigmatism characteristics. The fundamental reason is that the constraint mechanism of the laser in the vertical junction direction and the parallel junction direction is different, which makes the wavefront at the near-field position, that is, the cleavage plane, different. In the direction perpendicular to the junction plane, the beam is a plane wave, and the Gaussian beam waist is at the cleavage plane. The gain is unevenly distributed in parallel to the junction plane, and the beam shows a "gain focusing" phenomenon towards the axis during longitudinal propagation, causing the wavefront of the lasing light field to become curved. When designing the collimating lens for this kind of beam, even if all kinds of lens aberrations are ignored, that is, in the case of ideal imaging, the beam shows that it emit output collimated light in the fast axis and slow axis directions at the same time. If a short-focus lens is placed at a position further than its focal length, two focus points will obviously appear. This phenomenon has been observed in Toptica's laser diodes, but because the astigmatism is very small, this phenomenon can almost be ignored. Theoretically, this asymmetric waveguide structure definitely brings about the separation of the fast and slow axis beam waists, but if the separation amount is at the micron level, the lens focal length is near the 4 mm range, the aforementioned phenomenon is almost unobservable during collimation. However, if the separation is on the order of tens of micrometers, it severely disrupts the collimation effect of the light beam and affects the efficiency of the external cavity feedback. Table 5.2 shows the beam quality results of collimated qualified lasers.

(3) *Polarization direction of laser diodes*

Faraday lasers are atomic frequency selected lasers using atomic filters as the frequency selection elements. The frequency selection principle is to obtain a high signal-to-noise ratio narrowband transmission spectrum through two orthogonal polarizing prisms, using the Zeeman effect and the magneto-optical rotation effect of atoms placed in a magnetic field. Therefore, to reduce the polarization loss in the cavity, the light emitted by the laser diode needs to be linearly polarized light with a high extinction ratio. Due to the waveguide structure of the laser diode, it determines that its emitted light can only be transverse electric mode (TE) or transverse magnetic (TM) mode, and cannot be perfect linearly polarized light.

2. Atomic filters

The frequency and transmittance of the atomic filter transmission peak are achieved by adjusting the temperature and magnetic field of its internal atomic cell. In order to ensure the long-term stability of the atomic filter transmission spectrum, to make the laser work at the transmission peak frequency of the atomic filter for a long time, it is necessary to precisely control the temperature and magnetic field of the atomic cell, and study the working parameter requirements of different atomic filters. A large amount of experimental data is recorded, such as the influence of the temperature, magnetic field, buffer gas density in the atomic cell and the influence of factors such as the material of the atomic cell, the flatness of the transmission window, optical loss, and the angle of incidence of light. Through theoretical calculations and experimental tests to obtain the optimal temperature and magnetic field conditions, the research on the atomic filter is more in-depth and meticulous, providing reliable data for the required application scenarios. Since Chap. 4 has conducted an in-depth theoretical research and a large amount of experimental testing on the transmission spectrum of the atomic filter with the cell length, temperature, magnetic field, and PBS polarization angle, only the mechanical design and assembly process of the atomic filter are analyzed and introduced here. The integrated Faraday atomic filter is shown in Fig. 5.27.

In the entire design, as shown in Fig. 5.27, in order to reduce the impact of external temperature fluctuations on the temperature of the atomic cell, three layers of thermal

Table 5.2 Diode beam quality test

Distance	Wavelength 780nm	Wavelength 852nm
30cm	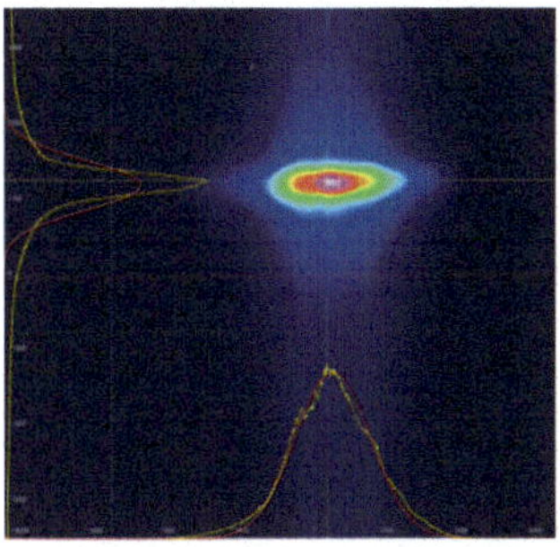	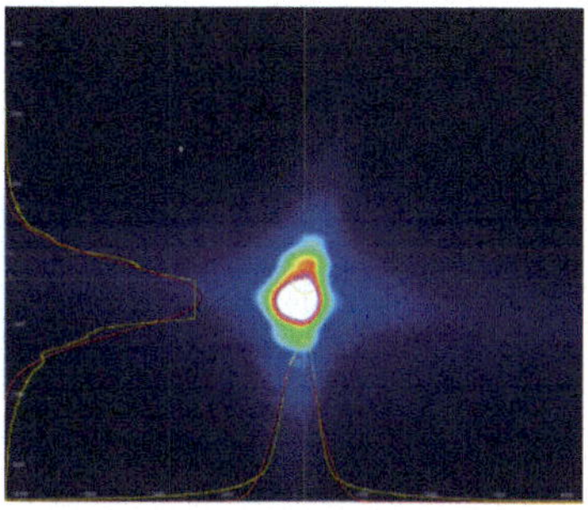
50cm	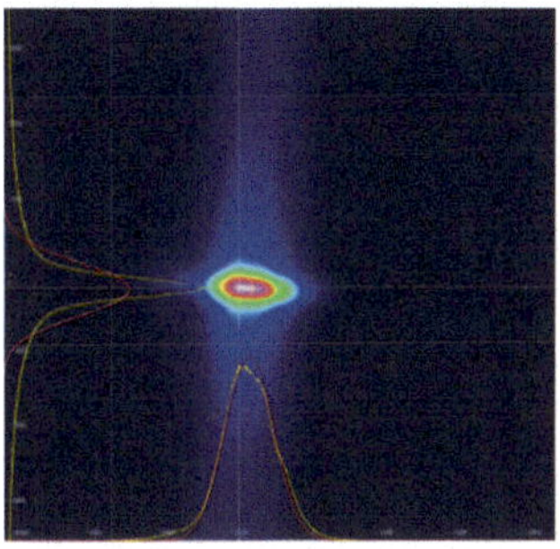	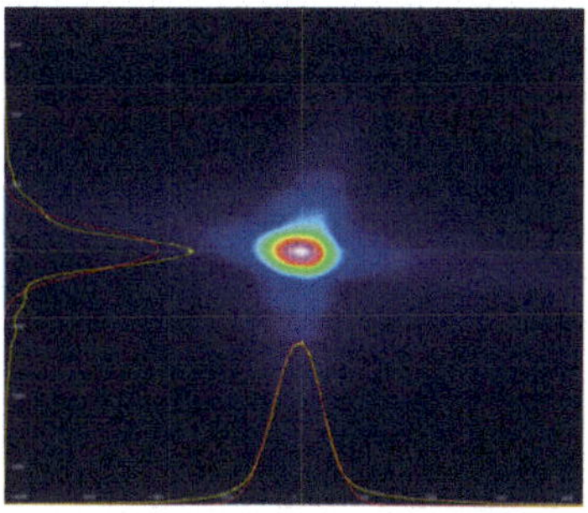
100cm	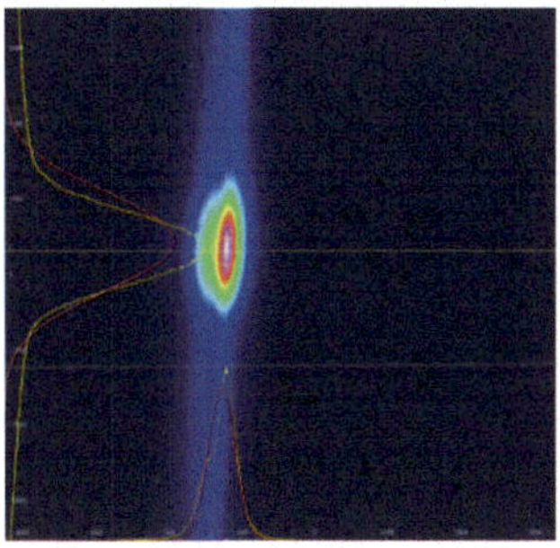	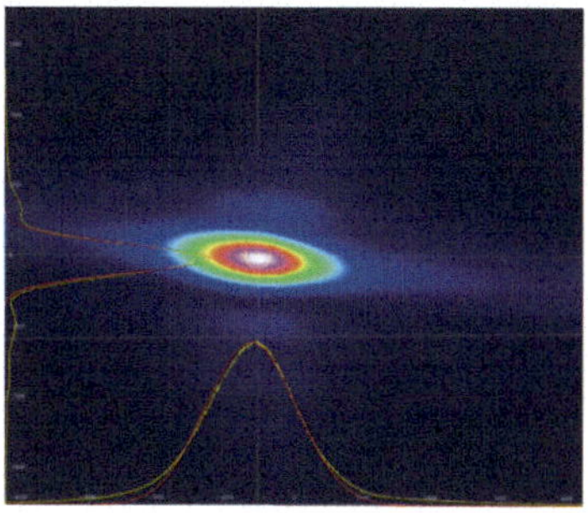

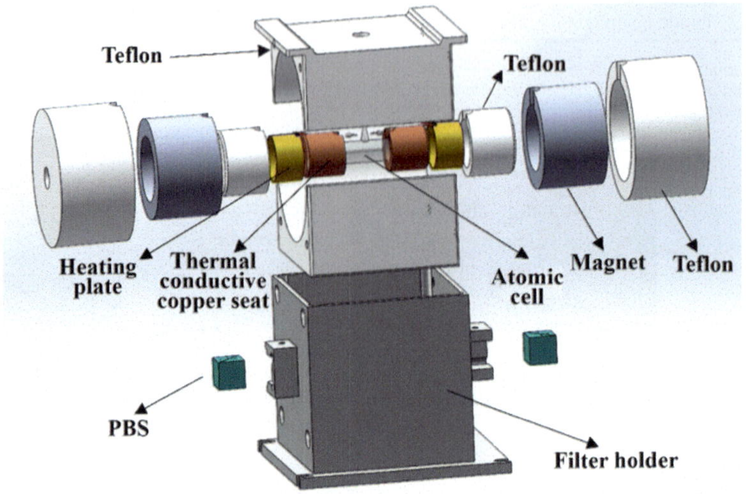

Fig. 5.27 Integrated atomic filter mechanical design assembly diagram

insulation composed of polytetrafluoroethylene material are wrapped around the exterior of the cell. Afterward, the entire assembly is placed and securely fastened within a housing made of 6061 aluminum alloy material. The overall volume of the shell is $(100 \times 90 \times 85)$ mm^3. The transmission peak position and transmission of atomic filters are affected by factors such as magnetic field, temperature, atomic cell composition, and buffer gas number density. On the one hand, relevant theoretical calculations are carried out to design reasonable experimental parameters. On the other hand, different atomic cell compositions and buffer gas number densities are tried experimentally to obtain the best transmission spectrum. The out-of-band suppression ratio is mainly determined by the extinction ratio of the polarization beam splitters, and the dispersion and other non-ideal characteristics of optical devices in the optical path reduce the out-of-band suppression ratio. Therefore, when selecting optical devices, their optical characteristics should be tested, and the polarization beam splitters are required to have a high transmission and parallelism $< 5'$. The flatness of the atomic cell surface is $< \lambda/6$, coated with an anti-reflection film.

3. Laser external cavity feedback mirror

In addition to the cat's eye cavity structure, there are two types of cavity mirrors for Faraday lasers, plane mirrors and corner-cube retroreflector. The plane mirror, as a feedback element of the laser resonator, is sensitive to the angle of incident light. As shown in Fig. 5.28, assuming the light-emitting surface size of the laser diode to be 1 μm × 4 μm, and the cavity length of the Faraday laser resonator to be 160 mm, then the maximum allowable deflection angle of the plane mirror comes out to be $< 1.2'$. During the movement of the laser or environmental vibration, it is easy to cause resonance cavity detuning. By replacing the plane mirror with a corner-cube

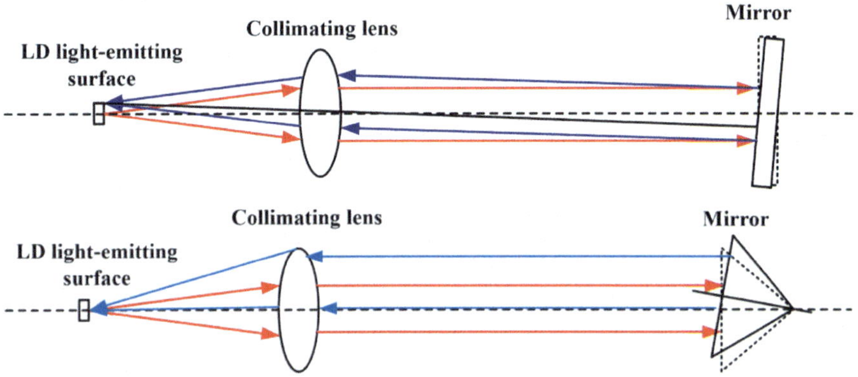

Fig. 5.28 Analysis of detuning angles of different feedback cavity mirrors

retroreflector, according to the size of the effective aperture of the lens, the detuning angle can be < 2°, greatly improving the anti-detuning performance of the laser.

There are three types of corner-cube retroreflector, namely, glass corner-cube retroreflector using total reflection, glass corner-cube retroreflector using mirror reflection, and hollow corner-cube retroreflector, as shown in Fig. 5.29. Theoretical analysis and experiments are carried out on them respectively. Because total reflection greatly changes the polarization state of light, linearly polarized light is incident and elliptically polarized light is emitted. When the light passes through the PBS again, it causes an increase in cavity loss and makes it difficult for the laser to start oscillating. The glass corner-cube retroreflector using mirror reflection as the cavity mirror has about 10% single-pass loss, and the laser threshold is relatively increased. The hollow corner-cube retroreflector avoids the above situation, and the structure of the hollow corner-cube retroreflector is more conducive to assembly with PZT, using cavity length tuning to achieve laser tuning.

Before introducing the corner-cube adjustment mechanical device, a simple introduction to the propagation of light in the corner-cube is made. As shown in Fig. 5.31, the laser beam is incident at the center of the corner-cube, and then reflected back to the second polarizing prism. For conventional corner-cube use, as shown in Fig. 5.31, light is incident from any sector of the corner-cube (1, 2, 3, 4, 5, 6), and then emitted

Fig. 5.29 Three different types of corner-cube retroreflectors: (left) total reflection glass corner-cube; (middle) mirror reflection glass corner-cube; (right) hollow corner-cube

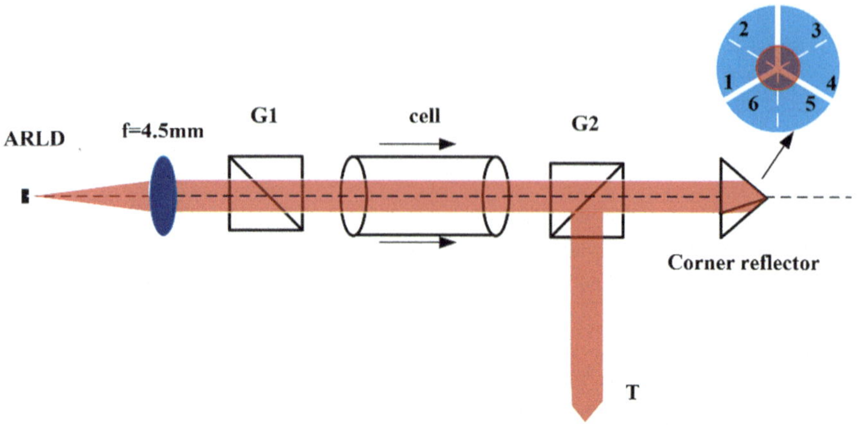

Fig. 5.30 Schematic diagram of light incident on the corner-cube in the Faraday laser. *ARLD* laser diode with anti-reflection coating, $f = 4.5$ mm: collimating lens focal length, *G1 and G2* a pair of orthogonally polarized Glan prisms

from the symmetrical (4, 5, 6, 1, 2, 3) direction. Moving left and right causes lateral displacement of the light spot, and moving up and down causes vertical displacement of the light spot. If the corner-cube is placed unreasonably, the light spot is found to be incident on the edge of the corner-cube, damaging the quality of the light spot or reducing the feedback efficiency. Combining Figs. 5.30 and 5.31, in addition to paying attention to the left and right and up and down displacements, the position where the light is incident on the corner-cube in the Faraday laser also needs to pay attention to the pitch and yaw adjustment. If the light emitted from the collimating lens has no directional error, it greatly reduces the angular adjustment of the corner-cube. The main impact of the adjustment error of the corner-cube is the feedback efficiency, that is, the slope efficiency of the laser. Furthermore, whether the efficiency of feedback affects the output wavelength of the laser requires further theoretical analysis and systematic experimental design to determine.

As shown in Fig. 5.32, considering the optical stability of the laser, while designing the mechanical fixture of the corner-cube, the XY direction translation adjustable knob is removed to avoid the instability of the laser performance caused by the elastic deformation of the screw or spring during the long-term operation of the laser. In order to pursue the best working state of the laser, an external auxiliary adjustment device is pre-adopted for the corner-cube, as shown in Fig. 5.33. Through the up, down, left, and right knobs, the position of the light incident on the corner-cube is adjusted to ensure the best feedback. After the laser output is stable, it is glued to the alumina adapter of the fixed PZT with a small deformation and a long curing time. During the curing process of the glue, the stability of the laser output power and wavelength is monitored throughout the process to prevent the performance of the laser from declining.

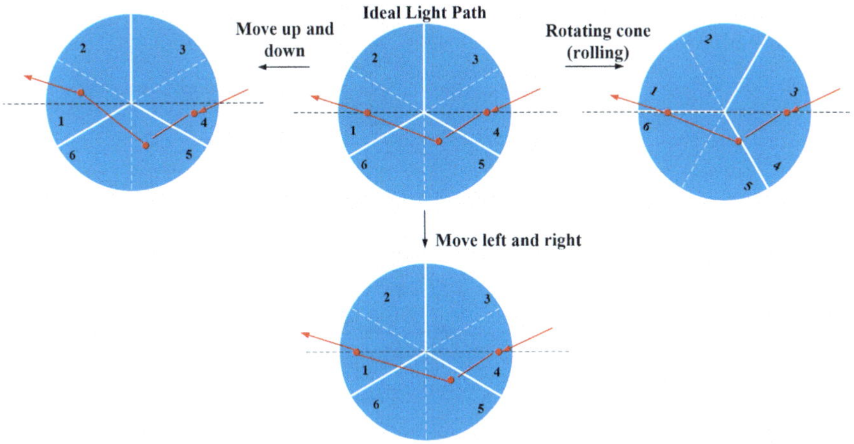

Fig. 5.31 Relationship between incident light and outgoing light positions in the corner-cube

Fig. 5.32 Mechanical design of corner-cube fixture

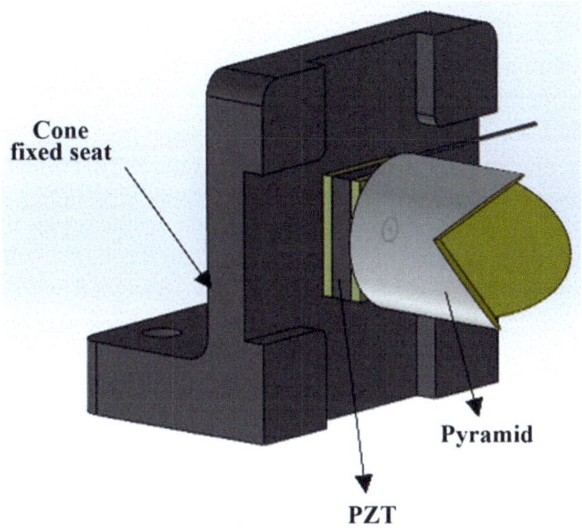

5.3 Circuit System

5.3.1 *Temperature Control Circuit System*

1. Temperature control circuit of the Faraday laser

The temperature control circuit of the Faraday laser plays a crucial role in the operation of the laser. It can ensure the stable operation of the Faraday laser, extend its service life, and improve its output power, etc. The temperature control circuit system

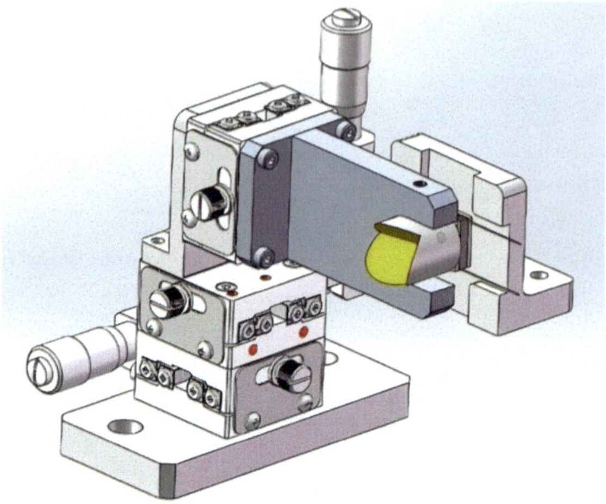

Fig. 5.33 Auxiliary adjustment design of corner-cube

is briefly described. Diode lasers generally have built-in semiconductor thermoelectric coolers (TEC) and temperature sensors and other related temperature control components to ensure that the temperature of the laser diode core is controllable. The temperature sensor (commonly used negative temperature coefficient thermistor) and the laser diode are installed on the same heat sink, which plays a role in real-time monitoring of the laser diode temperature. This system case is mainly composed of three units: setting unit, comparison operation unit, and output unit (Fig. 5.34).

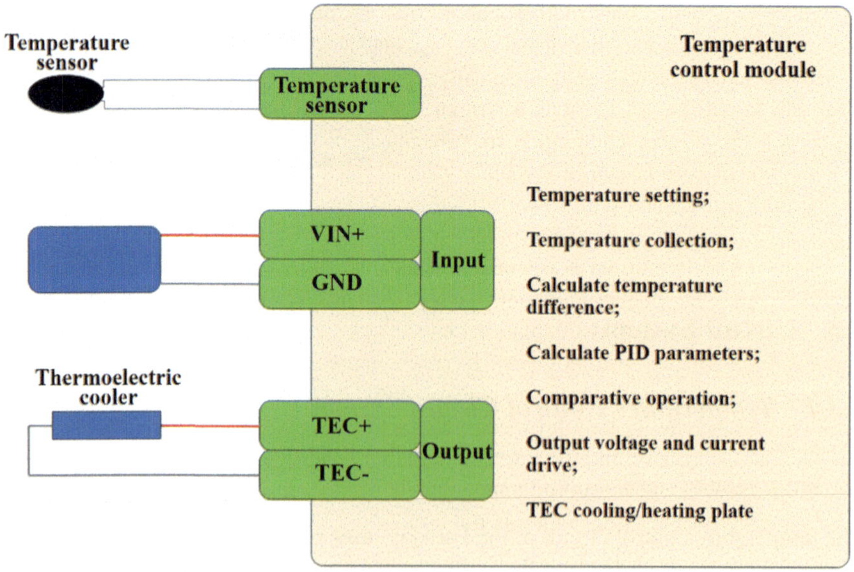

Fig. 5.34 Temperature control circuit block diagram

2. Analysis of the principle of temperature control circuit

Temperature setting: The unit uses LM399H as a voltage reference source, which has the characteristics (Table 5.3).

The operational amplifier OP07 is used as the follower device, with the characteristics (Table 5.4).

Reference circuit: LM399H (Fig. 5.35), OP07 chip internal schematic (Fig. 5.36), application circuit (Figs. 5.37 and 5.38) in the reference source U01, input 12 V, output 6.95 V standard voltage. R06 and R08 are adjustable potentiometers, by changing the voltage value at TR1P2 to set the target temperature, the operational amplifier OP07 acts as a voltage follower, this op-amp is used to reduce output impedance and improve load capacity.

Table 5.3 LM399H characteristic parameters

Serial number	Performance	Indicator
1	High precision	Typical output voltage stability is 0.0005%
2	Low temperature drift	Typical temperature coefficient is 2 ppm/°C
3	High stability	Long-term stability is 0.5 ppm per 1000 h
4	Wide operating temperature	− 55 to + 125 °C
5	Low noise	Typical noise is 0.1 μVp-p

Table 5.4 Operational amplifier OP07 characteristic parameters

Serial number	Performance	Indicator
1	Ultra-low offset	Maximum 150 mV
2	Low input bias current	1.8 nA
3	Low offset voltage drift	0.5 mV/°C
4	Ultra-stable time	Maximum 2 mV/month

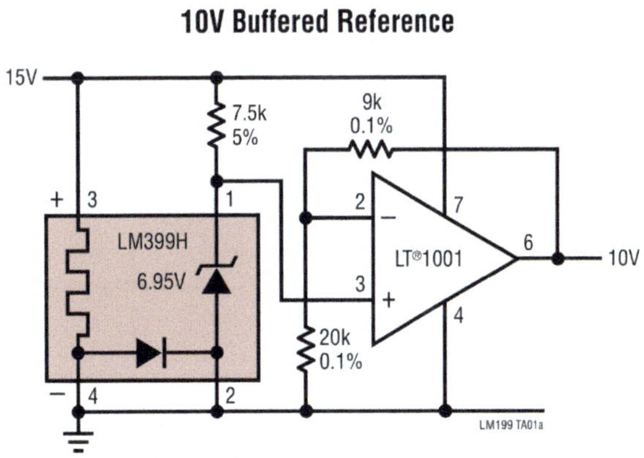

Fig. 5.35 LM399H application schematic

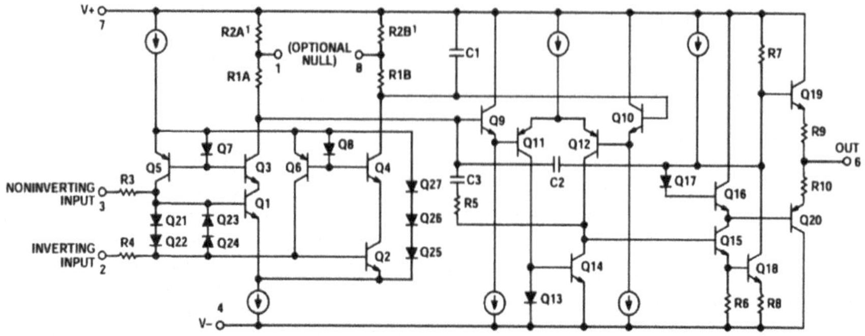

¹R2A AND R2B ARE ELECTRONICALLY ADJUSTED ON CHIP AT FACTORY FOR MINIMUM INPUT OFFSET VOLTAGE.

Fig. 5.36 OP07 chip internal schematic

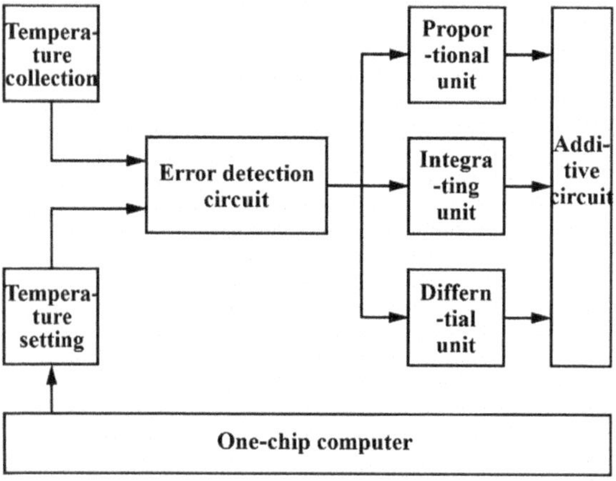

Fig. 5.37 Schematic diagram of temperature control principle

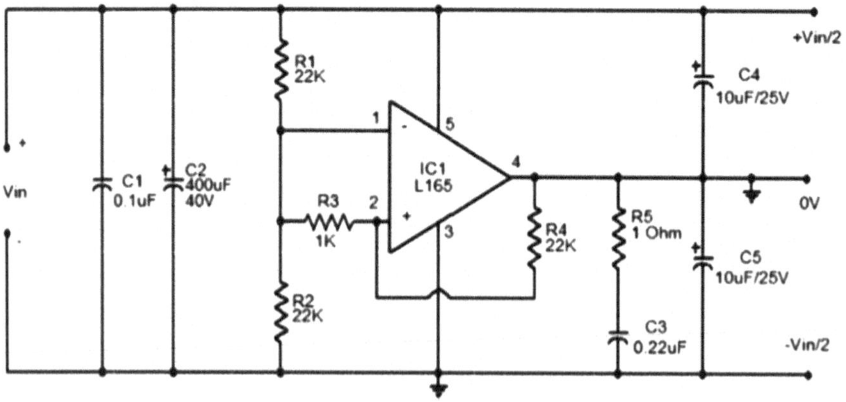

Fig. 5.38 L165CV application schematic

The principle of temperature control is shown in Fig. 5.38. A microcontroller is used as the main controller, and the temperature is set by setting the analog voltage of the digital-to-analog converter (DAC). A negative temperature coefficient thermistor (NTC) is used for temperature collection. The collected voltage value is compared with the set voltage value to obtain the temperature difference signal, complete the error signal extraction, and input it into the PID closed-loop control link. Finally, it is sent to the TEC through the addition circuit to control the temperature of the Faraday laser (see Fig. 5.37).

Output unit: It is built using the L165CV power operational amplifier and current limiting power resistor, and output to the semiconductor refrigerator (TEC). The L165CV can provide up to 3.5 A of current, with internal SOA protection, thermal protection, etc. The L165 is connected as a voltage amplifier, amplifying the available voltage at the junction between R1 and R2. The output voltage is always half of the input voltage (Vin). For example, if a 24 V voltage is provided at the input end (Vin), the output end gets + 12 V/− 12 V voltage. C2 is the input filter capacitor, C1 is the input decoupling capacitor. C4 and C5 are used to improve the symmetry of the output voltage (Fig. 5.38).

5.3.2 *Current Control Circuit System*

1. Current control circuit

The output characteristics of the laser are closely related to the performance of its driving power supply. The fluctuation and noise of the current affect the stability of the laser output wavelength and power. A constant current output mA-level current control circuit is designed according to the needs of the laser. It is roughly divided into several parts for unit analysis, voltage regulation circuit, current limiting unit, current driving part. The circuit design schematic is shown in Fig. 5.39.

2. Current control circuit principle analysis

Voltage regulation circuit design: The linear voltage regulator uses LM317. LM317 is one of the most widely used power integrated circuits. It does not only has the simplest form of fixed three-terminal voltage regulator circuit, but also has the feature of adjustable output voltage. In addition, it also has the advantages of wide voltage regulation range, good voltage regulation performance, low noise, and high ripple suppression ratio. The LM317 is an adjustable 3-terminal positive voltage regulator that can provide more than 1.5 A of current when the output voltage range is 1.2–37 V. Therefore, the voltage required by the Faraday laser can be adjusted adaptively using the sampling operation circuit according to the needs.

LM317 characteristic parameters and reference circuit: as shown in Table 5.5 and Fig. 5.40.

The voltage reference source is built with LM399, reference circuit: Fig. 5.41.

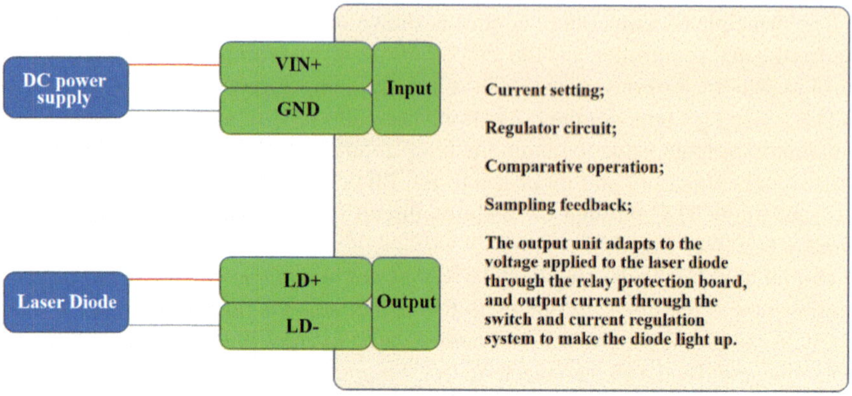

Fig. 5.39 Current control circuit schematic

Table 5.5 LM317 characteristic parameters

Serial number	Characteristic parameters	Serial number	Characteristic parameters	Serial number	Characteristic parameters
1	Adjustable output voltage down to 1.2 V	4	Typical load adjustment rate 0.1%	7	Overcurrent, overheat protection
2	Output current 1.5 A	5	20 dB ripple suppression ratio	8	Protection of the safe working area of the adjustment tube
3	Typical line regulation rate 0.01%	6	Output short circuit protection	9	Standard three-terminal transistor package

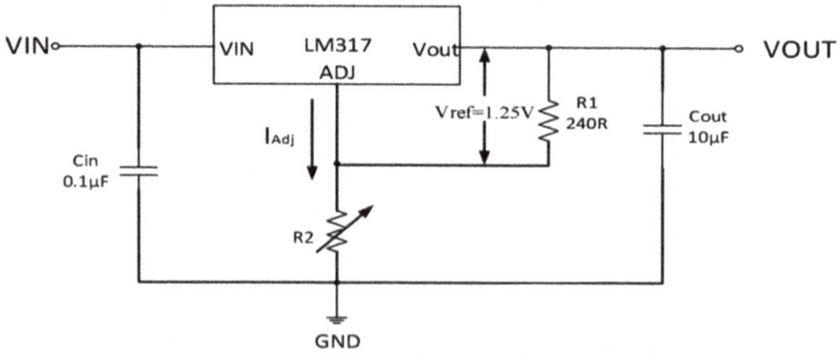

Fig. 5.40 LM317 application schematic

Fig. 5.41 LM399H reference circuit

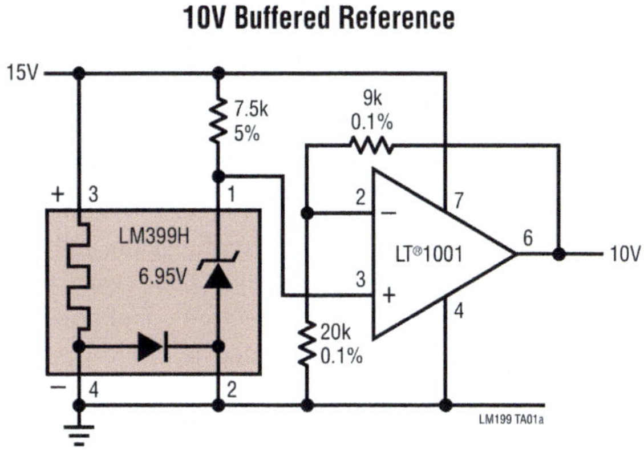

Current limiting unit: built with operational amplifier OPA2192, OPA2192 has excellent DC accuracy and AC performance, including rail-to-rail input/output, low offset (typical value: ± 5 μV), low temperature drift (typical value: ± 0.2 μV/°C) and 10 MHz bandwidth. OPA2192 characteristic parameters as shown in Table 5.6, reference circuit see Fig. 5.42.

Table 5.6 OPA2192 characteristic parameters

Serial number	Characteristic parameters	Serial number	Characteristic parameters	Serial number	Characteristic parameters
1	Low offset voltage: ± 5 μV	5	Low bias current: ± 5 pA	9	Low quiescent current: 1 mA per amplifier
2	Low offset voltage drift: ± 0.2 μV/°C	6	Rail-to-rail input and output	10	Wide power supply voltage: ± 2.25 to ± 18 V, 4.5–36 V
3	Low noise: 5.5 nV/at 1 kHz $\sqrt{Hz}$	7	High bandwidth: 10 MHz GBW	11	Input filtered for electromagnetic interference (EMI)/radio frequency interference (RFI)
4	High common-mode rejection: 140 dB	8	High slew rate: 20 V/μs	12	High capacitive load drive capability: 1 nF

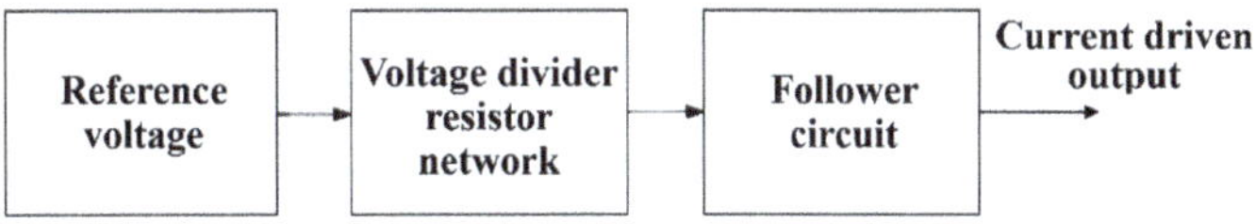

Fig. 5.42 Current setting principle diagram

Current drive design: Built with the operational amplifier AD8042, the AD8042 is a low-power, high-speed voltage feedback amplifier, powered by + 3 V, + 5 V, or ± 5 V. It has single power supply capability, with an input voltage range from 200 mV below the negative supply rail to within 1 V of the positive supply rail.

The output voltage swing can be extended to within 30 mV of each supply rail to provide a larger output dynamic range. In addition, its 0.1 dB gain flatness bandwidth is 14 MHz, and the differential gain and phase error while using a 5 V single power supply are 0.04% and 0.06° respectively. Therefore, the AD8042 is suitable for professional video electronic equipment applications, such as cameras, video switchers, or any high-speed portable devices. Its low distortion and fast settling characteristics make it an ideal buffer for single power supply, high-speed analog-to-digital converters (ADCs).

The power supply current of the AD8042 is low, with a maximum of only 12 mA, and can be powered by a 3.3 V single power supply. These features are very suitable for portable and battery-powered applications with strict size and power consumption requirements.

When the AD8042 is powered by a 5 V single power supply, it has a bandwidth of 160 MHz and a slew rate of 200 V/μs, so it can be used for many general-purpose high-speed applications that require a + 3.3 to + 12 V single power supply and up to ± 6 V dual power supply.

The characteristics of the AD8042 are shown in Table 5.7, and the reference circuit is shown in Figs. 5.43 (simplified schematic), 5.44 (capacitive load and closed-loop gain), and 5.45 (current drive principle diagram).

Table 5.7 AD8042 characteristic parameters

Serial number	Feature parameters	Serial number	Feature parameters
1	Rated power supply voltage: + 3 V, + 5 V and ± 5 V	5	Input exceeding power supply voltage by 0.5 V will not invert
2	Output swing extends to within 30 mV range of any power rail	6	Low power consumption: 5.2 mA per amplifier
3	Input voltage range extends 200 mV below ground voltage	7	Slew rate: 200 V/μs
4	0.1% settling time: 39 ns	8	Drives 50 mA 0.5 V from power rail

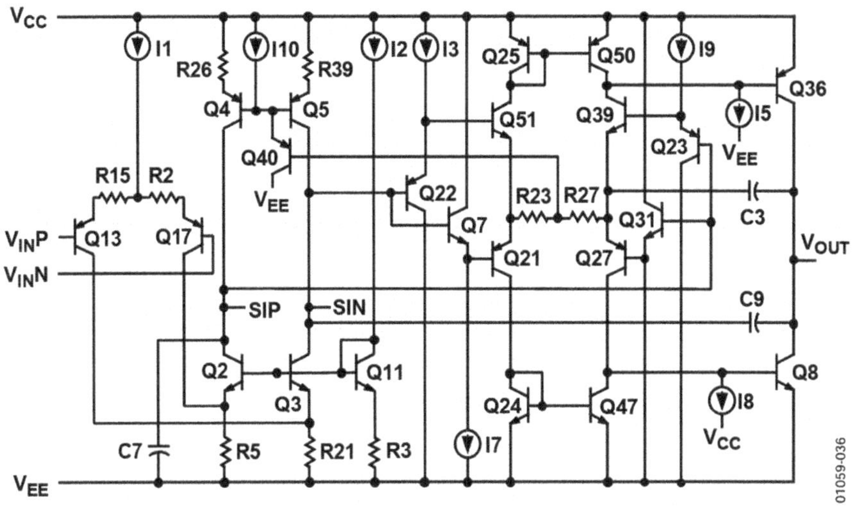

Fig. 5.43 Simplified schematic of AD8042

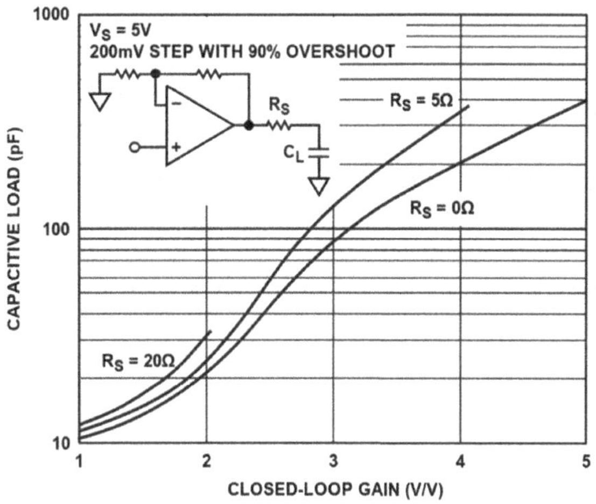

Fig. 5.44 Capacitive load and closed-loop gain of AD8042

5.4 Laser Stabilization Power Supply

5.4.1 Composition of Integrated Stabilization Circuit

Detailed design of the integrated stabilization circuit for the Faraday laser. This stabilization circuit can be divided into 10 modules, including power interface circuit design, SPI (Serial peripheral interface) topology circuit design, main control circuit

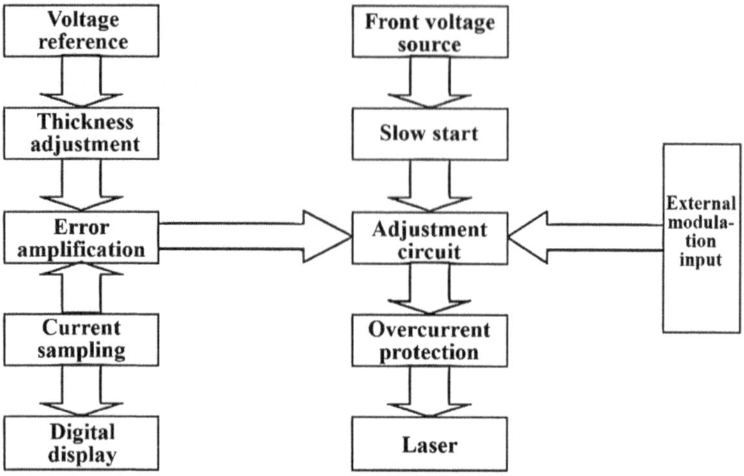

Fig. 5.45 Schematic diagram of current drive circuit

design, laser diode current source circuit design, LD (Laser diode) primary temperature control circuit design, LD secondary temperature control circuit design, PZT (Piezoelectric ceramics) drive circuit design, PID (Proportional integral differential) feedback circuit design, ADC (Analog-to-digital converter) acquisition circuit design, signal source circuit design. The circuit designs are as follows:

(1) Power interface circuit design, as shown in Fig. 5.46.
(2) SPI topology circuit design, as shown in Fig. 5.47, using multi-channel optocoupler isolation to drive different SPI chips.

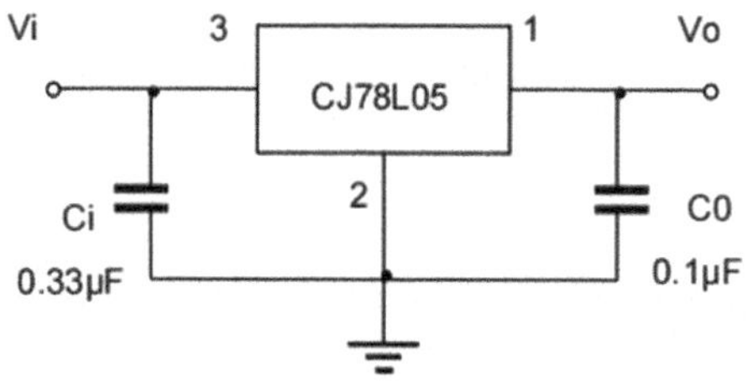

Fig. 5.46 Power interface circuit

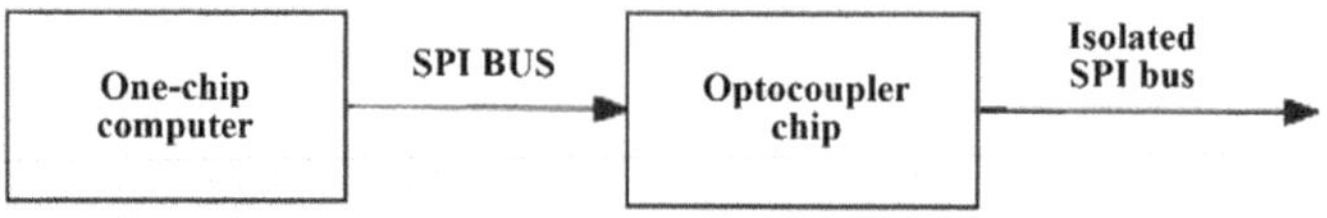

Fig. 5.47 SPI topology circuit design

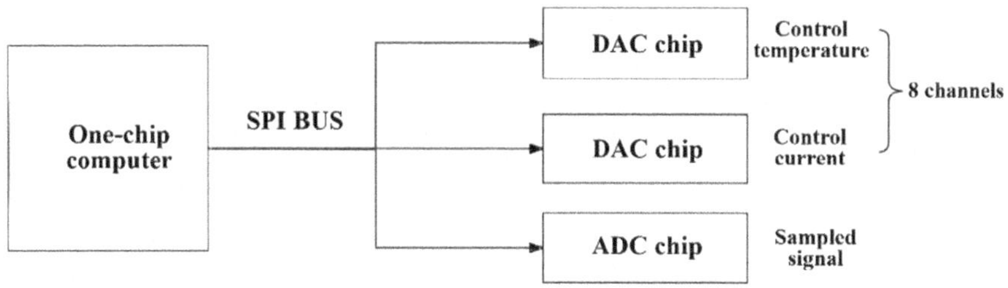

Fig. 5.48 Main control circuit block diagram

(3) Main control circuit design, as shown in Fig. 5.48. The currently used Arduino system development board (an open-source electronic prototyping platform, including software and hardware) needs to control the external circuit through the interface brought out by the development board.

The main control part of the original Arduino system digital board includes a set of SPI interfaces that are led out, with MISO directly connected to the AD_OUTA network of the ADC, including other ADC control ports, + 12 V power supply, the development board outputs + 5 and + 3.3 V, 8-channel relay control interface, 8-channel SPI control interface.

(4) Design of the laser diode current source circuit, as shown in Fig. 5.49. AD5541 uses a + 14 V power supply, uses the reference source LM399, obtains an output voltage of 0–5 V through voltage division, the output voltage passes through a voltage follower composed of OP07, the low output impedance of the voltage follower can enhance the driving ability, and the high input impedance can stabilize the output voltage of the reference source, and then modulate through the 1 V/1 mA constant current source driving circuit.
(5) The LD temperature control circuit design is shown in Fig. 5.50. The LD temperature control circuit employs a reference voltage source. After the temperature setting is reduced in input impedance through a primary operational amplifier loop, it is input into a secondary operational amplifier loop together with the reference voltage to generate a temperature difference signal. This difference signal enters the feedback network loop to compensate the temperature difference to the input port of the primary operational amplifier, achieving the purpose of controlling temperature fluctuations.
(6) Design of the LD secondary temperature control circuit, as shown in Fig. 5.51. The circuit scheme is the same as the primary temperature control.
(7) PZT driver circuit design, as shown in Fig. 5.52. The reference source used is still the low noise high stability reference source LM399.
(8) PID feedback circuit design, as shown in Fig. 5.53. Among them, JP16 is the PZT integral feedback channel, JP4 is the fast reverse current channel, and JP8 is the slow reverse current channel.
(9) ADC acquisition circuit design, as shown in Fig. 5.54.

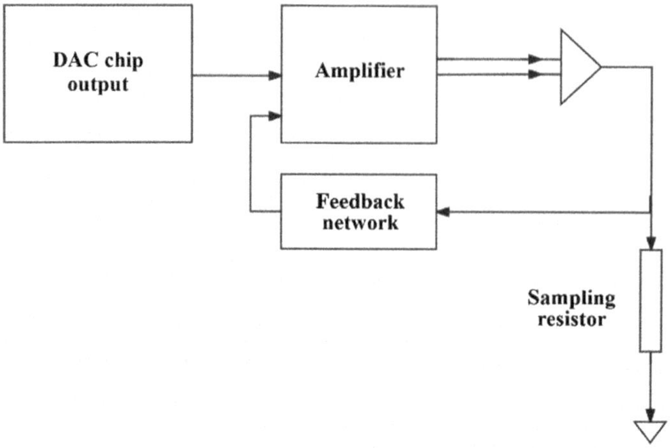

Fig. 5.49 LD current source circuit design

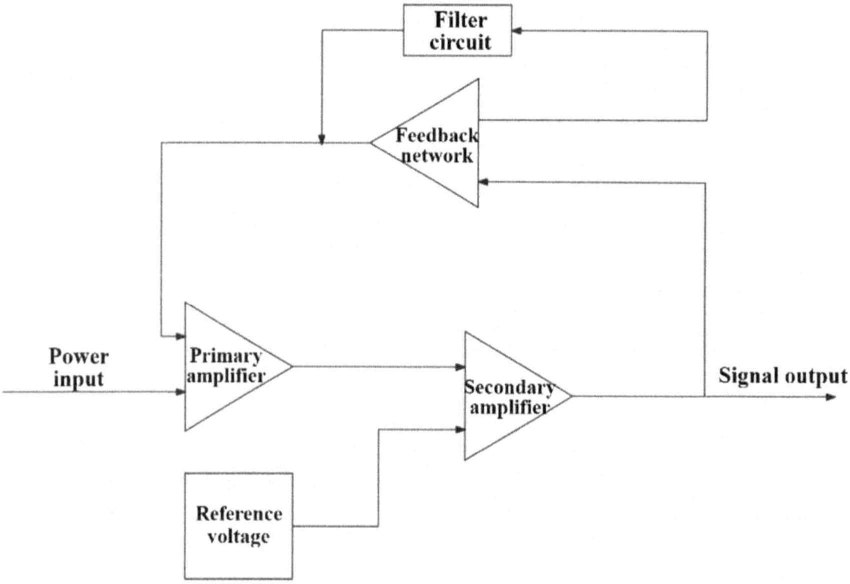

Fig. 5.50 Design of LD temperature control circuit

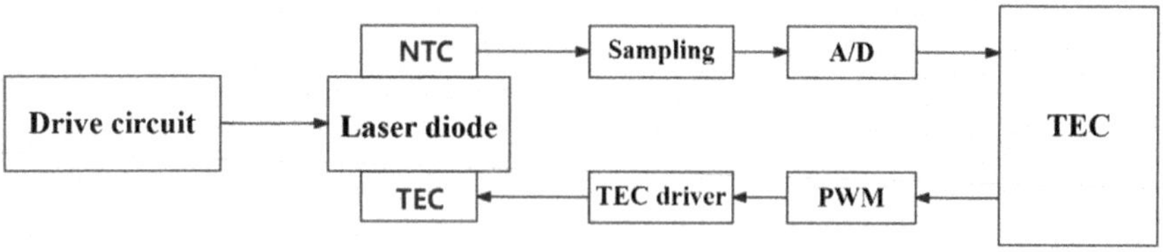

Fig. 5.51 LD secondary temperature control circuit design

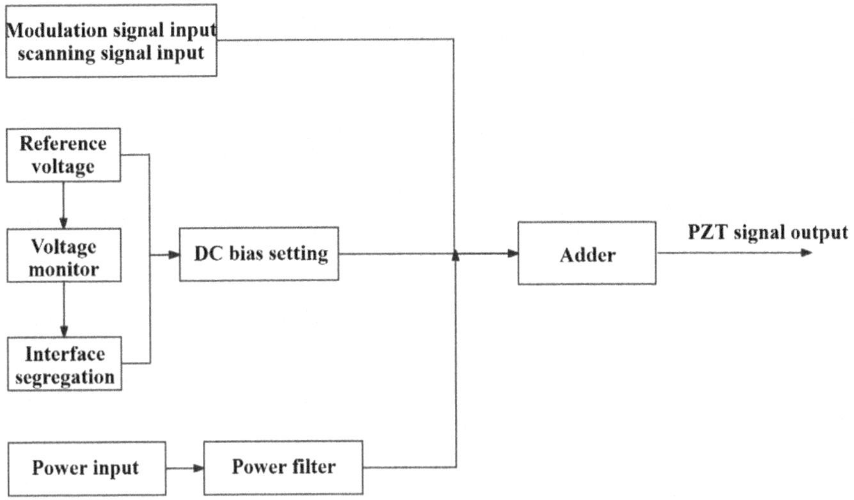

Fig. 5.52 PZT driver setting circuit block diagram

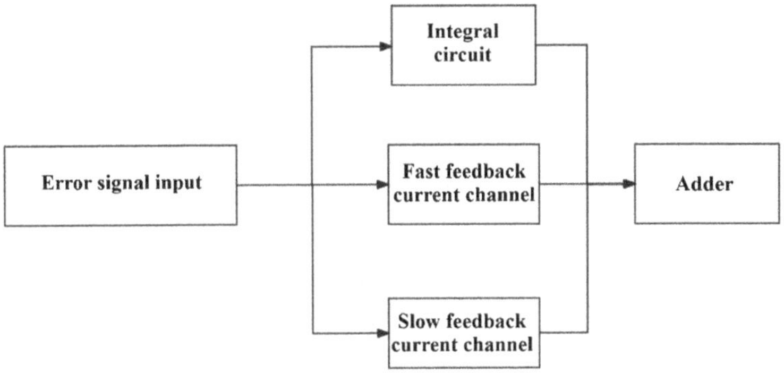

Fig. 5.53 PID feedback circuit block diagram

The input power supply of the ADC, the analog power supply uses + 5 V power supply, and the digital power supply uses + 3.3 V power supply. It uses serial read and write mode, the peripheral configuration is PAR/SER/BYTE SEL = 1, and DB15 = 0. STBY is standby mode, low level effective, default normal work, pull up to + 3.3 V_VDD power supply, the analog input range is set to ± 10 V. And the reference selection internal reference REF SEL = 1.

(10) Signal source circuit design. The system block diagram of the signal source circuit is shown in Fig. 5.55, including DDS chip, attenuation circuit, bandpass filter, power amplifier, etc., as well as other peripheral power supplies. It mainly generates the required sine wave through the DDS chip, adjusts the power at the output end, passes through the 3–6 MHz bandpass filter, then goes through

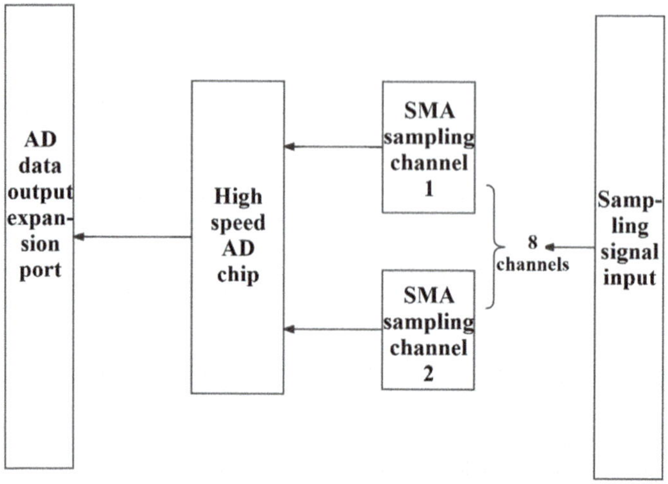

Fig. 5.54 AD7606 peripheral circuit block diagram

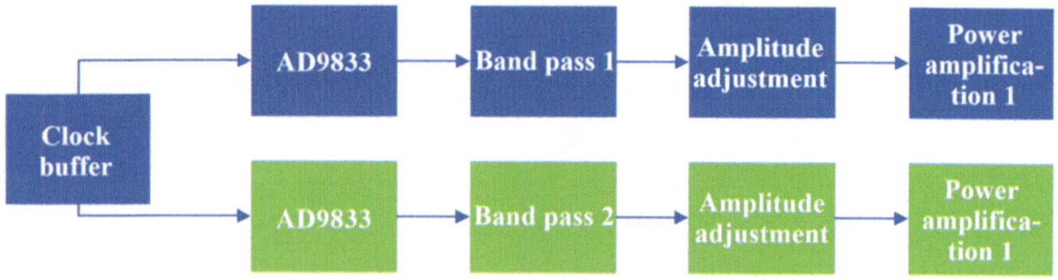

Fig. 5.55 Circuit system block diagram

power amplification, and outputs to the intrinsic signal of the mixer and drives EOM, respectively.

5.4.2 Integration Key Component of the Frequency Stabilization Circuit—Constant Current Source

In addition to the 10 modules that make up the integrated frequency stabilization circuit mentioned above, a separate constant current source design is also required to ensure the stable and reliable output of the laser frequency stabilization circuit. The performance of the current source directly determines the performance of the Faraday laser frequency stabilization system, so it is very necessary to deeply study the current source in the frequency stabilization system, and it is particularly critical to ensure the low noise and long-term temperature of the current source in the frequency stabilization system. For the low noise, long-term high stability design of the constant current source, precision signal chain, low noise regulator and switch converter, precise monitoring and reliable protection are essential. Higher system-level anti-interference and anti-noise performance is achieved on the power supply through high power supply suppression ratio (PSRR) low dropout regulator (LDO) and on-chip filtering. The selection of high stability, high precision IC, sampling resistor, etc., are all technical means to ensure the realization of low noise and high stability. The constant current source topology chooses to use high precision resistors as sampling resistors, operational amplifiers as negative feedback, and MOS tubes as current expansion forms. The error of the resistor will only affect the value of the output current, but will not affect the output characteristics. In actual use, VMOS tubes replace MOS tubes. VMOS tubes have smaller gate currents and larger drain currents, so that the operational amplifier does not need too much driving ability, and the circuit can work normally. VMOS tubes have the characteristics of good temperature stability and low noise, which make up for the shortcomings of transistors and help to improve the temperature stability of the constant current source. The selected operational amplifier has higher gain, lower input offset voltage and offset current, as well as low temperature drift and low noise voltage. In the actual circuit design, one should pay attention to the temperature rise caused by the layout of local area power devices, which can also play a good role in reducing the overall temperature coefficient. Using NI Multisim 14.0 simulation software, the basic circuit of the pre-adopted constant current source is shown in Fig. 5.56.

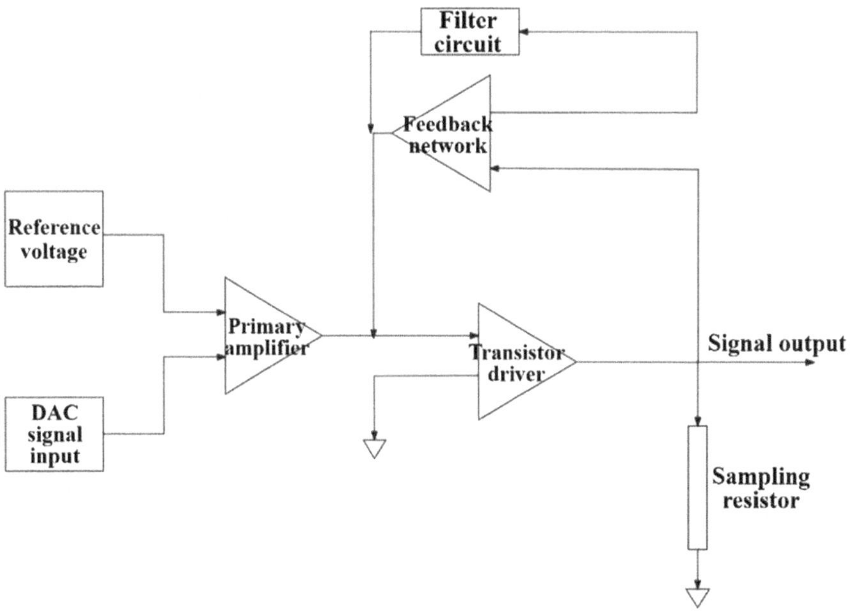

Fig. 5.56 Schematic diagram of constant current source circuit

During the simulation, LED is used to replace the laser diode (LD), the forward voltage drop of the LED is 2.18 V, and the current is 100 mA. The simulation result is that the voltage noise is ≤ 100 fV. In the actual circuit, due to the noise of the power supply, the noise of the voltage reference source, the noise of the resistor and the noise of the operational amplifier, the total noise of this constant current source sums up to 0.1 μA. The temperature drift of this constant current source circuit depends on the temperature stability of the sampling resistor R10 in the circuit. The actual use of ± 0.1 ppm precision non-inductive resistor is used to ensure the long-term stability of the constant current source. At the same time, multiple identical sampling resistors are used in parallel, such as five 50 Ω in parallel to 10 Ω, which reduces the current of each sampling resistor and also reduces the temperature rise of the sampling resistor to ensure the long-term stability of the constant current source.

Common measures for optimizing the noise of the constant current source include: ① overall thermal stability improvement, constant temperature insulation treatment of sensitive devices; ② use of low noise electronic components to reduce the inherent noise caused by the device itself; ③ use of effective electromagnetic shielding materials to reduce the pickup of spatial clutter and avoid the generation of interference current. The test results are shown in Figs. 5.57, 5.58, 5.59 and 5.60.

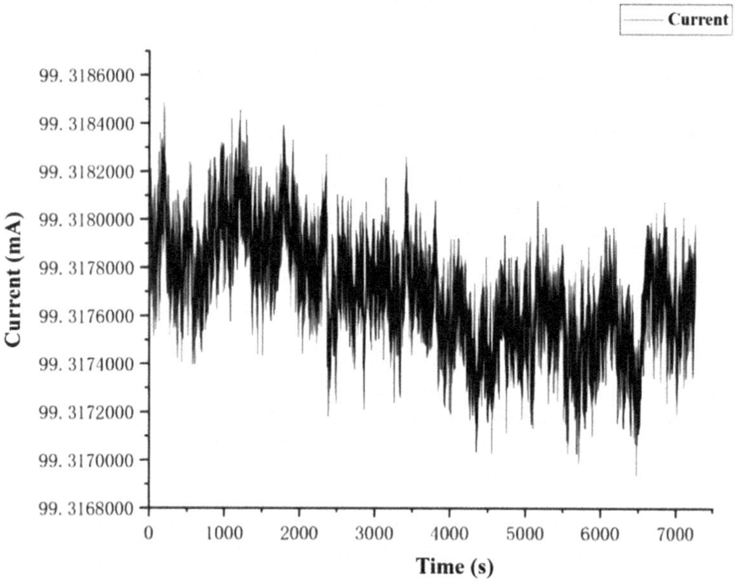

Fig. 5.57 Constant current source module current noise test

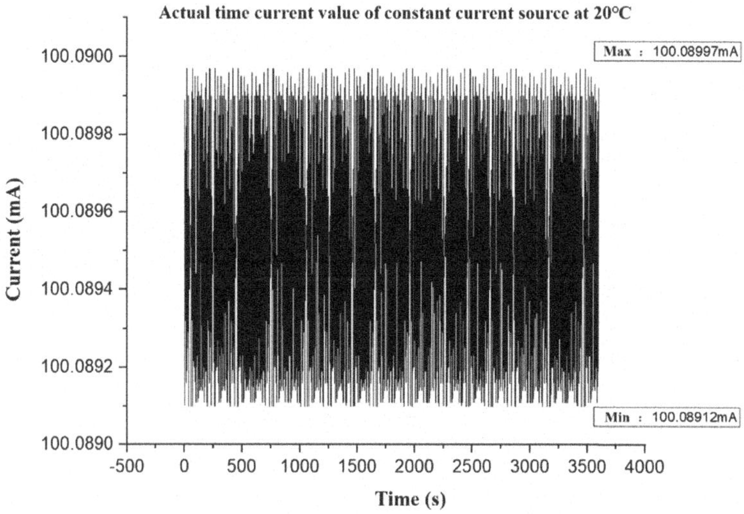

Fig. 5.58 Constant current source 20 °C test

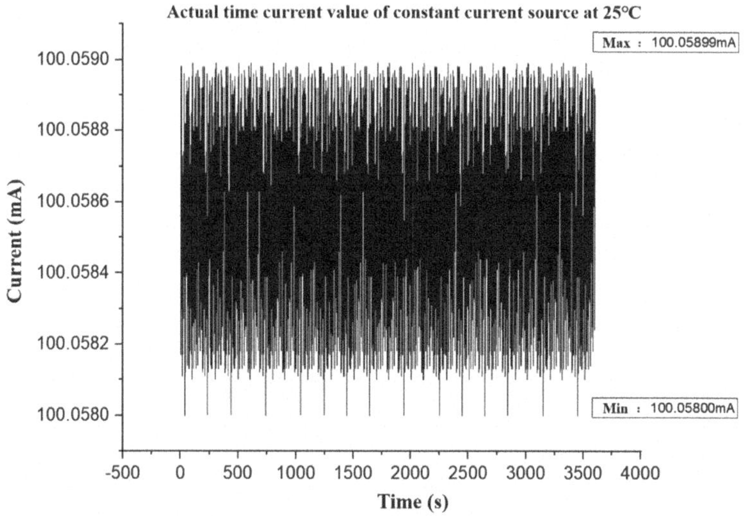

Fig. 5.59 Constant current source 25 °C test

5.4.3 Key Component of Integrated Frequency Stabilization Circuit-High Precision Temperature Control Module

To improve temperature stability, it is necessary to optimize the results of long-term temperature control. Since the Faraday laser frequency is greatly affected by

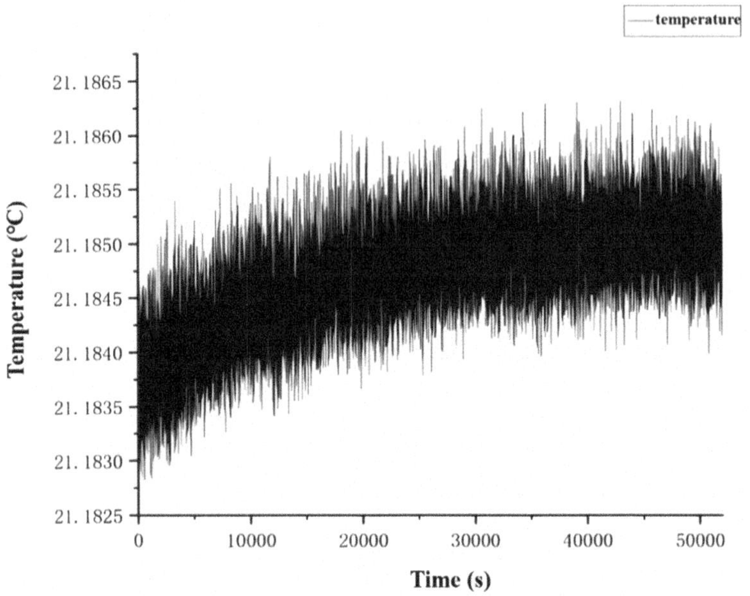

Fig. 5.60 Constant current source room temperature—temperature drift test

parameters such as current, temperature, and cavity length, one of the measures to minimize the frequency jitter and drift of the Faraday laser itself is to have a highly stable temperature control system. The current high-precision digital temperature control system still has defects such as insufficient temperature stability and insufficient resolution. The temperature control of the commonly used frequency-stabilized laser is still in the form of an analog PID module, the basic principle is shown in Fig. 5.61. The basic principle of the circuit is to use a thermistor of 10 kΩ@25 °C in the laser diode. This reference voltage source drives the thermistor and converts the temperature signal into a voltage signal through a voltage divider. After differential amplification with the set voltage, it is processed by PI and output to the power output stage combined with current negative feedback to drive the TEC, thereby controlling the temperature of the laser diode.

The advantage of this method is that it is implemented by pure analog circuits, with fast adjustment speed and simple structure. The main disadvantage is that it needs to manually replace the resistor and capacitor components for different thermal loads, which does not meet the requirements of digital systems and can only be implemented in analog. The current commercial diode laser temperature controllers have many models, such as the LDT-5900C 32 W laser diode temperature controller from Newport, USA, as shown in Fig. 5.62, and the TED200C laser diode temperature controller from Thorlabs, USA.

The LDT-5910C product is a separate system with a large size and cannot be integrated. It has no corresponding control interface. The advantage that we can draw from the product is the fully programmable PID loop. In addition, it supports

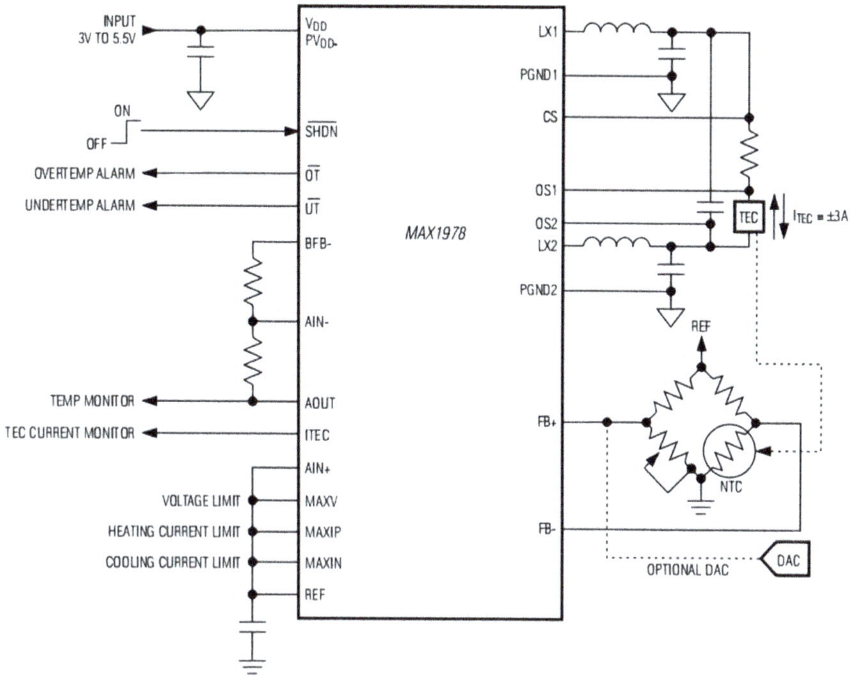

Fig. 5.61 Analog temperature control circuit of the laser

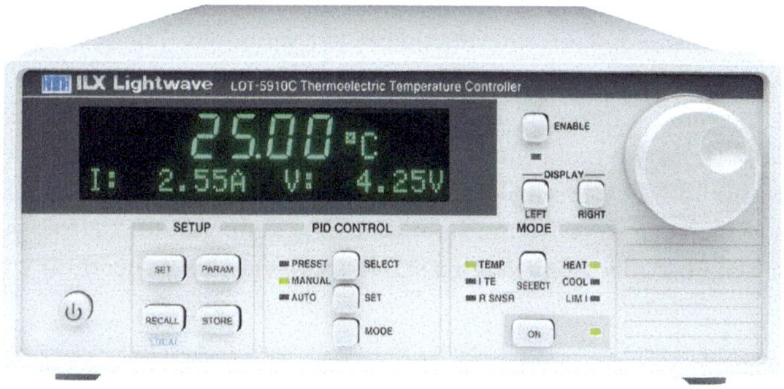

Fig. 5.62 Newport's LDT-5900C laser diode temperature controller

almost all common temperature sensors, such as thermistors, RTDs, IC sensors, etc., and can adapt to different thermal loads.

Thorlabs' TED200C is a precision temperature controller designed to drive thermoelectric cooler (TEC) components, with a current of up to ± 2 A. TED200C is a separate digital PID, P, I, D can adjust performance parameters separately, and the product size cannot be integrated, as shown in Fig. 5.63.

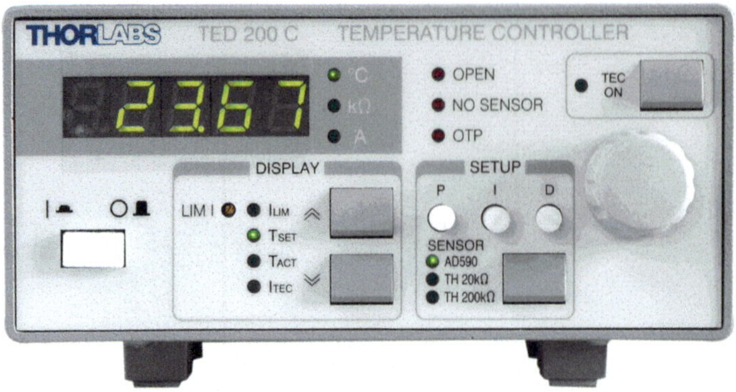

Fig. 5.63 Thorlabs' TED200C laser diode temperature controller

The temperature control system of the fully automatic laser frequency stabilization system jointly developed by Peking University and Zhejiang Faraday Laser Technology Co., Ltd., as shown in Fig. 5.64, this system uses a high-resolution DAC to set the temperature value to achieve a setting step better than 50 μK. In addition, it adopts an incremental digital PID, which can control the values of P, I, and D separately in a digital way to ensure that the temperature stability is less than ± 5 mK.

Due to the significant impact of external environmental temperature disturbances on the laser, in order to achieve long-term temperature stability within ± 5 mK, we have implemented a filtering process on the acquired temperature data to improve its accuracy. The Kalman filter is the optimal filter directly designed in the time domain. In 1961, Kalman extended this filtering theory to continuous time systems, forming a complete system of Kalman filtering. For systems where the system noise is Gaussian distributed, the minimum mean square estimate can be recursively derived. For the temperature system, the main interference to the system measurement accuracy is Gaussian white noise, so the Kalman filter can be introduced into the temperature application, which can improve the stability of the temperature control system. The temperature data is collected through a stable frequency laser spectrum weak signal

Fig. 5.64 Fully automatic laser frequency stabilization system jointly developed by Peking University and Zhejiang Faraday Laser Technology Co., Ltd.

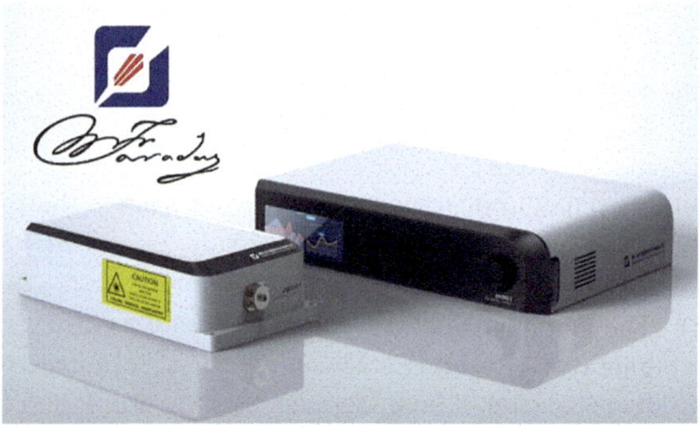

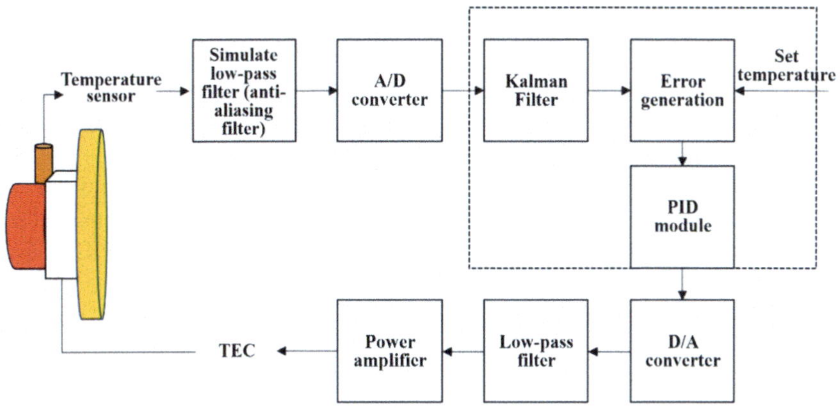

Fig. 5.65 Digital temperature control system block diagram

multi-channel high-speed high-precision data acquisition system, sent to FPGA or DSP for Kalman filter data processing, and then incremental PID data processing is used to control the TEC heating or cooling laser.

The system mainly includes digital circuits and analog circuits. The digital circuit is used for data processing, such as Kalman filtering and incremental PID data processing. The analog circuit is used for data collection and DAC digital-to-analog conversion, analog low-pass filtering, and finally power amplification. The workflow of each part of the circuit is as shown in Fig. 5.65.

The digital temperature control process shown in Fig. 5.65 includes the following 5 steps:

(1) The temperature sensor signal first passes through an analog low-pass filter (anti-aliasing filter). Since the temperature signal is a slow variable, the cut-off frequency of the low-pass filter can be set to below 10 Hz to filter out high-frequency components that affect the signal quality and reduce high-frequency components from aliasing into the digital signal;
(2) The temperature voltage signal required is collected by the stable frequency laser spectrum weak signal multi-channel high-speed high-precision data acquisition system to the ARM or FPGA internal, and then the collected temperature voltage value is used for Kalman filtering to improve the temperature control accuracy and reduce errors;
(3) Subtract the set value to obtain the error relative to the set value, input the incremental PID module, and get the control increment $\Delta u(k)$ after PID operation;
(4) Then it is converted into an analog voltage by a high-precision DAC, and then low-pass filtered again to filter out the high harmonics of the DAC;
(5) The output analog voltage signal is power amplified, usually by controlling the gate voltage applied to the FET to control the current value loaded on the TEC,

used to drive the semiconductor cooling piece TEC, thereby controlling the temperature of the object.

The temperature control uses NTC10K as the temperature sensor to characterize the temperature, so the measurement of temperature fluctuations can be converted into the measurement of thermistor resistance. The temperature control circuit can be divided into analog temperature control and digital temperature control from the control method. The current commercial solutions generally use analog temperature control. The existing MAXIM's MAX1978 and ADI's ADN8831 are both PID adjustment loops composed of internal operational amplifiers. The power output stage is PWM adjustment implemented by MOS tubes. The current on the TEC is adjusted by the duty cycle. The specific circuit implementation form is shown in Fig. 5.66, this method has high efficiency, but may introduce PWM frequency interference.

Therefore, considering the form of temperature control, analog temperature control is used. In order to reduce the interference that TEC current may introduce, the power output stage is changed to power amplifier output, sacrificing efficiency, requiring a larger heat sink, reducing ripple, and reducing the interference of temperature control on other modules. The specific power amplifier circuit form is planned to be adopted as shown in Fig. 5.67.

The temperature stability of temperature control is a key indicator. By testing the voltage on the thermistor and by calculating the temperature stability through the corresponding relationship between the resistance of the thermistor and the temperature, the variation range does not exceed 60 mV. According to the relationship that the resistance of the NTC thermistor corresponds to the temperature near 25 °C,

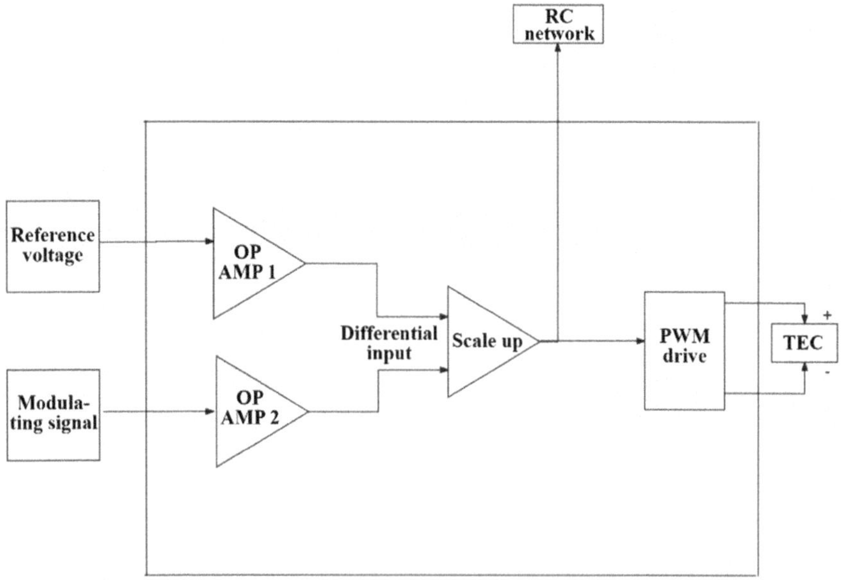

Fig. 5.66 MAX1978 circuit form

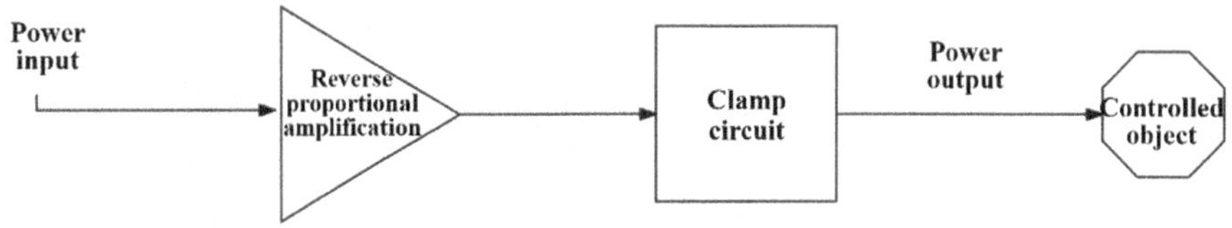

Fig. 5.67 Power amplifier output stage

1.5 mK corresponds to 100 mV, so the maximum temperature change is ± 0.3 mK, as shown in Fig. 5.68.

At the same time, Peking University and Zhejiang Faraday Laser Technology Co., Ltd. jointly developed a dual-channel 0–5 MHz digital signal source module. The dual-channel DDS frequency and phase can be independently configured, supporting high-stability sine wave, triangle wave and other waveform outputs. The signal processing module was developed, using a 4-layer board design to retain a complete inner layer plane to ensure the continuity of the impedance of the local oscillator signal and fluorescence signal loop, and can emit a high signal-to-noise ratio error signal by mixing frequencies. The frequency locking module is developed, the frequency locking PID part is digitally controlled by the MCU, and it is combined with the signal source module, signal processing module, and temperature control module to achieve automatic frequency locking. As shown in Figs. 5.69, 5.70 and 5.71.

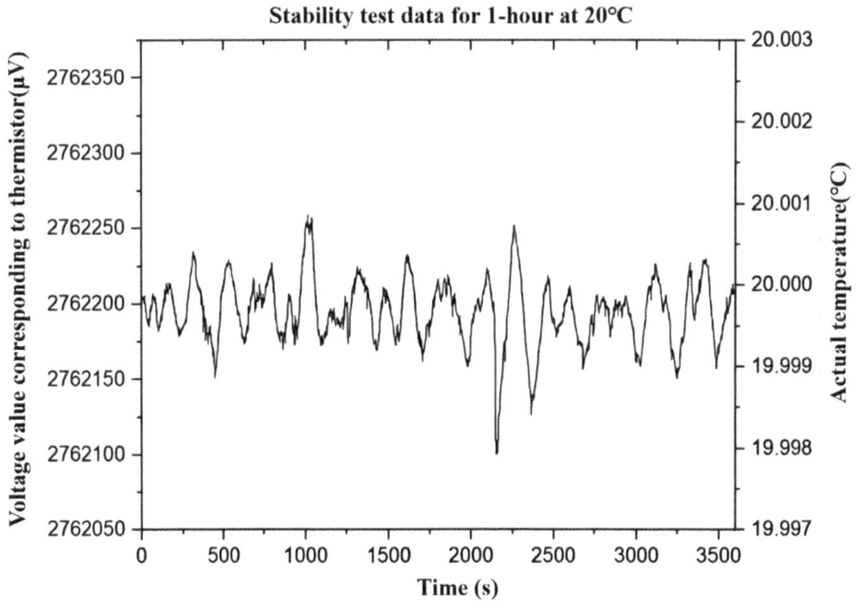

Fig. 5.68 20 °C temperature control 1 h test indicator

Fig. 5.69 Signal source module

Fig. 5.70 Frequency stabilization module

Fig. 5.71 Frequency locking module

5.4.4 Integration Key Component of the Frequency Stabilization Circuit—Power Stabilization Module

The stability of Faraday laser power is particularly critical for related research in the field of quantum precision measurement. For example, for the frequency stability of cold atomic clocks, it is necessary to enter 10^{-15} at an integration time of 10^4 s, which requires the relative power fluctuation of the laser ($\Delta I/I$) to be at the level of 10^{-5}. In addition, the measurement accuracy/resolution of many light absorption type measurement systems is directly related to the stability of laser power. For this reason, automatic laser power compensation technology is used. According to the laser power monitored by the photodetector, the amplitude of the radio frequency voltage injected into the acousto-optic crystal is adjusted, thereby changing the diffraction efficiency of the acousto-optic crystal, achieving the purpose of controlling the stability of the output light power, and making the laser power output meet the requirements.

There are many factors that affect the long-term stability of laser power. On the one hand, there are power fluctuations caused by the drive current from the laser itself. All laser drivers inevitably have current drift and noise, which cause the laser power to change in the same way as the current. The quantum deficit of diode lasers turn the laser diode into a heat source, and the temperature changes caused by heat also affect the laser output power. Frequency-stabilized lasers have high-precision temperature and current control circuits, but the process of laser frequency stabilization uses the characteristics of the injection current related to the laser frequency and response speed, and stabilize the laser frequency by changing the current. Long-term test results show that the fluctuation of laser current is around $\pm$ 8 μA. For lasers with an output power of several mW, this causes a power fluctuation of one in ten thousand; on the other hand, external optical paths, including those for laser transmission and dispersion processes, can affect power stability. Polarization components (e.g., wave plates, single-mode fibers) in lasers and optical paths may undergo changes in their optical characteristics due to external factors such as locking point drift and airflow temperature variations. For example, the polarization plane of the laser rotates with the increase of operating time. Although polarization-independent spectroscopic devices are used, it still causes changes in the spectroscopic ratio; fluctuations in environmental temperature directly cause changes in the optical path direction, etc. These factors cause the effect of power stability to deteriorate.

In order to solve the problem of long-term stability of output light, the method of external modulation of the laser is adopted, and the laser power fluctuation information is obtained by spectroscopic sampling. In the current laser system, to avoid sampling interference, a linear thin film polarizer with a high extinction ratio (> 40,000) is added before the laser is emitted from the single-mode polarization-maintaining fiber and enters the spectroscopic prism, and its transmission polarization direction is fixed at an appropriate position. In this way, the rotation of the laser polarization plane can be converted into power fluctuations, and these fluctuations can be corrected by the subsequent power stabilization loop. As shown in Fig. 5.72.

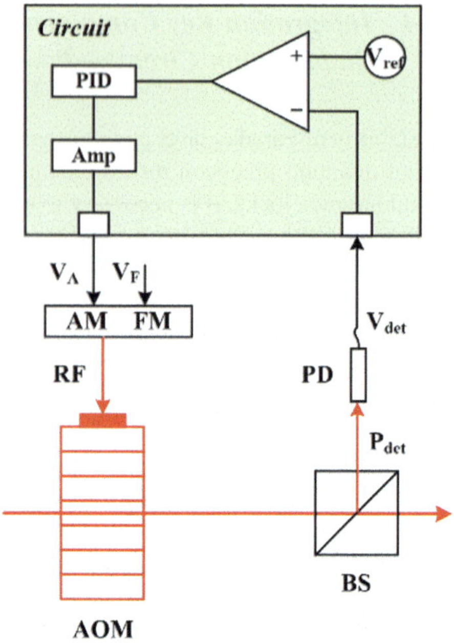

Fig. 5.72 Schematic diagram of power stabilization optical path

The laser power sampling and stabilization circuit uses precision-matched components. The power stabilization servo circuit and the trans-impedance amplifier of the monitoring light photodiode monitors the light photodiode using a PIN silicon photodiode (Hamamatsu S2386-44K), and the photoelectric conversion efficiency is 0.5 A/W. Since the monitoring light power is at the μW level, the photoelectric current after the photodiode conversion is at the μA level. The trans-impedance amplifier uses a precision operational amplifier OP27 and a 100 kΩ low-temperature drift resistor to convert the photoelectric current signal into a voltage signal. The current/voltage conversion factor is 10^6, and the voltage is amplified 10 times after conversion. The trans-impedance amplifier circuit has a filter bandwidth of 100 kHz, which has a certain filtering effect on noise. The voltage reference source of the servo control circuit selects the MAX6341 chip (+ 4.096 V output, temperature drift coefficient 1 ppm/°C), and the output voltage after proportional integral operation is injected into the subsequent acousto-optic driver after low-pass filtering (bandwidth is about 15.9 kHz). The acousto-optic crystal acts as the final power adjustment for the laser, mainly by changing the amplitude voltage of the acousto-optic driver to control its microwave output power, thereby changing the light diffraction efficiency of the acousto-optic crystal to achieve power adjustment. Therefore, the noise level of its driver directly affects the quality of the detection light.

A dedicated acousto-optic crystal drive circuit is designed according to system requirements (frequency shift fixed at 125 MHz). The main idea of this drive circuit is to generate the required frequency signal from a low phase noise crystal oscillator (125 MHz) and a fine-tuning circuit, which is injected into the acousto-optic crystal

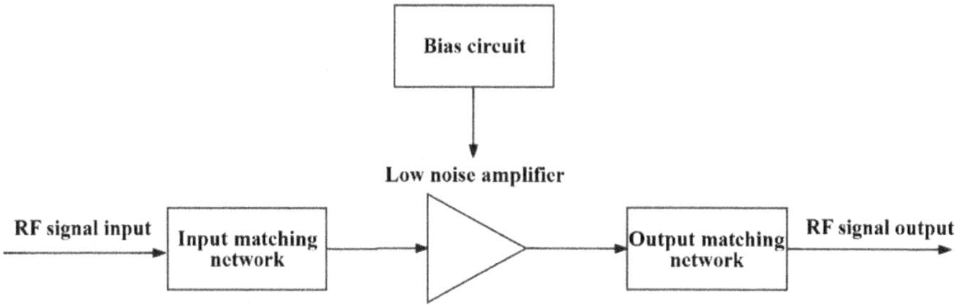

Fig. 5.73 RF circuit principle diagram

to produce acousto-optic diffraction after attenuation and power amplification. The circuit principle is shown in Fig. 5.73.

By packaging the stable power system into an independent module, it not only facilitates integration and use but also reduces external interference to the circuit, improving system stability.

After adopting the above measures, long-term monitoring and recording of laser power changes are conducted. After calculation, the Allan deviation of laser power fluctuations is shown in Fig. 5.74. It can be seen from the figure that after the above optimization, the Allan deviation, the long-term stability of laser power has been greatly improved.

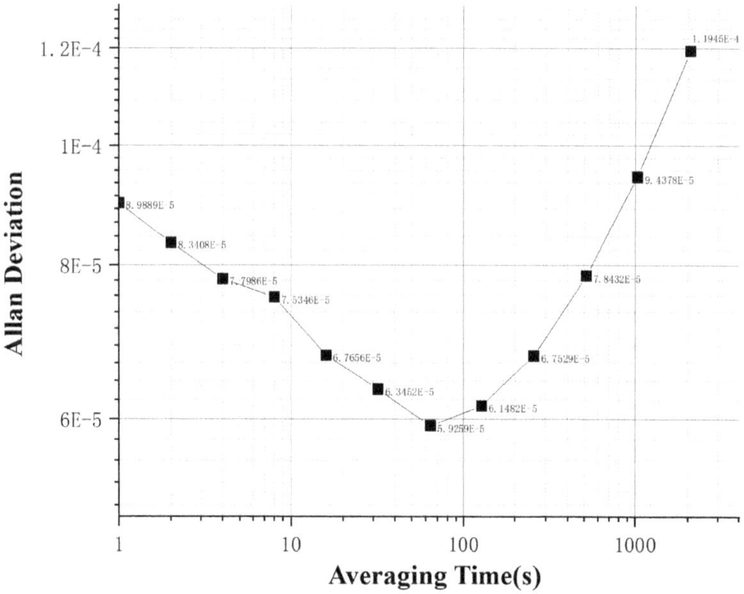

Fig. 5.74 Allan deviation of optical power fluctuations

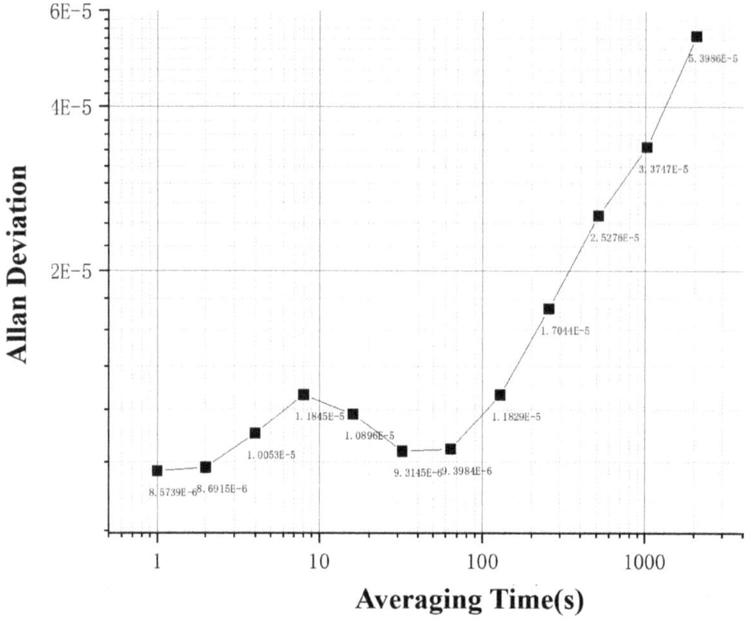

Fig. 5.75 Final stable power index

In addition, by developing a stable power drive circuit board and optical system, optical power stability tests are conducted. The design scheme of the circuit drive board and the optical system are continuously optimized, the parameters of the components in the fine-tuning circuit PID feedback part are finely adjusted, and the temperature drift devices that are greatly affected by temperature are upgraded and replaced. Low temperature drift, low jitter coefficient time–frequency devices are selected, achieving optical power second-sampling stability reaching 8.57×10^{-6}, and the thousand second sampling stability reaching 3.37×10^{-5}, as shown in Fig. 5.75.

5.4.5 Laser Frequency Stabilization Circuit Parameter Characteristics

In summary, the laser frequency stabilization circuit has the following four main characteristics:

(1) Stability

During operation, the laser may be affected by various external factors such as temperature, pressure, vibration, etc., causing its output frequency to drift, thereby affecting the applicability of the laser. To maintain the constancy of the laser output frequency, this stability is achieved by adopting specific control mechanisms and techniques, such as Frequency Lock Loop (FLL) or Phase Lock Loop (PLL), etc.

(2) Anti-interference

The laser frequency stabilization circuit usually has strong anti-interference ability and can maintain the stability of the laser output frequency in the presence of external interference. This characteristic allows the frequency stabilization circuit to work normally even in complex electromagnetic environments.

(3) Accuracy

The laser frequency stabilization circuit can control the laser to provide very accurate frequency output. By adopting high-precision control mechanisms and techniques, the laser frequency stabilization circuit can achieve very high frequency accuracy, meeting various precise control needs.

(4) Fast response

The laser frequency stabilization circuit usually has the ability to respond quickly and can adjust and correct the laser frequency changes in a short time. This characteristic allows the laser frequency stabilization circuit to adapt to rapidly changing environmental conditions and application needs.

The laser frequency stabilization circuit has characteristics such as stability, anti-interference, accuracy, and fast response. These characteristics make the laser frequency stabilization circuit play an important role in various applications that require precise laser frequency control.

5.5 Main Performance Features and Indicators

5.5.1 Overview of Faraday Laser Performance Features

The Faraday laser, as a new type of atomic frequency selection diode laser, essentially uses FADOF as an atomic frequency selection component. Through the special properties of the frequency selection component, the output laser wavelength can be limited within the Doppler broadening range of the atomic transition spectrum. At the same time, compared to traditional external cavity diode lasers, Faraday lasers have good robustness to diode drive current and working temperature fluctuations, and the output wavelength is always limited within the transmission bandwidth of the atomic filter, corresponding to the atomic transition spectrum. Faraday lasers have broad application prospects in areas such as optical frequency standards and laser cooling. With the development of Faraday laser-related science and technology, it inevitably plays an important role in more important fields, providing important support for theoretical innovation, technological innovation, and application innovation.

5.5.2 *Power-Current (P-I) Curve*

When the laser diode current is changed, the output frequency of the output laser drifts within the transmission bandwidth of the highest transmission peak of the atomic filter. As shown in Fig. 5.76, the transmission spectrum of the cesium atom FADOF, the transmission peak within the bandwidth range of the highest transmission spectrum is not flat, and the transmission changes from 0.65 to 0.75. Therefore, when changing the current, the total loss in the laser resonator also changes accordingly, causing fluctuations in the output power of the laser. Therefore, when the current is increased, the output power of the laser not only increases due to the increase in pump power, but also fluctuates due to the change in the transmission of the atomic filter. This fluctuation is strongly related to the longitudinal mode frequency of the laser output. Figures 5.77 and 5.78 show the changes in output wavelength and output power of the 852 nm Faraday laser when the current is increased. It can be seen that the output power generally shows an upward trend, but when the output laser undergoes mode hopping, the output power also jumps accordingly. Figure 5.79 shows the power change of the Toptica 780 nm corner-cube Faraday laser with current, and Fig. 5.80 shows the relationship between the output power of the dual-frequency 780 nm laser and the drive current.

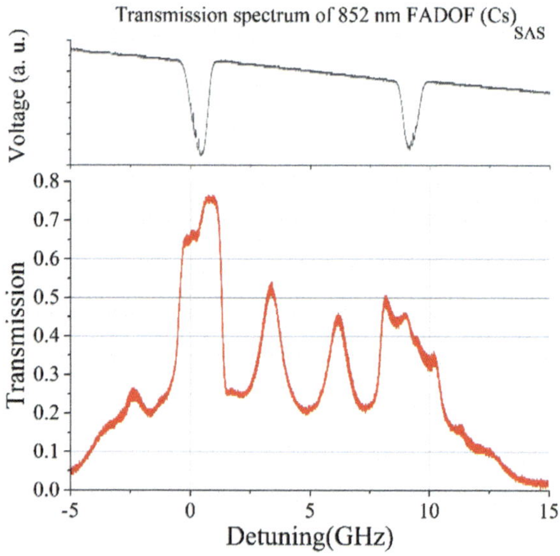

Fig. 5.76 Transmission spectrum of cesium atom FADOF

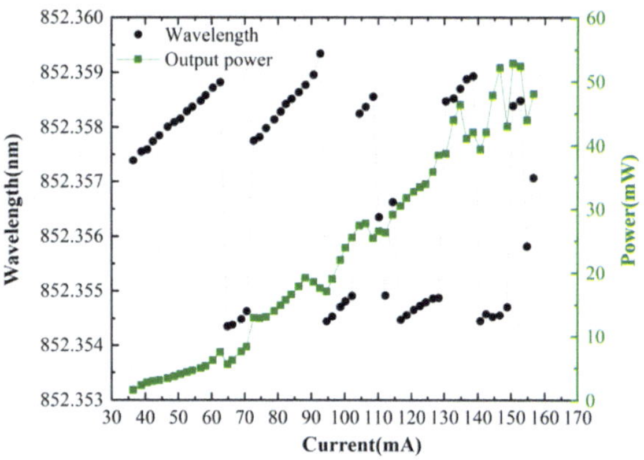

Fig. 5.77 Test results of cesium Faraday laser

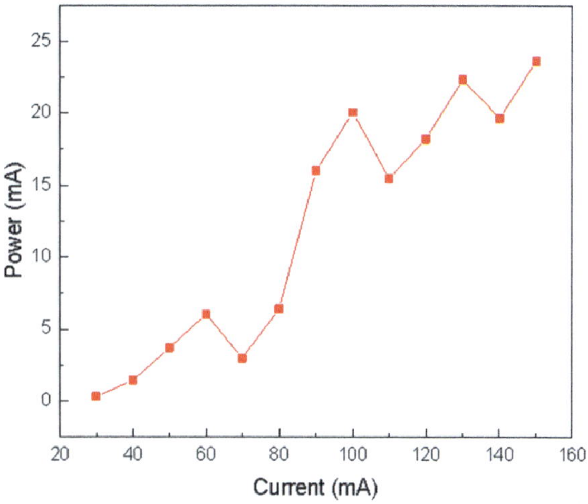

Fig. 5.78 Power change of Toptica 852 nm corner-cube Faraday laser with current

5.5.3 Temperature Change Resistance

In the overview of the Faraday laser, we mentioned that the main advantage of the Faraday laser compared to traditional external cavity diode lasers is its ability to intrinsically correspond with the target spectral line (usually the atomic spectral line) and its excellent robustness to the temperature and current of the laser diode. The key to achieve this ability is to reasonably design the structure and performance parameters of the atomic filter, so that the peak of its transmission spectrum corresponds to the target spectral line. The smaller the bandwidth of the transmission peak of the atomic filter, the smaller the frequency fluctuation of the output laser of the Faraday laser built, but correspondingly, it becomes more prone to laser mode hopping during free running. Therefore, while choosing the parameters of the atomic filter in the actual construction of the laser, one cannot choose the parameters of the

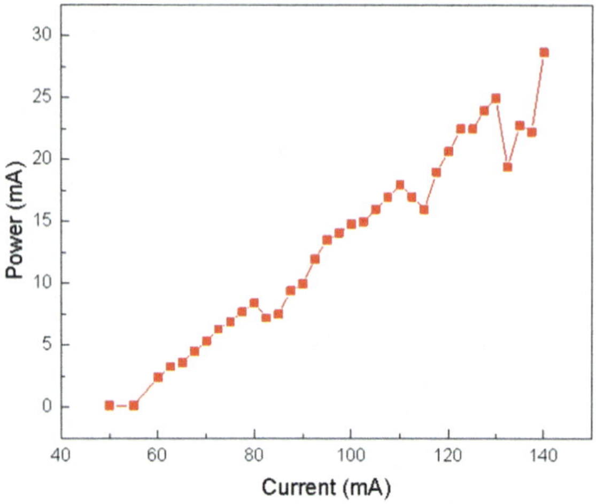

Fig. 5.79 Power change of Toptica 780 nm corner-cube Faraday laser with current

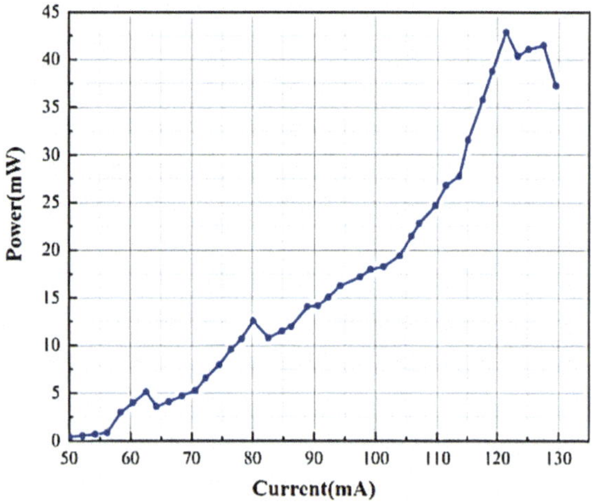

Fig. 5.80 Relationship between output power of dual-frequency 780 nm laser and drive current

atomic filter in the field of optical communication, the smaller the transmission bandwidth, the better. In actual use, one needs to consider the performance indicators of the laser, balance the range of output laser frequency fluctuations and the importance of mode hopping probability to the application, and then make a decision. Here, we take the 780 nm Faraday laser and the 780 nm Voigt laser as examples to analyze their resistance to changes in the temperature of the laser diode. Figure 5.81 shows the transmission spectrum of the atomic filter of the 780 nm Faraday laser and the relationship between the output wavelength and the temperature of the laser diode. It can be seen that when the ^{87}Rb Faraday atomic filter works at different temperatures, its transmission spectrum changes significantly. When the working temperature is 50 °C, the highest transmission peak corresponds to the Doppler broadening of the

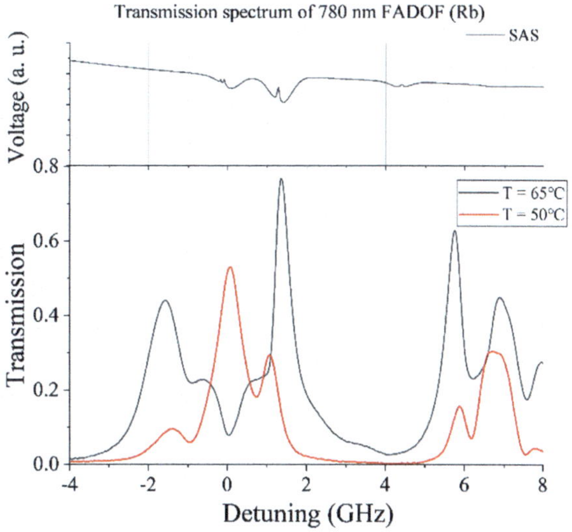

Fig. 5.81 Transmission spectrum of the atomic filter of the 780 nm Faraday laser

^{87}Rb atomic D_2 transition line, with a transmittance of about 0.55, and when the working temperature is 65 °C, the highest transmission peak corresponds to the Doppler broadening of the ^{85}Rb atomic D_2 transition line, with a transmittance of about 0.78.

While using this Faraday atomic filter to build a Faraday laser, the working temperature of the FADOF also causes the Faraday laser to have different output performance. When the working temperature is 50 °C, the output frequency of the laser should correspond to the Doppler spectrum of the ^{87}Rb atomic D_2 transition line, that is, the output wavelength fluctuates within the range of 1 ~ 2 pm near 780.246 nm, and the specific fluctuation range depends on the transmission bandwidth of the atomic filter, as shown in Fig. 5.81. At this time, if the temperature of the laser diode is changed, it causes the gain of the laser diode to change, and it also affects the length of the laser cavity, causing the output wavelength of the Faraday laser to drift, but no matter how the output wavelength drifts, it does not exceed the bandwidth range of the atomic filter, achieving excellent resistance to changes in the temperature of the laser diode. If the working temperature of the FADOF is set to 65 °C, according to the transmission spectrum of the atomic filter, the output frequency of the Faraday laser should be near the ^{85}Rb atomic D_2 transition line at this time, and the corresponding output wavelength is around 780.244 nm. The fluctuation is within the range of 1 ~ 2 pm near nm. Since the transmission bandwidth at the working temperature of 65 °C is narrower and the transmittance is higher than those at the working temperature of 50 °C, so the fluctuation range of the output wavelength becomes smaller and the output power becomes larger when the working temperature is 65 °C. Figure 5.82 shows the relationship between the output wavelength and the temperature of the laser diode when the working temperature of FADOF is 50 °C.

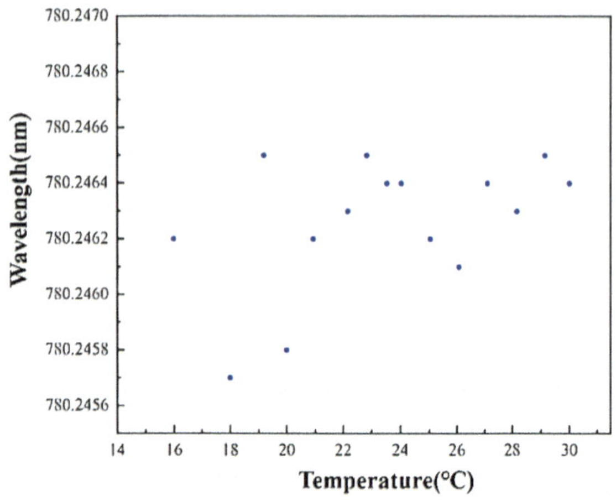

Fig. 5.82 Relationship between output wavelength and laser diode temperature (FADOF working temperature is 50 °C)

Not only does the Faraday laser have the ability to resist changes in the temperature of the laser diode, but the Voigt laser also has the ability to resist changes in the temperature of the laser diode. In fact, this resistance is an inherent attribute of diode lasers with atomic filters as frequency selection devices. The transmission spectrum of the atomic filter in the Voigt laser can be seen in Chap. 4. By selecting different temperatures and magnetic fields, the position and shape of the highest transmission peak of the atomic filter can be controlled, thus controlling the output performance of the Voigt laser. When the working temperature and magnetic field of VADOF are determined, the output wavelength of the laser also becomes limited within the transmission bandwidth of the highest transmission peak of the atomic filter. The relationship between the output of the Voigt laser and the temperature of the laser diode is shown in Fig. 5.83, and the output wavelength always fluctuates within the range of 1 pm near 780.2557 nm.

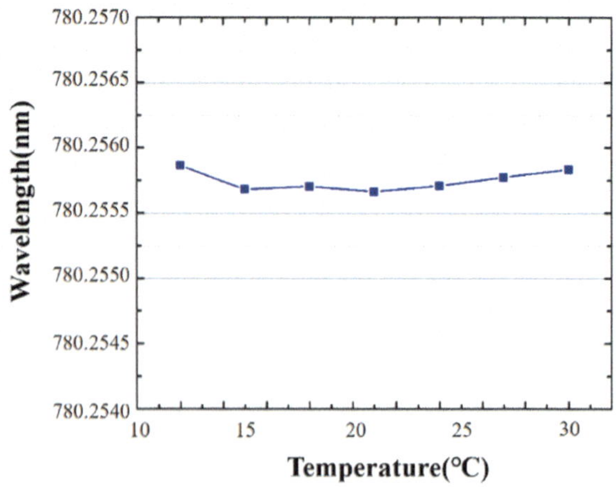

Fig. 5.83 Relationship between Voigt laser output wavelength and laser diode temperature

5.5.4 Resistance to Current Changes

In the previous section, we discussed the resistance of the 780 nm Faraday laser and the 780 nm Voigt laser corresponding to the changes in the temperature of the laser diode, considering this as an inherent attribute of diode lasers with atomic filters as frequency selection devices. In addition, for the same reason, diode lasers with atomic filters as frequency selection devices also have the ability to resist changes in the current of the laser diode. The combination of the two means that this type of Faraday laser has excellent resistance to changes in the parameters of the laser diode.

Continuing from the previous section, we introduce the resistance of the 780 nm Faraday laser and Voigt laser to current changes. As shown in Figs. 5.84 and 5.85, the figures depict the relationship between the output wavelength and the laser diode current for a Faraday laser and a Voigt laser, respectively, at a wavelength of 780 nm.

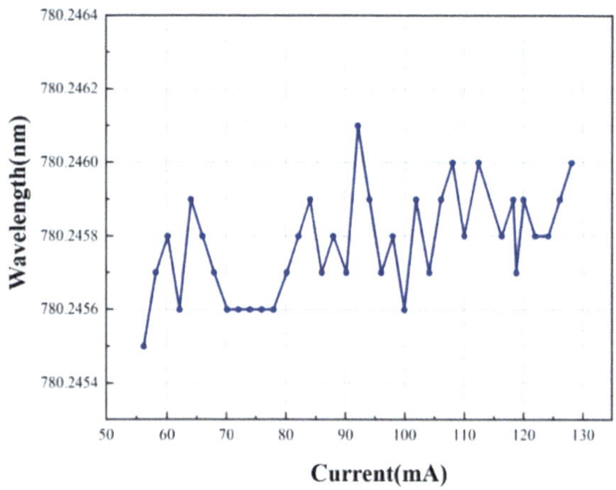

Fig. 5.84 Relationship between Faraday laser wavelength and drive current

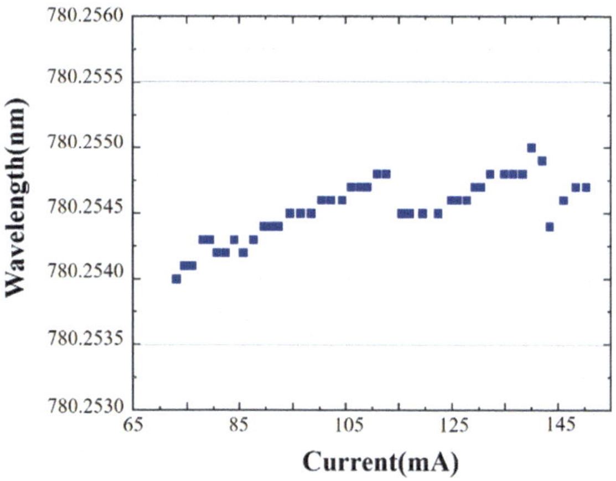

Fig. 5.85 Relationship between Voigt laser wavelength and drive current

For the Faraday laser, when the drive current changes from 55 to 129 mA, the center wavelength of the laser operation is 780.2458 nm, corresponding to the Rb atomic transition spectrum line, and the change range is within 0.6 pm. From this, it can be concluded that the output wavelength of the Faraday laser is almost unaffected by current changes.

For the Voigt laser, when the drive current changes from 70 to 150 mA, the center wavelength of the laser operation is 780.2545 nm, and the change range is within 1 pm, as shown in Fig. 5.85. From this, it can be concluded that the output wavelength of the Voigt laser is almost unaffected by current changes. The output wavelength of the Voigt laser can be changed by changing the magnetic field and working temperature, and the appropriate output wavelength can be selected by adjusting the parameters for different application scenarios.

In addition to the 780 nm Faraday laser, the cesium-based 852 nm Faraday laser has also become very mature, with excellent performance in diode lasers with an output wavelength of 852 nm. Here, we take the 852 nm Faraday laser as an example to introduce its resistance to temperature changes in laser diodes. As shown in Fig. 5.86, it is the transmission spectrum of the atomic filter of the 852 nm Faraday laser. It can be seen that the highest transmission peak of the transmission spectrum corresponds to the Doppler spectrum of the cesium atom D_2 transition line, with a transmittance of about 0.83 and a transmission bandwidth of 2.2 GHz. The transmission bandwidth is greater than the transmission bandwidth of the 780 nm laser in the previous section, so the fluctuation range of the output wavelength of the Faraday laser built with this atomic filter should be greater than the fluctuation range of the output wavelength of the 780 nm laser in the previous section.

As shown in Fig. 5.87, it is the curve of the output wavelength of this Faraday laser with the change of the laser diode current. The output frequency of the laser should

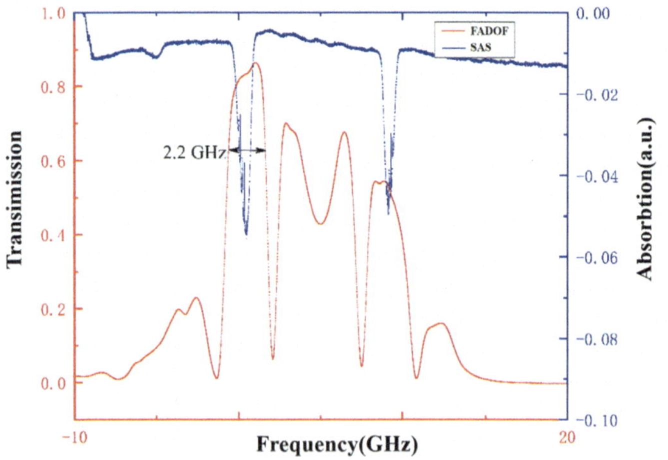

Fig. 5.86 Transmission spectrum of cesium atom 852 nm atomic filter

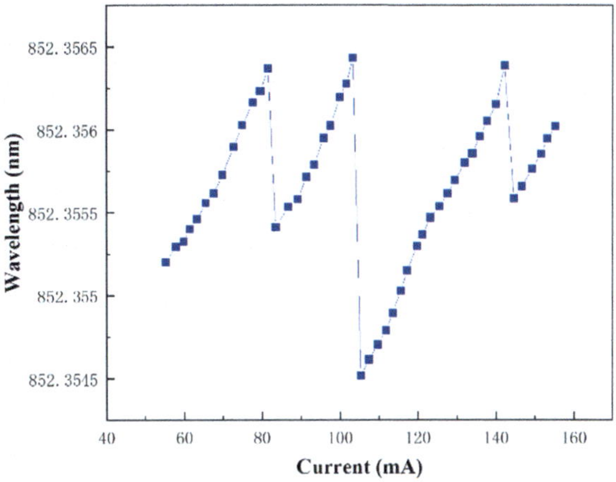

Fig. 5.87 Change of 852 nm Faraday laser wavelength with current

correspond to the Doppler spectrum of the cesium atom D_2 transition line, that is, the output wavelength is around 852.356 nm. At this time, if the temperature of the laser diode is changed, it increases the pump power of the laser and also increases the overall power consumption of the laser, which affects the cavity length of the laser to a certain extent, causing the output wavelength of the Faraday laser to drift. However, no matter how much the output wavelength drifts, it does not exceed the bandwidth range of the atomic filter. Currently, the output wavelength range of this laser is less than 2 pm, reflecting its excellent resistance to temperature changes in laser diodes.

5.5.5 *Resistance to Mechanical Vibration*

The Faraday laser uses an atomic filter to replace the interference filter and grating in traditional diode lasers as frequency selection devices. The biggest advantage is that it uses atomic frequency selection to intrinsically align with the atomic transition line, and it has excellent robustness to changes in laser diode parameters. In addition, considering that the incident angle of the input laser has little effect on the frequency selection ability of the atomic filter, it mainly depends on the working temperature and magnetic field size of the atomic filter, while the frequency selection effect of the interference filter and grating completely depends on the incident angle of the input laser. Therefore, when the laser is disturbed by mechanical vibration, compared with the diode laser with the interference filter and grating as frequency selection devices, the Faraday laser shows better mechanical robustness.

In addition to the inherent anti-mechanical vibration capability that may exist in Faraday lasers, the anti-mechanical vibration capability of the laser can also be

Fig. 5.88 Relationship between laser output wavelength and current, corner-cube rotation angle

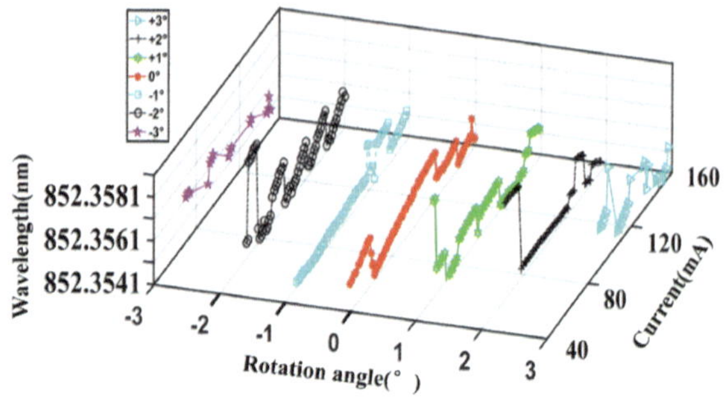

improved by using a corner-cube retroreflector instead of a plane mirror as the reflective cavity mirror of the laser. As shown in Fig. 5.88, the angle of the rotating corner-cube retroreflector, between − 3 and 3°, the Faraday laser can always maintain oscillation, and always has robustness to the current changes of the laser diode. If a plane mirror is used as the feedback cavity mirror, even if the rotation angle is less than 0.01°, the laser fails to oscillate. In comparison to the plane mirror, the corner-cube retroreflector as the reflective cavity mirror is insensitive to the angle of the incident laser, and it is easy to make the laser return along the original path to achieve laser oscillation. Therefore, the corner-cube Faraday laser has better anti-mechanical vibration capability.

5.5.6 *Free-Running Wavelength Stability Performance (Including λ-t Curve)*

During the long-term free-running process, environmental disturbances (temperature, vibration, etc.) cause changes in the position and posture of the optical components in the laser resonator, which brings about drift of the cavity mode of the laser, which is unavoidable. Generally, it is possible to limit the drift of the laser cavity mode and prevent the output frequency of the laser from deviating from the target working frequency during the long-term free-running process by using materials with low thermal expansion coefficients to customize the laser cavity, designing appropriate temperature control systems, and designing appropriate mechanical structures to enhance the anti-mechanical vibration capability. However, this method still has its limitations. Even if the anti-interference ability of the laser system is done well, once the laser runs for a long time, due to the aging of the material, the output frequency of the laser eventually drifts away from the target frequency and needs human intervention to compensate for this drift.

One of the unique advantages of the Faraday laser compared to other types of lasers is that, its frequency is always limited within the transmission bandwidth of the atomic filter, which indicates an excellent frequency stability. The wavelength of Faraday laser is always corresponding to the atomic transition line even if the time flies. Figures 5.89 and 5.90 show the wavelength changing more than one day of 852 nm and 780 nm Faraday laser, respectively. Figure 5.91 shows the Allan deviation of the 852 nm free-running Faraday laser [22]. Except for time changing, the wavelength would stay within the bandwidth of FADOF transmission with the change of laser current and temperature, as shown in Fig. 5.92. Generally, the transmission bandwidth of the atomic filter is on the order of hundreds of MHz, which is the same order of magnitude as the Doppler broadening of the atomic transition line. Therefore, when the target working frequency is the atomic transition spectrum line, the wavelength of Faraday laser will always align with the Doppler spectrum of the atomic transition line.

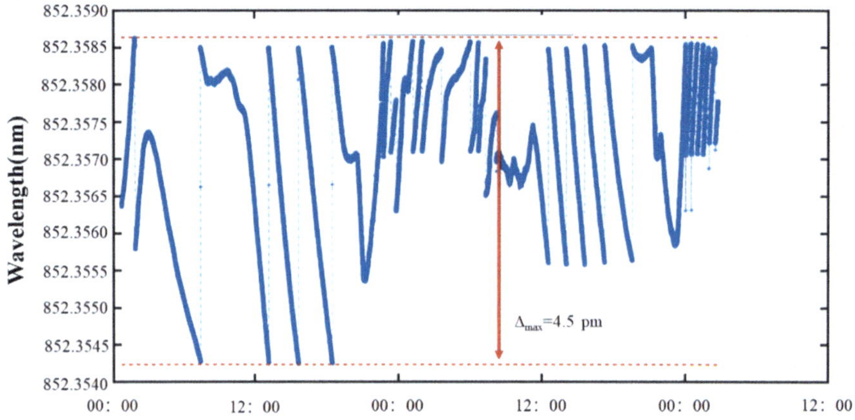

Fig. 5.89 Cesium atom 852 nm Faraday laser wavelength stability

Fig. 5.90 Wavelength of Rb atom 780 nm Faraday laser with the change of time

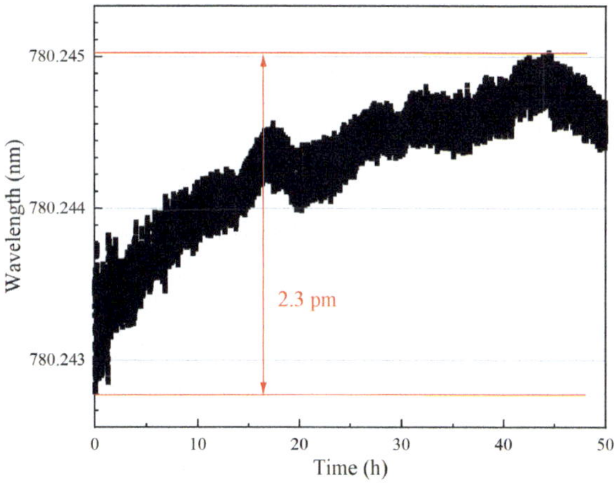

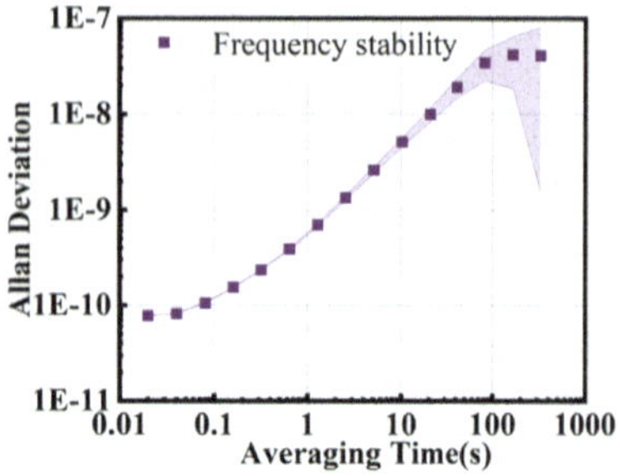

Fig. 5.91 Allan deviation of the 852 nm Faraday laser frequency stability [22]

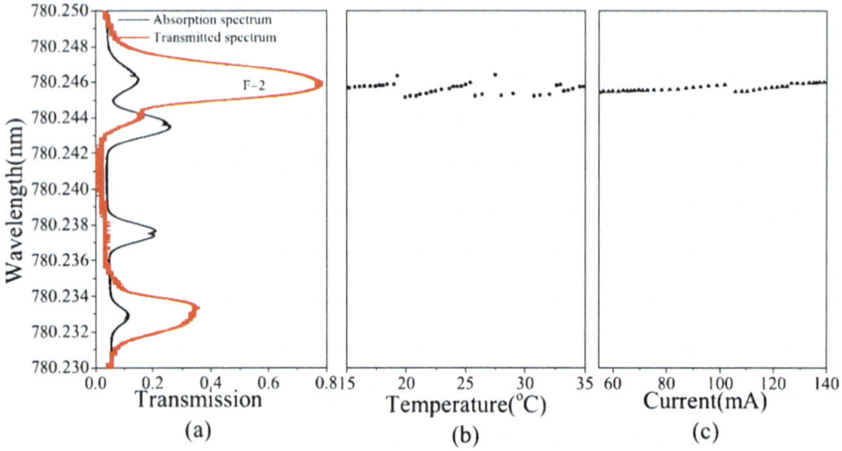

Fig. 5.92 Wavelength stability of Rb atom 780 nm Faraday laser. **a** Transmission spectrum; **b** wavelength changes with diode temperature; **c** wavelength changes with diode current [11]

5.5.7 Power Performance

In the existing research work on Faraday lasers, exploring the power of Faraday lasers mostly focuses on changing the laser diode operating current and temperature. For low-power Faraday lasers, the output power is generally around 30 mW. For high-power Faraday lasers, through the use of high-power semiconductor laser chips and FADOF parameter design, it is possible to realize high-power Faraday lasers with output power of the order of W or even 100 W. The power stability of Faraday laser is affected by several factors, such as the temperature of the laser diode, the temperature of the vapor cell in FADOF, and the tiny vibration of the cavity mirror caused by the vibration of the external environment, resulting in its power stability is a little worse than that of the interference filter configuration external cavity diode laser. However, according to the experimental data, its power stability at 1 s can also reach 10^{-5} order of magnitude under the stable working condition.

For the relative intensity noise (RIN), there is a study [17] comparing 852 nm dal-frequency (DF) Faraday laser with an interference-filter external-cavity diode laser (IF-ECDL). The optical signals of the DF Faraday laser and the IF-ECDL are converted to electric signals by the photodiode, then the intensity noises of the electric signals are separately measured by a signal source analyzer. As depicted in Fig. 5.93, with the tested frequency range from 1 Hz to 10 MHz, the RIN measurement results of the DF Faraday laser and the IF-ECDL are shown by the red and black curves. It demonstrates that the RIN of the DF Faraday laser is comparable to that of IF-ECDL, even if it operates in dual-frequency mode. The RIN performances of the DF Faraday laser and the IF-ECDL are measured to be − 134 and − 137 dBc/Hz at 10 kHz offset when the optical power entering the photodiode are both 1 mW. For comparison, Fig. 5.94 shows the RIN of a single-mode Faraday laser. This indicates that the RIN is − 141.1 dBc/Hz at 10 kHz, which is better than that of the IF ECDL.

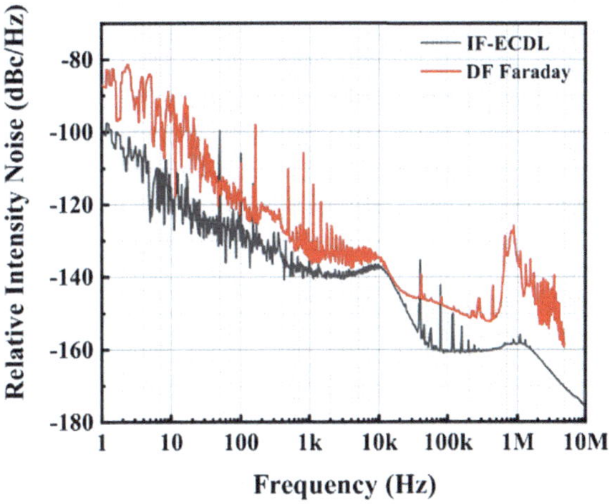

Fig. 5.93 The measured RIN of the 852 nm DF Faraday laser (red curve) and the homemade 852 nm IF-ECDL (black curve) [17]

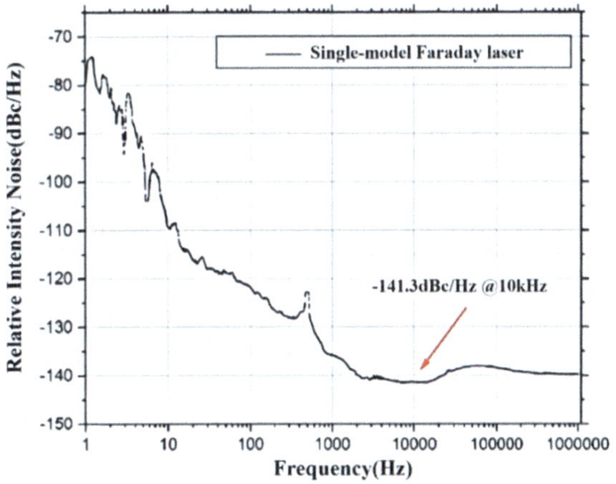

Fig. 5.94 The measured RIN of the 852 nm single-mode Faraday laser

Overall, the power characteristics of Faraday lasers are affected by a variety of factors, including the absorption of the magneto-optical element, the thermal lens effect, the Faraday effect, the thermal expansion, the design and geometry, and the magnetic field strength. To meet the application needs of different situations, the power characteristics of Faraday lasers can be optimized in experiments by optimizing the design of Faraday atom filters, using high reflectivity cavity mirrors to reduce the laser loss in the cavity.

5.5.8 Technical Noises of Faraday Laser

The noise source of the Faraday laser after frequency locking exhibits a highly intricate nature. Given our primary focus on the frequency characteristics of the Faraday laser, it is imperative to thoroughly test and evaluate its frequency noise. The inherent noise of the Faraday laser can be effectively mitigated through implementation of a phase-locked loop during the process of frequency locking. However, suppressing residual amplitude modulation caused by parasitic standard effect and birefringence effect in the locked light path remains challenging, as it significantly impacts the post-locking frequency noise of the Faraday laser. By employing a phase noise analyzer to measure baseline noise at far detuning from phase detection signal when the laser is locked, we can mathematically convert this into frequency noise $S(f)$, as depicted in Fig. 5.95.

The low-frequency and high-frequency components of frequency noise correspond to the long-term and short-term instabilities of the phase-locked Faraday laser, respectively. Formula 5.5 can be utilized to convert frequency noise into equivalent Allan deviation, as illustrated in Fig. 5.96.

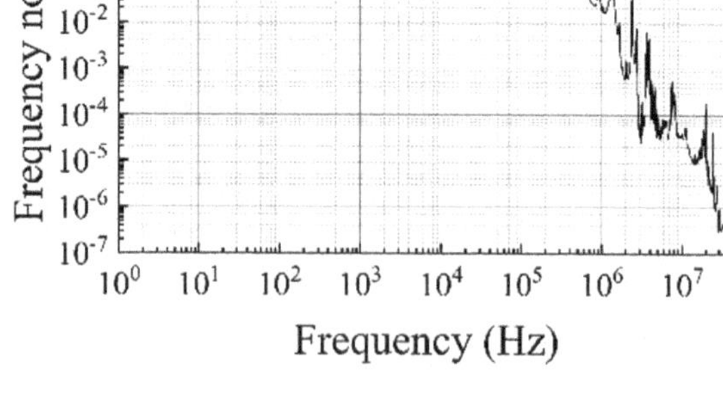

Fig. 5.95 Frequency noise spectrum of Faraday laser

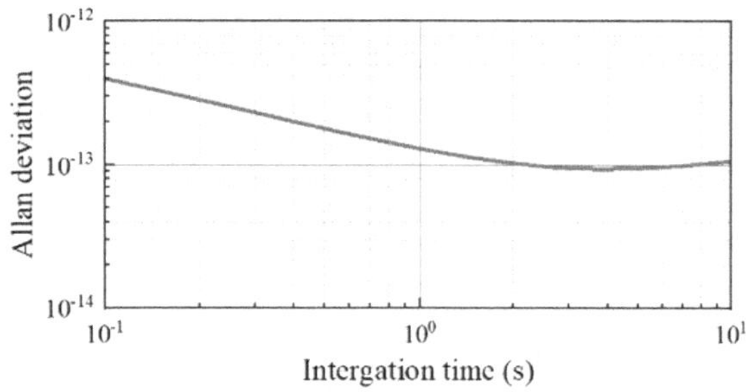

Fig. 5.96 Equivalent Allan deviation from frequency noise of Faraday laser

$$\sigma^2(\tau) = 2\int S^2(f)\frac{\sin^4(\pi\tau f)}{(\pi\tau f)^2}\frac{1}{\nu^2}df \tag{5.5}$$

where ν is the laser frequency. Frequency noise is the fundamental factor that limits the frequency instability of the Faraday laser, with its optimal stability achieving 1.3 $\times$ 10^{-13} at 1 s. To further enhance this performance, it is imperative to optimize the structural configuration of components within the locked optical path in order to minimize residual amplitude modulation resulting from standard specific effects and birefringence effects induced by devices such as EOM crystal and atomic vapor cell. Moreover, fine-tuning parameters of individual components and adjusting input laser power also contribute to reducing noise levels. Based on current estimations, it is anticipated that the frequency stability of the Faraday laser can be pushed below a coefficient level of 10^{-14} at 1 s.

It is worth noting that frequency noise analysis can not only be utilized for the analysis of short-term stability, as shown in Fig. 5.96, but also enables an accurate assessment of long-term stability. Unfortunately, due to limitations in the measurement instrument, it is not feasible to obtain data for extended periods. Therefore, a scheme employing two phase-locked lasers can be implemented to examine the impact of different parameters on instability. Primarily, residual amplitude modulation significantly influences long-term instability; thus, the structure and parameters of related components such as EOM crystal and atomic vapor cell play a crucial role. Additionally, power instability within the laser itself can partially transfer into frequency instability; hence it becomes necessary to stabilize laser power prior to locking its frequency. By optimizing these aforementioned steps, it is expected that Faraday lasers' long-term stability could be improved at a small coefficient level around 10^{-13} at 10,000 s or even at a larger coefficient level around 10^{-14} at 10,000 s.

The phase noise spectrum of the Faraday laser was obtained through frequency locking tests conducted using two Faraday lasers, as shown in Fig. 5.97. It is evident that the phase noise of the Faraday laser is significantly suppressed upon frequency locking. A linear regression analysis performed on the phase noise curve reveals a slope of approximately − 2 within the Fourier frequency range of 1–10 kHz, indicating a predominant presence of white frequency noise. Furthermore, examination of the lock-in curve highlights a servo peak near 100 kHz, representing our servo

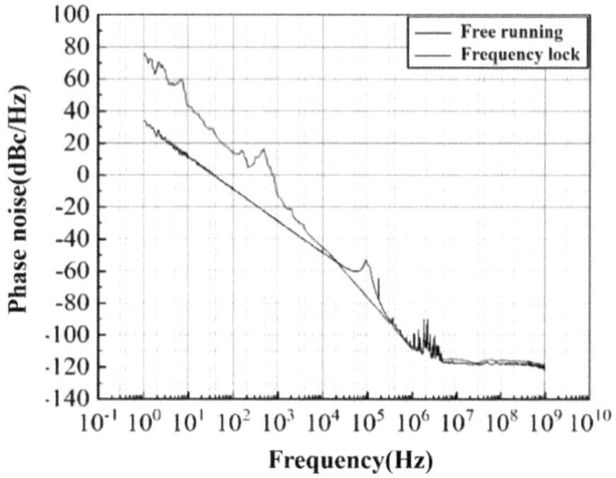

Fig. 5.97 Phase noise spectrum, wherein the red line depicts free running and the black line represents frequency locking

circuit's bandwidth. Expanding this bandwidth further would be advantageous in achieving enhanced suppression of system noise.

Furthermore, we acquired the widely used commercial laser Toptica TA pro to conduct a comparative analysis with our self-running Faraday laser. The obtained test results reveal that both lasers exhibit similar phase noise levels at approximately 1 Hz; however, as the frequency increases, the commercial laser exhibits higher phase noise relative to the Faraday laser. Beyond 1 MHz, the Faraday laser demonstrates a remarkable superiority over its commercial counterpart by exhibiting 40 dBc/Hz lower phase noise in high-frequency regions. Given the inherent capability of Faraday lasers to autonomously align atomic spectrum lines and the enhanced frequency performance resulting from atomic selectivity, Faraday lasers are poised to play a pivotal role in quantum-related disciplines.

References

1. J.N. Holonyak, S.F. Bevacqua, Coherent (visible) light emission from Ga ($As_{1-x}P_x$) junctions. Appl. Phys. Lett. **1**(4), 82–83 (1962)
2. R.N. Hall, G.E. Fenner, J.D. Kingsley et al., Coherent light emission from GaAs junctions. Phys. Rev. Lett. **9**(9), 366 (1962)
3. M.I. Nathan, W.P. Dumke, G. Burns et al., Stimulated emission of radiation from GaAs p–n junctions. Appl. Phys. Lett. **1**(3), 62–64 (1962)
4. V.L. Velichanskll, A.S. Zibrov, V.S. Kargopol'tsev et al., Minimum line width of an injection laser. Sov. Tech. Phys. Lett. (Engl. Transl.) (United States) **4**(9) (1978)
5. X.Y. Miao, L.F. Yin, W. Zhuang et al., Note: demonstration of an external-cavity diode laser system immune to current and temperature fluctuations. Rev. Sci. Instrum. **82**(8), 086106 (2011)
6. Z. Liu, X. Guan et al., An atomic filter laser with a compact Voigt anomalous dispersion optical filter. Appl. Phys. Lett. **123**(13) (2023)
7. M. Faraday, On the magnetization of light and the illumination of magnetic lines of force. Philos. Trans. R. Soc. Lond. **136**, 1–20 (1846)

8. Y. Öhman, A tentative monochromator for solar work based on the principle of selective magnetic rotation. Stockholms Obs. Ann. **19**(4), 9–11 (1956)
9. P.P. Sorokin, J.R. Lankard, V.L. Moruzzi et al., Frequency-locking of organic dye lasers to atomic resonance lines. Appl. Phys. Lett. **15**(6), 179–181 (1969)
10. W.D. Lee, J.C. Campbell, Optically stabilized $Al_xGa_{1-x}As$/GaAs laser using magnetically induced birefringence in Rb vapor. Appl. Phys. Lett. **58**(10), 995–997 (1991)
11. Z.M. Tao, Y.L. Hong, B. Luo et al., Diode laser operating on an atomic transition limited by an isotope 87 Rb Faraday filter at 780 nm. Opt. Lett. **40**(18), 4348–4351 (2015)
12. Z.M. Tao, X.G. Zhang, D. Pan et al., Faraday laser using 1.2 km fiber as an extended cavity. J. Phys. B At. Mol. Opt. Phys. **49**(13), 13LT01 (2016)
13. J. Keaveney, W.J. Hamlyn, C.S. Adams et al., A single-mode external cavity diode laser using an intra-cavity atomic Faraday filter with short-term linewidth < 400 kHz and long-term stability of < 1 MHz. Rev. Sci. Instrum. **87**(9), 095111 (2016)
14. P.Y. Chang, H.F. Peng, S.N. Zhang et al., A Faraday laser lasing on Rb 1529 nm transition. Sci. Rep. **7**(1), 8995 (2017)
15. M.D. Rotondaro, B.V. Zhdanov, M.K. Shaffer et al., Narrowband diode laser pump module for pumping alkali vapors. Opt. Express **26**(8), 9792–9797 (2018)
16. P.Y. Chang, Y.L. Chen, H.S. Shang et al., A Faraday laser operating on Cs 852 nm transition. Appl. Phys. B **125**(12), 230 (2019)
17. T.T. Shi, X.L. Guan, P.Y. Chang et al., A dual-frequency Faraday laser. IEEE Photonics J. **12**(4), 1–11 (2020)
18. H. Tang, H.Z. Zhao, R. Wang et al., 18 W ultra-narrow diode laser absolutely locked to the Rb D_2 line. Opt. Express **29**(23), 38728–38736 (2021)
19. P. Chang, H. Shi, J. Miao et al., Frequency-stabilized Faraday laser with 10^{-14} short-term instability for atomic clocks. Appl. Phys. Lett. **120**(14), 141102 (2022)
20. H. Tang, H. Zhao, D. Zhang et al., Polarization insensitive efficient ultra-narrow diode laser strictly locked by a Faraday filter. Opt. Express **30**(16), 29772–29780 (2022)
21. H. Shi, P. Chang, Z. Wang et al., Frequency stabilization of a cesium Faraday laser with a double-layer vapor cell as frequency reference. IEEE Photonics J. **14**(6), 1–6 (2022)
22. Z.Y. Wang, Z.J. Liu, J.X. Miao et al., A corner-cube-feedback Faraday laser with 8 kHz linewidth. arXiv:2309.04485 (2024)

Chapter 6
Frequency Stabilization Techniques for Faraday Lasers

Diode lasers are very important in many research fields such as quantum precision measurement, quantum optics, atomic physics, etc. Traditional diode lasers usually use gratings and interference filters to adjust the output frequency, and the output frequency is finely mechanically adjusted to be near the atomic transition line, and then the laser frequency is locked by precision spectroscopy, such as saturated absorption spectroscopy, modulation transfer spectroscopy, etc. At the same time, Pound–Drever–Hall (PDH) technology is also often used for laser frequency locking and laser linewidth narrowing. When using precision spectroscopy technology to lock the laser frequency, the frequency of traditional diode lasers drifts or mutates in the case of long-term operation or changes in working parameters, causing the system to lose lock. However, the Faraday laser can keep the laser frequency continuously resonant with the atomic transition line during long-term operation when the driving current and working temperature of the laser diode fluctuate greatly, greatly improving the convenience and efficiency of the initial frequency locking and re-locking process after losing lock. This chapter considers the modulation transfer spectroscopy technology in precision spectroscopy locking as an example to show the process to achieve the locking of the Faraday laser frequency. At the same time, it demonstrates the process to lock the output frequency of the Faraday laser through PDH technology.

J. Chen et al., *Faraday Laser*, https://doi.org/10.1007/978-981-97-8023-5_6

6.1 Frequency Stabilization of Faraday Lasers with Saturated Absorption Spectroscopy

6.1.1 Basic Principles

Saturated absorption spectroscopy is a general technique used to measure the Doppler-free transition spectra of atoms and molecules, with an important value in the field of laser frequency stabilization. By using the narrow linewidth saturated absorption peak or crossover peak as the quantum reference of the laser frequency, and through the feedback system, the frequency of the laser can be locked at a specific absorption peak position, thereby achieving the precise positioning of the laser output frequency at a certain atomic transition frequency. The characteristics of simple application, reliability, and high resolution of saturated absorption spectroscopy made it the first to be used in laser locking, and it stands as one of the cornerstones of frequency stabilization technology.

1. Classical theory and power saturation effect of absorption spectroscopy

In the basic principles of light and atomic interactions, when the photon frequency equals the transition frequency of the stationary ground state atom, the atom transition occurs from ground state to the excited state by absorbing photons, generally, known as resonance absorption. By passing a laser through an atomic vapor cell and scanning its frequency, when the frequency matches the atomic transition frequency, the laser transmittance decreases, the absorption rate increases, forming an absorption signal. This signal reflects the atomic absorption spectrum, which is widely used in material composition analysis. In quantum mechanics, the finite lifetime of energy levels leads to uncertainty in transition frequencies, forming a natural linewidth, and the Lorentzian linewidth reflects the lifetime of the energy level. Under high light intensity, the absorption signal saturates with increasing light intensity, a phenomenon known as power saturation effect, indicating that all atoms participate in the interaction, and the sample's absorption of light no longer increases.

2. Doppler broadening and saturation absorption spectrum

In practical applications, the interaction between atoms and photons is affected by the Doppler effect, as atoms are usually in a state of thermal motion. For example, when atoms move towards photons, the perceived photon frequency is higher. Conversely, when atoms move away from photons, the perceived frequency is lower. This phenomenon causes the resonance absorption of atoms depending upon the speed of atomic motion. In an ideal gas state, the speed distribution of atomic vapor follows the Maxwell–Boltzmann distribution, leading to a Gaussian distribution of photon absorption frequencies, known as the Doppler broadening effect. The Doppler linewidth, i.e., the full width at half maximum of the Gaussian distribution, is greater than the natural linewidth of atomic transitions, affecting the resolution of absorption spectral lines.

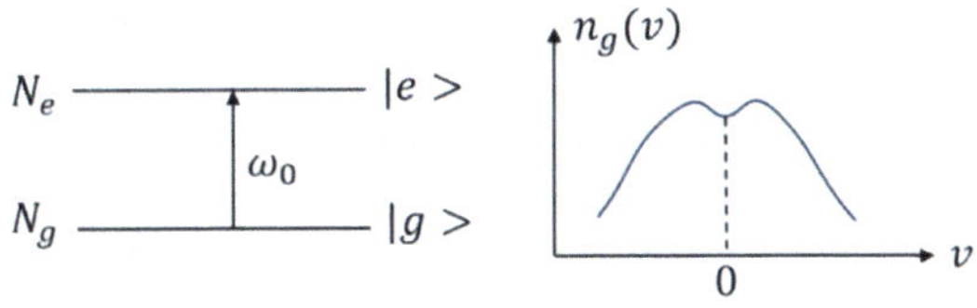

Fig. 6.1 Schematic diagram of a two-level system and Bennet hole under strong resonant light

To reduce the impact of Doppler broadening, a saturation absorption peak can be constructed through the power saturation effect. In this process, a strong laser beam is used to irradiate atomic vapor, pumping atoms from the ground state to the excited state, achieving power saturation. This reduction in the number of ground state atoms manifests as an inverted Lorentzian dip (known as a Bennet hole or hole burning) [1], the width of which is determined by the natural linewidth, typically much smaller than the Doppler broadening, thus improving the accuracy of the measurement. Figure 6.1 is a schematic diagram of a two-level system and a Bennet hole under strong resonant light.

In the early 1970s, Schawlow and Hänsch developed an experimental method to detect the Bennett hole, with the detected spectrum known as saturated absorption spectroscopy [2, 3]. The linewidth of this spectrum is unaffected by Doppler broadening and is much narrower than the Doppler linewidth. Consequently, atomic spectral lines closer to the natural linewidth can be obtained in the experiment. This method utilizes the interaction between two overlapping, counter-propagating laser beams and atoms to generate a spectrum with saturated absorption peaks. The laser beam emitting from a single laser source is divided into two beams of different intensities: the stronger one is designated as the "pump" laser, while the weaker one is referred to as the "probe" laser. The pump and probe lasers have the same frequencies, i.e., $\omega_{\text{pump}} = \omega_{\text{probe}} = \omega$. The two beams of light pass through the atomic vapor from opposite ends of the atomic vapor cell, and their optical paths are overlapping each other, as shown in Fig. 6.2.

In the setup where the pump and probe lasers propagate in opposite directions, the atomic vapor is simultaneously affected by two lasers. If the velocity component of the atom on the z-axis is v_z, then the frequencies of the pump and probe lasers it perceived by the atoms are adjusted to match the transition frequency ω_0,

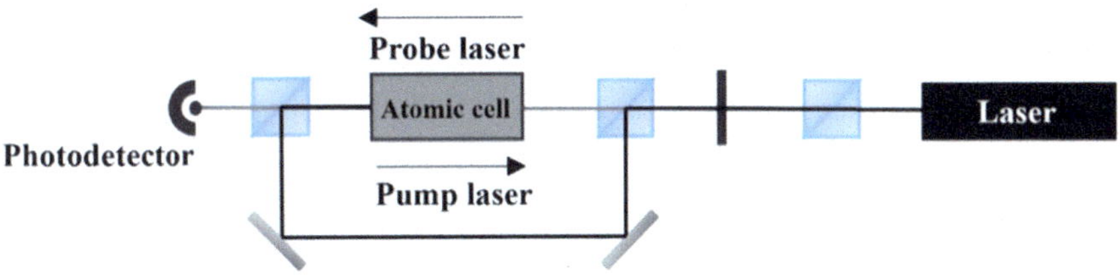

Fig. 6.2 Schematic diagram of the saturated absorption spectroscopy device

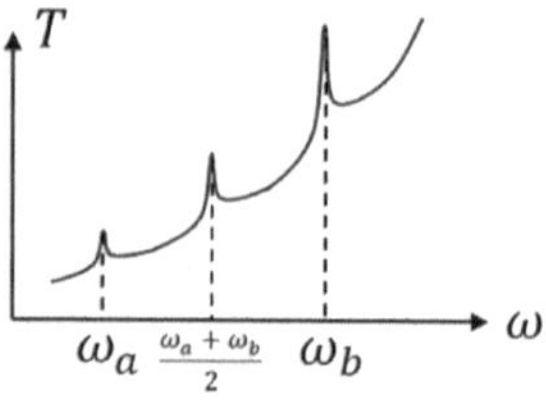

Fig. 6.3 Saturated absorption crossover peaks

thus achieving resonant absorption. Specifically, when $v_z = 0$, the atom resonates with both lasers simultaneously, effectively "filtering" out atoms unaffected by the Doppler effect.

In a multi-level system, such as a three-level model, two excited states share a ground state, and the difference in their transition frequencies is less than the Doppler linewidth. In this setup, when the pump and probe lasers have the same frequency, there will be two groups of atoms resonating with the pump and probe lasers respectively. One group of atoms absorbs pump photons and the atomic transitions to an excited state in the redshift direction occur, while the other resonates with the probe laser in the blueshift direction. This results in the formation of four Bennet holes in the velocity distribution, appearing as saturated absorption crossover peaks at specific frequencies. These peaks represent the saturated absorption features formed between two adjacent transitions. These crossover peaks provide a deep understanding of atomic transitions and spectral characteristics, and are helpful for precise control and application in high-precision spectral measurements. Figure 6.3 shows the saturated absorption crossover peaks.

6.1.2 *Saturated Absorption Spectroscopy of Faraday Laser*

The output frequency of the Faraday laser is always limited within the transmission peak of the atomic filter, and can be fine-tuned by changing parameters such as cavity length. The red curve in Fig. 6.4 is the saturated absorption spectroscopy obtained while scanning the cavity length of the single-frequency mode Faraday laser working at cesium atom 852 nm transition. It shows four absorption peaks, but $6S_{1/2}F = 4 \rightarrow 6^2P_{3/2}F' = 3$ transition and crossover peaks are not observed. This owes to the fact that the transmission of the Faraday atomic filter decreases near these two peaks, and the mode at this frequency does not form feedback oscillation and laser output, so the Faraday laser cannot be tuned to this range. If the transmission bandwidth and peak transmittance of the Faraday atomic filter are further increased to cover the crossover peak frequency range, it becomes possible to scan out the complete saturated absorption spectroscopy.

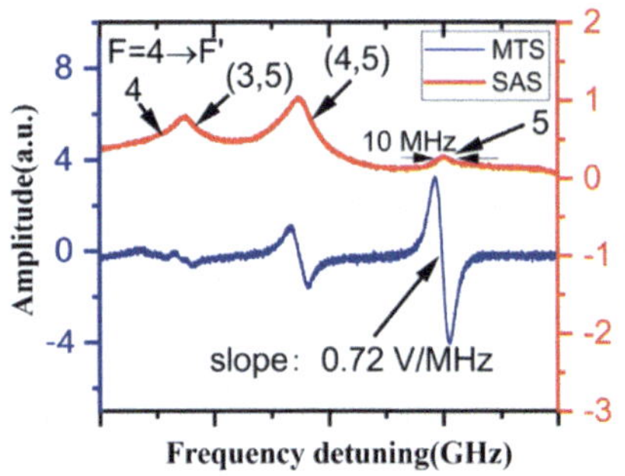

Fig. 6.4 Single-frequency Faraday laser saturated absorption spectroscopy and modulation transfer spectroscopy

6.2 Frequency Stabilization of Faraday Lasers via Modulation Transfer Spectroscopy

6.2.1 Basic Principles

There are mainly two explanations for the basic principles of modulation transfer spectroscopy [4]: one is based on the four-wave mixing nonlinear effect, whereas the other is based on the hole burning effect. Among them, the four-wave mixing effect is the most widely used method to explain modulation transfer spectroscopy.

Four-wave mixing is a third-order nonlinear effect and is a common method to achieve optical phase conjugation. The specific model can be described as: in a nonlinear medium, two beams of light propagate in opposite directions, and the third input light wave propagates in the same direction or at a small angle with one of the beams, interacting in the nonlinear medium to produce a fourth beam. If the frequencies of the three input lights are the same, it is degenerate four-wave mixing; if the frequencies are different, it is non-degenerate four-wave mixing [5, 6].

Taking a two-level atomic system as an example, the principle of four-wave mixing interaction between light and a two-level atom is shown in Fig. 6.5 [7]. The upper and lower energy levels of the two-level atomic system are $|b\rangle$ and $|a\rangle$, respectively. Two beams of light with frequency ω are shot at each other in the two-level atomic system. One of them is modulated, the modulation frequency is δ, and first-order sidebands are generated on both sides of the center frequency, which are $\omega + \delta$ and $\omega - \delta$ frequency components. According to the principle of four-wave mixing, any three beams of light undergoes four-wave mixing to produce new frequency components of light. The only ones that meet the conditions are the new sidebands $\omega + \delta$ and $\omega - \delta$. When these two frequency component lights enter the photodetector together with the unmodulated light ω, after beat frequency filtering, only the electrical signal with frequency δ is left. It is mixed with the reference signal with the same frequency δ, that is, phase-sensitive demodulation to obtain the modulation transfer spectrum signal.

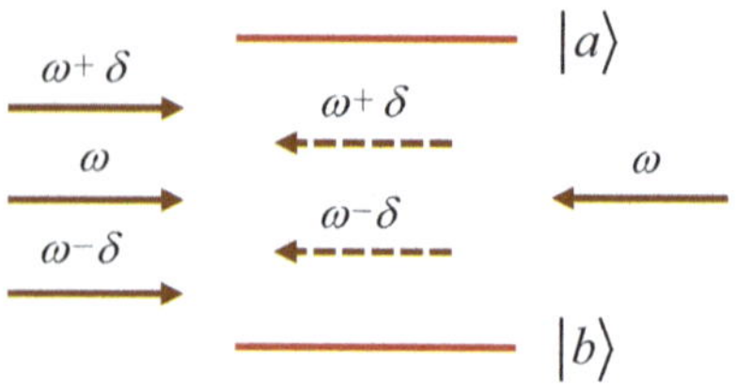

Fig. 6.5 Four-wave mixing principle of a two-level atomic system

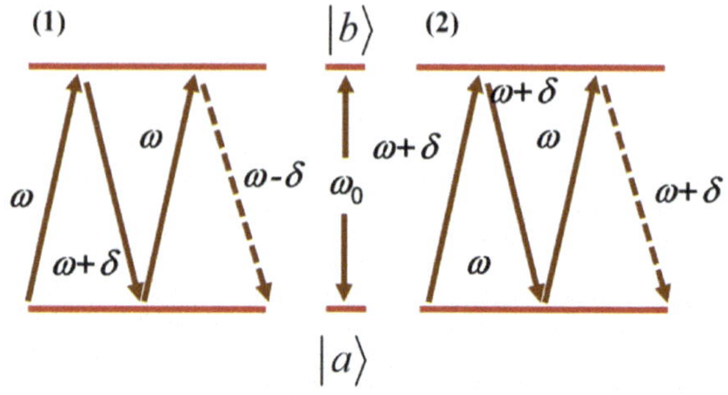

Fig. 6.6 Four-wave mixing resonance absorption process

The four-wave mixing resonance absorption process is shown in Fig. 6.6 [7]. As can be seen from photon resonance absorption, in Fig. 6.6(1), $\omega - \mathrm{k}v = \omega_0$, $\omega - \mathrm{k}v - (\omega + \delta - kv) + \omega + kv = \omega_0$, we get: $\omega = \omega_0 + \delta/2$, where kv is the Doppler frequency shift. In Fig. 6.6(2), from $\omega + \delta - \mathrm{k}v = \omega_0$, $\omega + \delta - \mathrm{k}v - (\omega - kv) + \omega + kv = \omega_0$, we get: $\omega = \omega_0 - \delta$.

The modulation transfer process occurs when the unmodulated probe light and the modulated pump light interact with the nonlinear medium together, and the probe light obtains the modulated sidebands. The modulation method for the pump light can be phase modulation and amplitude modulation, among which the modulated phase information can be extracted through phase-sensitive detection to obtain a high-resolution modulation transfer spectrum. The modulation transfer process of the specific two-level atom phase modulation can be explained by Fig. 6.7. When only phase modulation is applied to the pump laser, the modulated pump laser and the unmodulated probe laser interact with the atoms together, and the modulation of the pump light is transferred to the probe light through the atoms. The probe light after modulation transfer also has the characteristics of phase modulation, but the intensity of the modulated sidebands of the probe light (i.e., $\pm$ 1 order sidebands) is related to the detuning amount $\Delta = (\omega_L - \omega_0)$, which shows a clear absorption line shape relationship with the frequency detuning, that is, the intensity of the modulated sidebands is the largest when the laser resonates with the atoms.

The modulation transfer spectrum line is obtained by phase-sensitive demodulation of the beat signal of the probe light sidebands and the main frequency. This process uses high-frequency modulation and demodulation to transfer the signal of laser frequency to the high-frequency for detection. According to the noise spectrum law, the high-frequency noise is much smaller than the low-frequency, so extracting

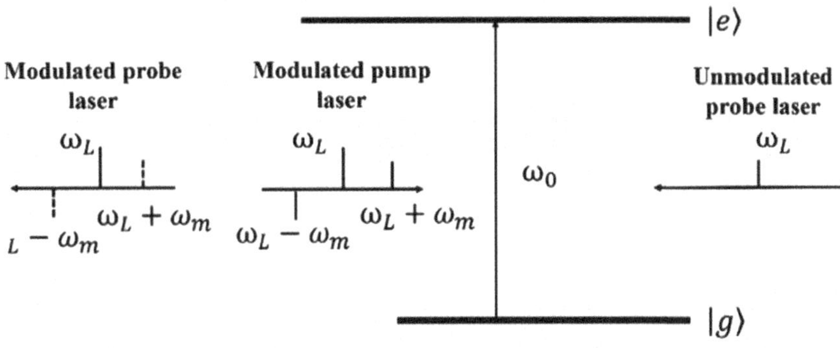

Fig. 6.7 Schematic diagram of the modulation transfer process of phase modulation of a two-level atom

the laser frequency information at high frequency can significantly improve the signal-to-noise ratio of the spectrum line and effectively reduce the impact of low-frequency noise in the loop. Additionally, when the detuning is too large, the intensity of the modulated sidebands of the probe light decreases rapidly. Consequently, the modulation transfer spectrum exhibits the characteristic of no Doppler background. The heterodyne detection method can effectively eliminate noise that is independent of the laser frequency (phase) and cannot be phase modulated, and can achieve high-sensitivity spectrum line detection.

6.2.2 *Single-Frequency MTS Frequency Stabilization System of Cesium Atom 852 nm Faraday Laser*

In 2023, Shi and colleagues from Peking University realized the single-frequency modulation transfer spectroscopy frequency stabilization of the cesium atom 852 nm Faraday laser [8]. This research group built a modulation transfer spectroscopy based on a single-frequency Faraday laser, and improved the long-term frequency stability through the application of a double-layer atomic cell. This section provides a detailed introduction to this work.

1. Single-frequency MTS frequency stabilization system of cesium atom 852 nm Faraday laser

The schematic diagram of the integrated frequency stabilization system of the Faraday laser modulation transfer spectroscopy is shown in Fig. 6.8. Figures 6.9 and 6.10 are the mechanical diagram and physical diagram of the integrated frequency stabilization system of the Faraday laser modulation transfer spectroscopy, respectively. In the system, the Faraday laser first passes through the optical isolator to prevent stray light feedback, and then is divided into two beams by the half-wave plate and the polarizing beam splitter. The transmitted light serves as the frequency-stabilized laser output, and the reflected light is used for MTS frequency stabilization. The reflected light is further divided into two beams by the half-wave plate and the polarizing beam splitter, one of which serves as the pump light with a power of about 1.2–1.5 mW, and the other serves as the probe light with a power of about 0.2–0.3 mW. These intensities can be adjusted by a combination of a half-wave plate and a polarizing beam splitter. The pump light is phase modulated by a resonant electro-optic modulator, with a tunable range of 3.5–7 MHz. When the modulation depth is 1 rad, the most efficient resonance frequency point is 4.6 MHz (− 3 dB). The EOM is driven by the RF signal generated by the signal generator, producing two first-order modulation sidebands. The pump light modulated by the EOM is reflected into the cesium atomic vapor cell by the polarizing beam splitter, and the probe light is also reflected into the cesium atomic vapor cell by the mirror, interacting with the modulated pump light and cesium atoms in the atomic vapor cell. By adjusting the angles of the mirror and the polarizing beam splitter, the pump light and the probe light are made to strictly overlap in space and pass through the center of the atomic vapor cell. After four-wave mixing, the probe light is transmitted into the high-speed photodetector by the polarizing beam splitter. The signal detected by the PD, mixed with the reference signal consistent with the frequency and EOM driving frequency, is then demodulated to obtain the modulation transfer spectrum signal, i.e., the error signal. This error signal is then filtered and amplified, and input to the PID feedback locking module. The PID outputs three feedback signals connected to the piezoelectric ceramic end, current end, and laser head of the Faraday laser driver power supply. Finally, the frequency of the Faraday laser is stabilized to the atomic transition frequency corresponding to the quantum frequency reference cesium atom, achieving frequency stability of the Faraday laser.

2. Cesium atom 852 nm Faraday laser MTS stabilization system test principle

Using MTS stabilization technology, when the Faraday laser operates in single-frequency mode, the laser frequency is locked on the quantum reference cesium atom D_2 transition line 852 nm $6S_{1/2}F = 4 \rightarrow 6P_{3/2}F' = 5$ cyclic transition peak, and the error signal is fed back to the laser head, current, and piezoelectric ceramic ports of the Faraday laser.

When the Faraday laser is locked by MTS, the frequency stability range is less than a hundred Hz, which is significantly suppressed compared to the frequency fluctuation range of several hundred MHz when running freely. Here, the frequency stability is derived from the residual error signal. To clearly explain the conversion process from the residual error signal to frequency fluctuation, Fig. 6.11 shows the MTS signal and the residual error signal after locking.

By recording the voltage fluctuations of the error signal, the frequency fluctuation $\delta\nu$ can be derived from the relationship $\delta\nu/\Delta\nu = \delta V/\Delta V$, where $\delta\nu$ and δV are the frequency fluctuation and voltage fluctuation, respectively, and $\Delta\nu$ and ΔV are the linewidth and voltage amplitude of the MTS error signal. From this, the frequency fluctuation can be derived from the remaining error signal voltage fluctuation. In addition, the frequency stability of the frequency standard can be expressed

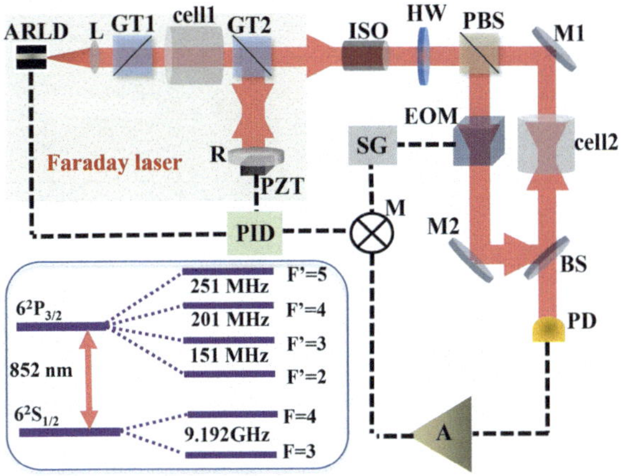

Fig. 6.8 Schematic of the modulation transfer spectroscopy stabilization Faraday laser device. *ARLD* laser diode with anti-reflective coating, *L* lens, *GT1* Glan–Taylor prism 1, *cell1* cell 1, *GT2* Glan–Taylor prism 2, *R* cavity mirror, *PZT* piezoelectric ceramic, *PID* proportional-integral-derivative feedback circuit, *PBS* polarizing beam splitter, *EOM* electro-optic modulator, *SG* signal generator, *A* amplifier

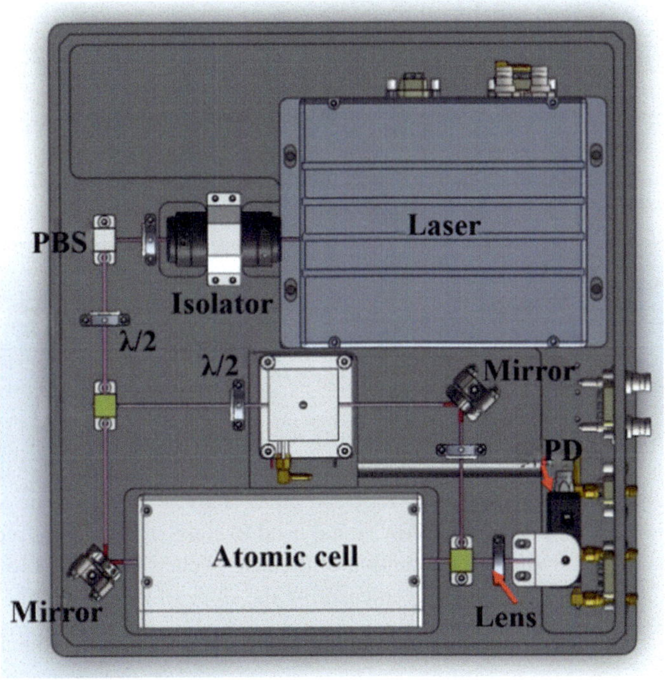

Fig. 6.9 Mechanical diagram of the integrated system of the modulation transfer spectroscopy stabilization Faraday laser

Fig. 6.10 Integrated system of the modulation transfer spectroscopy stabilization Faraday laser instrument setup

as follows:

$$\sigma_y(\tau) = \frac{1}{K}\frac{1}{Q}\frac{1}{S/N}\frac{1}{\sqrt{\tau}} \tag{6.1}$$

where K is a constant of the order of unity and can be simplified as $K = 1$, $Q = \nu_0/\Delta\nu$ is the quality factor of the atomic resonance line, S/N is the signal-to-noise ratio determined by the signal $S = I_{\max}$ and noise N by σ^I $(\tau = 1)$. The amplitude fluctuation of the demodulation error signal used to stabilize the laser frequency is proportional to the frequency fluctuation of the laser. When the laser is locked near the zero point of the error signal, by measuring the slope of the error signal near the resonant frequency, the amplitude fluctuation described by σ^I can be converted into the frequency fluctuation described by σ^ν, that is

$$\sigma^{\nu} = \frac{\sigma^{I}\Delta\nu}{I_{\max}} \tag{6.2}$$

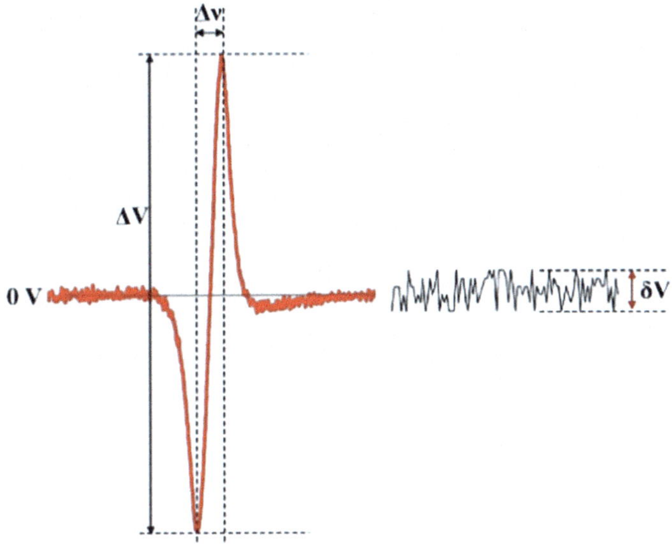

Fig. 6.11 MTS signal (red line) and residual error signal after locking (black line)

Δv is the frequency interval between the maximum and minimum values of the error signal, $I_{\max}$ is the amplitude of the error signal. According to the known parameters $I_{\max}$, σ^I, Δv, v_0, the frequency stability expression of the laser system can be obtained as:

$$\sigma_y(\tau) = \frac{1}{Q}\frac{1}{S/N}\frac{1}{\sqrt{\tau/s}} = \frac{\sigma^I \Delta\nu}{I_{\max}\nu_0}\frac{1}{\sqrt{\tau/s}} = \frac{\sigma^\nu}{\nu_0}\frac{1}{\sqrt{\tau/s}} \tag{6.3}$$

In a system with only one frequency standard, measuring stability through residual error signals is one of the common methods in the field of frequency metrology to evaluate frequency standards. This method converts the voltage fluctuations of the residual error signal into frequency fluctuations to measure the Allan deviation of frequency fluctuations, which can reflect the ability of the local oscillator laser to track the quantum reference.

The scanning acquisition and spectral line locking of the quantum reference spectral line both require tuning and controlling the frequency of the laser, which is usually accomplished by the laser drive circuit and the electronic servo feedback loop. For diode lasers, the laser frequency can be tuned by changing the drive current (corresponding to changing the refractive index of the laser diode gain medium), operating temperature (corresponding to changing the carrier density of the semiconductor gain), and resonator length. Among them, the tuning of the resonator length is achieved by changing the drive voltage of the piezoelectric ceramic (PZT) on the diode laser. The operating temperature is tuned with the help of heating or cooling devices (such as semiconductor coolers, TEC) and temperature sensing devices (such as thermistors). At the same time, under free-running conditions, in order to keep the frequency drift and noise of the laser at a low level, it is necessary to precisely control the parameters that cause the laser frequency to change with a precision control circuit, including the drive current, operating temperature, and drive voltage. Such precision control circuits can usually output low-noise stable voltage or current signals, respectively achieving control of the above parameters. Considering that the tuning of the laser frequency by the operating temperature is relatively slow, in diode lasers, the locking of the laser frequency is usually achieved by externally controlling the drive current and drive voltage. At the same time, the laser center frequency that has drifted can be pulled back by tuning the operating temperature slowly.

3. Cesium atom 852 nm Faraday laser single frequency MTS stabilization results

The experimental setup is shown in Fig. 6.12. The Faraday laser is used as the local oscillator. The FADOF used for frequency selection is composed of two polarizing beam splitters (PBS), a cesium atom vapor cell, and NdFeB magnets. The atomic vapor cell is 3 cm long and 1.5 cm in diameter, filled with cesium atoms and 10 torr Ar buffer gas, heated by a copper coil and wrapped by a heat-insulating PTFE layer. The 10 torr Ar buffer gas can widen the full width at half maximum (FWHM) of the FADOF transmission spectrum, thereby increasing the mode-hop-free range of the Faraday laser, and can suppress the spurious peaks of the transmission spectrum. The temperature of the vapor cell is controlled at 72 °C, with temperature fluctuations within 0.1 °C, thus maintaining the stability of the cesium saturated vapor pressure and the number of atoms in the vapor cell. The NdFeB magnet provides an axial magnetic field of 900 G. The laser is emitted by a laser diode with an anti-reflective coating, which eliminates the cavity mode. The first PBS is used to generate linearly polarized light. Due to the Faraday effect, the polarization of the light rotates after passing through the cesium vapor cell. Only light that has rotated 90° in polarization can pass through the second orthogonal PBS. The light is reflected at the high reflection mirror (R) for optical feedback. It then passes through the FADOF used for frequency selection again, returning to the ARLD to form continuous optical feedback, frequency selection, and light amplification, thereby improving the coherence of the laser and narrowing the laser linewidth. The stability of the Faraday laser frequency stabilization system is mainly limited by the temperature fluctuations of the cell used as the quantum reference for the modulation transfer spectroscopy. Temperature fluctuations can cause changes in the collision broadening and collision frequency shift of the atomic system, and affect the atomic density in the vapor cell, thereby causing fluctuations in the transmittance of the atomic vapor cell, which affects the error signal and feedback signal. Therefore, the research group at Peking University proposed using a double-layer vapor cell as the quantum frequency reference. The space between the two layers of quartz glass in the double-layer atomic vapor cell is evacuated, which can effectively suppress the heat conduction between the vapor cell and the environment, helping to solve the problem of temperature fluctuations in the atomic vapor cell. The effects of the detection light power, pump light power, modulation frequency, and vapor cell temperature on stability are shown in

Figs. 6.13, 6.14, 6.15 and 6.16. Based on this rule, the system parameters can be optimized individually. The final system operating conditions are selected as follows: a pump light power of 0.9 mW, a detection light power of 0.7 mW, a vapor cell temperature of 25 °C, and a modulation frequency of 4.6 MHz.

After measured by the residual error signal evaluation method, the system's stability reaches 5.8×10^{-15}/s, and decreases to 2.0×10^{-15} at 50 s. As the temperature fluctuates, the atomic density and atomic velocity distribution in the vapor cell also change, resulting in drift of the atomic frequency and broadening of the spectral

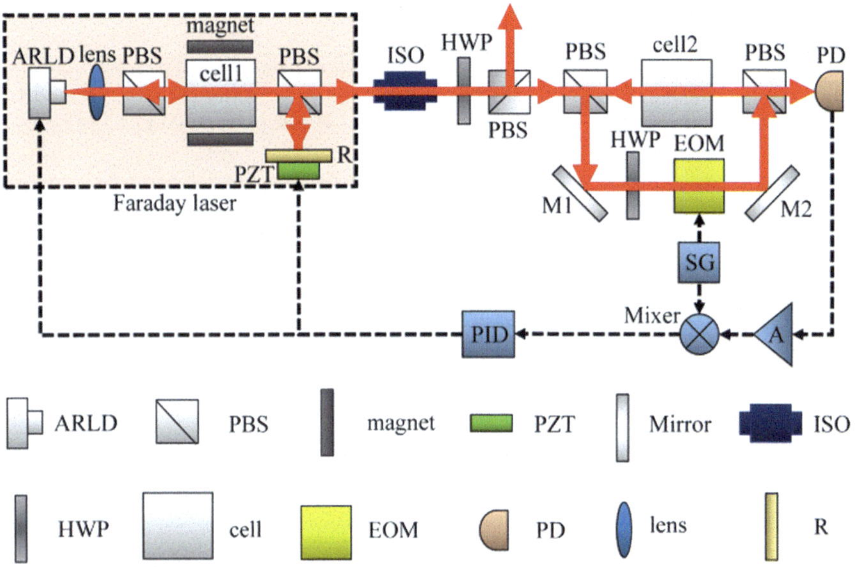

Fig. 6.12 Schematic diagram of the experimental setup of the high-stability Faraday laser

Fig. 6.13 The effect of pump light power on the slope of the MTS signal

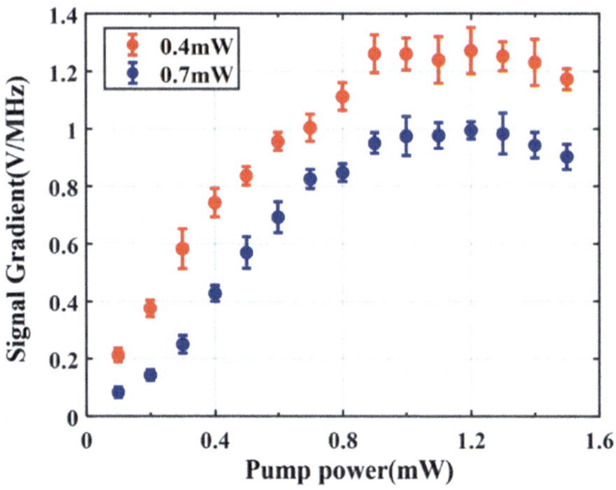

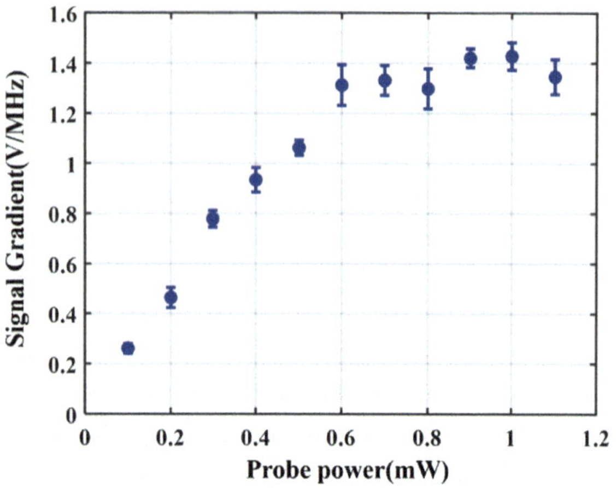

Fig. 6.14 The effect of detection light power on the slope of the MTS signal

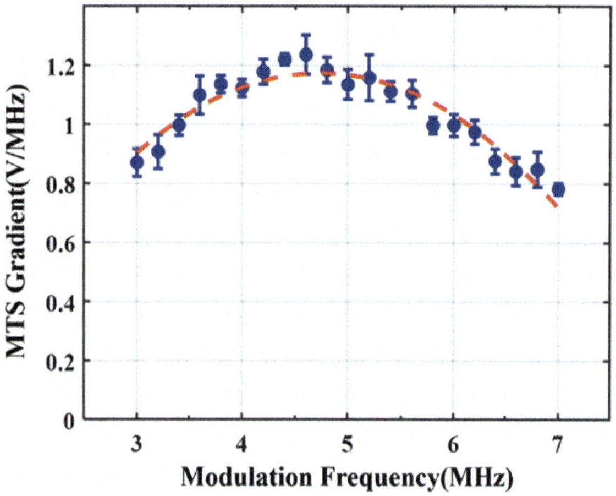

Fig. 6.15 The effect of modulation frequency on the slope of the MTS signal

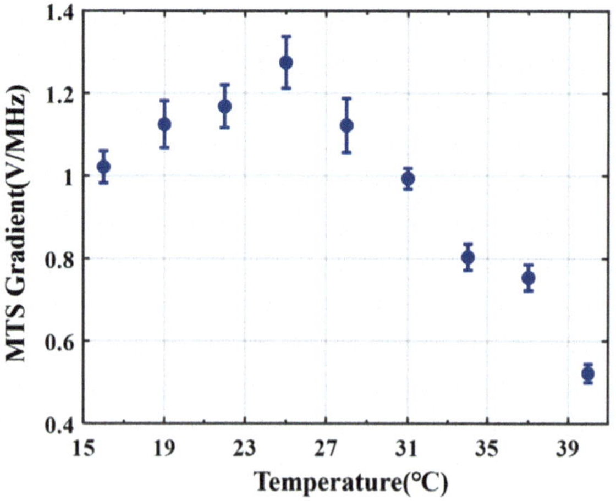

Fig. 6.16 The effect of vapor cell temperature on the slope of the MTS signal

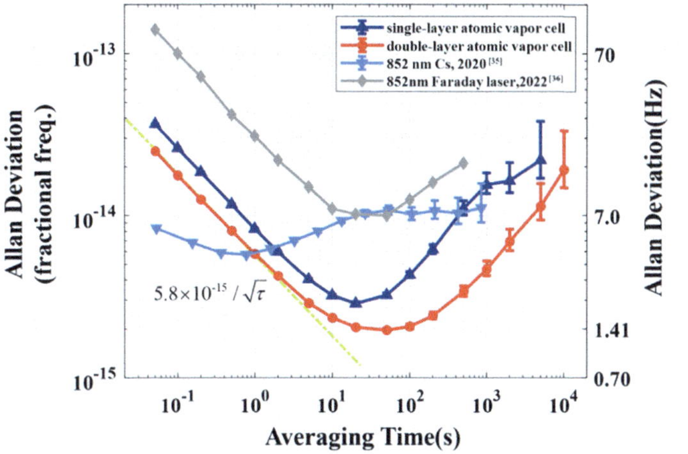

Fig. 6.17 Stability diagram of high-stability Faraday laser

linewidth. Because temperature drift occurs primarily at low frequencies, it mainly affects the long-term stability of the system. Since the space between the two quartz layers of the double-layered cell is evacuated, the heat conduction between the interior of the cell and the external environment can be basically isolated, resulting in a very slow temperature change with time. In other words, the temperature of the cesium atoms inside the vapor cell cannot keep up with the changes in the external temperature, which greatly reduces the temperature fluctuations inside the vapor cell. Therefore, using a double-layer vapor cell greatly improves the long-term stability of the system. Experimental results show that compared with using a single-layer atomic vapor cell, the stability of the laser frequency is significantly improved, as shown in Fig. 6.17.

6.2.3 Cesium Atom 852 nm Single-Frequency Mode and Switchable Dual-Frequency Mode Faraday Laser Frequency Stabilization

The operating mode of the Faraday laser depends on the number of transmission peaks and their relative transmittance of the Faraday atomic filter (FADOF) used as a frequency selector. Therefore, a multi-frequency Faraday laser can be realized by controlling the transmission spectrum of the Faraday atomic filter. In 2022, Chang and others from Peking University [9] achieved a cesium atom Faraday laser that can switch between single-frequency and dual-frequency modes, and stabilized the output frequency to the hyperfine transition spectrum line by modulating the transfer spectroscopy.

Fig. 6.18 Transmission spectrum of 852 nm FADOF of cesium atoms

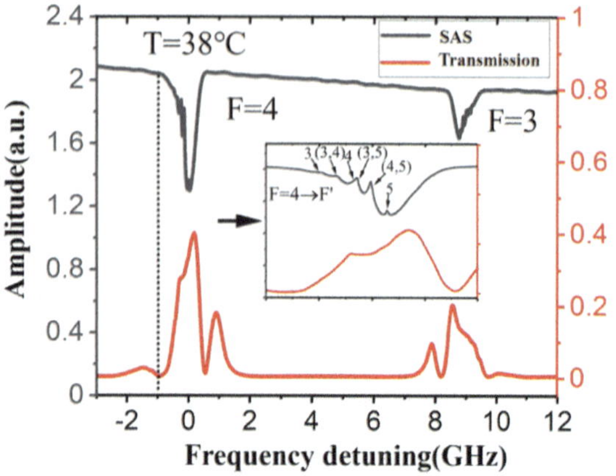

Figure 6.18 shows the transmission spectrum of the FADOF inside the Faraday laser. From the figure, it can be seen that there are mainly two larger transmission peaks, and with the change of the FADOF temperature, the maximum transmittance of these two transmission peaks also changes. When there is a significant difference in the transmittance of the two transmission peaks, the frequency component with a larger transmittance forms a resonance, and hence, the laser works in a single-frequency mode. When the transmittance of the two transmission peaks are comparable, the corresponding two frequency components simultaneously form a resonance in the cavity, and the Faraday laser works in a dual-frequency mode at this time.

The internal magnetic field of the FADOF is 330 G. By changing the working conditions of the Faraday laser, such as current and temperature, the Faraday laser can work in single-frequency mode or dual-frequency mode. This depends on the size relationship of the peak transmittance of the two transition lines of cesium atom $6S_{1/2}$ ($F = 4$) and $6S_{1/2}$ ($F = 3$) to $6P_{3/2}$. When in single-frequency mode, the laser works on the transition peak from the ground state $F = 4$ to the excited state $6P_{3/2}$, and the saturation spectroscopy and modulation transfer spectroscopy are shown in Fig. 6.19a. The modulation transfer spectroscopy line width is 10 MHz, and the maximum slope is 0.72 V/MHz when the modulation frequency is taken as 5.4 MHz. The Faraday laser is locked to the $6S_{1/2}F = 4 \rightarrow 6P_{3/2}F' = 5$ cycling transition peak through the modulation transfer spectroscopy, and the second stability obtained by the residual error signal evaluation method is $3 \times 10^{-14}/\sqrt{\tau}$. The stability decreases to 9×10^{-15} within an average sampling time of 50 s. In the dual-frequency mode, the output frequency of the free-running Faraday laser corresponds to $6S_{1/2}$ ($F = 4$) and $6S_{1/2}$ ($F = 3$) to $6P_{3/2}$ transitions, the detected dual-frequency saturation spectroscopy and modulation transfer spectroscopy are shown in Fig. 6.19b. After the laser is locked, the beat frequency signal between the two modes of the Faraday laser is detected, the absolute value of this signal corresponds to the microwave frequency of the hyperfine energy level transition of the cesium atom ground state. Since the two frequency modes share the same resonant cavity, the common mode noise brought

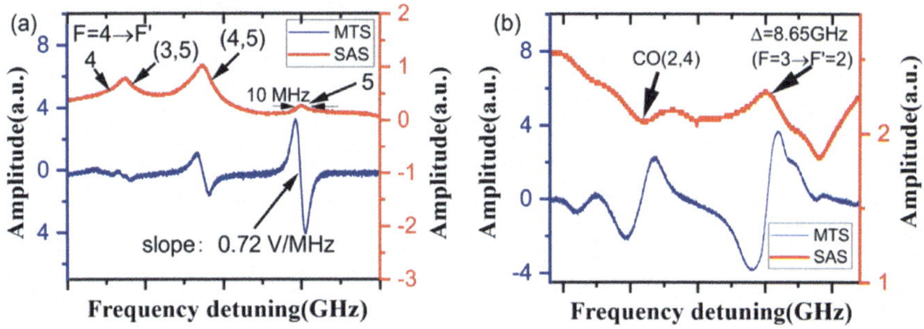

Fig. 6.19 Single frequency (**a**) and dual frequency (**b**) Faraday laser saturation spectroscopy and modulation transfer spectroscopy

Fig. 6.20 Dual frequency Faraday laser beat signal

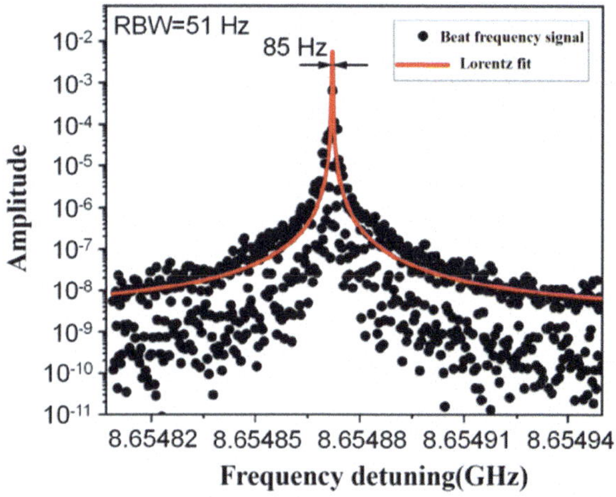

by the cavity length jitter is suppressed, and the linewidth of this microwave signal is narrowed to 85 Hz, as shown in Fig. 6.20. By further improving the performance of this microwave signal, it is expected to achieve more stable optically generated microwaves, and apply them in fountain atomic clocks, CPT atomic clocks, and other systems.

6.2.4 MTS Frequency Stabilization System of Rb Atom 780 nm Faraday Laser

For the Rb atom Faraday laser, the biggest difference from the cesium atom Faraday laser is that Rb has two stable isotopes, ^{85}Rb and ^{87}Rb, while cesium only has one stable isotope. Therefore, there are some differences in the implementation process. Specifically, cesium has a simpler energy level structure, the setting of FADOF parameters is easier, and it can correspond to the wavelength of the target output laser more easily. However, the 780 nm Rb atom Faraday laser have more diverse situations due to the presence of dual isotopes. For natural Rb, considering the complexity of the dual isotope spectral lines existing at the same time, it is generally difficult to generate a laser frequency that directly corresponds to the atomic spectral line [10]. However, this kind of laser still has applications, for example, in atomic magnetometers, a laser source with a certain detuning amount (GHz) is needed. When developing a Faraday laser with a frequency corresponding to the atomic transition spectral line, researchers often use a single isotope FADOF to achieve the target frequency selection, for example: the frequency corresponding to ^{85}Rb or ^{87}Rb-D_2 line transition can be achieved by setting the parameters of FADOF in a single isotope (^{85}Rb or ^{87}Rb). An interesting phenomenon is that by choosing FADOF parameters, the transmission peak can correspond to the transition frequency of another isotope, that is, using ^{85}Rb (or ^{87}Rb)-FADOF can achieve ^{87}Rb (or ^{85}Rb)-D_2 transition frequency output. In addition, Qin and colleagues from Peking University reported a switchable Faraday laser that can switch between the D_2 transition frequencies of the two isotopes ^{85}Rb and ^{87}Rb using a single isotope FADOF (^{87}Rb-FADOF). This research mainly includes the 780 nm Faraday laser locked to the Rb atom hyperfine transition line realized by Gao and others from Peking University in 2024. In 2024, Qin from Peking University [11] and colleagues achieved the locking to the ultrafine transition line of Rb atoms with a 780 nm single/dual-frequency switchable Faraday laser. These systems stabilize the output frequency of the Faraday laser on the cycling transition of the Rb atom D_2 line through the modulation transfer spectroscopy method. This section provides a detailed introduction to these works.

1. Single-frequency mode 780 nm stabilized Faraday laser

Peking University research group used a 5 torr mixed Ar buffer gas ^{85}Rb FADOF as a frequency selection element to build a single-frequency Faraday laser. Using MTS technology, the output frequency of the laser is locked on the ^{87}Rb D_2 line's $5^2S_{1/2}F = 2 \rightarrow 5^2P_{3/2}F' = 3$ cycling transition peak.

The experimental setup for locking the Faraday laser to the ^{87}Rb D_2 line using MTS technology is shown in Fig. 6.21. The single-frequency Faraday laser consists of an anti-reflective coated laser diode, a collimation lens (CL), FADOF, a hollow corner cube metal reflector, and a piezoelectric ceramic (PZT). The 780 nm beam emitted from the ARLD is collimated by an aspheric lens with a focal length of 4.5 mm and a numerical aperture of 0.55, and then enters the FADOF system for frequency selection. Inside the FADOF system, the PZT is attached to the cone to adjust the cavity length, allowing the laser frequency to be tuned. The light emitted by the Faraday laser passes through a 30 dB isolator to avoid harmful stray light feedback, and then is split into two beams by a half-wave plate (HWP) and a polarizing beam splitter (PBS). The transmitted light serves as the stabilized laser output, and only 2 mW of reflected light enters the MTS system for stabilization. The reflected light is further split into two beams by the half-wave plate and polarizing beam splitter, one of which serves as the detection light, and the other as the backward transmitted pump light. The vapor cell for the MTS system is a double-layer vapor cell, with a vacuum area between the two quartz layers, which has been proven to improve the frequency stability of the laser system [12]. The pump beam passes through an electro-optic modulator (EOM) for phase modulation, with a modulation frequency of 5.8 MHz and a modulation depth of $\delta = 0.4$. One channel of the signal generator (SIG) drives the EOM, while the other serves as the local oscillator signal for the mixer. The pump light modulated by the EOM is reflected into the Rb atomic vapor cell by the polarization beam splitter, where it interacts with the probe light and Rb atoms. After four-wave mixing, the probe light is transmitted into the high-speed photodetector through the polarization beam splitter and is detected. The signal detected by the PD is demodulated by the mixer, generating a dispersion-like modulated transfer spectrum signal, i.e., the error signal. The error signal is then processed by the Proportional Integral Derivative (PID) servo control module. The PID output feedback signal is fed back to the current and PZT ports of the Faraday laser driver, ultimately locking the frequency of the Faraday laser to the atomic transition frequency corresponding to the Rb atoms, achieving frequency locking of the Faraday laser.

Figure 6.22 shows the Saturated Absorption Spectroscopy (SAS) and Modulated Transfer Spectroscopy (MTS) signals of this system. From left to right, the SAS signal corresponds to the ^{87}Rb $5^2S_{1/2}F = 2 \rightarrow 5^2P_{3/2}F' = 3$, 3 and 2 as well as 1 and 3 crossover line transitions. Due to the ultra-narrow transmission spectrum of FADOF and the combined effect of laser mode hopping, the output frequency of the Faraday laser only shows three saturated absorption peaks during the PZT scanning

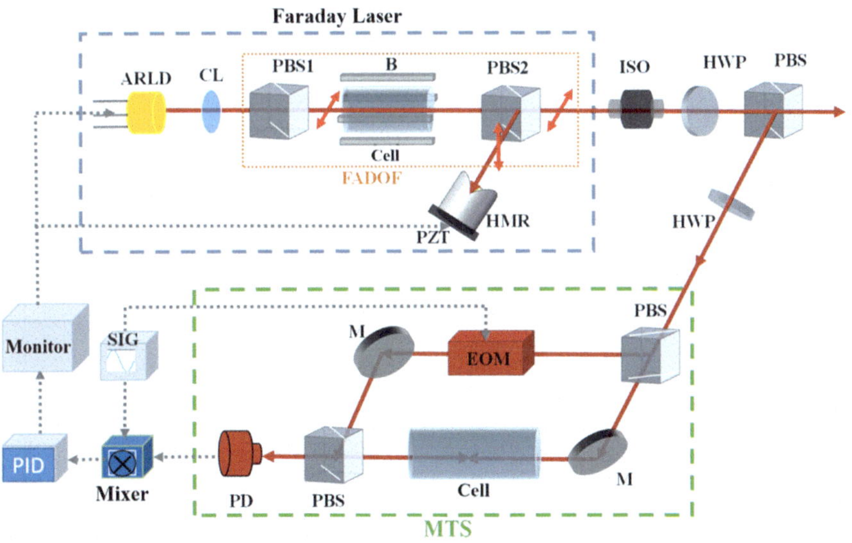

Fig. 6.21 Rb atom 780 nm single-frequency Faraday laser MTS stabilization system. *ARLD* anti-reflective coated laser diode, *CL* collimation lens, *PBS* polarizing beam splitter, *ISO* isolator, *HWP* half-wave plate, *M* mirror, *PZT* piezoelectric ceramic, *HMR* hollow conical metal reflector, *EOM* electro-optic modulator, *PD* photodetector, *SIG* signal generator, *PID* proportional integral derivative feedback circuit

period. At the ^{87}Rb $5^2S_{1/2}F = 2 \rightarrow 5^2P_{3/2}F' = 3$ transition, the linewidth of the SAS spectrum line and the peak-to-peak value of the MTS spectrum line are 13.5 MHz and 11.4 V, respectively, and the maximum slope of the MTS is 0.84 V/MHz.

By using a multimeter to record the voltage fluctuations of the error signal, the frequency fluctuations are converted using the aforementioned relationship $\delta v/\Delta v = \delta V/\Delta V$. When this Faraday laser is locked, the Allan deviation measured by the residual error signal method reaches 2.87×10^{-14} at 1 s, then decreases to 1.25×10^{-14} at 100 s, as shown in Fig. 6.23. Unlike other diode lasers, this Faraday laser can scan the saturated absorption peak within a current range of 75 ~ 140 mA, although the slope of the MTS signal obtained under these conditions is slightly different. By using MTS technology, this Faraday laser system can quickly and accurately lock onto the atomic hyperfine transition spectrum within a larger injection current range, achieving a "plug and play" system effect.

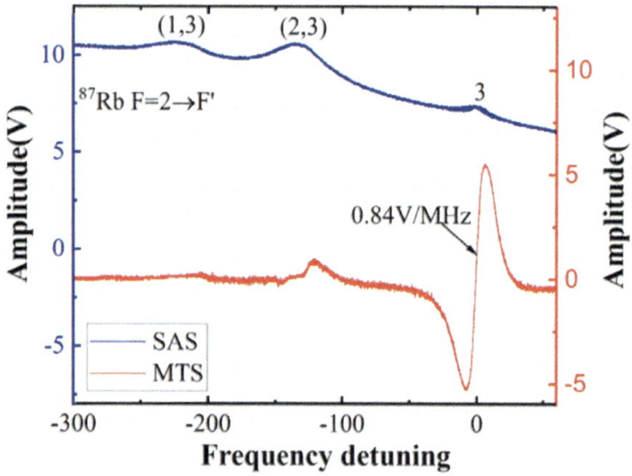

Fig. 6.22 Rb 780 nm single-frequency Faraday laser's saturated absorption spectroscopy (SAS) and modulated transfer spectroscopy (MTS)

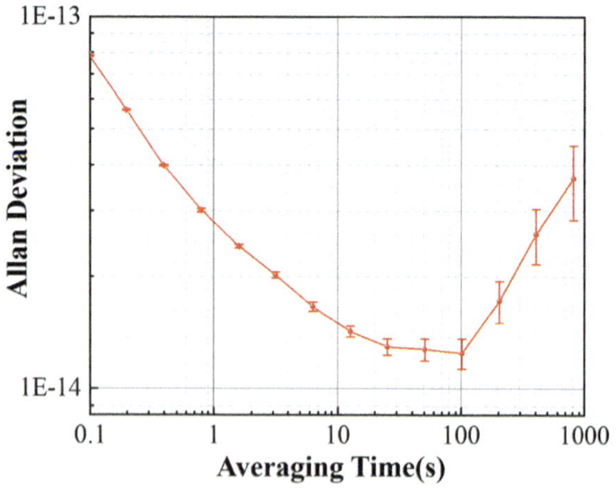

Fig. 6.23 Stability measurement results of the 780 nm single-frequency Faraday laser for Rb atoms

2. Dual-frequency switchable 780 nm stabilized Faraday laser

The research group used a single isotope ^{87}Rb-FADOF to construct a dual-frequency switchable 780 nm Faraday laser, the frequency of which corresponds to the cycling transitions in the ^{85}Rb and ^{87}Rb D_2 lines.

To achieve specific frequency selection functions, the research group studied the effects of different temperatures and laser intensities on the transmission spectrum of ^{87}Rb-FADOF, as shown in Fig. 6.24. The solid and dashed lines in the figure represent the transmission curves under strong laser (66.37 mW/mm^2) and weak laser (4.42 mW/mm^2) respectively. The peaks 1, 2, 3, and 4 of the saturated spectrum are marked as ^{87}Rb D_2 line, ^{85}Rb D_2 line, ^{85}Rb D_1 line, and ^{87}Rb D_1 line transition peaks respectively. At temperatures of 60 and 65 °C, the FADOF transmission peaks corresponding to positions 1 and 2 have similar transmittance. At the same time, at low light intensity (i.e., low drive current), the transmittance of peak 2 is the best, corresponding to the ^{85}Rb D_2 line transition peak. At high light intensity (i.e., high drive current), the transmittance of peak 1 increases so as to become comparable to peak 2. At this point, the frequency corresponding to the ^{87}Rb D_2 line transition can be achieved. Therefore, at an appropriate temperature, by changing the size of the drive current, the output laser frequency of this system can be switched between the ^{85}Rb D_2 line and ^{87}Rb D_2 line transitions. At the same time, the research group also found that the Faraday laser system has robustness to fluctuations in laser diode current and temperature, and the output wavelength changes with current intensity and temperature as shown in Fig. 6.25.

The research group has set up a modulation transfer spectroscopy device that can stabilize the Faraday laser frequency to the cycling transitions of the ^{85}Rb or ^{87}Rb D_2 lines (i.e., ^{85}Rb $5^2S_{1/2}$ $(F = 3) \rightarrow 5^2P_{3/2}$ $(F' = 4)$ and ^{87}Rb $5^2S_{1/2}$ $(F = 2) \rightarrow 5^2P_{3/2}$ $(F' = 3)$). The experiment is illustrated in Fig. 6.26.

In the Faraday laser, ^{87}Rb-FADOF consists of a pair of polarizing beam splitters (PBS), a cylindrical single isotope ^{87}Rb vapor cell with a length of 30 mm and a diameter of 15 mm, a 300 G axial magnetic field formed by four permanent magnets, and a temperature control system with an accuracy of 0.1 °C. The output frequency can be switched by changing the temperature of the vapor cell in the FADOF or the driving current.

After obtaining the modulation transfer spectroscopy signal, the system has three feedback ports, including a slow feedback port that controls the laser current, a fast feedback port that controls the ARLD, and a feedback port that controls the PZT, thereby achieving better noise suppression and frequency stabilization effects, and achieving frequency stability for noise suppression.

To characterize the frequency stability of the locking system, the research group used both the inner loop and the beat note measurement methods, as shown in Fig. 6.27. The inner loop evaluation method only uses a set of Faraday laser systems. When the system is locked on the cycling transitions of ^{85}Rb and ^{87}Rb D_2 lines respectively, the system's Allan deviation is better than $3 \times 10^{-14}/\sqrt{\tau}$. In addition, a Faraday laser MTS system with the same FADOF parameters locked on the cycling transition of ^{87}Rb D_2 line, and another one locked on the cycling transition of ^{85}Rb

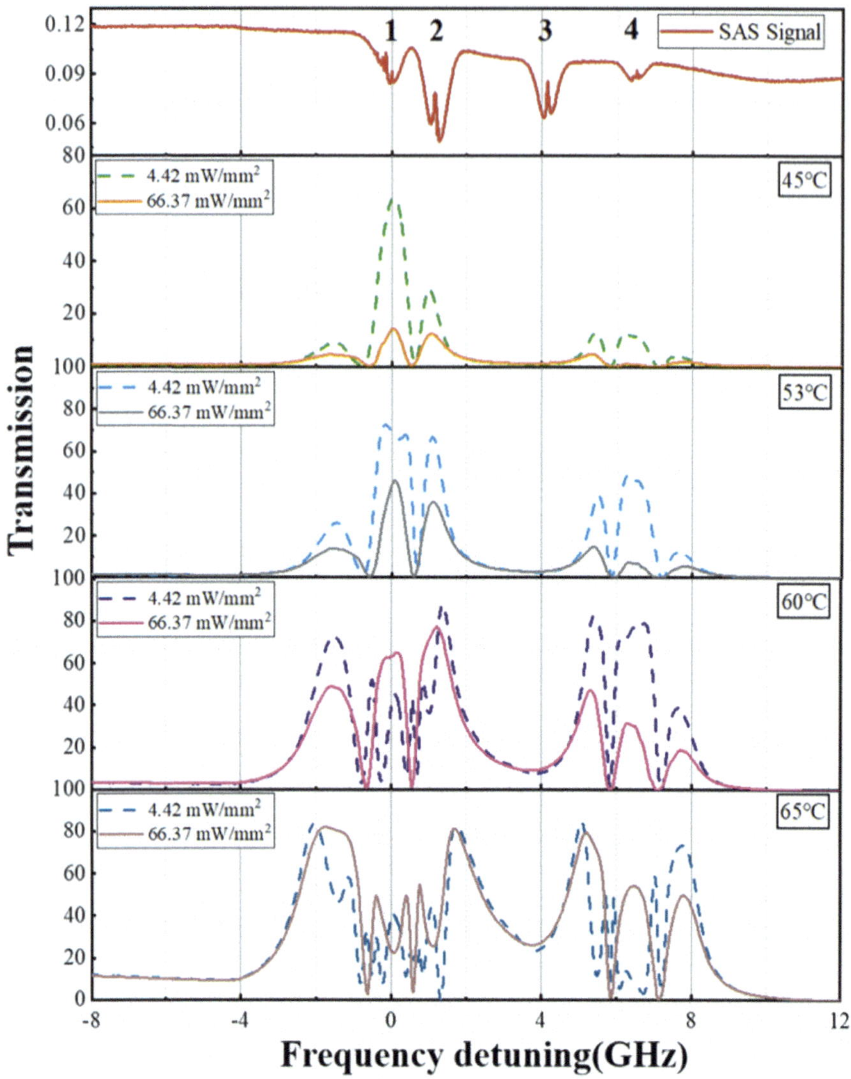

Fig. 6.24 Transmission spectrum of ^{87}Rb-FADOF at different temperatures and laser intensities

D_2 line, the second stability obtained by beat note is 2.8×10^{-12}, with a thousand second stability of 8.2×10^{-12}.

This research has expanded the application range of Faraday lasers, providing key laser sources for cold atomic physics, atomic interferometers, compact optical clocks, and other atomic physics applications. This scheme can also be extended to atoms such as cesium and potassium, using the aforementioned FADOF to achieve switchable Faraday lasers and Voigt lasers of different frequencies.

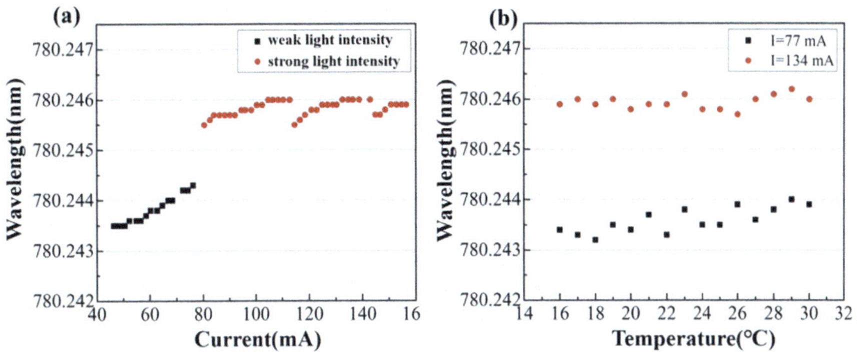

Fig. 6.25 Changes in the system's output wavelength under different current intensities and temperatures

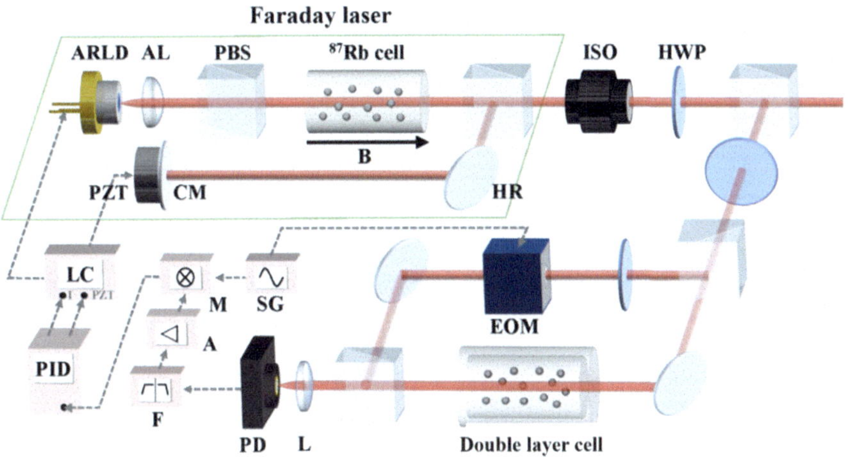

Fig. 6.26 MTS frequency stabilization system of the 780 nm dual-frequency switchable Faraday laser of Rb atom. *ARLD* anti-reflective coated laser diode, *AL* lens, *PBS* polarizing beam splitter, *^{87}Rb cell1* ^{87}Rb cell, *ISO* isolator, *HWP* half-wave plate, *HR* high reflection mirror, *PZT* piezo-electric ceramic, *EOM* electro-optic modulator, *PD* photodetector, *SG* signal generator, *M* mixer, *A* amplifier, *PID Proportional-Integral-Derivative feedback circuit*

6.2.5 A Switchable Potassium Faraday Laser Between 770 and 766 nm

Being Group I elements, Rb, Cs and K possess the similar properties. Potassium has multiple isotopes, the most common of which are potassium-39 (^{39}K) and potassium-41 (^{41}K). ^{39}K is the most abundant isotope, accounting for about 93.3% of the natural existence of potassium, while ^{41}K accounts for about 6.7%. In addition, there exist

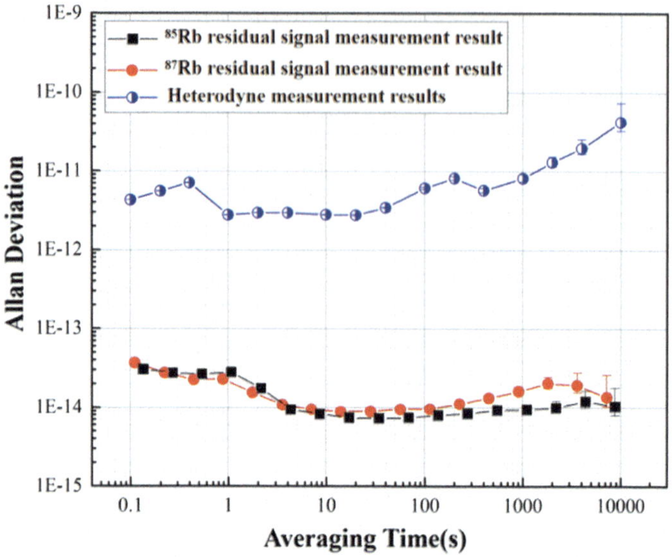

Fig. 6.27 Frequency stability of the 780 nm dual-frequency switchable Faraday laser system of Rb atom

a radioactive isotope—potassium-40 (^{40}K), which is widely used in cold atomic physics experiments.

The potassium Faraday laser uses a natural potassium atomic vapor cell within the potassium-based FADOF as a quantum reference. Figure 6.28 shows the related energy levels of ^{39}K. In comparison to rubidium and cesium, potassium has a narrower energy level splitting, with the gap between its ground states $F = 1$ and $F = 2$ being smaller than the Doppler width. Therefore, there is an equivalent energy level X$(1, 2)$. The wavelength of the ^{39}K D_1 line is about 770.108 nm, and the D_2 line wavelength is about 766.701 nm. These two wavelengths are widely used in potassium cooling and trapping, Bose–Einstein condensation (BEC), potassium magnetometer, optical frequency standard, high-resolution spectroscopy, and other fields.

The wavelengths of the potassium D_1 and D_2 lines are close, covered by the fluorescence spectrum of a single laser diode, which makes it possible to switch its wavelength between 770 and 766 nm in a single laser system. Figure 6.29 shows the influence of the operating temperature of vapor cell and the magnetic field on the transmitted spectrum of potassium FADOF. We find that when the magnetic field is set at 1500 G, 105 °C is the optimal temperature for ^{39}K D_1 line and 95 °C is the optimal temperature for D_2 line. That is, at optimal working parameters, the potassium Faraday laser operate at 770 nm when the vapor temperature of FADOF set at 105 °C and once adjusting the vapor temperature to 95 °C, it will switch to 766 nm immediately.

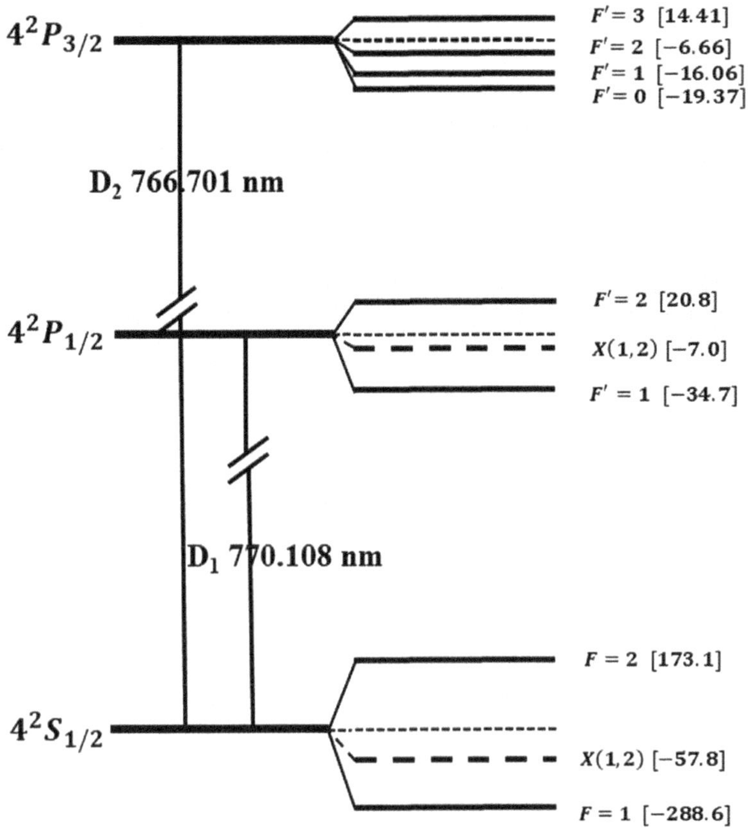

Fig. 6.28 Energy level diagram of potassium atom D_1 and D_2 transition lines. F and F' represent the total (electronic plus nuclear) angular quantum numbers of the ground states and excited states, respectively. The values in the parentheses represent the hyperfine shifts in MHz

Figure 6.30 shows the schematic diagram of the potassium Faraday laser with MTS frequency stabilization technique. The design of the system includes a home-made potassium atom Faraday laser, which emits a broad fluorescence spectrum. After passing through an isolator, the light beam passes through a half-wave plate and a polarizing beam splitter, and is divided into two parts: transmitted light and reflected light. The transmitted light serves as the frequency-stabilized laser output, while the reflected light is used for modulation transfer spectroscopy (MTS) frequency stabilization. The reflected light is further divided into two beams by another set of half-wave plates and polarizing beam splitters, which serve as the pump light and probe light in the modulation transfer spectroscopy. The light intensity of these two beams can be adjusted by adjusting the combination of the half-wave plate and the polarizing beam splitter. The pump light undergoes electro-optic phase modulation through a resonant electro-optic modulator, and then undergoes four-wave mixing with the probe light in a double-layer atomic vapor cell, further transferring the

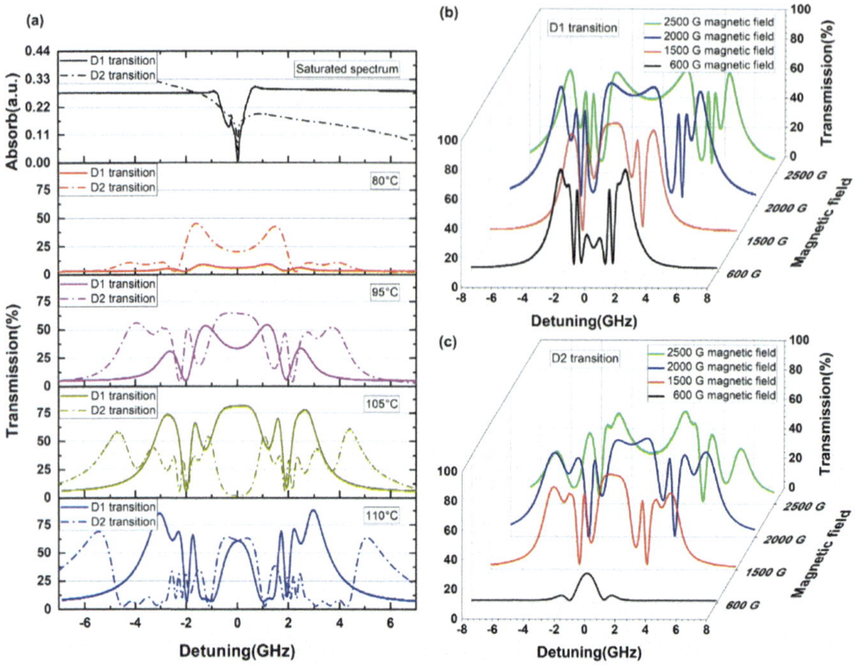

Fig. 6.29 The transmitted spectrum of FADOF on the D_1 and D_2 lines at different operating parameters. **a** The transmitted spectrum of FADOF at different vapor temperatures. The top black curves are the SAS signals of ^{39}K. The red, pink, green, blue curves represent the transmitted spectrum at the temperatures of 80, 95, 105, and 110 °C. The solid curves relate to the D_1 line and the dashed lines relate to the D_2 line. The magnetic field is maintained at 1500 G. **b** The transmitted spectrum of FADOF operating at D_1 line at different magnetic fields of 600 G (black line), 1500 G (red line), 2000 G (blue line), and 2500 G (green line). The vapor temperature is kept at 105 °C. **c** The transmitted spectrum of FADOF operating at D_2 line at different magnetic fields of 600 G (black line), 1500 G (red line), 2000 G (blue line), and 2500 G (green line). The vapor temperature is kept at 95 °C

modulation information to the unmodulated probe light, which is then received by a photodetector. After demodulation and processing of the detected signal, three feedback signals are output through the PID controller, which are connected to the piezoelectric ceramic end, current end, and laser head of the Faraday laser, thereby achieving precise stabilization of the Faraday laser frequency and ensuring its frequency stability on the potassium quantum frequency reference.

The potassium saturated absorption spectroscopy and modulation transfer spectroscopy obtained from the experiment are shown in Fig. 6.31, corresponding to the potassium D_1 and D_2 lines, respectively. The two excited states of the ^{39}K D_1 transitions, $F' = 1$ and $F' = 2$, differ by about 55.5 MHz, which is significantly larger than the natural linewidth. Therefore, we can clearly distinguish the transitions from the ground state to each excited state. Figure 6.31a shows the saturated absorption spectroscopy (SAS) and modulation transfer spectroscopy (MTS) signals of the ^{39}K

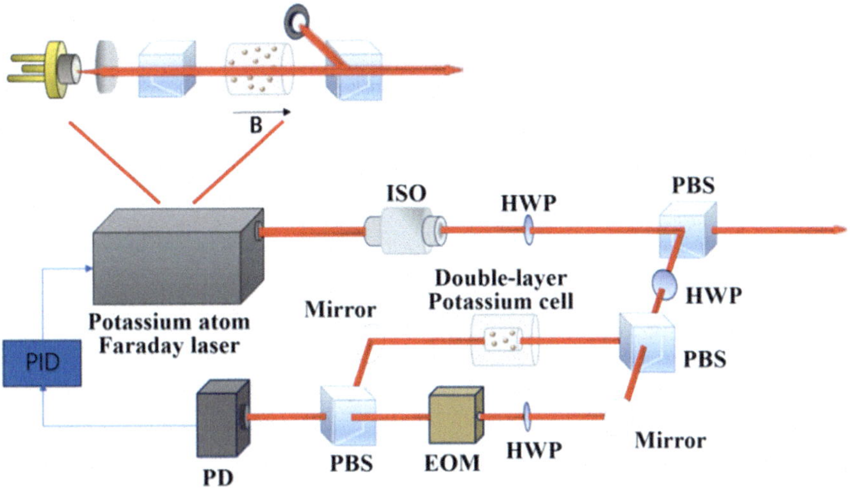

Fig. 6.30 Schematic diagram of the wavelength switchable potassium Faraday laser between 770 nm D_1 and 767 nm D_2 transitions

D_1 transitions. Since the energy level intervals of the potassium D_2 transitions are small and close to the natural linewidth, it is difficult for us to distinguish the transitions from the ground state to each excited state. Figure 6.31b shows the saturated absorption spectroscopy (SAS) and modulation transfer spectroscopy (MTS) signals of the ^{39}K D_2 transitions.

Then, the laser is locked to ^{39}K $4S_{1/2}F = (1, 2) \rightarrow 4P_{1/2}F' = 2$ and $4S_{1/2}F = (1, 2) \rightarrow 4P_{3/2}F'$ transitions, operating at 770.108 nm and 766.701 nm, respectively.

By optimizing the modulation frequency of the EOM, the power of the pump light and the probe light, and the working temperature of the double-layer potassium atom

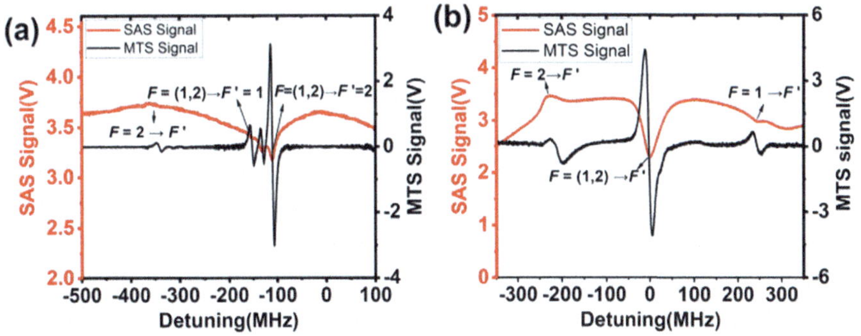

Fig. 6.31 **a** The SAS (red) and MTS (black) signals at the D_1 line. From left to right correspond to the transitions from the ground state $F = 2$ and $F = (1, 2)$. **b** The SAS (red) and MTS (black) signals at the D_2 line. From left to right correspond to the transitions from the ground state $F = 2$, $F = (1, 2)$ and $F = 1$

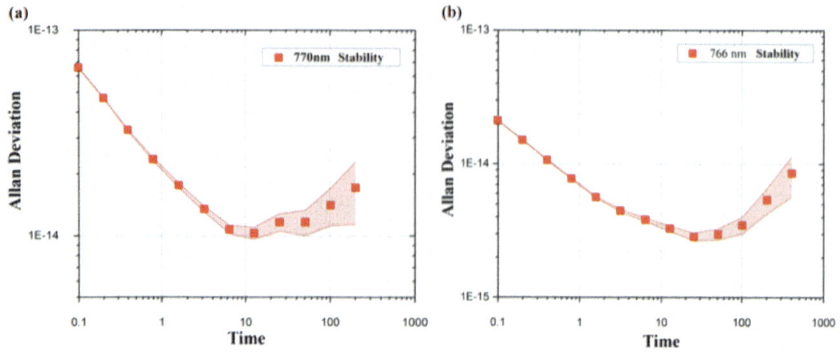

Fig. 6.32 **a** In-loop stability of the potassium Faraday laser operating at 770 nm. **b** In-loop stability of the potassium Faraday laser operating at 766 nm

vapor cell, a large frequency discrimination signal slope, about 1.2 V/MHz at 766 nm and 0.58 V/MHz at 770 nm, is obtained. Furthermore, the locking of the switchable potassium Faraday lasers at both 770 and 766 nm is achieved. The in-loop stabilities of 2.2×10^{-14}/s for 770 nm and 7.0×10^{-15}/s for 766 nm are achieved and are demonstrated in Fig. 6.32a, b, respectively.

We utilized two identical potassium Faraday lasers to conduct a beat frequency experiment. The frequency of one laser is shifted by 200 MHz. Figure 6.33 shows the frequency stability of the beat signals at 770 nm and 767 nm, which are 2.23×10^{-12}/s and 1.24×10^{-12}/s, respectively. Consider two lasers contribute the same portion to the beat signal, the frequency stability of a single laser is 1.57×10^{-12}/s for 770 nm and 8.77×10^{-13}/s for 766 nm. In the future, to improve the medium- and long-term stability, we can use a wedged EOM and a vapor cell to suppress the RAM noise caused by the etalon effect. Besides, since the power fluctuation of the laser causes energy level shift, stabilizing the laser power further enhances the long-term stability.

6.3 Frequency Stabilization of Faraday Laser Using PDH (Pound–Drever–Hall) Method

6.3.1 Basic Principles

1. Spectral characteristics of Fabry–Pérot resonator

Ultra-stable optical resonators are used to generate lasers with high phase coherence and frequency stability, and are widely used in laser frequency stabilization [13, 14]. A thorough analysis of the characteristics of the optical FP cavity is necessary. This section derives the exact solutions for the finesse and FWHM of the FP cavity under

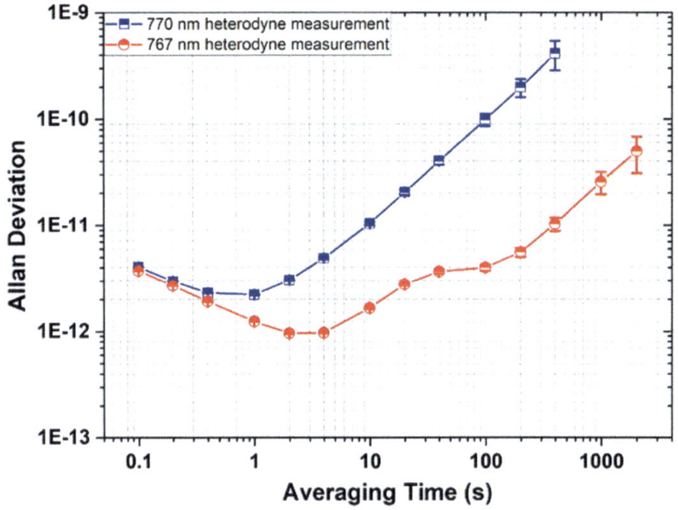

Fig. 6.33 Allan deviations of the beat signals by two identical lasers at 770 nm (the blue line) and 767 nm (the red line)

any mirror reflectivity, which differs from the results given in traditional textbooks for high reflectivity FP cavities [15, 16]. Moreover, it analyzes the generalized expression of the cavity enhancement factor under any reflectivity, and proves that under the same mirror reflectivity, the intra-cavity light intensity enhancement caused by the resonant FP cavity and the intra-cavity light intensity suppression introduced by the anti-resonant FP cavity are symmetrically distributed.

2. Airy function and power reflection factor

Using the typical model of the optical cavity, let us assume an FP cavity composed of two plane mirrors M_1 and M_2, with a cavity length of L. Further, the reflection coefficients and transmission coefficients of the two cavity mirrors are r_1, t_1 and r_2, t_2, respectively the reflectance and transmittance of the two cavity mirrors are $R_1 = {r_1}^2$, $T_1 = {t_1}^2$ and $R_2 = {r_2}^2$, $T_2 = {t_2}^2$, the path of the light field in the FP cavity is shown in Fig. 6.34. Assuming that the complex amplitude of the electromagnetic field incident on M_1 is E_0, E_T and E_R are the complex amplitudes of the transmitted and reflected electromagnetic waves through the FP cavity, respectively.

The ratio of the light intensity transmitted through the FP cavity to the incident light intensity is called the Airy function, as shown in the following formula:

$$T(\omega) = \frac{E_T E_T^*}{E_0^2} = \frac{t_1^2 t_2^2}{1 + r_1^2 r_2^2 - 2r_1 r_2 \cos(\omega 2L/c)} \tag{6.4}$$

where, $\Delta\varphi = \omega 2L/c$ represents the phase shift of the reflected field in the cavity, ω is the angular frequency of the incident light, c is the speed of light. Corresponding to the transmitted light, the power reflected by the FP cavity can be expressed as:

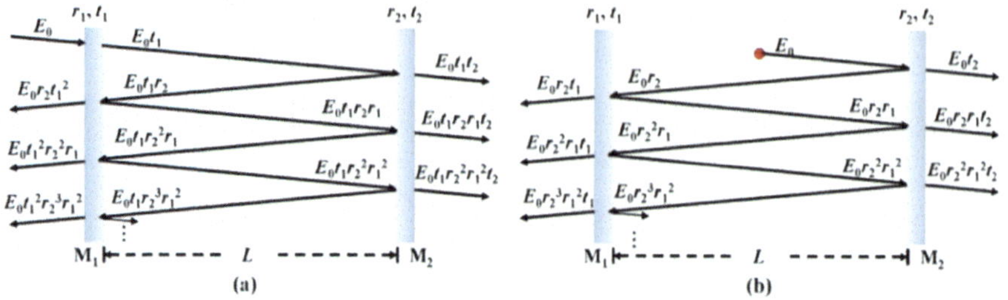

Fig. 6.34 The amplitude of light transmitted and reflected through the FP cavity. **a** The initial incident wave is emitted from one of the mirrors of the FP cavity. **b** The initial incident wave is generated by a radiation source located at the center of the FP cavity

$$R(\omega) = \frac{E_R E_R^*}{E_0^2} = \frac{2r_1 r_2[1 - \cos(\omega 2L/c)]}{1 + r_1^2 r_2^2 - 2r_1 r_2 \cos(\omega 2L/c)} \tag{6.5}$$

From the above two formulas, it can be seen that both transmission and reflection depend on the phase shift $\Delta\varphi$, the transmission factor $T(\omega)$ and the reflection factor $R(\omega)$ vary with the phase shift as shown in Fig. 6.35, and satisfy the condition $T(\omega) + R(\omega) = 1$. It is worth noting that the Airy function applies not only to the case where the incident light enters the FP cavity from one side of the cavity mirror, but also to the case where the radiation source is located in the center of the two FP cavities. These two cases correspond to Fig. 6.34a, b, respectively.

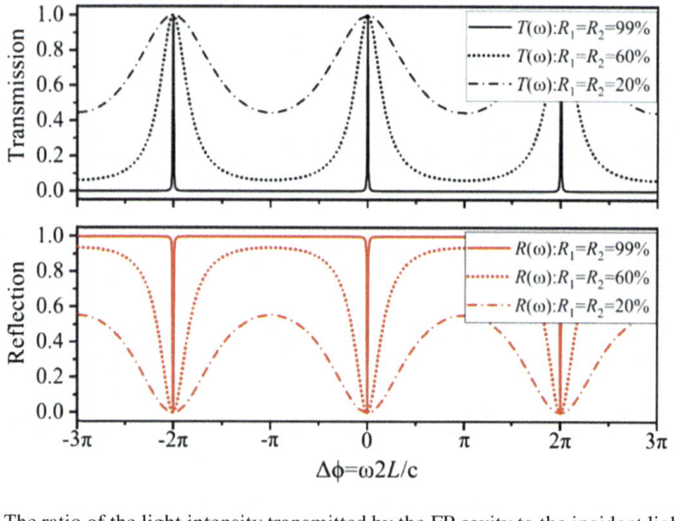

Fig. 6.35 The ratio of the light intensity transmitted by the FP cavity to the incident light intensity varies with phase shift, showing periodic transmission peaks (above), and the ratio of the light intensity reflected by the FP cavity to the incident light intensity varies with phase shift, showing periodic reflection peaks (below)

Obviously, as the reflectivity of the cavity mirror increases, the power transmission of the FP cavity in the anti-resonance region gradually decreases, and so as the FWHM of the resonance region. Reflection and transmission are in contrast to each other, as shown in Fig. 6.35.

When the light frequency coincides with the cavity resonance frequency, the transmittance reaches its maximum value, and when the light frequency is located in the middle of two adjacent cavity modes, the transmittance reaches its minimum. The situation of reflection is exactly the opposite of the transmission situation. The upper and lower figures share a horizontal axis, and the solid line, dotted line, and dotted-dashed line respectively represent the case where the mirror reflectivity is 99%, 60%, and 20%.

3. Cavity finesse and FWHM exact solution

According to formula (6.4), for an empty FP cavity, the maximum and minimum transmittance can be represented as:

$$T_{\max} = \frac{t_1^2 t_2^2}{1 + r_1^2 r_2^2 - 2r_1 r_2} \tag{6.6}$$

and

$$T_{\min} = \frac{t_1^2 t_2^2}{1 + r_1^2 r_2^2 + 2r_1 r_2} \tag{6.7}$$

The above results of cavity mode frequency correspond to the vicinity of the qth resonance frequency ω_q, where the cavity mode frequency is exactly in the middle of two adjacent resonance cavity modes. Since the minimum transmission T_{min} can be ignored under high reflectivity conditions, the FWHM is usually expressed as a frequency interval at half of the maximum transmittance of a transmission spectrum [15, 16]. However, as the mirror reflectivity decreases, T_{min} gradually increases to the point where it cannot be ignored when calculating FWHM. Therefore, in order to obtain a universal expression, this section uses the half of the sum of T_{max} and T_{min} to solve for FWHM, as shown below:

$$T(\omega) = \frac{1}{2}(T_{\max} + T_{\min}) \tag{6.8}$$

According to formula (6.4) and formula (6.8), the exact solution of the FWHM (i.e., cavity mode linewidth) of a periodic transmission spectrum of the resonant cavity is obtained, as shown in the following formula:

$$\Delta\nu_r = \frac{\Delta\omega_r}{2\pi} = \frac{FSR}{\pi}\cos^{-1}\left(\frac{2r_1 r_2}{1 + r_1^2 r_2^2}\right) \tag{6.9}$$

where, FSR is free spectral range given by $FSR = c/2L$. Consequently, the cavity finesse is

$$F_r = \frac{FSR}{\Delta \nu_r} = \frac{\pi}{\cos^{-1}\left(\frac{2r_1 r_2}{1+r_1^2 r_2^2}\right)} \tag{6.10}$$

It should be pointed out that when the reflectivity of the cavity mirror approaches 1, formulas (6.9) and (6.10) can be simplified to traditional expressions:

$$\Delta \nu_r = FSR \frac{1 - r_1 r_2}{\pi \sqrt{r_1 r_2}} \tag{6.11}$$

and

$$F_r' = \frac{\pi \sqrt{r_1 r_2}}{1 - r_1 r_2} \tag{6.12}$$

However, when the reflectivity is very low and approaches zero, formula (6.10) can be expressed by Taylor expansion,

$$F_r'' = \frac{\pi}{\frac{\pi}{2} - \frac{2r_1 r_2}{1+r_1^2 r_2^2}} = 2 + \frac{8r_1 r_2}{\pi\left(1 + r_1^2 r_2^2\right) - 4r_1 r_2} \tag{6.13}$$

Obviously, as the reflectivity R gradually approaches zero, the cavity finesse gradually approaches the minimum value of 2, which is different from the result obtained by formula (6.12).

Normally, in order to achieve laser oscillation and produce a highly coherent laser source, the cavity mode frequency is tuned to the exact resonance position. This counter-intuitive anti-resonance cavity, i.e., the situation where the cavity mode frequency is in the middle of two adjacent cavity modes, is rarely considered. However, this does not mean that anti-resonance lasers cannot be realized; rather, they possess unique laser characteristics and hold invaluable research significance. To distinguish the anti-resonance cavity from the traditional resonance cavity, the linewidth and finesse of the anti-resonance cavity are shown in the following formula:

$$\Delta \nu_{antires} = FSR - \Delta \nu_r = FSR\left(1 - \frac{1}{\pi}\cos^{-1}\left(\frac{2r_1 r_2}{1 + r_1^2 r_2^2}\right)\right) \tag{6.14}$$

and

$$F_{antires} = \frac{FSR}{\Delta \nu_{antires}} = \frac{\pi}{\cos^{-1}\left(\frac{2r_1 r_2}{1+r_1^2 r_2^2}\right)} \tag{6.15}$$

Similarly, under the high reflection approximation, formulas (6.14) and (6.15) can be simplified to the following forms:

$$\Delta v'_{antires} = FSR\left(1 - \frac{1 - r_1 r_2}{\pi\sqrt{r_1 r_2}}\right) \tag{6.16}$$

and

$$F'_{antires} = \frac{1}{1 - \frac{1 - r_1 r_2}{\pi\sqrt{r_1 r_2}}} \tag{6.17}$$

When the reflectivity is extremely low to the point of approaching zero, formula (6.15) can be expressed as

$$\begin{aligned} F''_{antires} &= \frac{1}{1 - \frac{1}{\pi}\left(\frac{\pi}{2} - \frac{2r_1 r_2}{1 + r_1^2 r_2^2}\right)} \\ &= 2 - \frac{8 r_1 r_2}{\pi\left(1 + r_1^2 r_2^2\right) - 4 r_1 r_2} \end{aligned} \tag{6.18}$$

and approaches the maximum value of 2, which coincides with the result obtained in the resonant case where R approaches 0.

Furthermore, according to formula (6.5), the average of R_{max} and R_{min}, i.e., the FWHM of the anti-resonant cavity, can also be solved by the power reflection factor. The FWHM of a reflection spectrum obtained using this method is consistent with formula (6.14).

Figure 6.36a, b respectively show the ratio of the cavity mode linewidth to the FSR, and the variation of the cavity finesse with the cavity reflectivity R. The black dashed-dotted line (approximate value) and the red solid line (accurate value) represent the calculation results of the resonant FP cavity, while the blue short dashed-dotted line (approximate value) and the green dashed line (accurate value) represent the calculation results of the anti-resonant FP cavity. The inset shows the results for reflectivity R between 0 and 0.5, and the black stars and black circles represent the results achieved in previous work [17, 18]. As shown in Fig. 6.36, when the reflectivity is below a certain value, the high-reflection approximation result is no longer accurate and cannot be used to describe the characteristics of the FP cavity. As the reflectivity R increases, the accurate and approximate values of the cavity mode linewidth (cavity finesse) tend to coincide. Figure 6.36a, b shows that the FWHM [cavity finesse] at the resonance and anti-resonance points obtained from formulas (6.9) and (6.14) [formulas (6.10) and (6.15)] coincide at $R = 0$, unlike the results obtained from the approximate values, which separate at $R = 0$. This also reflects to a certain extent that the high-reflection approximation result is not applicable when the reflectivity is very low. This section presents the accurate expressions of FWHM and cavity finesse, which can effectively solve the singularity problem existing in traditional expressions [19].

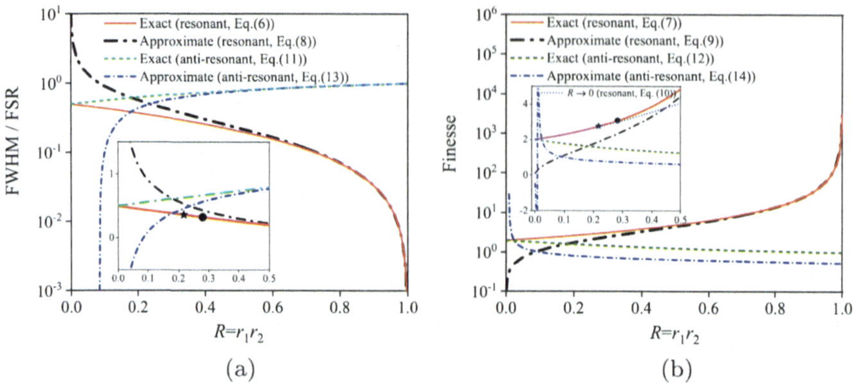

Fig. 6.36 **a** The ratio of the half-width to the free spectral range as the mirror reflectivity of the cavity changes when the FP cavity is in resonance and anti-resonance with the incident light. **b** The finesse changes with the mirror reflectivity of the cavity when the FP cavity is in resonance and anti-resonance with the incident light

In the figure, the main graph's vertical axis is a logarithmic coordinate system, and the inset is a linear coordinate system. The red solid curve and green dashed line represent the exact solution, and the black dotted-dashed line and blue short dotted-dashed line represent the high-reflection approximation results. The results show that when the mirror reflectivity is greater than or equal to 70%, whether the FP cavity is working in the resonance or anti-resonance region, the results obtained from the exact and approximate expressions tend to coincide, and separate below 70%. This phenomenon can be more clearly represented in the inset. At the same time, the inset also proves that the results of this work are credible, because at high reflectivity, the resonance and anti-resonance results obtained from the exact solution coincide. The black star and circle represent typical values that have been experimentally verified. Using formulas (6.9) and (6.14), when the reflectivity approaches zero, the cavity mode linewidths of the resonant cavity and the anti-resonant cavity both approach FSR/2, not infinity. And when the reflectivity approaches 1, the cavity mode linewidths of the resonant cavity and the anti-resonant cavity approach zero and FSR, respectively. Similarly, the cavity finesse equals 2 at $R = 0$, not zero. These results have strong reference significance for the research and development of ultra-low finesse FP cavities.

4. Symmetry of cavity enhancement factor and cavity suppression factor

Based on the above results, the cavity enhancement factor is analyzed. Due to the Purcell effect, compared to free space, the spontaneous emission rate of a two-level atom in a resonant cavity can be enhanced by $\eta_c = 3Q_c\,\lambda^3/4\pi^2 V$ times, where $Q_c = \nu_0/\Delta\nu$. Represents the cavity quality factor, which is defined as the ratio of the atomic transition frequency to the linewidth of the cavity mode, λ represents the atomic transition wavelength, V represents the equivalent cavity mode volume. Therefore, the Purcell factor is proportional to the cavity quality factor, which is proportional to the cavity finesse. The resonant cavity enhances the spontaneous radiation rate, on the contrary, the anti-resonant cavity suppresses the spontaneous radiation rate, that is, the suppression factor is inversely proportional to the cavity finesse [20].

Assume that the atomic dipole is located in a symmetric confocal cavity, as shown in Fig. 6.37, the curvature radii of cavity mirrors M_1 and M_2 are equal, and their reflectivities are respectively $R_1 = R_2 = R$. In order to calculate the total power radiated by the atomic dipole, assume that the power radiates within a sphere, which can be divided into three parts: S_1 is the part radiated by cavity mirror M_1; S_2 is the part radiated by cavity mirror M_2; S_{side} is the remaining part of the sphere. S_1, S_2, and S_{side} contain solid angles $\Delta\Omega_1$, $\Delta\Omega_2$, and $\Delta\Omega_{side}$, respectively, and satisfy the relation $\Delta\Omega_1 + \Delta\Omega_2 + \Delta\Omega_{side} = 4\pi$. The solid angle of the cavity can be expressed as $\Delta\Omega_c = \Delta\Omega_1 + \Delta\Omega_2$.

As known from Ref. [21], the power through S_{side} can be regarded as the total power radiated into free space minus the power radiated into $\Delta\Omega_c$, expressed as:

$$P_{side} = \left(1 - \frac{3}{8\pi}\Delta\Omega_c\right)P_{free} \tag{6.19}$$

By calculating the power transmitted through S_1 and S_2, we get:

$$P_c = \frac{1 - R^2}{1 + R^2 - 2R\cos(\omega 2L/c)}P_{free} \tag{6.20}$$

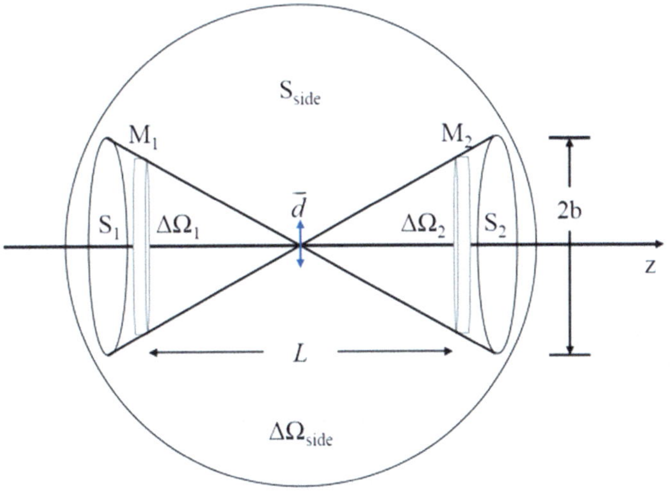

Fig. 6.37 Atomic dipole in a symmetric confocal cavity. Composed of two plane mirrors M_1 and M_2, the distance between M_1 and M_2 is $L = 2a$, the radii of curvature of M_1 and M_2 are respectively $+a$ and $-a$

where, $R = r_1 r_2$ represents the reflectivity of the cavity mirror, which indicates that the power entering the cavity from the dipole radiation follows the Airy function linearity of the cavity.

Ultimately, the total power of the dipole radiation can be represented as the sum of the total power of the boundary and cavity mirror radiation:

$$P_{total} = \left[1 + \left(\frac{1 - R^2}{1 + R^2 - 2R\cos(\omega 2L/c)} - 1\right)\frac{3}{8\pi}\Delta\Omega_c\right]P_{free} \tag{6.21}$$

According to formula (6.21), the power radiated into $\Delta\Omega_{\rm side}$ is fixed. When the phase of the reflected light field changes, only the power radiated into $\Delta\Omega_{\rm c}$ changes. Therefore, the total power also follows the cavity absorption linearity. For ease of analysis, this work studies the relationship between the power radiated into the cavity and the phase shift.

The ratio of the power radiated into the cavity to the power radiated into free space is defined as $\alpha = P_c/P_{free}$, and is related to the phase shift. When the cavity is resonant, it reaches the maximum value $\alpha_{\max} = (1 + R)/(1 - R)$, whereas when the cavity is anti-resonant, it reaches the minimum value $\alpha_{\min} = (1 - R)/(1 + R)$. Taking the resonant case as an example, the relationship between the cavity enhancement factor and the cavity finesse when $R \to 1$ and $R \to 0$ is analyzed below:

(1) For $R \to 1$,

$$\alpha_{\max} = \frac{1+R}{1-R} \simeq \frac{2}{1-R} \tag{6.22}$$

And

$$F_r' = \frac{\pi\sqrt{R}}{1-R} \simeq \frac{\pi}{1-R} \tag{6.23}$$

Therefore,

$$\alpha_{\max} = \frac{2F_r'}{\pi} \tag{6.24}$$

(2) For $R \to 0$, using Taylor expansion: $1/(1 - R) = 1 + R + R^2 + \cdots$, therefore,

$$\alpha_{\max} = \frac{1+R}{1-R} = \frac{2}{1-R} - 1 \simeq 1 + 2R \tag{6.25}$$

According to formula (6.13),

$$F_r'' \simeq 2 + \frac{8R}{\pi - 4R} \tag{6.26}$$

From this, we can obtain

$$\alpha_{\max} = 1 + \frac{\pi}{2F_r''}(F_r'' - 2) \tag{6.27}$$

Since when $R \to 0$, $F_r'' \to 2$, the cavity enhancement factor tends to 1 at this time, which further proves the correctness of the above analysis.

In summary, for the resonant cavity, when $R \to 1$, $\alpha_{\max}$ can be simply represented by $2F_r'/\pi$; for the anti-resonant cavity, $\alpha_{\max}$ is simplified to $1/(2F_r'/\pi)$. This result shows that for high-precision FP cavities, the cavity enhancement effect is very obvious. On the contrary, for ultra-low precision cavities ($R \to 0$, even without cavity mirror), a at both the resonant and anti-resonant cavities is close to 1, which is obvious because light can pass through the FP cavity with ultra-low reflectivity almost unimpeded. This work provides a universal expression for the cavity enhancement factor, rather than the traditional high-reflection approximation result. P_c/P_{free} The relationship between the change and the cavity phase shift is shown in Fig. 6.38a. To make this result easier to understand, Fig. 6.38b shows the relationship between P_c/P_{free} and the cavity phase shift in the logarithmic coordinate system. As can be seen from the figure, as the reflectivity decreases, this ratio gradually approaches 1, indicating that the cavity enhancement factor at $R = 0$ is 1. This result is different from the result obtained from the traditional expression $2F_r'/\pi$, which is only suitable for high-reflection approximation results, and is not suitable for high-loss FP cavities. This is because when the reflectivity approaches zero, the cavity enhancement factor should approach 1, and the traditional $2F_r'/\pi$ expression is clearly inaccurate. In addition, the enhancement factor of the resonant cavity and the suppression factor of the anti-resonant cavity are symmetrical, as shown in Fig. 6.38b.

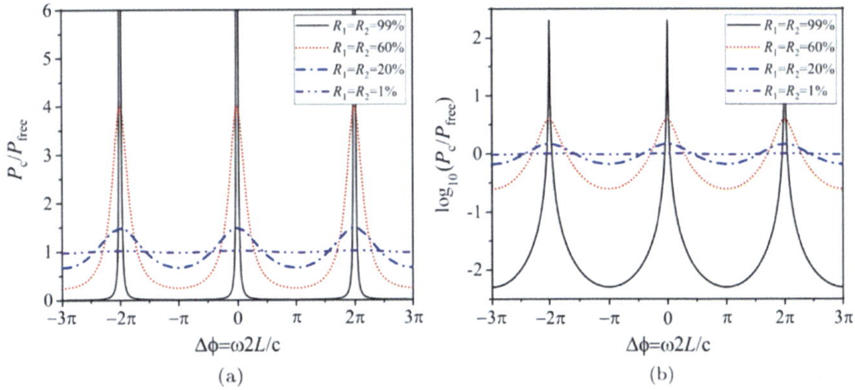

Fig. 6.38 **a** In the linear coordinate system, the change of the ratio of the power radiated to the cavity to the power radiated to free space with phase shift. **b** In the logarithmic coordinate system, the change of the ratio of the power radiated to the cavity to the power radiated to free space with phase shift

The results show the symmetry of the enhancement factor of the resonant cavity and the suppression factor of the anti-resonant cavity. The black solid curve, red dashed curve, blue dotted-dashed curve, and purple short dotted-dashed curve represent the results of the cavity mirror reflectivity $R = 99\%, 60\%, 20\%, 1\%$, respectively. As the reflectivity decreases, P_c/P_{free} gradually approaches 1.

5. The principle and characteristics of PDH frequency stabilization technology

PDH frequency stabilization technology was first proposed by Dr. Drever and Hall in 1983 [22] and is used to lock the laser frequency to the F-P reference cavity to narrow the linewidth. It uses the resonant frequency of the optical resonant cavity as the frequency reference standard, and first uses a crystal to perform radio frequency phase modulation on the frequency emitted by the laser, and then incident on the resonant reference cavity. By detecting the sidebands generated by radio frequency modulation with optical heterodyne, high-sensitivity resonance information is obtained, and a high signal-to-noise ratio error signal is obtained after demodulation. The PDH frequency stabilization method can quickly respond to phase changes in the system, and the bandwidth of its servo system is not limited by the response time of the resonant cavity, and can achieve high-speed locking of the laser frequency to the resonant frequency of the reference cavity. The basic principle of PDH frequency stabilization technology is shown in Fig. 6.39.

The output of the laser, after being adjusted by the polarizer (P), passes through the electro-optic phase modulator (EOM) to obtain a pair of sidebands with equal amplitude and opposite phase. The carrier and sidebands together are incident on the Fabry–Pérot (F-P) resonator. Since the intrinsic mode linewidth of the F-P resonator used for frequency stabilization is relatively narrow, when the laser frequency deviates from the resonant frequency of the cavity, only a very small part of the sideband frequency in the transmitted light, with very weak power, can be detected. However, the power of reflected light is stronger, as it contains light directly reflected by the cavity mirror without entering the resonator, as well as leaked light after oscillation in the resonator. Even if the laser frequency is not in the center of the resonant

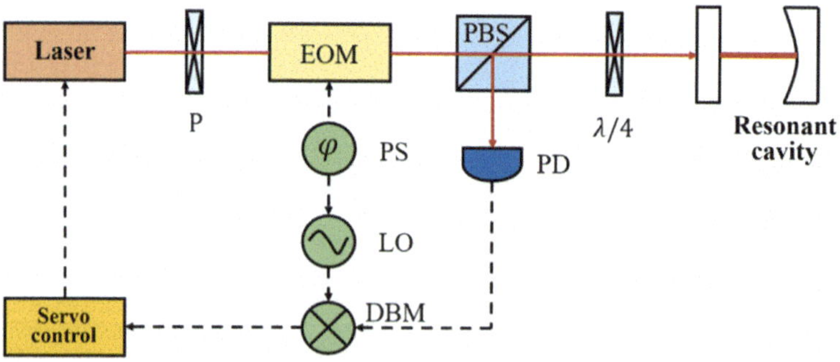

Fig. 6.39 Schematic diagram of the implementation principle of PDH frequency stabilization technology. *P* polarizer, *EOM* electro-optic modulator, *PBS* polarizing beam splitter, *λ/4* quarter-wave plate, *PD* photodetector, *DBM* double-balanced mixer, *LO* local oscillator, *PS* phase shifter

frequency of the cavity, the sideband frequency can be reflected back, so the signal is stronger and easier to detect. Therefore, PDH frequency stabilization technology generally detects the reflected signal of the F-P resonator. Moreover, the composite phase of these two parts of light in the reflected light near the resonance peak of the cavity is a quasi-dispersion curve, which is very suitable for phase demodulation as an error signal for frequency stabilization.

6. The principle of error signal generation in PDH frequency stabilization technology

PDH frequency stabilization laser technology is an active feedback regulation technology [22, 23], the core of which is the generation of error signals. PDH frequency stabilization laser technology needs to use an electro-optic modulator (EOM) to modulate the light, a detector (PD) to collect the reflected light of the cavity, and then demodulate the frequency detuning information in the reflected light to generate an error signal. The error signal, after being processed by a proportional integral differential circuit, is fed back to the laser's piezoelectric ceramics or acousto-optic modulators and other frequency responsive devices for frequency compensation.

The laser field strength $E(\mathrm{t})$ is represented as a function of time:

$$E(t) = E_0 e^{i\omega_c t} \tag{6.28}$$

The laser output is incident on the resonator after phase modulation by the EOM, and the incident light field strength $E_\mathrm{i}(\mathrm{t})$, $E_i(t)$ is represented as

$$E_i(t) = E_0 e^{i(\omega_c t + \beta \sin \Omega t)} \tag{6.29}$$

where, E_0 is the initial light field intensity, ω_c is the laser carrier frequency, Ω is the radio frequency modulation frequency, β is the modulation coefficient. When β is relatively small (the modulation depth of the PDH locking method is generally small, only the first-order modulation sideband appears), we ignore the high-order modulation sidebands, expand (6.29) using the Bessel function, and finally the incident light field strength $E_i(t)$, $E_i(t)$ can be expressed as:

$$E_i(t) = E_0\left[J_0(\beta)e^{i\omega_c t} + J_1(\beta)e^{i(\omega_c+\Omega)t} - J_1(\beta)e^{i(\omega_c-\Omega)t}\right] \tag{6.30}$$

where $J_\mathrm{n}(\beta)$ is the Bessel function of the first kind of order n, the formula only gives the carrier ω_c and the first-order modulated sideband $\omega_\mathrm{c} + \Omega$. After the laser frequency is phase modulated, the amplitude in the frequency domain is shown in Fig. 6.40.

After the laser is modulated by the EOM with a radio frequency drive frequency of Ω, two modulated sidebands with equal amplitudes and opposite phases (differing by π) appear at the carrier's sides, with a frequency interval of Ω. When the modulation depth is very low, considering only the first-order modulated sideband, the expression for the reflected light after the modulated light incident on the F-P cavity is:

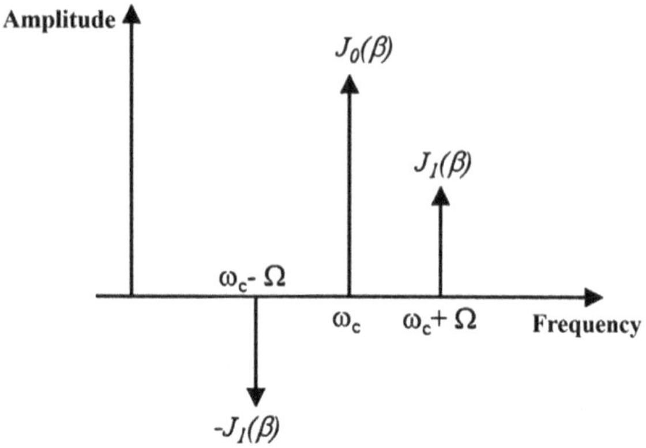

Fig. 6.40 Sidebands formed by phase modulation of laser frequency

$$E_r(t) = E_0\left[F(\omega_c)J_0(\beta)e^{i\omega_c t} + F(\omega_c + \Omega)J_1(\beta)e^{i(\omega_c+\Omega)t} - F(\omega_c - \Omega)J_1(\beta)e^{i(\omega_c-\Omega)t}\right] \tag{6.31}$$

In the formula, $F(\omega_c)$ is a reflection function of the cavity, related to the reflectivity of the endoscope and the phase of the laser traveling back and forth in the cavity. In order to demodulate the phase information of the reflected light, the output frequency of the local oscillator signal (LO) is the same as the modulation frequency Ω, but there is a phase difference with the modulation frequency, denoted as φ, expressed as:

$$E_L(t) = E_{LO}e^{i(\Omega t+\varphi)} \tag{6.32}$$

The LO signal and the cavity reflection signal detected by the detector are mixed on the mixer (DBM). Only considering the case where the carrier frequency is close to the resonance of the resonant cavity, and the above formula chooses the appropriate phase shift ($\varphi = \pi/2$), after demodulating formula (6.31), the output signal of the mixer is the error signal containing phase information, expressed as:

$$\begin{aligned}\varepsilon &= -2J_0(\beta)J_1(\beta)E_0^2\mathrm{Im}\left[F(\omega_c)F^*(\omega_c+\Omega) - F^*(\omega_c)F(\omega_c-\Omega)\right] \\ &\approx -\frac{4}{\pi}J_0(\beta)J_1(\beta)E_0^2\frac{\Delta\omega}{\Delta\nu_c}\end{aligned} \tag{6.33}$$

where $\Delta\omega = \omega_c - \omega_0$, ω_0 is the resonant frequency of the F-P cavity, $\Delta\nu_c$ is the intrinsic linewidth of the F-P cavity, and it satisfies $\Delta\omega << \Delta\nu_c$. Figure 6.41 shows a schematic diagram of the corresponding relationship between the transmission peak of the resonant cavity observed when the laser frequency is near the resonant frequency of the cavity and the frequency scanning range is greater than the free spectral range of the cavity. The error signal contains phase information after demodulation.

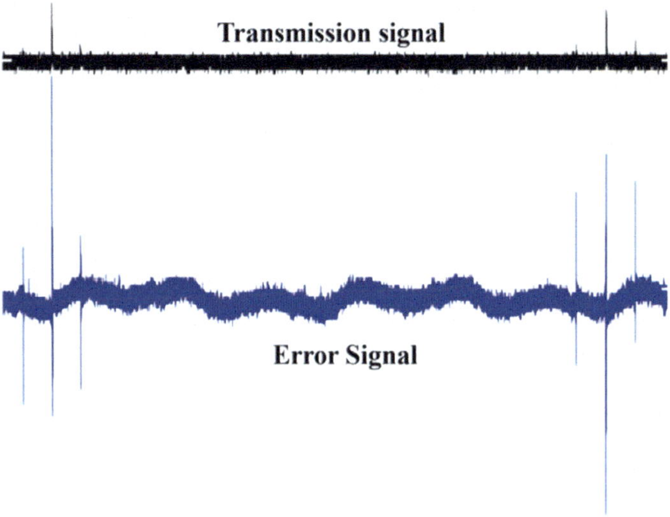

Fig. 6.41 Transmission peak, reflection peak, and error signal obtained by scanning the laser near the resonant frequency of the cavity

6.3.2 Frequency Stabilization of Faraday Laser with PDH

In 2024, a research group from Peking University locked the Faraday laser to a portable optical reference cavity developed independently using the PDH frequency stabilization method, effectively narrowing the linewidth of the Faraday laser and reducing its short-term frequency instability. This section provides a detailed discussion to the work.

1. Design and fabrication of F-P optical resonator

The core component of the PDH ultra-stable resonator is the F-P optical resonator. The F-P optical resonator is composed of two high-reflectivity mirrors. The incident laser reflects between the two mirrors, forming multiple interferences in the cavity. When a stable in-phase interference occurs, it indicates that resonance occurs in the cavity, as shown in Fig. 6.42. At this time, the optical frequency instability of the ultra-stable cavity is equal to the optical length instability of the F-P cavity, which can be expressed as:

$$d\nu/\nu = -d(nL)/nL \tag{6.34}$$

where, $d\nu$ is the frequency jitter, $d(nL)$ is the jitter of the equivalent optical length of the cavity, n is the refractive index of the medium, L is the cavity length.

Expanding formula (6.34) can yield

$$d\nu/\nu = -(dn/n + dL/L) \tag{6.35}$$

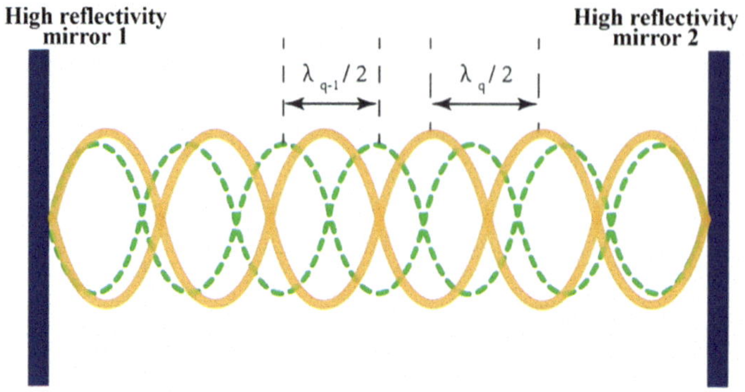

Fig. 6.42 Principle of F-P resonant cavity

As can be seen from the formula, the frequency instability of the cavity is the sum of two instabilities: the instability of the refractive index of the medium dn/n and the instability of the cavity length dL/L. There are two ways to reduce the instability of the refractive index of the medium: one is to create a ultra-low pressure environment by vacuuming, and the other is to control the environmental temperature at the temperature corresponding to the zero point of the thermal expansion coefficient of the cavity material or to use a refrigerator to create a ultra-low temperature environment to reduce thermal noise. The former is easier to achieve in practice, and hence, the common method is to place the optical reference cavity in a vacuum environment. So, it is necessary to design a vacuum cavity to ensure that the reference cavity is in a high vacuum environment. The instability of the cavity length is affected by temperature, vibration, and thermal noise. The fluctuation of temperature causes the material to expand or contract, so it is necessary to control the temperature and use ultra-low expansion coefficient microcrystalline glass or single crystal silicon as the cavity material. Vibration causes corresponding stress in the cavity and causes deformation, so active and passive vibration isolation platforms, soundproof boxes, and designs insensitive to vibration can be used. Thermal noise refers to the fluctuation in cavity length caused by Brownian motion thermal noise of particles within the cavity, cavity mirrors, and coatings. The magnitude of the thermal noise in equivalent cavity length primarily depends on properties such as the cavity's material, shape, and temperature. Therefore, we need to control the temperature of the reference cavity, choose the appropriate material, and design the appropriate shape of the cavity.

(1) *Design of F-P optical resonator*

In order to make the optical reference cavity insensitive to inertial forces, the physical structure and support structure of the reference cavity should have a high degree of symmetry. At the same time, to satisfy portability, the reference cavity should be completely constrained in all spatial degrees of freedom. In view of this, a cubic optical reference cavity can be used. The eight vertices of the cube are cut, the cutting depth is parameter dc, and the deformation of the two cavity mirrors under the action of squeezing force can be weakened by changing the cutting depth. Support objects are used to squeeze support at the cut corners, and the support force is set as parameter Fa. The cavity is made of ultra-low expansion coefficient glass (ULE), and each of the three pairs of parallel planes in the cubic cavity has a through hole in the center, one of which is a light transmission hole, and the other two are exhaust holes. The length of the light transmission hole is the cavity length, set as parameter *L*. Ultra-high reflectivity fused silica cavity mirrors are used at both ends of the light transmission hole, the diameter of the cavity mirror is parameter d_m, and the thickness of the cavity mirror is parameter T_m.

The vibration sensitivity of the reference cavity refers to the relative change in the length of the optical reference cavity due to changes in the acceleration of the environment in which it is located, formulated as:

$$S \propto (L - L_i)/L \tag{6.36}$$

where, *S* represents the vibration sensitivity of the reference cavity, *L* represents the original cavity length, L_i represents the cavity length after deformation.

Figure 6.43 depicts the design drawing of the optical reference cavity, with a cavity length of 50 mm and a light transmission aperture of 6 mm. Figure 6.44 is the structural model of the optical reference cavity built in conjunction with the design drawings, which includes the ULE cavity and a pair of cavity mirrors. Here, the cubic cavity length is 50 mm, the cutting depth is 10 mm, the mirror diameter is 1 in., and the light transmission aperture is 6 mm.

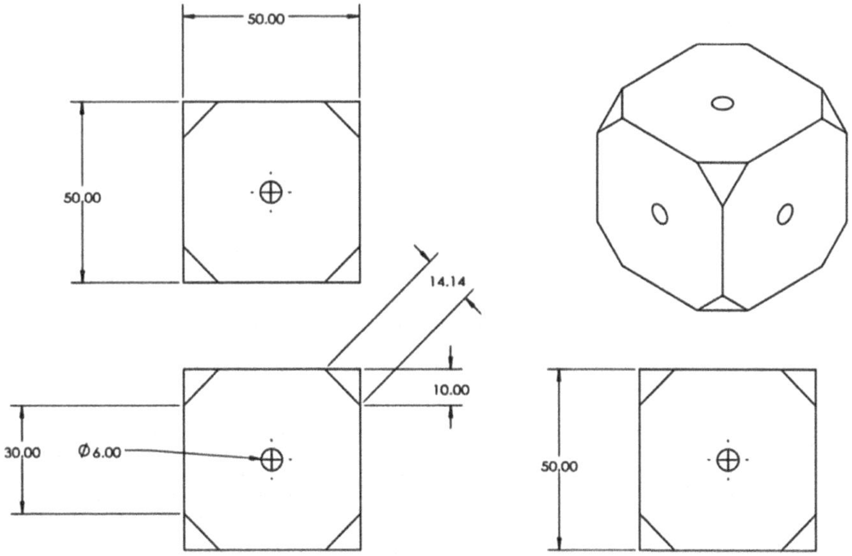

Fig. 6.43 Design drawing of the optical reference cavity

Fig. 6.44 Modeling diagram of the optical reference cavity

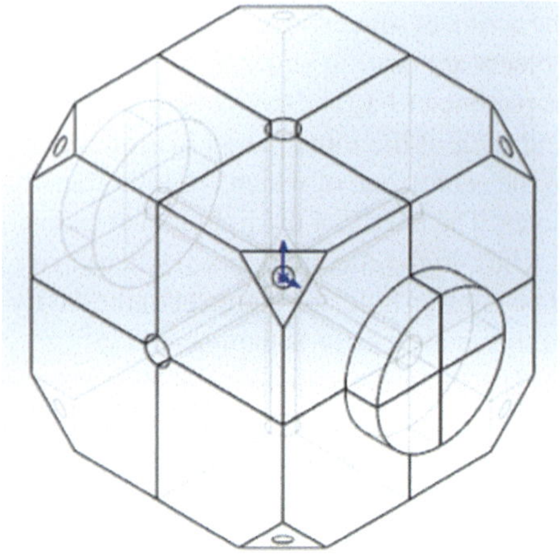

2. Simulation analysis of F-P optical resonator design

(1) *Changes in the pressure sensitivity of the reference cavity with cutting depth under different pressures*

The squeezing force is set to 1 N, 2 N, 3 N, 4 N, and 5 N, respectively, and the changes in the pressure sensitivity of the reference cavity with the cutting depth obtained by simulation are shown in Fig. 6.45. The results show that the pressure sensitivity of the reference cavity under different pressures is different. With other conditions remain unchanged, there is an optimal cutting depth, at which the pressure used for fixation is insensitive. In Fig. 6.45, the unit of force Fa is Newton (N), and M1 and M2 represent the positions of the two cavity mirrors.

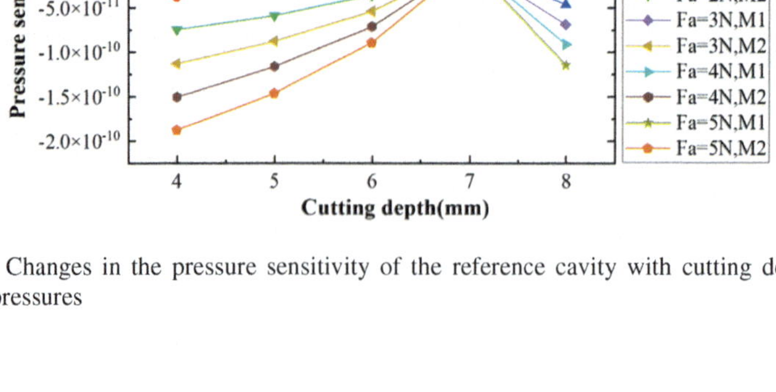

Fig. 6.45 Changes in the pressure sensitivity of the reference cavity with cutting depth under different pressures

(2) *Changes in the pressure sensitivity of the reference cavity with cutting depth under different aperture diameters*

The changes in the pressure sensitivity of the reference cavity with the cutting depth under different light transmission aperture diameters are shown in Fig. 6.46. The light transmission aperture values are considered to be 4, 5, and 6 mm, with 1 mm the step length of the cutting depth, a total of 24 simulation points are obtained, and 6 curves are drawn. The optimal cutting depth under different light transmission aperture diameters is obtained, so that the deformation of the cavity mirror is zero. At a cutting depth of 7 mm, a light transmission aperture of 5 mm can be selected, and the pressure sensitivity of the cavity is close to zero.

3. Fabrication and system setup of F-P optical resonator

Based on the aforementioned theoretical research and finite element simulation analysis, the research group and its collaborators designed and fabricated a cubic cavity. Figure 6.47 shows the physical image of the fabricated reference cavity, which includes the ULE material cavity and cavity mirror. The reference cavity adopts a cubic structure with a cavity length of 50 mm. The cubic cavity has 8 cut corners with a cutting depth of 10 mm. To maintain structural symmetry and stability as much as possible, in addition to the light transmission aperture in the horizontal direction, the reference cavity also has two exhaust apertures with the same diameter as the light transmission aperture, forming three apertures with lengths equal to the cavity length. Considering the manufacturing cost, a pair of high-reflectance mirrors with a diameter of 1 in. and a thickness of 4 mm, featuring a 6 mm clear aperture for light transmission, are selected as the cavity mirrors.

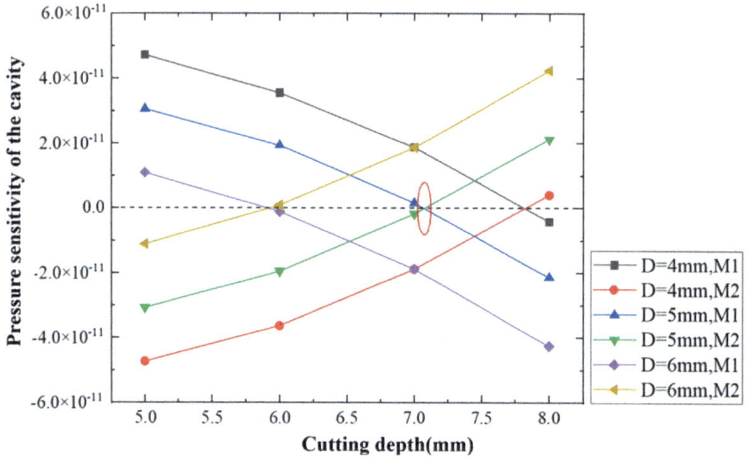

Fig. 6.46 Changes in the pressure sensitivity of the reference cavity with cutting depth under different aperture diameters

Fig. 6.47 Physical image of the optical reference cavity

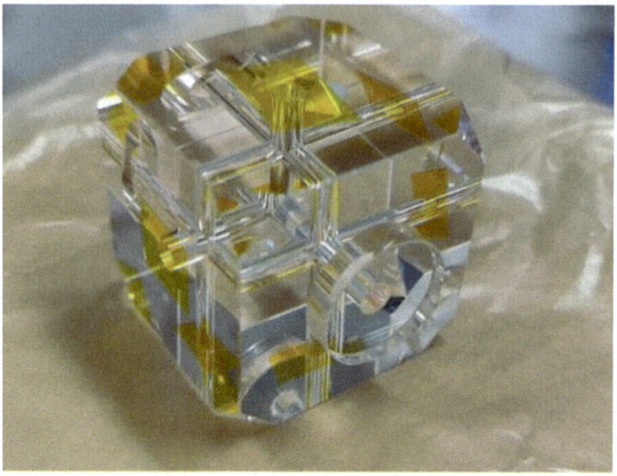

The ultra-stable resonator mirror chooses a one-inch ULE material cavity mirror, with a coating range covering 780.243 ± 0.001 and 852.356 ± 0.001 nm wavelengths, and a reflectivity greater than 99.999%. According to formula (6.30), the finesse of the cavity is greater than 300,000.

$$F = \frac{\pi\sqrt{R}}{1-R} \tag{6.37}$$

The domestically coated cavity mirror is tested using a PE_Lambda950 model UV–visible–NIR spectrophotometer produced by PerkinElmer. The working principle of the UV–visible–NIR spectrophotometer is a double-beam double-monochromator spectrometer, using a halogen lamp and deuterium lamp as light sources, with automatic switching and automatic collimation controlled by software. The detector in the UV–visible region is a highly sensitive photomultiplier tube, and in the near-infrared region, it is a PbS detector. The detectable wavelength range is 175 ~ 3300 nm, with a wavelength accuracy of $\pm$ 0.08 nm.

The test uses normal incidence, with an incident wavelength range of 320 ~ 855 nm and a wavelength interval of 1 nm. The reflectivity test results of the cavity mirror obtained are shown in Fig. 6.48. Among them, the transmittance at 780 nm is 0.000557%, the reflectivity is 99.999443%, the transmittance at 852 nm is 0.000689%, and the reflectivity is 99.999311%, with the reflectivity all greater than 99.999%.

According to formula (6.29), in a normal laboratory open environment, the instability of laser frequency caused by temperature and air refractive index changes is too large. It is necessary to control the temperature and refractive index within the optical reference cavity. The common method is to place the optical reference cavity in a vacuum environment.

At room temperature, a pressure change of 1 Pa causes a refractive index change of 2.8×10^{-9}. Therefore, to achieve a laser frequency stability of the 10^{-15} order

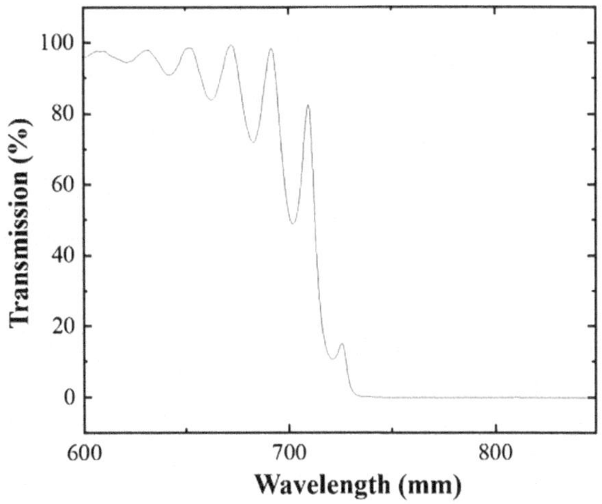

Fig. 6.48 Cavity mirror reflectivity test

of magnitude, the rate of pressure change should be less than 2.8×10^{-7} Pa. This requires the reference cavity to be in a vacuum environment of the 10^{-5} Pa order of magnitude. The vacuum extraction requires a mechanical pump and a molecular pump to first extract the internal pressure of the vacuum cell to the vacuum level where the ion pump can be started, then cooperate with the ion pump to further reduce the vacuum level, and finally disconnect the mechanical pump and the molecular pump, and let the ion pump work alone to maintain the vacuum level inside the vacuum cell. Therefore, the design of the vacuum cavity needs to reserve the interface of the molecular pump. Figure 6.49 is the design and simulation model of the vacuum cavity.

The physical object of the vacuum cavity is shown in Fig. 6.50. The shell of the vacuum cavity is an aluminum shell, which includes the interfaces of the ion pump and the molecular pump. Considering the volume of the vacuum cavity, an ion pump with a pumping speed of 5 L/s can maintain a vacuum level of 10^{-6} Pa or lower.

After the reference cavity provides a vacuum environment, the refractive index of the medium inside the reference cavity can be considered to be approximately equal to the refractive index of the vacuum, and the rate of change is also maintained at approximately acceptable. At this time, the instability of the laser frequency is basically equivalent to the instability of the length of the reference cavity:

$$d\nu/\nu = -dL/L \tag{6.38}$$

Due to the thermal expansion effect of the reference cavity, it has a significant impact on the length of the reference cavity. Ultra-low expansion (ULE) glass has very small thermal expansion coefficient, and its thermal expansion coefficient at room temperature is close to zero. The thermal expansion coefficient of designed ULE near room temperature basically changes linearly between − 30 and 30 ppb (part per billion, 10^{-9}), and the zero point of the thermal expansion coefficient is

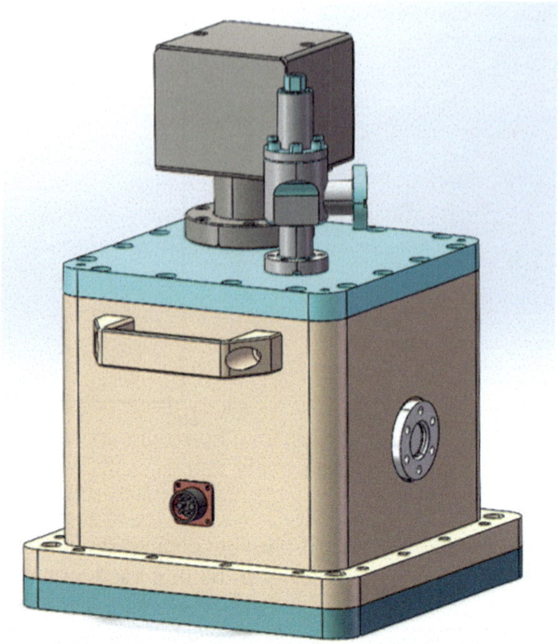

Fig. 6.49 Vacuum cavity simulation model

around 20 °C. For the thermal changes in the reference cavity, heat conduction and convection are blocked in the vacuum cavity, so thermal radiation is the main factor in heat transfer.

Figure 6.51 is a 3D cross-sectional view of the thermal shield design. Using two layers of brass with a very low radiation coefficient (gold-plated) as the insulation material, the square optical reference cavity is placed between the two insulation layers, which can ensure a large time constant for thermal radiation. Under stable laboratory temperature conditions, the temperature of the reference cavity can be kept relatively stable. The actual thermal shield layer is shown in Figs. 6.52 and 6.53.

Figure 6.54 is a structural diagram of the square PDH ultra-stable resonator, which includes the F-P optical reference cavity, thermal shield layer, vacuum cavity, and 5 L/s pumping speed ion pump. The left figure is a left view of the ultra-stable resonator, and the right figure is a cross-sectional view, where the structure of the optical reference cavity, two layers of thermal shield, vacuum structure, and ion pump can be clearly seen.

The ultra-stable resonator mirror with a coating range covering 780.243 ± 0.001, 852.356 ± 0.001 nm wavelengths is glued to both sides of the light transmission hole of the F-P optical reference cavity, wrapped in an oxygen-free copper shield layer and placed in the vacuum system. Quartz windows with a coating range covering 780.243 ± 0.001, 852.356 ± 0.001 nm wavelengths are glued to both sides of the vacuum

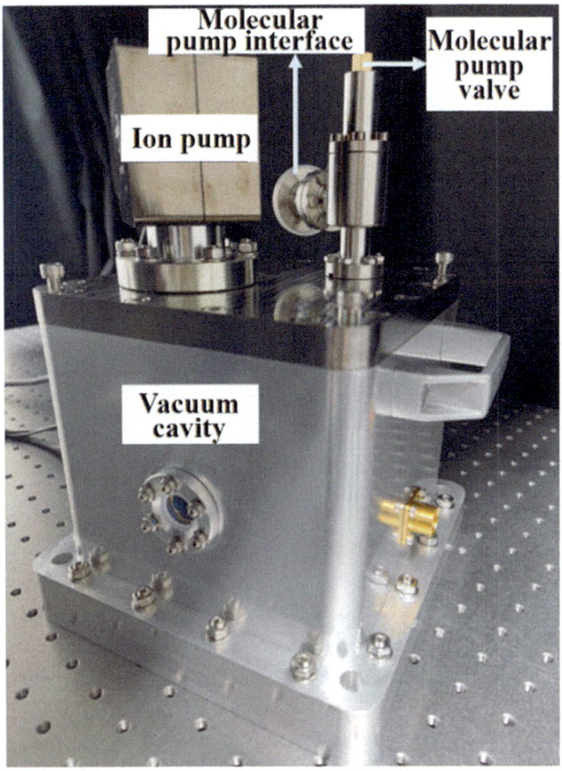

Fig. 6.50 Vacuum cavity physical object

system, requiring a light transmittance greater than 99.5%. A standard vacuum pump-ion pump is used for vacuuming, with a maximum voltage of 7 kV for the ion pump, and the ultimate vacuum of the ion pump is better than 10^{-8} Pa. The minimum start-up pressure is better than 10^{-4} Pa. The system's ultimate vacuum is better than 5×10^{-6} Pa, and the leakage rate is better than 1×10^{-10} Pa m^3/s. The overall system structure is shown in Figs. 6.55 and 6.56.

4. PDH frequency stabilized Faraday laser

The optical path unit of the PDH frequency-stabilized Faraday laser includes a laser, a transmission optical path (including fiber part and free space light parts), a PDH ultra-stable resonator, and a transmission/reflection signal receiver, as shown in Figs. 6.57 and 6.58. The light emitted by the laser is split by the fiber, one output, one can be connected to the wavelength meter, and the other is emitted as free space light through the fiber coupler. The free space light enters the ultra-stable cavity through optical components such as a 1/2 wave plate, a mirror, a polarization beam splitter prism, a 1/4 wave plate, etc. The transmitted signal is received by the camera and photodetector 2, and the reflected signal is received by photodetector 1.

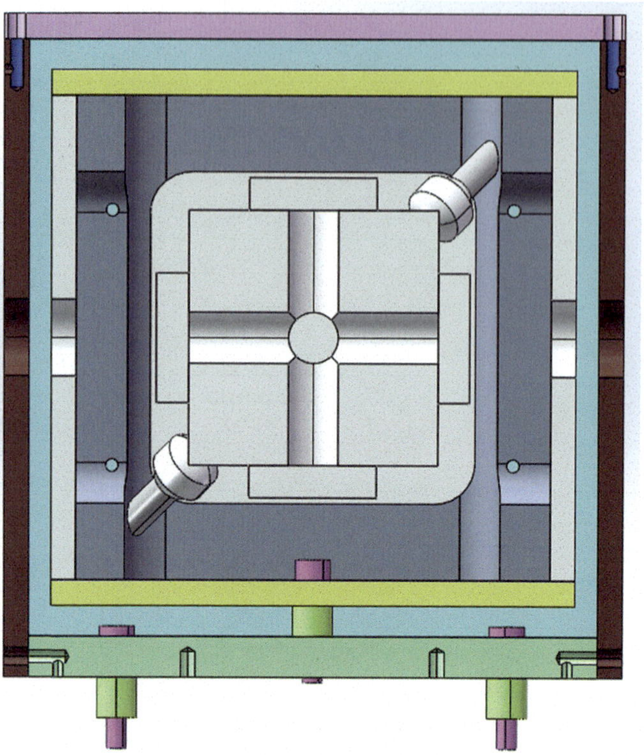

Fig. 6.51 Cross-sectional design of the thermal shield layer

The physical unit of the PDH frequency-stabilized Faraday laser includes an F-P optical resonator and its supporting temperature control structure and vacuum structure, as shown in Fig. 6.59.

The locking bandwidth of the Faraday laser when locked to the PDH system is shown in Fig. 6.60a, b shows the bandwidth display after the Toptica laser is locked. It can be seen that the bandwidth generally required for the Faraday laser PDH lock is about 220 kHz. Considering the further increase in bandwidth and low-frequency gain, it may further improve the performance after the Faraday laser is locked to the PDH ultra-stable resonator. The maximum bandwidth of the Faraday laser PDH lock can currently reach 400 kHz.

To measure the characteristics of the Faraday laser and Toptica TA laser during free running, another Toptica DL pro laser locked to PDH is used as a reference. Figure 6.61a shows the stability of the Faraday laser during free running reference to the Toptica DL pro laser locked to the PDH super cavity system, and the stability of the Toptica TA laser during free running reference to the Toptica DL pro laser

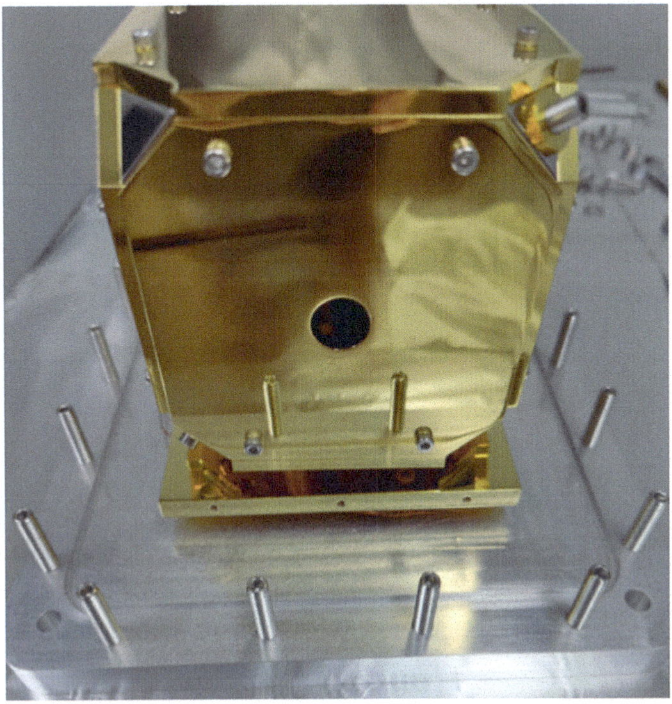

Fig. 6.52 First layer of the thermal shield layer processing physical diagram

locked to the PDH super cavity system. It can be seen that the short-term frequency stability of the free-running Faraday laser is about 4.5×10^{-10}/s, and the short-term frequency stability of the free-running Toptica TA is about 4.1×10^{-10}/s.

In order to evaluate the characteristics of the Faraday laser after PDH locking, the method of the triangle hat is used to measure the short-term (1s) frequency stability of the Faraday laser PDH lock, Toptica TA PDH lock, and Toptica DL pro PDH lock. Figure 6.61b shows the stability result of the triangle hat beat frequency, where both Faraday and Toptica TA use a 5 cm cavity, and Toptica DL pro uses a 10 cm cavity. The short-term (1s) frequency stability of the Faraday laser is 5.29×10^{-15}/s, the short-term frequency stability of Toptica TA is 5.4×10^{-15}/s, and the short-term frequency stability of Toptica DL pro is 5.56×10^{-15}/s, all at the same performance level. It can also be seen that the cavity of 10 cm has a serious drift.

Figure 6.62a shows the phase noise results of the beat frequency of the Faraday laser locked to the PDH ultra-cavity system and the Toptica TA laser also locked to the ultra-cavity system. From the phase noise, it can be seen that the locking bandwidth of the Faraday laser is around 200 kHz, and the locking bandwidth of the Toptica TA laser is around 700 kHz. Figure 6.62b shows the beat frequency linewidth measured by the FFT spectrum analyzer after locking, with RBW of 4 Hz, and the linewidth of the Faraday laser after fitting is about 8 Hz.

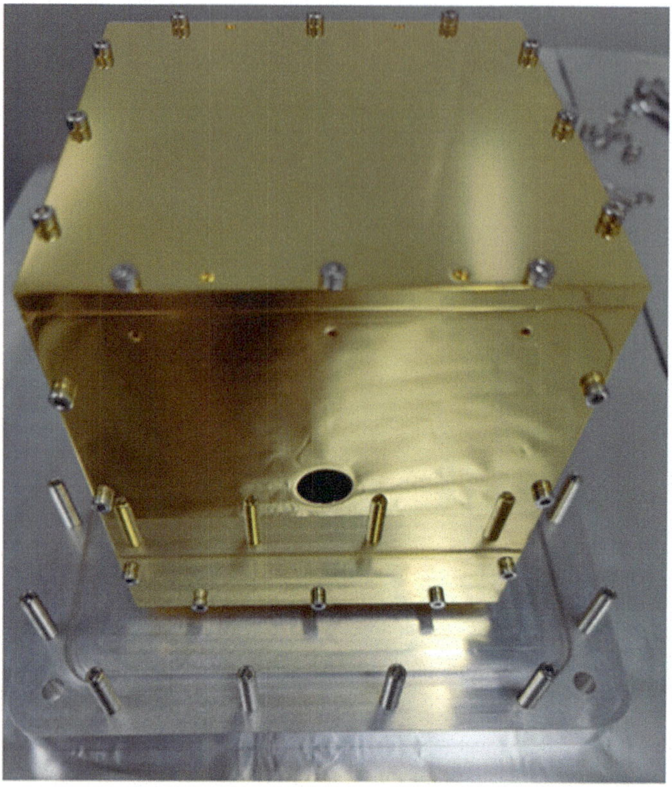

Fig. 6.53 Second layer of the thermal shield layer processing physical diagram

6.3.3 PDH Frequency Stabilization of Whispering Gallery Mode Microcavity Based on Faraday Laser

Traditional Fabry–Pérot (F-P) cavities can achieve high Q values and are widely used in many fields. However, their large cavity size is not easy to integrate, and the high cost of high reflectivity cavity mirrors and the need for complex stabilization devices limit the application of F-P cavities. Whispering gallery mode optical microcavities have high quality factors, small mode volumes, and the inner walls are generally very smooth, allowing light to circulate multiple times in the cavity, thus easily confining the light beam in a specific area, thereby increasing the energy density of the light in the cavity, which is beneficial for the occurrence of nonlinear optical effects. The following is PDH frequency stabilization based on Faraday laser.

1. Working principle of microcavity

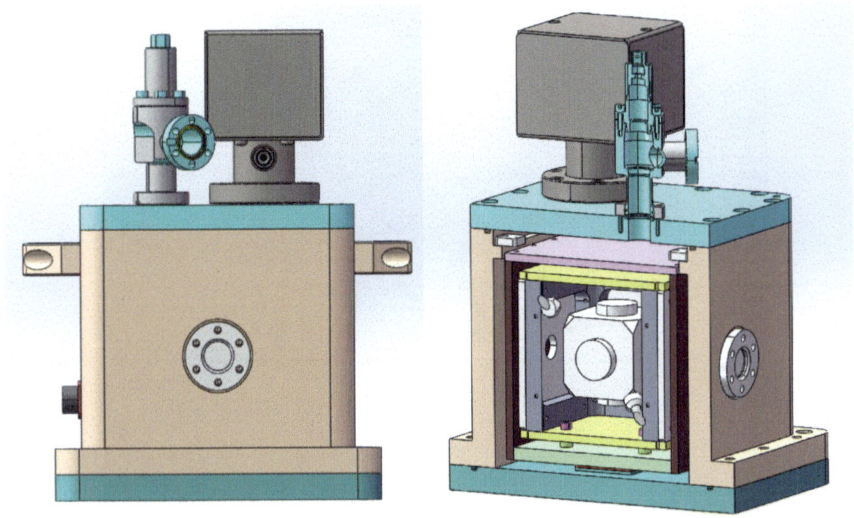

Fig. 6.54 Structural diagram of the square ultra-stable resonator

Fig. 6.55 PDH ultra-stable resonator square cavity structure decomposition physical diagram

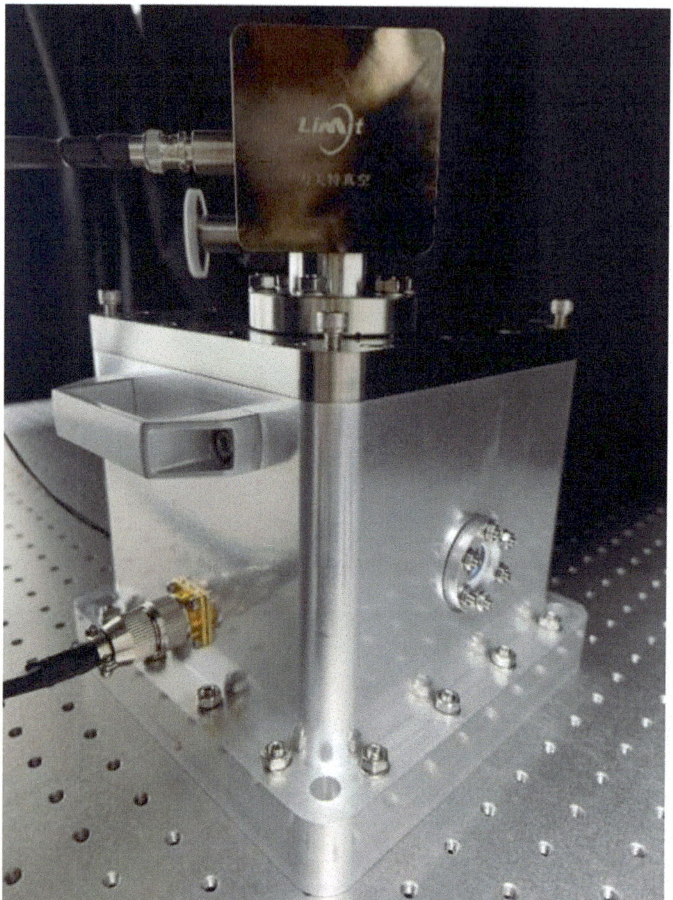

Fig. 6.56 Ultra-stable resonator square cavity overall physical diagram

Figure 6.63a, b are the echo walls of the Temple of Heaven in Beijing and St. Paul's Cathedral in London, respectively. Both have a similar structure of a circular "whispering gallery". Their common feature is that when two people stand close to the inner wall, if one person whispers at one end of the gallery, even if they are far apart, the person at the other end of the gallery can clearly hear it, as if whispering in the ear. This phenomenon gives the gallery its name, "whispering gallery." In 1910, Rayleigh published a paper detailing its physical principles [24]. Sound waves can reflect off the curved smooth wall with very little loss, so sound can travel a long distance along the wall. This effect is known as the whispering gallery mode (WGM), which we usually call the "echo wall mode". A similar effect in a circular dielectric can lead to an analogous effect in optics. Similar to the reflection of sound waves on the wall, when light is incident from an optically denser medium to an optically rarer medium and the incident angle is large enough, it can also undergo total internal reflection at

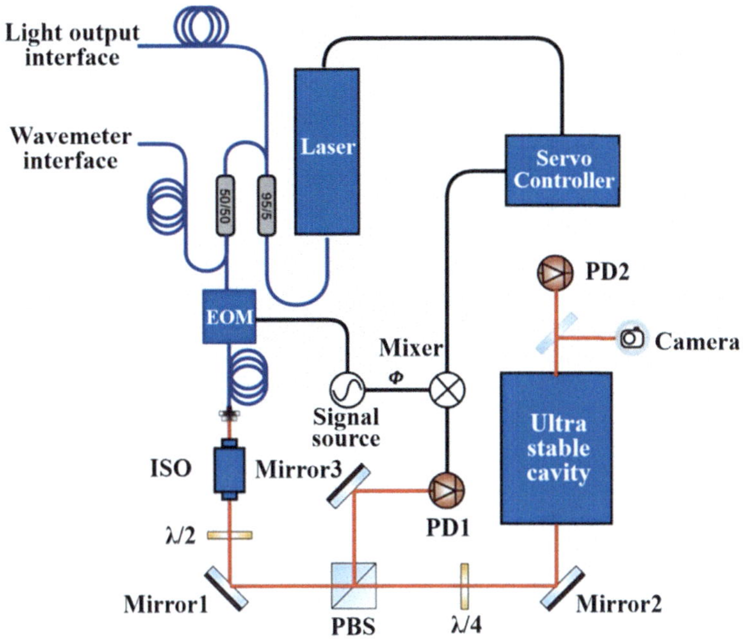

Fig. 6.57 Design of the optical path unit of the PDH frequency-stabilized Faraday laser

the interface of the two media. Therefore, there is an optical echo wall mode at the curved high refractive index interface. Within the boundary of a closed cavity, light can be trapped inside the cavity to maintain a stable wave transmission whispering gallery mode [25].

In Fig. 6.64, it is assumed that a beam of light enters the resonator on the left and moves along the circumference of the resonator. If the angle of the light is very steep, according to Snell's law, the light partially reflects and refracts out at the same angle. When the light enters the surface at an angle θ_{inc} greater than the total internal reflection angle $\sin^{-1}\frac{n_0}{n}$, all the light is reflected and confined within the resonator. If the resonator is circular or elliptical, due to the symmetry of the system, the light repeatedly reflects from the surface at θ_{inc} and overlaps with its starting position. As shown on the right side of Fig. 6.64, if the light is in the same phase after such a round trip, which depends on the frequency of the light, a standing wave resonance occurs. This is the so-called whispering gallery mode, where the annular structure used to confine the light field is called the whispering gallery mode optical microcavity (WGM microcavity) [26, 27].

If N_R is very large, then the complete round trip path is close to the circumference of the circle, and the phase difference produced after one round trip is an integer multiple of 2π. For a resonator with a radius of a, the phase difference is: $k \cdot 2\pi a \cdot n$, where k is $\frac{2\pi}{\lambda}$, n is the refractive index, l is the mode number of the resonant cavity, leading to the condition for resonance: $\frac{2\pi}{\lambda} \cdot 2\pi a \cdot n = l \cdot 2\pi$, simplified to

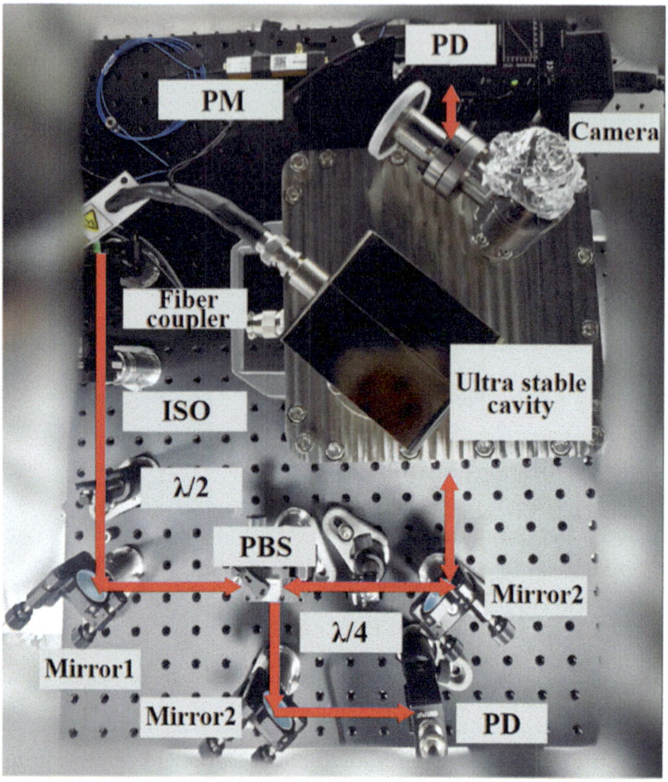

Fig. 6.58 Physical picture of the optical path unit of the PDH frequency-stabilized Faraday laser

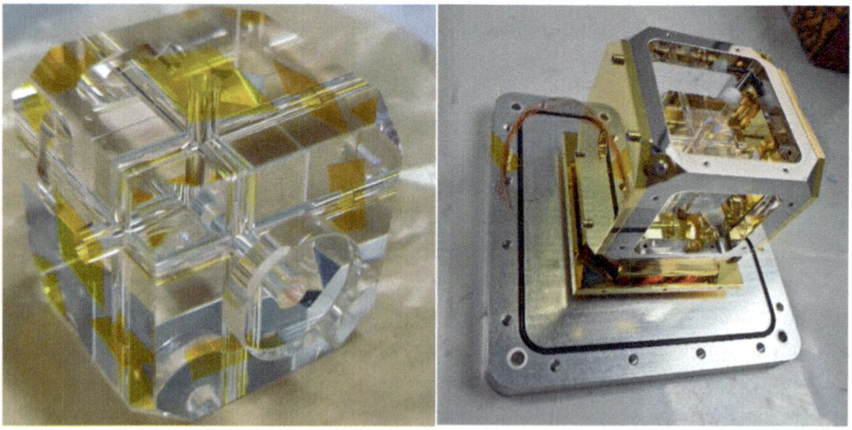

Fig. 6.59 Physical picture of the optical resonator (5 cm square cavity)

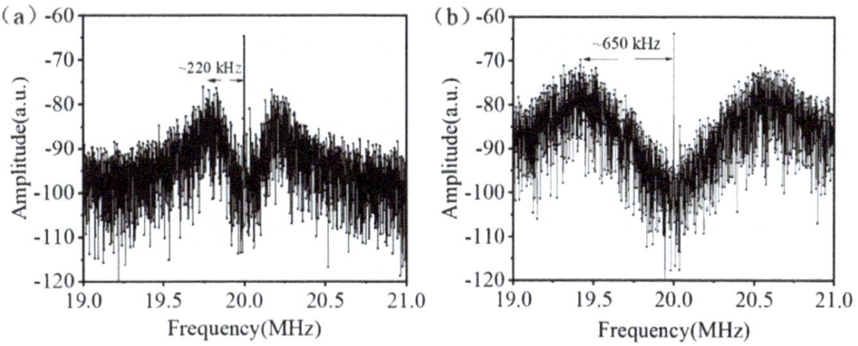

Fig. 6.60 Locking bandwidth information. **a** Locking bandwidth information displayed by the reflected signal after the Faraday laser is locked. **b** Locking bandwidth information displayed by the reflected signal after the Toptica laser is locked

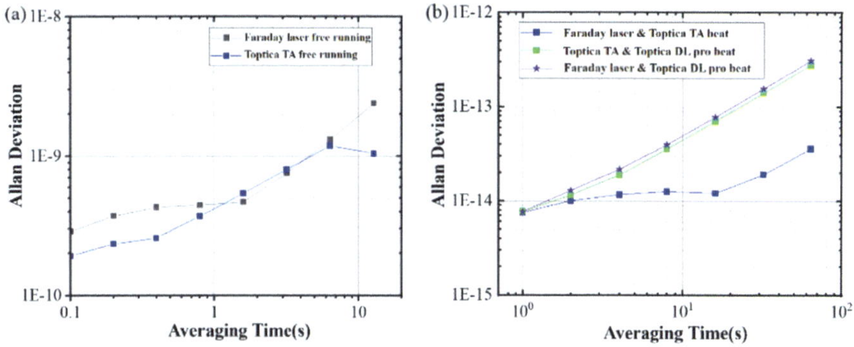

Fig. 6.61 **a** Stability of free-running frequency of Faraday laser and Toptica laser. **b** Beat frequency stability results of Faraday laser locked to PDH system and Toptica laser locked to PDH system

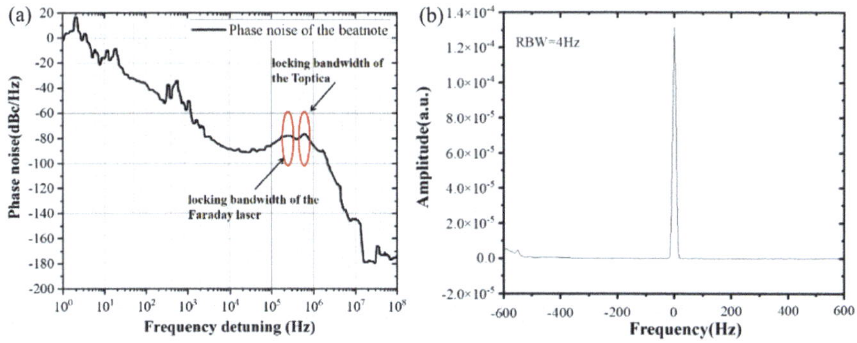

Fig. 6.62 **a** Phase noise obtained from the beat frequency after the Faraday laser and Toptica laser are locked to PDH. **b** Beat frequency linewidth of Faraday laser locked to PDH system and Toptica laser locked to PDH system measured by FFT spectrum analyzer

Fig. 6.63 **a** Temple of Heaven Echo Wall. **b** St. Paul's Cathedral

Fig. 6.64 Two-dimensional diagram of WGM resonator

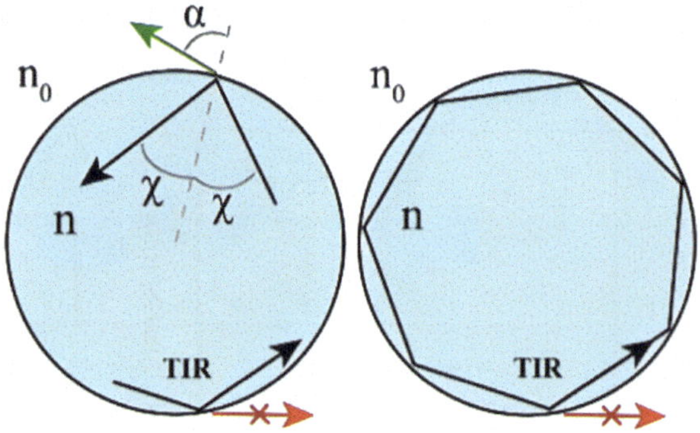

$$n \cdot 2\pi a = l \cdot \lambda \tag{6.39}$$

2. Basic parameters of WGM microcavity

The whispering gallery mode of an optical microcavity is usually described by several characteristic parameters, the most important of which are the quality factor (Q) and the mode volume (V). In addition, there is the free spectral range (FSR). In the field of sensing, it is often required that the microcavity has a high quality factor and a small mode volume to increase the sensitivity of the sensor. If the movement and broadening of the mode spectrum are to be measured, the free spectral range and finesse are of particular interest [28].

(1) *Quality factor Q*

An important parameter for measuring the quality of a resonant cavity is its quality factor (Q value), defined as follows:

$$Q = \omega \frac{E_{stored}}{P_{loss}} = \frac{2\pi \vartheta_0 T_{rt}}{l} \tag{6.40}$$

where ω is the angular frequency, E_{stored} is the energy stored in the cavity per cycle, P_{loss} is the power loss per cycle, ϑ_0 is the light frequency, T_{rt} is the round-trip time in the cavity, l is the relative power loss per round trip.

Therefore, the Q value of the whispering gallery mode microcavity is inversely proportional to the energy loss, where the total intrinsic loss is mainly composed of radiation loss, absorption loss, and scattering loss, all of which are intrinsic losses of the microcavity:

$$\frac{1}{Q_{int}} = \frac{1}{Q_{rad}} + \frac{1}{Q_{abs}} + \frac{1}{Q_{sca}} \tag{6.41}$$

Q_{rad} refers to radiation loss. The radiation loss is related to the size parameter R/λ of the microcavity, as well as the refractive index contrast between the material composing the microcavity and the external environment. Q_{rad} increases exponentially with the increase of R. When the cavity is large enough, the radiation loss can be ignored.

Q_{abs} refers to absorption loss, which comes from the absorption of electromagnetic waves by the dielectric material constituting the microcavity and the surrounding environment. The absorption loss varies with different working bands.

Q_{sca} refers to scattering loss, which comes from the uneven undulation of the surface of the actual prepared microcavity and the defects inside the dielectric, causing scattering light to cause loss. It can be reduced by improving optical material preparation technology and microcavity processing technology.

In addition, in the experiment, the introduction of external coupling devices to excite and collect whispering gallery modes also brings additional loss Q_{ext}, the Q value measured in practice is:

$$\frac{1}{Q} = \frac{1}{Q_{ind}} + \frac{1}{Q_{ext}} \tag{6.42}$$

In general, when coupling devices can ensure effective coupling, the introduced losses are very small, and a high Q value can still be maintained.

In an ideal microcavity, when there is no loss in the cavity, the Q value is infinite. However, light actually contains many losses when it propagates in the microcavity. The dissipation factor in the cavity is related to the photon lifetime and the resonance linewidth, and the expression for Q is:

$$Q = \omega_r \tau_p \approx \frac{\vartheta_r}{\Delta\vartheta_r} = \frac{\lambda}{\Delta\lambda} \tag{6.43}$$

where, τ_p is the photon lifetime within the microcavity, $\Delta\vartheta_r$ is the linewidth of the full-width half-maximum (FWHM), λ is the wavelength of the resonant wave in the cavity, $\Delta\lambda$ is the FWHM of the transmission spectrum. The quality factor Q of the microcavity can be obtained by measuring the wavelength entering the cavity and the FWHM of the transmission spectrum.

(2) *Mode volume of the microcavity*

The mode volume of the microcavity is also an important parameter. The size of the mode volume usually represents the size of the light confinement space in the microcavity. The mode volume V can be expressed as:

$$V = \frac{\int \rho dxdydz}{\max(\rho)} \tag{6.44}$$

In the formula, ρ is the energy density of the light field, and $\rho = \varepsilon E^2$. The mode volume strongly depends on the size of the microcavity. Generally, the smaller the radius of the microcavity, the smaller the mode volume of the light confined inside it, and the stronger the interaction between the light in the cavity and the medium, making it easier for various nonlinear interactions to occur.

(3) *Free spectral range of the microcavity*

Free Spectral Range (FSR) refers to the wavelength interval between two adjacent resonant modes, which is mainly determined by the size of the WGM microcavity and the refractive index of its own material, and can be expressed as:

$$f_{FSR} = \frac{c}{2\pi nR} \quad \lambda_{FSR} = \frac{\lambda^2}{2\pi nR} \tag{6.45}$$

where, f_{FSR} and λ_{FSR}, respectively represent the frequency interval and wavelength interval, c is the speed of light, n is the refractive index of the WGM material, R is the radius of the WGM microcavity, λ is the resonant wavelength. As can be seen from the formula, the size of the FSR is inversely proportional to the radius of the microcavity. The smaller the microcavity, the larger is its FSR. In the actual process, the radius of the prepared microcavity can be increased or decreased as needed to control the FSR of the microcavity.

3. Microcavity coupling

The mechanism of light coupling into the cavity is a key part of the experimental device. The commonly used method is to use near-field coupling devices outside the microcavity to effectively couple light into the microcavity. The more commonly used coupling devices can be divided into three types: conical fiber coupling, angled fiber coupling, and prism coupling [28], as shown in Fig. 6.65. The principle of the three methods is the same, the evanescent field of the light beam overlaps with the evanescent field of the WGM. When the phase matching conditions are met, the resonant mode of the cavity can be efficiently excited [29].

Conical fiber coupling is currently the most efficient method, and high-transmittance conical fibers can be stably prepared. The conical fiber coupling method is simple and low-cost, so it is commonly used in laboratories to measure the Q value of microcavities.

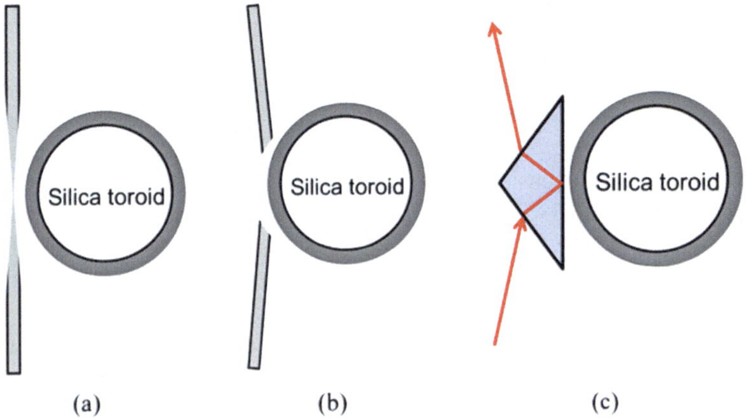

Fig. 6.65 Microcavity coupling methods **a** conical fiber coupling; **b** angled fiber coupling; **c** prism coupling

The angled fiber coupling is similar to prism coupling. The end face of the angled fiber is polished into a reflective surface through precise cutting and polishing technology, and the light beam is coupled into the microcavity through total reflection. However, this method has high processing requirements, high production difficulty, and low coupling efficiency.

Prism coupling can achieve different coupling states and different coupling efficiencies by adjusting the distance between the microcavity and the prism. The input light is focused on the coupling surface of the prism and the microcavity using an optical lens, and the input light is fully reflected at the coupling interface and coupled into the microcavity in the form of an evanescent wave. After resonating in the microcavity for a period of time, the light is coupled out of the microcavity in the same way and enters the prism. Its feature is that the performance structure is stable, which is conducive to the final packaging and practical application of the microcavity. However, the efficiency of prism coupling is still low, up to about 80%, and requires more optical devices, so it is still not commonly used in laboratory tests of the Q value of microcavities.

4. Faraday laser-based WGM microcavity PDH locking

In order to test the resonant optical spectrum characteristics of the crystal microcavity used, a crystal microcavity Q value test system is built. Since this is a near-field interaction, the coupling efficiency depends on the gap between the resonator and the coupler, and requires nanometer-level positioning stability [30].

Figure 6.66a is a schematic diagram of the microcavity coupling test system. The light source uses a continuous tunable laser (Toptica, CTL1550) with a central wavelength of 1550 nm. The signal generator (Keysight, 33500B) generates a triangular wave with an amplitude of 1 V and a frequency of 20 Hz to scan the laser. The emitted light passes through a 90/10 beam splitter, one beam enters the wavelength meter (High Finesse, WSU10-IR2) to detect the wavelength of the scanning light, and the other beam enters the polarization controller to control the polarization state of the emitted light. The light is coupled into the crystal microcavity using a conical fiber, and the emitted light enters the oscilloscope (Keysight, DSOX2002A) through a photodetector (Thorlabs, PDA10CS2). Figure 6.66b is an experimental diagram of the microcavity coupling test system. The crystal microcavity and the conical fiber are placed on a three-dimensional precision displacement platform, and the displacement platform is adjusted with an optical microscope to gradually bring the fiber cone closer to the microcavity. It can be observed that the brightness in the middle area of the outer circular surface of the crystal microcavity is higher, indicating that the roughness of this area is very low, which is conducive to achieve efficient coupling. When the conical fiber is observed, the position of the microcavity is adjusted left and right while observing whether there are resonance lines on the oscilloscope to determine whether the conical fiber and the crystal microcavity are coupled [30]. As shown in Fig. 6.66c, it can be observed that the brightness in the middle area of the outer circular surface of the crystal microcavity is higher, indicating that the roughness of this area is very low, which is conducive to achieving efficient coupling. When the conical fiber is observed, the position of the microcavity is adjusted left and right while observing whether there are resonance lines on the oscilloscope to determine whether the conical fiber and the crystal microcavity are coupled.

The resonance spectrum obtained by coupling is shown in Fig. 6.67. It can be seen that the resonance spectrum is very dense, and a large number of high-order modes are excited in one scanning cycle, and almost all modes are above the baseline. This is due to the interference of the light delay in the microcavity with the wavelength offset and the output light of the laser to form damped oscillations, resulting in exponential decay oscillations. The speed of decay of the light field energy in the microcavity can be characterized by extracting the envelope of the decay oscillation spectrum line, which indicates that most of the modes in the crystal microcavity have very high Q values, which meet the requirements.

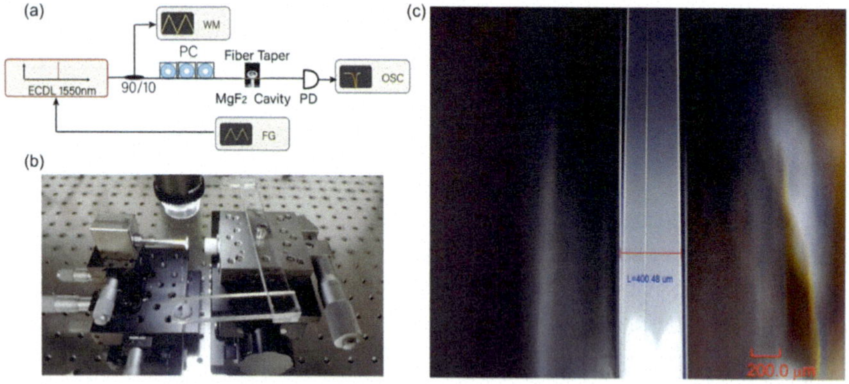

Fig. 6.66 **a** Schematic diagram of the crystal microcavity coupling test system. **b** Experimental diagram of the coupling test system. **c** Coupling of the crystal microcavity and the conical fiber

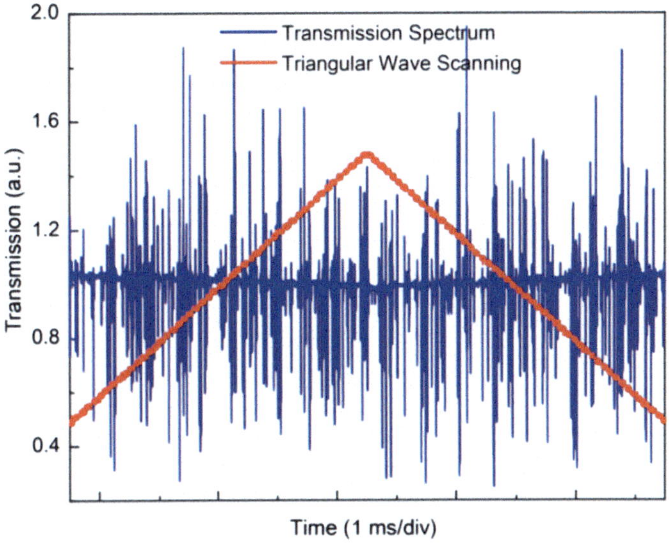

Fig. 6.67 Crystal microcavity coupling resonance spectrum

Figure 6.68 is a schematic diagram of the WGM microcavity PDH frequency stabilization system of the Faraday laser. The light emitted by the 780 nm Faraday laser is coupled into the fiber after passing through the isolator. The sinusoidal signal output by the signal source modulates the electro-optic modulator EOM to obtain the PDH sidebands. After the laser is modulated by the electro-optic modulator to produce sidebands, it is connected to the WGM microcavity that has been coupled by the conical fiber through the fiber flange. The transmitted light through the microcavity enters the detector. The signal output by the detector is split after passing through the electronic beam splitter. One path of the signal is input to the oscilloscope for monitoring, while the other path is input to the mixer to mix with the local oscillation signal. The error signal output by the mixer is input to the PID servo loop after low-pass filtering. The feedback signal is fed back to the laser current and piezoelectric ceramics to adjust the laser, finally achieving the WGM microcavity PDH frequency stabilization based on the Faraday laser. The MgF_2 crystal microcavity used in the experiment, the radius size R of the MgF_2 crystal microcavity is 4.5 mm, with a thickness $H = 400\ \mu m$, a wedge angle $\theta = 40°$, and the intrinsic Q value reaching as high as 9.24×10^9 [31].

The Faraday laser employs a Faraday anomalous dispersion optical filter (FADOF) as the frequency-selective element. And the laser output wavelength intrinsically trace to the atomic transition. Using PDH frequency stabilization technology, the output laser frequency of the Faraday laser is stabilized to the WGM, thus suppressing laser frequency drift and achieving a narrow linewidth laser source with a level of tens of Hz.

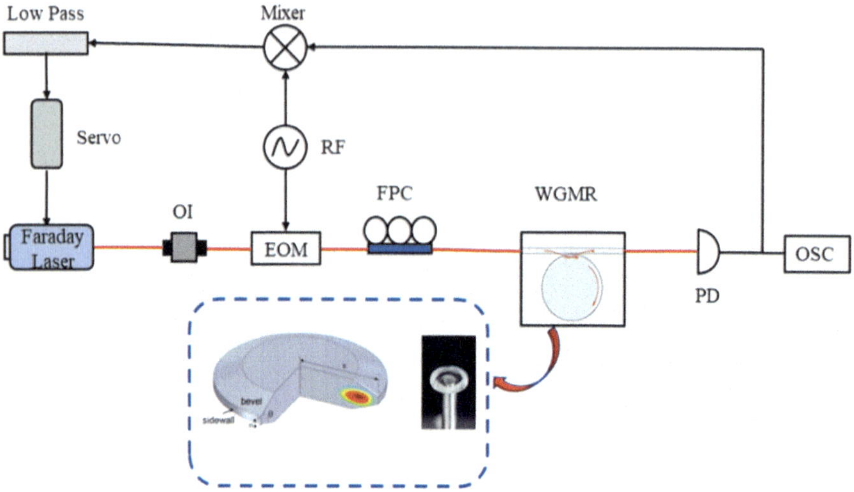

Fig. 6.68 Schematic diagram of the WGM microcavity PDH frequency stabilization system based on the Faraday laser

6.3.4 Combination of Faraday Laser PDH and Atomic Frequency Stabilization

Because the output wavelength of the Faraday laser can directly correspond to the atomic transition spectrum line, and it has the characteristic of being immune to changes in the working temperature and driving current of the laser diode, it has great potential in many practical applications represented by quantum precision measurement. In the field of quantum precision measurement, due to the differences in specific applications, there are different requirements and emphases on the short-term, medium and long-term stability, of the laser frequency after locking. Taking the atomic clock used for time–frequency measurement as an example, atomic clocks that independently serve as timekeepers are more concerned with long-term stability, but if they are only used to provide local oscillators (replacing quartz oscillators) for other atomic clocks (timekeepers or standard clocks) with good long-term stability, they are more concerned with short-term stability.

PDH frequency stabilization technology and atomic and molecular ultra-fine transition reference frequency stabilization technology are currently the two most commonly used laser frequency stabilization technologies. PDH frequency stabilization technology uses an ultra-high precision F-P cavity as a frequency reference, which can provide extremely high short-term frequency stability and has a very wide range of applications. When using PDH frequency stabilization, its unique technical characteristics need to be considered. First, the frequency reference used by PDH frequency stabilization is not an absolute frequency reference. Secondly, the ultra-high precision F-P cavity used as a reference is affected by external environmental changes such as temperature, which degrades the long-term frequency stability (generally refers to > 100 s time scale) of PDH frequency stabilization technology to a certain extent. Unlike PDH frequency stabilization technology, atomic and molecular ultra-fine transition reference frequency stabilization technology uses the transition frequency between atomic or molecular energy levels as an absolute frequency reference. The transition frequency has lower environmental sensitivity and can provide better long-term frequency stability. However, compared to PDH frequency stabilization technology, generally due to the wider linewidth of the absolute frequency reference, the signal-to-noise ratio during locking is reduced to some extent, making its short-term frequency stability, in most cases, slightly inferior to PDH frequency stabilization technology. For some applications that require high short-term and long-term frequency stability of the laser, such as the new generation of satellite navigation positioning systems, while achieving satellite network synchronization through two-way optical links, the short-term stability (1s) of the laser is extremely high. During the subsequent autonomous operation of the constellation (time scale of several hours to several days), the laser needs to have good medium and long-term stability [32–34]. This section focuses on the combination of PDH frequency stabilization technology and atomic frequency stabilization technology, so that the Faraday laser, based on its original characteristics, also has excellent short-term, medium and long-term frequency stability, making it able to contribute to the development of the field of quantum precision measurement and the construction of global satellite navigation positioning systems.

The key to locking the Faraday laser frequency by combining PDH frequency stabilization with atomic frequency stabilization lies in the coordination between the two phase-locked loops. That is, through the setting of the phase-locked loop inflection frequency, it is ensured that the two phase-locked loops do not interfere with each other and can cooperate with each other. Together, they effectively suppress laser noise after the laser frequency is locked, making it both short-term frequency stable and long-term frequency stable. In an ideal situation represented in logarithmic coordinates, after using PDH locking, the frequency noise limit of the laser should be the thermal noise of the ultra-high precision F-P cavity, i.e., frequency flicker noise, generally manifested as f^{-3} on the phase noise power spectral density, and generally manifested as f^{-1} on the frequency noise power spectral density. When converted to Allan deviation, the slope is 0, which is a flat line. Similarly, in an ideal situation, after using atomic locking, the frequency noise limit of the laser should be white frequency noise, generally manifested as f^{-2} on the phase noise power spectral density, and generally manifested as f^{0} on the frequency noise power spectral density, which is a flat line. When converted to Allan deviation, the slope is $\tau^{-1/2}$. We characterize it with normalized frequency noise ($\mathrm{Hz}^{-1/2}$), as shown in Fig. 6.69, where $S = N_0$ represents the white frequency noise after only atomic molecule locking, which can generally be at the ~ 10^{-13} $\mathrm{Hz}^{-1/2}$ level. After only using PDH technology for locking, it can be represented as $S = C_0/\sqrt{f}$, where C_0 is generally at the 10^{-15} $\mathrm{Hz}^{-1/2}$ level [32]. When two phase-locked loops are built to achieve joint locking, both PDH locking characteristics and atomic molecule locking characteristics can be possessed at the same time, where f_c is the mixed locking intersection frequency, which can generally be calculated as $f_c = (C_0/N_0)^2$. f_c can also be adjusted according to actual

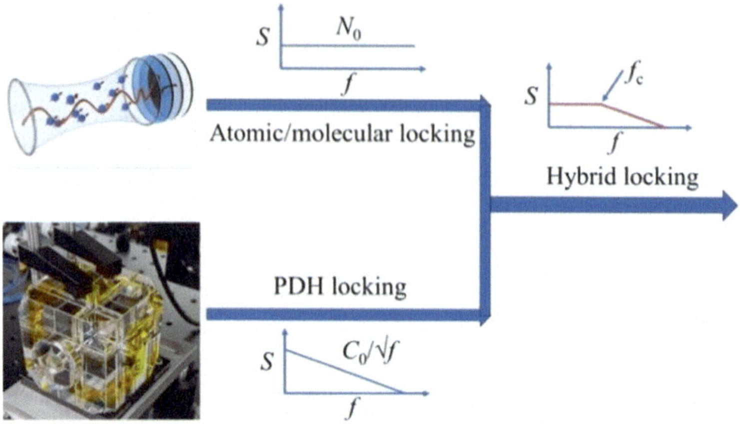

Fig. 6.69 Implementation of locking by mixing atomic molecule locking and PDH locking

application situations, which can be achieved by changing the PID parameters of the two phase-locked loops.

As shown in Fig. 6.70, it is a schematic diagram of atomic locking combined with PDH locking to form a Faraday laser joint locking. In atomic locking, we take the modulation transfer spectroscopy as an example for specific explanation. After the Faraday laser is locked by the modulation transfer spectroscopy, it is output to the PDH locking module for secondary locking. Although after MTS locking, compared to free running, the frequency noise of the laser has been greatly suppressed, but in the process of locking the laser that has been MTS locked into the PDH locking module, it still carries a certain frequency noise, denoted by ν_{MTS}. After using MTS to lock the Faraday laser frequency, the frequency of the laser has been fixed to a certain extent, and there is a high probability that it cannot resonate with the high-precision F-P cavity, so we need to shift the laser frequency to make it resonate with the cavity. Depending upon the size of the laser frequency moving range, an acousto-optic modulator or an electro-optic modulator can generally be selected as the specific frequency shifter. During the PDH locking process, a modulation and demodulation process is also needed to form an error signal as the input of servo feedback. The joint locking schematic diagram uses EOM as the frequency shifter, and the frequency shifting signal and modulation signal are input to the EOM together. Because the frequency shifting signal also needs to be the feedback object of PDH locking, and the frequency range is generally large, it is necessary to use VCO to provide the frequency shifting signal and act as the actuator of the feedback loop to achieve locking. The modulation signal is provided by a general signal generator. When using EOM, it is necessary to consider the impact of modulation depth on the overall locking loop, which can be calculated according to the Bessel function. The process of frequency shifting to achieve resonance with a cavity actually involves the 1st-order sideband resonating with the cavity (the specific situation determines whether a + 1st order or − 1st order frequency shift is more appropriate). The Bessel function is shown in

Fig. 6.71. In the actual frequency shifting process, ± 1 order sidebands and 0 order height similar points can be selected as the modulation depth selection points, which can not only ensure the size of the sidebands required for resonance, but also suppress the noise interference of other order sidebands as much as possible. The modulation and demodulation process of forming an error signal can also use the Bessel function to calculate and determine the modulation depth required for the maximum phase detection slope of the error signal, and then determine the driving power of the RF signal. Suppose the phase detection slope is D, which is proportional to the product of the 0th and 1st order of the Bessel function in the process of forming the error signal, that is, $D \propto J_0(\beta)\, J_1(\beta)$, generally $\beta = 1.08$, $J_0(\beta)\, J_1(\beta)$ is the maximum. For an EOM with a load of $R = 50\ \Omega$, the modulation depth $\beta = \pi V/V_\pi$, V is the driving voltage of the modulation, V_π is the half-wave voltage of the EOM, and then $\text{dBm} = 10\ \text{lg}\ [(V^2/2R)/1\ \text{mW}]$ to determine the corresponding values. Of course, there is some deviation between theory and experiment. After calculating the corresponding radio frequency power, appropriate adjustments need to be made to achieve the best experimental results.

From the schematic diagram of hybrid locking, we can get:

$$\nu_{MTS+PDH} = \frac{1}{1+G}(\nu_{MTS} + G\nu_{PDH}) \tag{6.46}$$

where G is the open-loop gain during PDH locking, which can be simply represented as the product of the phase detection slope D, the servo loop gain G_{servo} and the EOM (VCO) transfer function. Through the frequency noise analysis of the locking process jointly implemented by atomic locking and PDH locking in Fig. 6.69, it is known that when $f < f_c$, atomic locking is dominant, that is, $|G|$ is 0; when $f \geq f_c$, PDH

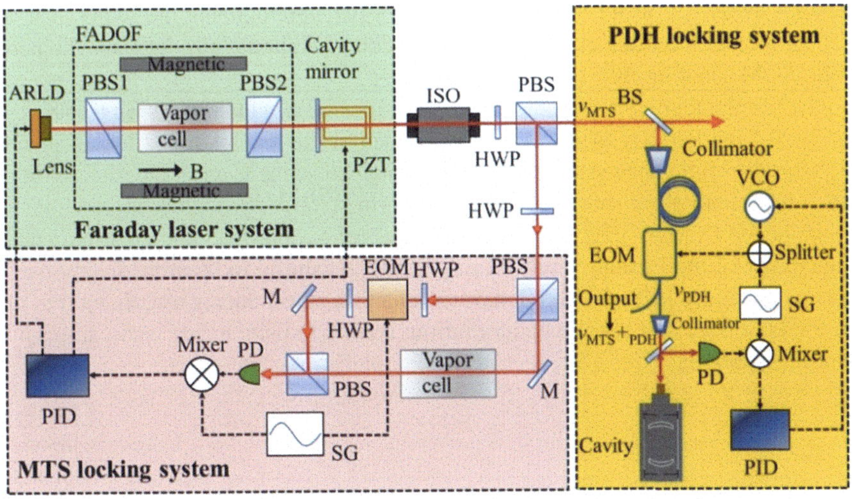

Fig. 6.70 Schematic diagram of hybrid locking

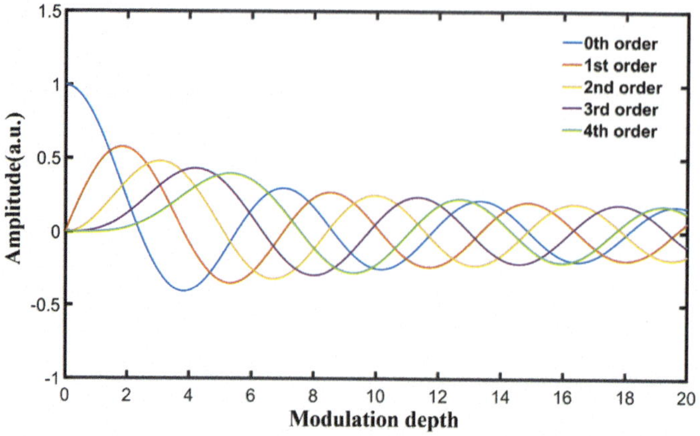

Fig. 6.71 Bessel function

locking plays a role, $|G| >> 1$. Generally speaking, the size of f_c is determined when $|G| = 1$. The value of G is closely related to the setting of the servo loop parameters. By changing the servo loop parameters, the f_c can be changed according to the actual application requirements.

During the PDH locking process of the Faraday laser, an AOM can be added before the EOM to scan the frequency, which is consistent with traditional laser frequency locking. The AOM can not only realize frequency scanning, but also serve as another actuator for PDH servo loop locking, used to eliminate frequency noise and adjust the size of f_c. It should be noted that after introducing the AOM, the size of the frequency noise level introduced by the AOM itself needs to be considered. It must be smaller than the frequency noise size after atomic locking, otherwise it damages the effect of the atomic locking loop. After introducing the AOM, if it is used as another feedback element for PDH locking, *G* should also increase the part of the AOM accordingly.

After implementing the joint locking, it is expected that the short-term frequency stability of the Faraday laser can be maintained at the level of PDH locking, and the long-term frequency stability can be maintained at the level of atomic locking, which have both good short-term and long-term stability, providing more choices for applications. More importantly, due to the strong robustness and environmental adaptability of the Faraday laser, it can provide the ability to be immediately used and automatically locked. Even if sudden situations occur during use, the output of the Faraday laser can always be around the atomic transition line, and can easily achieve re-locking.

References

1. W.R. Bennett, Hole burning effects in a He–Ne optical maser. Phys. Rev. **126**(2), 580–593 (1962)
2. T.W. Hänsch, I.S. Shahin, A.L. Schawlow, High-resolution saturation spectroscopy of the sodium D lines with a pulsed tunable dye laser. Phys. Rev. Lett. **27**(11), 707–710 (1971)
3. R.L. Barger, J.L. Hall, Pressure shift and broadening of methane line at 3.39 μ studied by laser-saturated molecular absorption. Phys. Rev. Lett. **22**(1), 4–8 (1969)
4. J.H. Shirley, Modulation transfer processes in optical heterodyne saturation spectroscopy. Opt. Lett. **7**(11), 537 (1982)
5. M. Ducloy, D. Bloch, Theory of degenerate four-wave mixing in resonant Doppler-broadened systems—I. Angular dependence of intensity and lineshape of phase-conjugate emission. J. Phys. **42**(5), 711–721 (1981)
6. M. Ducloy, D. Bloch, Theory of degenerate four-wave mixing in resonant Doppler-broadened media—II. Doppler-free heterodyne spectroscopy via collinear four-wave mixing in two- and three-level systems. J. Phys. **43**(1), 57–65 (1982)
7. R.K. Raj, D. Bloch, J.J. Snyder et al., High-frequency optically heterodyned saturation spectroscopy via resonant degenerate four-wave mixing. Phys. Rev. Lett. **44**(19), 1251–1254 (1980)
8. H. Shi, P. Chang, Z. Wang et al., Frequency stabilization of a cesium Faraday laser with a double-layer vapor cell as frequency reference. IEEE Photonics J. **14**(6), 1–6 (2022)
9. P. Chang, H. Shi, J. Miao et al., Frequency-stabilized Faraday laser with 10^{-14} short-term instability for atomic clocks. Appl. Phys. Lett. **120**(14), 141102 (2022)
10. X. Miao, L. Yin, W. Zhuang et al., Note: demonstration of an external-cavity diode laser system immune to current and temperature fluctuations. Rev. Sci. Instrum. **82**, 086106 (2011)
11. X. Qin, Z. Liu, H. Shi et al., Switchable Faraday laser with frequencies of ^{85}Rb and ^{87}Rb 780 nm transitions using a single isotope ^{87}Rb Faraday atomic filter. Appl. Phys. Lett. **124**, 161104 (2024)
12. H. Shi, P. Chang, Z. Wang et al., Frequency stabilization of a cesium Faraday laser with a double-layer vapor cell as frequency reference. IEEE Photonics J. **14**, 1561006 (2022)
13. R.W.P. Drever, J.L. Hall, F.V. Kowalski et al., Laser phase and frequency stabilization using an optical resonator. Appl. Phys. B **31**, 97–105 (1983)
14. J.M. Robinson, E. Oelker, W.R. Milner et al., Crystalline optical cavity at 4 K with thermal-noise-limited instability and ultralow drift. Optica **6**, 240 (2019)
15. A.E. Siegman, *Lasers* (University Science, Mill Valley, CA, 1986)
16. F. Riehle, *Frequency Standards: Basics and Applications* (Wiley, New York, 2004)
17. T.T. Shi, D. Pan, J. Chen, Realization of phase locking in good-bad-cavity active optical clock. Opt. Express **27**, 22040 (2019)
18. T.T. Shi, D. Pan, J. Chen, An inhibited laser. Commun. Phys. **5**, 208 (2022)
19. N. Ismail, C.C. Kores, D. Geskus, M. Pollnau, Fabry–Pérot resonator: spectral line shapes, generic and related airy distributions, linewidths, finesses and performance at low or frequency-dependent reflectivity. Opt. Express **24**(15), 16366–16389 (2016). https://doi.org/10.1364/OE.24.016366
20. D. Kleppner, Inhibited spontaneous emission. Phys. Rev. Lett. **47**, 233–236 (1981)
21. D.J. Heinzen, M.S. Feld, Vacuum radiative level shift and spontaneous-emission linewidth of an atom in an optical resonator. Phys. Rev. Lett. **59**, 2623 (1987)
22. R.W.P. Drever, J.L. Hall, F.V. Kowalski et al., Laser phase and frequency stabilization using an optical resonator. Appl. Phys. B **31**(2), 97–105 (1983)
23. R.L. Barger, M.S. Sorem, J.L. Hall, Frequency stabilization of a cw dye laser. Appl. Phys. Lett. **22**(11), 573–575 (1973)
24. L. Rayleigh, CXII. The problem of the whispering gallery. Lond. Edinb. Dubl. Philos. Mag. J. Sci. **20**, 1001–1004 (1910)
25. L. Baumgartel, *Whispering Gallery Mode Resonators for Frequency Metrology Applications* (ProQuest Dissertations Publishing, 2013)

26. B.E.A. Saleh, M.C. Teich, *Fundamentals of Photonics*, 2nd edn. (Wiley-Interscience, Hoboken, NJ, 2007)
27. B. Sprenger, *Stabilizing Lasers Using Whispering Gallery Mode Resonators* (ProQuest Dissertations Publishing, 2010)
28. B. Liu, *Experimental Study on Ultra-High Q Microcavity Lasers* (University of Electronic Science and Technology, 2022)
29. X.F. Jiang, Y.F. Xiao, C.L. Zou et al., Highly unidirectional emission and ultralow-threshold lasing from on-chip ultrahigh-Q microcavities. Adv. Mater. **24**(35), OP260–OP264 (2012)
30. Z. Zhai, *Research on the Fabrication and Application of High-Quality Factor Whispering Gallery Mode Crystal Microcavities* (Beijing Institute of Technology, 2023)
31. Z. Qu, X. Liu, C. Zhang et al., Fabrication of an ultra-high quality MgF_2 micro-resonator for a single soliton comb generation. Opt. Express **31**, 3005–3016 (2023)
32. J.K. Abich, L.L. Blümel et al., Simultaneous laser frequency stabilization to an optical cavity and an iodine frequency reference. Opt. Lett. **46**(2), 360–363 (2021)
33. C. Günther, Kepler–satellite navigation without clocks and ground infrastructure, in *Proceedings of the 31st International Technical Meeting of the Satellite Division of the Institute of Navigation (ION GNSS+ 2018)* (2018), pp. 849–856
34. G. Giorgi, T.D. Schmidt, C. Trainotti et al., Advanced technologies for satellite navigation and geodesy. Adv. Space Res. **64**(6), 1256–1273 (2019)

Chapter 7
Applications of Faraday Lasers

Faraday lasers with characteristics of narrow linewidth and high frequency stability have broad application prospects in basic physical research and quantum precision measurement. This chapter mainly introduces the applications of Faraday lasers in atomic clocks, atomic interference gravimeters, magnetometers, optical communication, as well as laser wavelength standards, ultra-narrow band weak light coherent amplification and other fields.

7.1 Application of Faraday Lasers in the Field of Atomic Clocks

7.1.1 Faraday Lasers Applied to Atomic Clocks

Faraday lasers use atomic filters as frequency selectors, so the output frequency can be intrinsically limited near the atomic transition wavelength, making it the preferred light source in the field of quantum precision measurement. As the core instrument of quantum precision measurement, the excellent performance of atomic clocks plays a decisive role in measurement of accuracy, and plays an important role in the current and redefinition of the International System of Units second, gravitational wave detection, dark matter detection, gravitational potential measurement, physical constant measurement, etc. Among them, high-performance laser sources are key components of almost all kinds of atomic clocks. For example, in the fields of microwave atomic clocks, cold atom fountain clocks, integrating sphere cold atom clocks, optically pumped atomic clocks, magnetically selected state optical detection atomic clocks, coherent population trapping (CPT) atomic clocks, etc., lasers as probe light, pump light, cooling light or repump light are needed; whereas in the fields of optical frequency atomic clocks, cold-atom optical lattice clocks, ion trapped optical

J. Chen et al., *Faraday Laser*, https://doi.org/10.1007/978-981-97-8023-5_7

clocks, cold-atom superradiant active optical clocks, miniaturized optical clocks and chip-scale optical clocks, etc., narrow linewidth, low frequency noise lasers as local oscillator lasers, as well as cooling light, repump light to achieve atomic cooling, probe light to achieve atomic spectroscopy detection are prerequisite. In summary, the performance of the laser directly affects the key performance parameters of the atomic clock and is the core component of the development of high-performance atomic clocks.

By utilizing the natural quantum frequency selection advantage of Faraday lasers and applying them to the above-mentioned atomic clocks, it is expected to promote a qualitative improvement in the key parameters (frequency stability) of existing atomic clocks. Taking the atomic clocks that have already been applied with Faraday lasers as an example, this section introduces the scheme in detail. As pump light and probe light, Faraday lasers are applied to optically pumped cesium beam clocks and magnetically selected state cesium clocks with optical detection. Faraday lasers are used as local oscillator light to achieve high-performance optical frequency standards, and are combined with optical combs to achieve portable miniaturized optical clocks. In addition, by combining high-power single-diode lasers or diode array lasers as gain medium, high-power lasers with power on the order of watts or even a hundred watts can be realized. Using atomic spectroscopy for frequency stabilization, Faraday lasers can be used for cooling light, repump light and pump light of cold atom clocks, especially important in the field of cold atom active optical clocks.

7.1.2 Application of Faraday Lasers in Optically Pumped Cesium Beam Clocks

In the field of time measurement, compact cesium beam atomic clocks occupy an important role, with excellent medium- and long-term frequency stability, which can be applied to timekeeping, positioning, navigation, and high-speed digital communication fields. Cesium beam atomic clocks can be divided into two types: magnetically selected state cesium beam atomic clocks and optically pumped cesium beam atomic clocks. The latter can achieve higher spectral line signal-to-noise ratio and better frequency uncertainty, and is easy to vacuum seal, becoming the current research hotspot of miniaturized cesium clocks.

1. Principle of optically pumped cesium beam clock system based on Faraday lasers

The optically pumped compact cesium clock mainly includes a cesium atomic beam tube (physical part), a cesium clock circuit, and a laser system. The cesium atomic beam tube provides a directional cesium atomic beam as a quantum frequency reference. The laser system provides pumping and detection lasers for particle pumping and quantum state detection, ultimately obtaining the clock transition spectrum. The cesium clock circuit locks the frequency of the microwave source or crystal oscillator, emitting a 10 MHz atomic clock signal. The optical pumping process introduces light shift, therefore, an ultra-stable laser system that can operate stably for a long time is crucial for achieving a high-performance compact optically pumped cesium beam clock. The schematic diagram of the compact optically pumped cesium beam clock is shown in Fig. 7.1. The heated cesium oven sprays out the cesium atomic beam. The cesium atomic beam interacts with the pumping light, pumping all cesium atoms from ground state $F = 4$ level to $F = 3$ level. After interacting with the 9.19 GHz microwave in the microwave action area, they return to $F = 4$. Then, after interacting with the detection laser, atoms keep on recycling between $F = 4$ and $F' = 5$. As shown in Fig. 7.2, finally, the fluorescence signals of these atoms are detected and Ramsey fringes are obtained. After processing by the servo system, feedback is given to the crystal oscillator to stabilize its frequency. The main factors limiting the performance of the optically pumped cesium atomic clock are the scattering noise of cesium atoms and the frequency noise of the laser. Increasing the number of interacting atoms can effectively reduce the scattering noise of cesium atoms, which can be achieved by increasing the temperature of the cesium oven. However, when the temperature of the cesium oven is determined, using a narrow linewidth, low-frequency noise laser can improve the signal-to-noise ratio of the Ramsey fringes, thereby effectively improving the performance of the cesium beam clock.

Compared with other types of external cavity diode lasers, such as interference-filter external cavity diode lasers and grating external cavity diode lasers, Faraday lasers have the advantages of narrow linewidth and automatic correspondence of laser frequency to atomic transition lines. By using modulation transfer spectroscopy for secondary frequency locking of Faraday lasers, high-frequency noise can be effectively suppressed, improving the signal-to-noise ratio of the optically pumped cesium beam clock spectrum line. By changing the temperature of the cesium oven, the atomic flux in the cesium beam tube can be increased. The temperature of the cesium furnace is generally set at 100 °C. When the temperature of the cesium oven is increased, the atomic flux increases, the signal-to-noise ratio of the Ramsey fringes increases, and the short-term stability of the cesium beam atomic clock is improved.

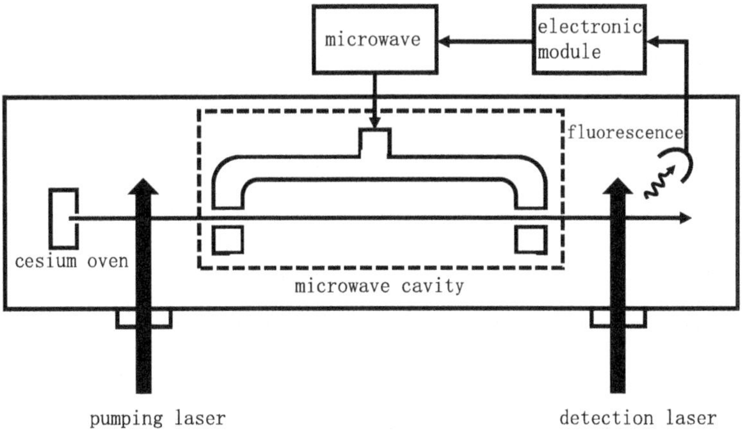

Fig. 7.1 Schematic diagram of the compact optically pumped cesium beam clock

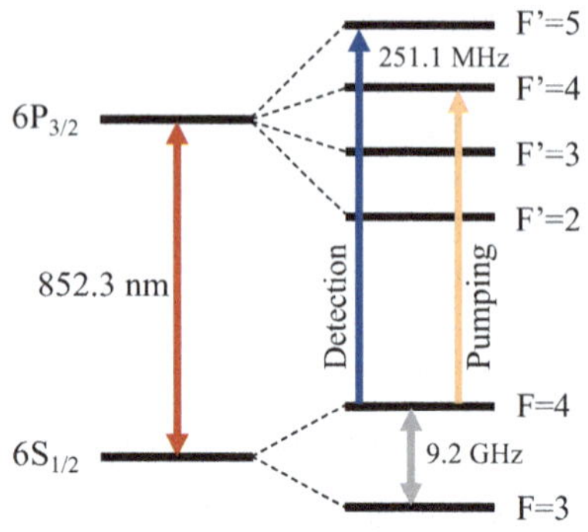

Fig. 7.2 Energy level diagram of cesium atoms

However, considering the service life of the cesium beam atomic clock, the beam tube temperature cannot be increased indefinitely. Considering all factors, the temperature of the cesium oven is set at 110 °C.

The frequency-stabilized Faraday laser is divided into two beams, serving as the detection light and pumping light of the optically pumped compact cesium clock. The Faraday laser is locked on the cesium atom $6s^2S_{1/2}\ |F=4\rangle - 6p^2P_{3/2}\ |F'=5\rangle$ transition, which can be used directly as the detection light. Because the pumping light corresponds to the cesium atom $6s^2S_{1/2}\ |F=4\rangle - 6p^2P_{3/2}\ |F'=4\rangle$ transition, therefore, a part of the stabilized Faraday laser needs to be frequency-shifted. The experimental setup is shown in Fig. 7.3. The stabilized Faraday laser is divided into two beams after passing through the half-wave plate and PBS, one beam is expanded through a lens and directly used as detection light, the other beam is reduced through a lens, then frequency-shifted by 251.1 MHz through AOM, and then collimated through a lens.

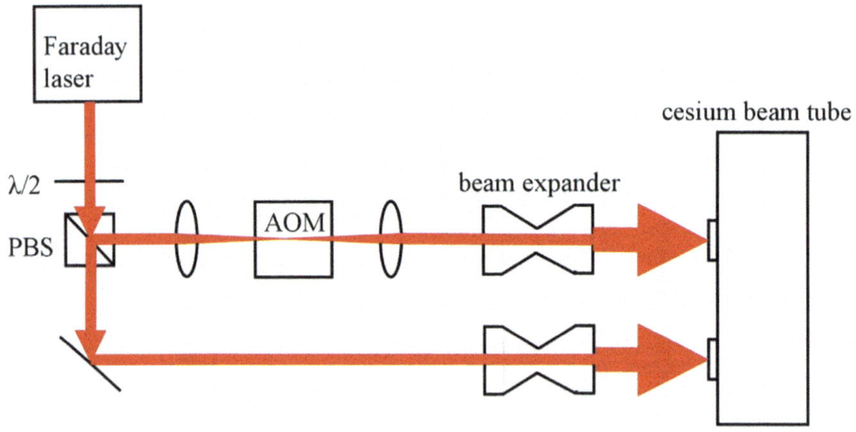

Fig. 7.3 Optical pumping cesium beam clock using Faraday laser as the detection light and pumping light. *λ/2* half-wave plate, *PBS* polarizing beam splitter, *AOM* acousto-optic modulator

2. Performance of the laser-pumped cesium beam atomic clock system based on Faraday laser

The performance of the Faraday laser is evaluated through optical heterodyne measurement. The frequency stability of the frequency stabilized Faraday laser is 2.7×10^{-12}/s, that is, the stability of a single system is 1.9×10^{-12}/s. For free running Faraday laser, the linewidth is 20.0 kHz, that is, the linewidth of a single system is 11.9 kHz. After MTS stabilization, the measured beat frequency linewidth is 3.5 kHz, that is, the linewidth of a single system is 2.3 kHz. Therefore, through MTS stabilization, the Faraday laser linewidth is significantly narrowed. Furthermore, using this Faraday laser as the detection light and pumping light of the optically pumped cesium beam clock, because the Faraday laser has a narrow linewidth and low frequency noise, and further increasing the cesium oven temperature to 110 °C to increase the atomic flux, can improve the signal-to-noise ratio of the Ramsey fringe, thereby improving the performance of the cesium beam clock. As shown in Fig. 7.4, the frequency stability of the cesium clock measured by frequency beating with the hydrogen maser is 7.9×10^{-13}/s, the stability is 1.3×10^{-13}/100 s, and reaching 2.6×10^{-14}/10,000 s. Internationally, the stability of the 5071A magnetic selected state cesium beam clock developed by Microsemi in the United States is 8.5×10^{-13}/100 s, and reaching 8.5×10^{-14}/10,000 s; companies such as Thales [1] in France and Oscilloquartz [2] in Switzerland have also achieved excellent performance optically pumped cesium beam clocks. The performance of the optically pumped cesium beam clock based on the Faraday laser is better than all compact cesium beam clocks achieved by other research institutes. In general, this optically pumped cesium atomic clock based on the narrow linewidth, high frequency stability Faraday laser can be used in timekeeping and time distribution, navigation and communication fields, and is conducive to promote the development of precise metrology fields.

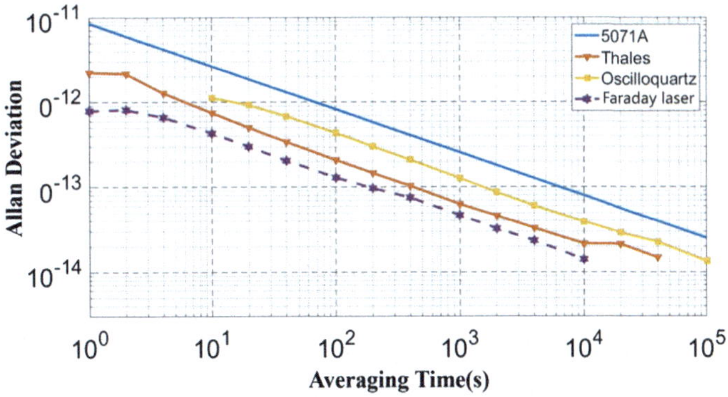

Fig. 7.4 Frequency stability of the compact cesium beam clock [1–3]

7.1.3 Faraday Lasers Applied in Dual-Frequency Optical-Microwave Atomic Clocks

Time/frequency is currently the most accurately measurable physical quantity, and atomic clocks, as high-precision instruments for time–frequency measurement, have important applications. For example, they are used for basic physical research through frequency comparison (verification of general relativity, dark matter detection, gravitational wave detection, changes in physical constants over time), the current definition and redefinition of seconds, the establishment of quantum sensing networks, and time–frequency transmission, etc. These wide-ranging applications have promoted the rapid development of atomic clock technology and types, from microwave clocks to optical clocks, and even nuclear clocks [4, 5] and pulsar astronomical clocks. Through optical frequency comparison, it has been proven that optical lattice clocks and ion optical clocks are currently the best performing atomic clocks. Even so, to cope up with different application scenarios, scientists are still striving to enrich the types of atomic clocks, such as highly charged ion clocks, single-molecule ion clocks, molecular lattice clocks, hydrogen atom lattice clocks, mercury atom lattice clocks, indium ion clocks, thorium-229 nuclear clock [4, 5] and even, millisecond wave pulsar astronomical clock. The types of atomic clocks are constantly being enriched, exploring the mysteries of nature, especially expanding the application range of quantum precision measurement.

Although there are many types of atomic clocks, the atomic clock currently used to define seconds is still the cesium atomic microwave clock, and it is expected that the roadmap for defining seconds with optical clocks will be determined in 2026. The cesium atomic clock, with its unique advantages, was successfully developed by the National Physical Laboratory of the United Kingdom in 1955 as the world's first atomic beam clock based on cesium atoms. Shortly after its realization, at the 13th International Metrology Conference in 1967, one second was defined on the transition frequency of the cesium atom's ground state hyperfine energy level, making cesium atoms the reference for primary time measurement standards. However, nowadays, the accuracy of microwave atomic clocks can no longer meet certain application needs, and the work of modifying the definition of "second" with optical clocks is in full swing. Optical lattice clocks and ion trapped clocks may become the basic clocks for the redefinition of seconds. However, since the three schemes [6, 7] for defining "second" have not yet been finally determined, and it is possible to use microwave clocks and optical clocks in parallel, by determining weights to achieve the transition and re-establishment of the definition of second. Therefore, the strategy to use the practical advantages of microwave clocks while combining the performance advantages of optical clocks is a highly challenging and widely applicable prospect.

Due to its unique performance advantages, the atomic clock realized with cesium atoms as the quantum reference is of great significance. Unlike optical lattice clocks and ion clocks, the optical clock based on thermal cesium atoms does not require ultra-stable oscillation and ultra-cold atomic quantum reference, and has the advantages of small size, simple system, and long-term continuous operation. Because the center transition frequency of optical frequency is much higher than that of microwave, its short-term frequency stability is better compared to microwave clocks, suitable for the current demand for high-performance optical frequency standards. In addition, applying the above-mentioned small-volume high-performance Faraday optical clock to the pumping light and detection light of the optically pumped cesium beam clock can obtain a microwave clock with excellent performance. In summary, this book proposes to use cesium atomic quantum reference to realize atomic clocks of two different bands of microwave and optics, and the two clocks can operate separately or simultaneously. In summary, this book proposes the utilization of a cesium atomic quantum reference to realize atomic clocks operating in two different wavebands: microwave and optical. These two wavebands can be switched between outputs freely, meaning that either one or both can be active at any given time. The microwave and the optical clocks, respectively have more excellent long-term and short-term frequency stability, realizing a dual-frequency optical-microwave atomic clock based on cesium atoms. And this scheme only needs one Faraday laser as the oscillation laser source [8], the output frequency intrinsically corresponds to the atomic transition. Further, the system structure is simple, easy to operate, has excellent performance, is highly practical, and expands the application range of microwave atomic clocks and miniaturized high-performance optical frequency atomic clocks based on Faraday lasers.

The working principle of the dual-frequency optical-microwave atomic clock is shown in Fig. 7.5. The frequency of the Faraday laser is locked to the hyperfine transition energy level of cesium atoms $6S_{1/2}$ ($F = 4$) to $6P_{3/2}$ ($F' = 5$) through modulation transfer spectroscopy, realizing a stable laser source. A beam of the stabilized laser is split off for frequency down-conversion, where the frequency of the laser is f_L, and it beats with a spectral component close to the frequency of stabilized laser, i.e., the Nth comb tooth f_N, to obtain the beat frequency $f_b = f_L - f_N$. The beat frequency f_b and the initial frequency f_0 of the optical comb are locked by microwave phase-locking technology. Therefore, the repetition frequency f_r of the optical comb is related to f_b, f_L, and f_0. According to the working principle of the optical comb, the frequency stability and accuracy of f_r are determined by the laser frequency f_L. This realizes the high-precision transfer of optical frequency to microwave frequency, forming a Faraday laser-based optical atomic clock.

The stabilized Faraday laser is used as the pumping light and detection light for the compact cesium beam clock, where the pumping light undergoes a frequency shift of 251 MHz, corresponding to the hyperfine transition energy level of cesium atoms $6S_{1/2}$ ($F = 4$) to $6P_{3/2}$ ($F' = 4$). All atoms at ground state $6S_{1/2}$ ($F = 4$) are pumped to the $6S_{1/2}$ ($F = 3$) state. Under the action of the Ramsey microwave excitation field, the cesium atoms are excited back from the ground state $6S_{1/2}$ ($F = 3$) to $6S_{1/2}$ ($F = 4$). The detection light frequency corresponds to the cesium atoms $6S_{1/2}$ ($F = 4$) to $6P_{3/2}$ ($F' = 5$), which is used to observe the change in the number of atoms on the ground state $6S_{1/2}$ ($F = 4$) under the action of the Ramsey microwave excitation field. After the frequency of the microwave source is locked, we obtain a microwave atomic clock based on the Faraday lasers.

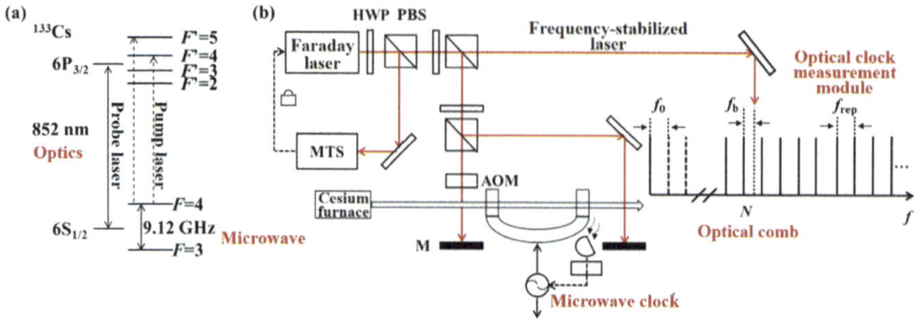

Fig. 7.5 **a** Relevant energy levels of the dual-frequency optical-microwave atomic clock; **b** working principal diagram. *HWP* half-wave plate, *PBS* polarizing beam splitter, *AOM* acousto-optic modulator, *M* reflected mirror, *MTS* modulation transfer spectroscopy stabilization, f_0 initial frequency of the optical comb, f_b beat frequency signal of the laser and one tooth of the optical comb, f_{rep} repetition frequency of the optical comb

In this way, a dual-frequency atomic clock with output signals working in both microwave and optical regions of the electromagnetic spectrum is built. Compared with optical lattice clocks and ion clocks, the optical clock based on thermal cesium atoms does not require ultra-stable local oscillators and ultra-cold atomic quantum references, and has the advantages of small size, simple system, and long-term continuous-wave operation; compared with traditional microwave optically pumped cesium beam clocks, applying the ultra-stable Faraday laser as the pumping and detection light of the optically pumped cesium beam clock can achieve a microwave clock with higher frequency stability. Moreover, this scheme requires only one laser as the local oscillator, namely the Faraday laser, whose output frequency intrinsically corresponds to the atomic transition wavelength. Furthermore, the dual-frequency optical-microwave atomic clocks based on the Faraday laser [8], the microwave atomic clock and the optical atomic clock have excellent long-term and short-term frequency stability, respectively, as shown in Fig. 7.6. The proposed dual-frequency optical-microwave atomic clock based on cesium atoms is a compact and high-performance atomic clock. From the viewpoint of scientific applications, this work connects multiple research fields, such as atomic clock and quantum metrology, and facilitates precision measurements for fundamental science. It demonstrates the feasibility of the commercialization and deployment of optical and microwave clocks. This work guides further development of integrated atomic clocks with better stability, thus broadening the horizon of quantum metrology and laser physics.

7.1.4 Faraday Laser Applied to Cold-Atom Active Optical Clock

The optical atomic clock has surpassed the microwave atomic clock in performance due to its higher working frequency, and has a wide range of applications in basic

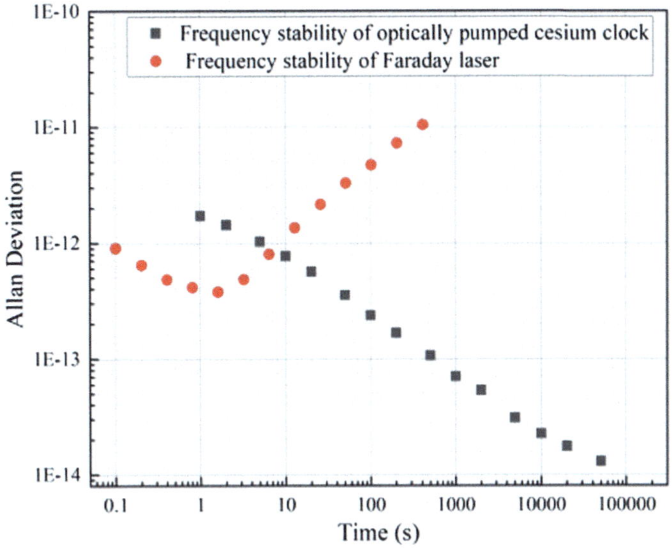

Fig. 7.6 Allan deviation of dual-frequency optical-microwave atomic clock based on cesium atoms

physical research, high-speed communication, high-precision positioning and other scientific research and engineering fields. Currently, the frequency stability and frequency uncertainty of the most advanced optical clock have both entered the 10^{-19} level [9–11]. According to the working mode, optical clocks can be divided into two categories: active optical clocks and passive optical clocks [12]. Passive optical clocks lock a pre-stabilized laser to an external, extremely narrow atomic spontaneous emission spectrum, such as a laser-cooled cesium atomic beam [13], or strontium atoms trapped in an optical lattice [14]. The local oscillator of a passive optical clock is usually locked to an ultra-stable cavity using the PDH method, and its system structure is complex and the working conditions are demanding. Moreover, since the clock laser of the passive optical clock works in the good cavity laser region, the frequency stability and the linewidth of the optical frequency standard are constrained by the cavity-length thermal noise, and its performance improvement faces great challenges [15].

In contrast, active optical clocks do not require external frequency references. Under weak feedback of the resonant cavity, active optical clocks directly use atoms as gain medium and use the stimulated emission of radiation signal generated by the gain medium atoms as the frequency standard. Active optical clocks work in the bad-cavity regime, where the gain linewidth is much smaller than the cavity-mode linewidth. Therefore, the optical frequency standard signal depends on the center frequency of the quantum reference, naturally suppressing the cavity-pulling effect brought by the cavity-length thermal noise [16]. The phase coherence of the optical frequency standard signal emitted by the active optical clock is excellent, and the linewidth can even surpass the quantum linewidth limit of spontaneous radiation [17,

18]. Overall, active optical clocks are expected to increase the stability by two orders of magnitude based on the best performing clocks. At the same time, the scheme of the active optical clock systems are simple and more conducive to miniaturization and integration [12].

Active optical clocks that use thermal atoms as the gain medium are influenced by the Doppler effect due to Brownian motion of the thermal atoms and the collisional broadening between the atoms, which can cause the frequency standard signal generated by the stimulated radiation to undergo a frequency drift, leading to a decrease in the frequency stability of the active optical clock system. Replacing the gain medium with a cold atom obtained by diffuse laser cooling can greatly suppress the Doppler and collision broadening phenomena, further improving the frequency stability of the active optical clock. This section considers the preparation of cold atomic gain medium by diffuse laser cooling technology as an example, and propose an application and experimental scheme of Faraday laser used in cold-atom active optical clock.

Diffuse laser cooling is a laser cooling scheme based on the Doppler cooling principle [19]. It has a simple structure and good performance. It uses the speed randomness of diffused photons to continuously compensate for the Doppler frequency shift required by atoms during deceleration, achieving multiple and continuous deceleration of atoms until the atoms are cooled to a very low temperature. In diffuse laser cooling technology, a high-power, narrow-linewidth, frequency-stable laser on the atomic transition spectrum is required to act as the cooling light and repumping light. The high-power frequency-stable Faraday laser meets these requirements well and has important application value in the direction of cold atom active optical clock.

Taking the rubidium atom 1367 nm four-level active optical clock as an example, the experimental scheme of Faraday laser applied to the cold-atom active optical clock and the energy level diagram of rubidium atoms are shown in Fig. 7.7. Two high-power 420 nm Faraday lasers serve as the cooling light and repumping light for laser cooling, and their output frequencies are locked to the corresponding cooling and repumping energy level transition frequencies through saturated absorption spectroscopy (or modulation transfer spectroscopy, etc.). The 420 nm cooling light is locked on the rubidium atom $5S_{1/2}F = 2 \rightarrow 6P_{3/2}F' = 3$ transitions, then generates red detuning through AOM frequency shifting, with the detuning amount Δ being twice the natural linewidth of rubidium atoms, approximately 10 MHz; the 420 nm repumping laser is locked to the rubidium atom $5S_{1/2}F = 1 \rightarrow 6P_{3/2}F' = 2$ transition. After the cooling light and repumping light are combined by the PBS, they are coupled into the optical fiber through the coupler. The optical fiber carrying the mixed cooling and repumping light is divided into multiple beams by the beam splitter, and the other end of the optical fiber is fixed on the side wall of the cold atom cell, and enters the cold atom tube from different positions. The cold atom tube is made of transparent quartz glass, and the side walls are wrapped with diffused reflection material. The mixed cooling and repumping light entering the cold atom tube will undergo diffuse reflection on the side walls. Ground state rubidium atoms moving against the cooling photons absorb these red-detuned photons, transiting to the $6P_{3/2}$ state, and then undergo spontaneous emission, releasing photons and falling back to the ground state. According to the conservation of momentum, when an atom absorbs a photon, its momentum undergoes a directional change; when it releases a photon through spontaneous emission, the direction of its momentum change is random. Therefore, statistically, the atom is subjected to a force that is always opposite to its direction of motion, i.e., it is decelerated. Since the diffuse reflection light has various directions, for rubidium atoms, a photon can always be found whose apparent red detuning amount relative to the atom just meets the requirement for resonant absorption. When the rubidium atom in the $6P_{3/2}$ state undergoes spontaneous emission transition to the lower energy level, it may fall to the $5S_{1/2}F = 1$ state. To prevent rubidium atoms from accumulating at $F = 1$, the action of the repump laser is needed to cause the rubidium atom to undergo a $5S_{1/2}F = 1 \rightarrow 6P_{3/2}F' = 2$ transition, and then fall back to the $5S_{1/2}F = 2$ state through spontaneous emission. Thus, under the action of the cooling light and repump light, this deceleration process can be repeated, achieving the Doppler cooling effect, and obtaining cold rubidium atoms uniformly distributed in the entire space of the cold atom tube, serving as the gain medium for the active optical clock.

One side window of the cold atom tube is flat, and the other side window is convex. Both windows are coated with a medium film that has a certain reflectivity to 1367 nm, serving as the front and rear mirrors of the plano-concave resonant cavity. Under the action of high-power cooling and repumping lasers, rubidium atoms are pumped to the $6P_{3/2}$ state. In the process of falling through spontaneous emission, a population inversion between $6S_{1/2}$ state and $5P_{3/2}$ state is established. The 1367 nm fluorescence signal generated by spontaneous radiation triggers stimulated radiation in the resonant cavity, is continuously amplified, and finally forms a 1367 nm stimulated radiation light signal. This serves as the optical frequency standard signal of the active optical clock, output from the window of the cold atom tube. This is the main physical process of the cold-atom active optical clock. After frequency beating with the optical comb, the optical frequency signal can be down-converted to the microwave domain, realizing the output of the time signal.

Similarly, if the gain medium of the active optical clock is replaced with cesium atoms, and the corresponding cooling and repumping lasers' wavelengths and locked transition lines are replaced, a cold-atom active optical clock with a 1470 nm wavelength optical frequency standard signal can be realized. The corresponding cesium atom energy level diagram is shown in Fig. 7.8. Two high-power 455 nm Faraday lasers are used to generate cooling light and repumping light, which are locked on the cesium atom $6S_{1/2}F = 4 \rightarrow 7P_{3/2}F' = 5$ and $6S_{1/2}F = 3 \rightarrow 7P_{3/2}F' = 4$ transition frequencies, respectively, to achieve diffuse laser cooling of cesium atoms in the cold atom tube. At the same time, due to the spontaneous radiation process, when a steady state is reached, a population inversion is achieved between the $7S_{1/2}$ and $6P_{3/2}$ states. The 1470 nm fluorescence signal generated by spontaneous radiation is continuously amplified in the resonant cavity formed by the two windows of the cold atom tube, forming a 1470 nm optical frequency standard signal and output from the window.

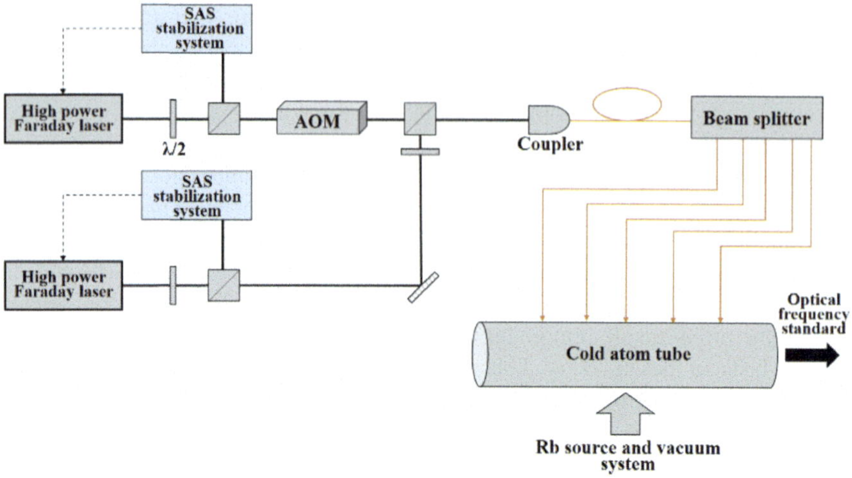

(a)

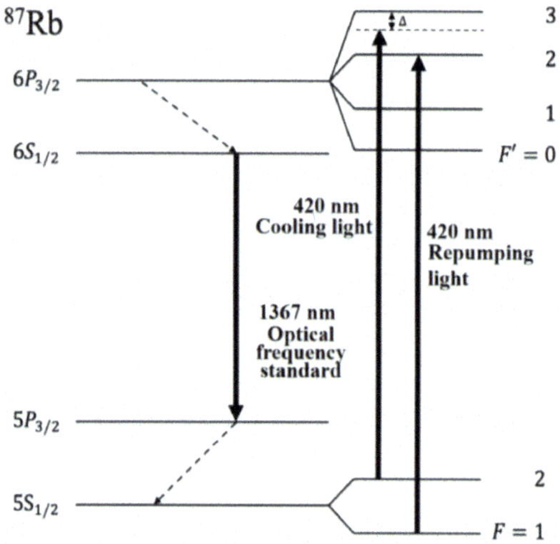

(b)

Fig. 7.7 **a** Experimental scheme of Faraday laser applied to cold-atom active optical clock; **b** Doppler cooling of Rb atoms and energy levels related to active optical clock

Taking the 1470 nm cesium atom clock laser transition as an example, for the thermal atom active optical clock, the working temperature of the cesium atom cell is around 130 °C. Due to the Doppler effect, the laser linewidth broadening is about 7.31 MHz, while the natural linewidth of the 1470 nm transition is only 1.81 MHz [20]. By replacing the gain medium of the active optical clock with cold atoms, the influence of the Doppler effect can be almost eliminated, thereby narrowing the

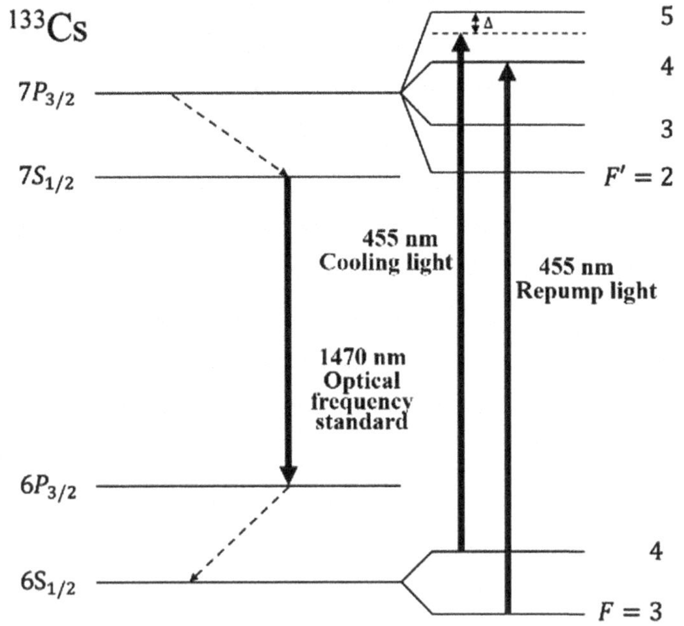

Fig. 7.8 Related energy levels of cesium atom cold-atom active optical clock

linewidth of the output optical frequency standard signal and improving the frequency stability. A narrower gain medium linewidth can further improve the bad-cavity factor (i.e., the ratio of the gain medium linewidth to the cavity-mode linewidth), thereby suppressing the influence of cavity-length thermal noise on the frequency stability.

7.2 Application of the Faraday Laser in Atomic Gravimeters

Acceleration due to gravity, usually denoted by "g", is the acceleration caused by the Earth or any other celestial body. It determines the rate of increase of velocity when an object is in free fall. On the surface of the Earth, g is approximately 9.8 m/s^2, and the direction always points towards the center of the Earth. When measuring "g", Galileo (Gal) is often used as a unit ($1\ m/s^2 = 100$ Gal). However, since the Earth is not a completely uniform sphere and its mass distribution is not completely uniform, the strength of "g" varies slightly at different locations.

Measuring high-precision acceleration due to gravity is of utmost importance for scientific research and technical applications [21–24]. This not only helps scientists gain a deeper understanding of the Earth's internal structure, but also plays a key role in multiple fields such as gravity mapping and navigation. For example, through precise gravity measurements, the distribution of underground mineral resources can be detected, and the accuracy of maritime and aviation navigation systems can be optimized. In addition, the high-precision value of acceleration due to gravity has also played a significant role in the definition of the kilogram in the International System of Units [25].

A gravimeter is a scientific instrument specifically designed for precise measurement of the acceleration due to gravity g. Its performance is mainly evaluated by indicators such as sensitivity, uncertainty, and stability. Sensitivity refers to the smallest change in gravity that the gravimeter can detect, which is used to measure the measurement noise of the gravimeter. Uncertainty is used to characterize the accuracy of the gravimeter's measurement of the acceleration due to gravity g. Stability generally refers to long-term stability, reflecting the ability of the gravimeter to maintain measurement consistency over a long period of time. A stable gravimeter can provide consistent measurement results over a long period of operation, even when environmental conditions change.

Gravimeters can be divided into relative gravimeters and absolute gravimeters. Relative gravimeters detect the relative change in the gravitational acceleration. This type of gravimeter usually has excellent sensitivity, for example, the LaCoste–Romberg zero-length spring gravimeter [26] has a sensitivity of up to 0.1 μGal/$\sqrt{\text{Hz}}$, and the iGrav portable superconducting gravimeter produced by the American GWR Company [27] has a sensitivity of 0.3 μGal/$\sqrt{\text{Hz}}$. In terms of long-term stability, the drift within a month is less than 0.5 μGal. Absolute gravimeters are a direct measurement of the gravitational acceleration at the Earth's surface, originally derived from the pendulum structure, with an uncertainty at the 10^{-4} g level [21]. With the development of laser technology, the laser interferometric absolute gravimeter is created. The laser interferometric absolute gravimeter generally adopts the Michelson interferometer structure, judging the distance of the falling object by observing the bright and dark stripes of interference, continuously recording the bright and dark changes of the interference stripe period and the corresponding clock signal, thereby obtaining the trajectory of the falling object and thus the gravitational acceleration. The mobile gravimeter developed by the University of Colorado and the Joint Institute for Laboratory Astrophysics (JILA) has an uncertainty of up to 10^{-8} g level [28]. Currently, the more mature commercial absolute gravimeters include the FG5 and A10 series laser interferometric absolute gravimeters from the United States. Taking the latest FG5/FG5X products as an example, their measurement uncertainty can reach 2 μGal, and the sensitivity can reach 15 μGal/$\sqrt{\text{Hz}}$ [29].

7.2.1 Principle and Development of Atomic Gravimeter

According to the de Broglie wave formula, matter also has wave-particle duality, so it is theoretically feasible to construct a gravimeter using the matter wave interference of atoms. After decades of development, various types of atomic gravimeters have been derived. According to the difference in the manipulation of atomic interference paths, they can be divided into Raman transition atomic gravimeters, Bragg diffraction atomic gravimeters, and Bloch oscillation atomic gravimeters, etc. Among them, the Raman transition atomic gravimeter is based on the principle of stimulated Raman transition of atoms to achieve atomic interference. This type of atomic gravimeter is more common, and this chapter mainly introduces this type of atomic gravimeter.

Analogous to the optical interferometer which needs to split, reflect, and combine the light beam (as shown in Fig. 7.9a), the atomic interferometer includes operations such as interference, superposition, and detection of atomic quantum states, using a three-pulse sequence of Raman lasers ($\pi/2 - \pi - \pi/2$) to manipulate the internal and external quantum states (as shown in Fig. 7.9b, this configuration of atomic gravimeter is also known as the Mach–Zehnder type atomic gravimeter). By properly adjusting the frequency and intensity of the laser, the splitting, reflection, and combination of atoms can be achieved, thereby forming a matter wave interference effect. This technology allows scientists to precisely control the quantum state of atoms, providing a foundation for the implementation of atomic interference.

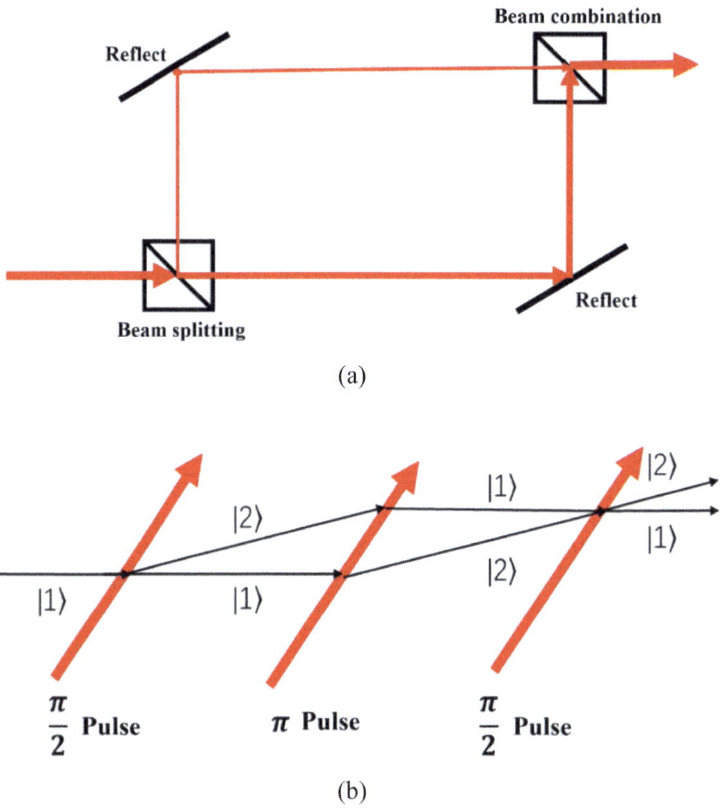

Fig. 7.9 **a** Optical interferometer; **b** atomic interferometer

Atomic gravimeters generally use alkali metal atoms (Rb, Cs, etc.) in ultra-high vacuum systems as control objects. This is because these types of atoms have relatively simple energy levels, which are easy to control, analyze, and study. First, the atoms are laser-cooled and trapped using magneto-optical trap technology, and then the temperature of the atomic cloud is further reduced by polarization gradient cooling, thereby improving the coherence of the atomic cloud. Then, the ultra-low temperature atomic cloud is allowed to fall freely under the action of gravity, and a microwave pulse or Raman π pulse laser is applied to select atoms with a magnetic sublevel of $m_F = 0$ which is insensitive to the magnetic field. Afterwards, three Raman pulses ($\pi/2 - \pi - \pi/2$) are applied at different times during the atom's fall to complete the splitting, state inversion, and merging of the atomic cloud. In this process, the atomic cloud not only undergoes transitions between internal states and changes in external state momentum due to the Raman pulses, but also experiences the effects of the gravitational field, leading to different phase accumulations on their respective paths after splitting. This ultimately manifests as interference phenomena in the population of internal state atoms and between external state matter waves. By scanning a variable (such as the chirp rate of the Raman light frequency α), interference fringes can be obtained. Since the atoms are freely falling in the gravitational field, information about gravity can be obtained from the interference fringes.

The first to realize a cold-atom interferometer and use it to measure gravitational acceleration was the research group of Steven Chu at Stanford University in the United States [30]. They utilized a cold atom fountain and a two-photon Raman transition process. In 1992, the group used a cold sodium atom fountain to measure the gravitational acceleration g, with a sensitivity of about 1.3×10^{-6} g/$\sqrt{\text{Hz}}$ [31]. Later in 1999, they upgraded the system to a cold cesium atom fountain, improving the sensitivity of gravity measurement to 2×10^{-8} g/$\sqrt{\text{Hz}}$ [32]. The Santos team at the Paris Observatory in France (LNE-SYRTE) has been conducting research on cold rubidium atom interferometer gravimeters since 2004 [33], using free-falling cold rubidium atoms to measure gravitational acceleration. This method reached a gravitational measurement sensitivity of 1.4×10^{-8} g/$\sqrt{\text{Hz}}$ [34]. The French Space Agency (ONERA) has been developing a compact atomic interferometer gravimeter since 2013. This type of miniaturized atomic gravimeter can reach a measurement sensitivity of 4.2×10^{-8} g/$\sqrt{\text{Hz}}$, with a stability close to 1×10^{-9} g after 1000 s [35]. The team subsequently reported on shipborne atomic gravimeters [36] and airborne atomic gravimeters [37]. In 2019, the University of California, Berkeley, reported on a mobile atomic gravimeter and conducted vehicle-mounted experiments, with a measurement sensitivity of 3.7×10^{-8} g/$\sqrt{\text{Hz}}$, a half-hour stability of 2×10^{-9} g, and a system uncertainty of 1.5×10^{-8} g [38].

In 2011, Zhou at the Institute of Precision Measurement Science and Technology Innovation of the Chinese Academy of Sciences (formerly the Wuhan Institute of Physics and Mathematics of the Chinese Academy of Sciences) reported an atomic gravimeter based on rubidium atom fountain, with a measurement sensitivity of 2 $\times 10^{-7}$ g/$\sqrt{\text{Hz}}$, and a stability of 2×10^{-9} g after 236 s [39]. In 2012, the research group of Huazhong University of Science and Technology led by Hu has developed a rubidium fountain atomic gravimeter based on active vibration isolation, with a measurement sensitivity reaching 4.2×10^{-9} g/$\sqrt{\text{Hz}}$, and a stability of 0.3 μGal/300 s [40]. In 2014, a team at Zhejiang University led by Lin developed a free-falling type atomic gravimeter, with a sensitivity of 1.0×10^{-7} g/$\sqrt{\text{Hz}}$, a stability of 1.1×10^{-8} g/200 s, and 5.7×10^{-9} g/1000 s [41]. The team subsequently developed a miniaturized atomic gravimeter, with a measurement sensitivity of 2.6×10^{-7} g/$\sqrt{\text{Hz}}$, a stability of 2.5×10^{-9} g/8000 s, and a system evaluation uncertainty of 2.0×10^{-8} g [42]. In 2018, the Gravity Standard Laboratory of the National Institute of Metrology, China, led by Wu developed the NIM-AGRb-1 free-falling type atomic gravimeter, with a measurement sensitivity reaching 44 μGal/$\sqrt{\text{Hz}}$, and a measurement uncertainty of 5.2 μGal [43].

7.2.2 Raman Laser Source Based on Faraday Atomic Filter

The atomic interference process is realized by the stimulated Raman transition of atoms through the interaction of Raman laser pulses with atoms. The Raman laser is essentially produced by two laser beams with a specific frequency difference. The traditional Raman laser is mainly produced by optical phase locking [44, 45] or electro-optic phase modulation [46–48] to achieve superior coherent characteristics. The latter is often used in compact, portable atomic gravimeters because the Raman laser implemented by the electro-optic phase modulation method only requires one laser source, thereby simplifying the optical system of the atomic gravimeter. In addition, since the two different frequencies of the Raman laser are modulated from one laser, they have excellent phase coherence. However, in the process of electro-optic phase modulation, in addition to generating two first-order sidebands, additional sidebands are also generated. These additional sidebands can bring system errors and directly affect the results of gravity measurement. To suppress these extra sidebands, in 2018, Zhu and others from the National University of Defense Technology used a single-sideband laser system based on a frequency-doubled I/Q modulator, which can suppress additional sidebands by more than 20 dB [49]. In 2022, Luo and others from Huazhong University of Science and Technology predicted and designed magic time intervals to further eliminate the impact of additional sidebands. This method can suppress additional sidebands by more than 20 dB but less than 30 dB, reducing the uncertainty caused by additional sidebands to 0.5 μGal [50]. In 2022, Wang and others from the National University of Defense Technology proposed a single-sideband modulation method based on FBG optical rectangular filtering, with a signal-to-noise ratio better than 19 dB [51]. Utilizing the optical filtering characteristics of FADOF, by setting the operating parameters of FADOF, only the frequencies in the first-order sideband of the Raman light that correspond to the atomic transition spectral lines in FADOF are allowed to pass through the optical filter, while the higher-order sidebands are unable to pass through the atomic optical filter. This results in the acquisition of Raman light with a single sideband.

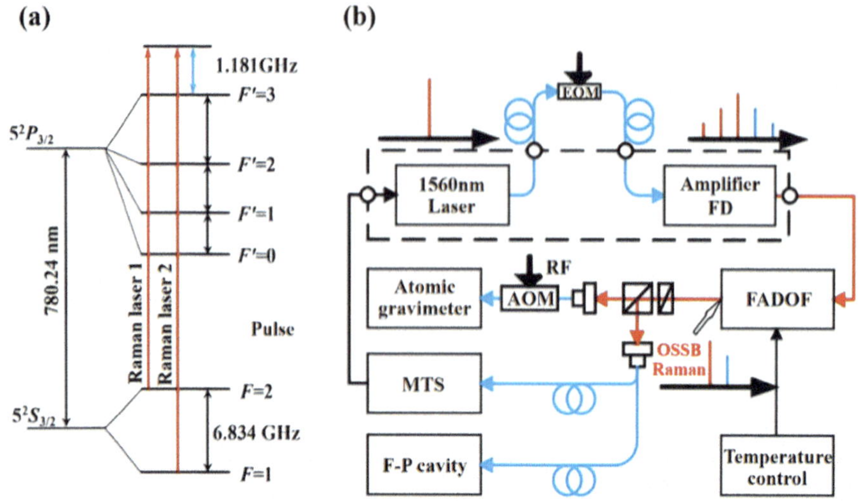

Fig. 7.10 **a** ^{87}Rb atom D_2 line energy level and Raman laser frequency; **b** experimental setup. *EOM* electro-optic modulator, *AOM* acousto-optic modulator

In 2023, Zhao from the National Institute of Metrology, China and Chen from the Peking University, China, used the optical filtering characteristics of FADOF and proposed a new method for generating optically single-sideband modulated Raman lasers [52]. The frequency of the Raman laser is shown in Fig. 7.10a, the lower frequency Raman light 1 is locked to the ^{85}Rb atom's $F = 3 \rightarrow F' = 4$ transition line through modulation transfer spectroscopy, and is detuned from the $5P_{3/2}$ $F' = 3$ of ^{87}Rb is approximately 1.18 GHz, and Raman light 2 is generated by electro-optic phase modulation.

As shown in Fig. 7.10b, the 1560 nm laser (Quantel, EYLSA780) output laser passes through an optical fiber electro-optic modulator (KG-PM-15) for electro-optic phase modulation, and the modulated signal is amplified and doubled to generate a 780 nm modulated laser. The carrier and modulation sidebands of the modulated laser pass through a FADOF optical filter to generate the required OSSB Raman light after filtering out the high-order sidebands. This research group applied the optical single-sideband Raman laser generated by FADOF to the ^{87}Rb free-falling atomic gravimeter for gravity measurement and for further experimental research.

The structure of FADOF is shown in Fig. 7.11a. The input power of FADOF is set to 300 mW, and the output power is about 120 mW. Using the Faraday magneto-optical rotation effect, by setting appropriate working parameters, the zero-order and positive first-order sidebands of the Raman light are passed through FADOF with a high transmission, and other sideband lights are optically filtered. In the experiment, the magnetic field is controlled by a magnetic ring, the temperature is controlled by a PI heater, the magnetic field is set to 300 G, and the temperature is set to 73.8 °C. The saturated absorption spectroscopy of rubidium and the transmission spectrum of FADOF under these working conditions are shown in Fig. 7.11b. The

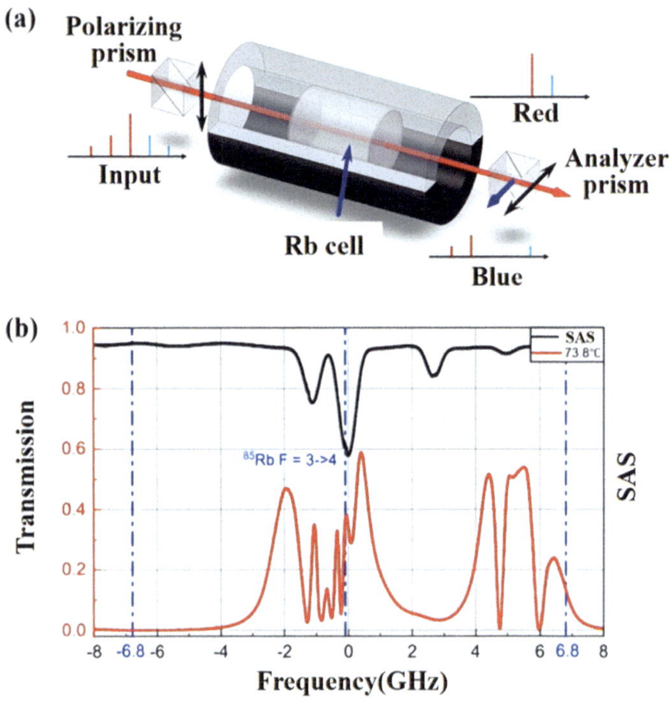

Fig. 7.11 **a** Experimental structure of FADOF; **b** saturated absorption spectroscopy of Rb atom, and transmission spectrum of FADOF under 300 G and 73.8 °C

blue dotted line in the figure represents the carrier of Raman light 1 and the ± 1st sidebands modulated by the electro-optic modulator, with a modulation frequency of 6.834 GHz. The red line transmission spectrum has a very low transmission at the − 1st order sideband. As shown in Fig. 7.11b, FADOF can only transmit 0th order and + 1st order components, and filters out the − 1st order and higher-order sidebands.

In order to evaluate the performance of the FA-OSSB (FADOF-based optical single-sideband Raman light) system, the research group studied the power distribution of the Raman spectrum without FADOF filtering (black line) and with FADOF filtering (red line), as shown in Fig. 7.12. When there is no FADOF filtering, the distribution of the carrier and each order of sidebands shows that the − 1st sideband is only slightly lower than the carrier, and the 2nd order sideband and even the 3rd order sideband are also clearly visible. After FADOF filtering, the excess sidebands are significantly suppressed by 50–60 dB. Based on previous research results, the effect of additional sidebands with a suppression ratio of more than 30 dB can be ignored [39–42], so it can be determined that the single-sideband Raman laser spectrum obtained by FADOF has a high purity and theoretically can eliminate the effect of additional sidebands.

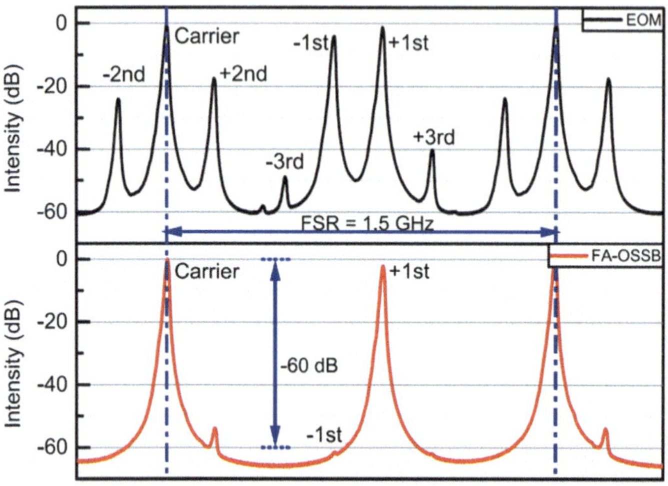

Fig. 7.12 Raman intensity spectrum without FADOF filtering (black line) and with FADOF filtering (red line)

To demonstrate the characteristics of the FA-OSSB Raman laser, the research group applied it to a free-fall atomic gravimeter, as shown in Fig. 7.13. They first measured the interference region from 10 to 83 mm from the center of the magneto-optical trap, and obtained the π transition spatial probability distribution under the comparison of electro-optic modulation and FA-OSSB, as shown in Fig. 7.13b. Under the condition of electro-optic modulation, there is an oscillation of the Raman π transition probability in space, with a half-period of 11 mm, which is consistent with the frequency interval of 6.834 GHz between the hyperfine energy levels of the ground state $F = 1$ and $F = 2$ [43], that is, the additional sideband effect in normal EOM modulation produces a spatial dependence of the effective Rabi frequency. However, the π transition probability in the FA-OSSB scheme is independent of position. It further proves that FADOF can effectively suppress sidebands.

In addition, because the phase shift is the largest when the π transition probability changes the most, two positions of 31 mm (peak) and 42 mm (valley) are selected in Fig. 7.13b, and the atomic interference fringes are measured with electro-optic modulation and FA-OSSB at two different free evolution time intervals (T) of 30 and 65 ms. The atomic interference fringes can be expressed as $P = \frac{1}{2}[1 - \cos(\Delta\Phi)]$, where the phase shift change $\Delta\Phi = \left(k_{eff}g - \alpha\right)T^2 + \Delta\Phi_0$, $\Delta\Phi_0$ includes the effect of phase shift with additional sideband effects. Figure 7.14a provides a Lissajous diagram to compare the phase shift between the 31 and 42 mm positions using EOM and FA-OSSB at T = 30 ms. It can be clearly observed that the phase and contrast of the stripes under the normal electro-optic modulation scheme have changed, while the FA-OSSB scheme has almost no change.

As shown in Fig. 7.14a, the phase measured at two different positions in the EOM scheme shifts by 358.8 mrad at 30 ms, and the stripe contrast drops from 59.7 to

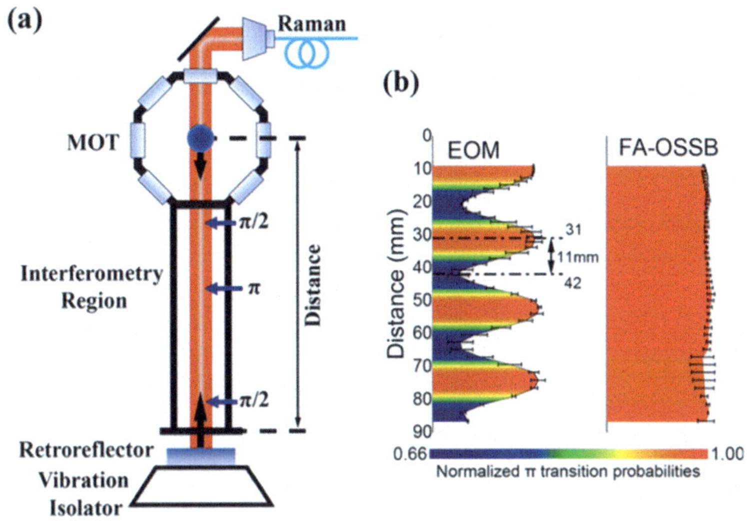

Fig. 7.13 **a** Structure of ^{87}Rb atom free-fall atomic gravimeter; **b** probability distribution of π transition

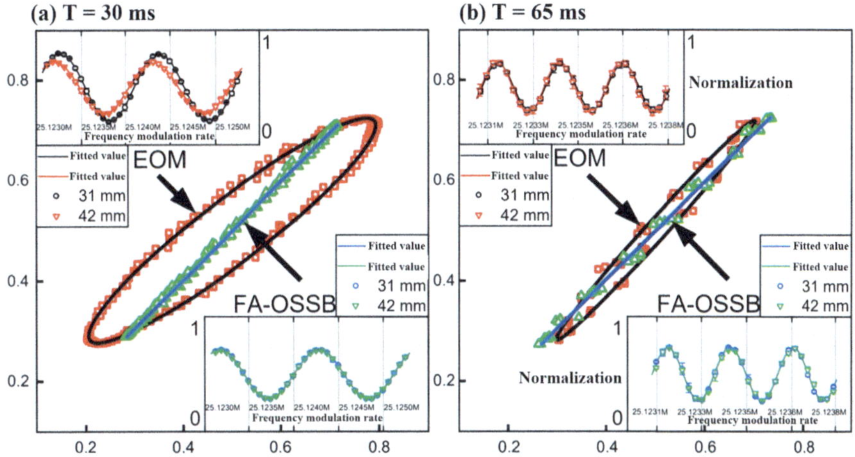

Fig. 7.14 **a** At T = 30 ms, the atomic interference stripes and Lissajous diagram under electro-optic modulation and the FADOF-based optical single-sideband Raman light; **b** when T = 65 ms, the atomic interference diagram and Lissajous diagram

45.5%. The FA-OSSB scheme reduces the phase shift to 2.2 mrad, and the contrast remains basically unchanged. Figure 7.14b is a comparison at T = 65 ms. In the EOM scheme, the phase shift is about 126.9 mrad, and in the FA-OSSB scheme, the phase shift is compressed to 2.3 mrad. This shows that the FA-OSSB scheme can suppress the impact of ASB on phase shift and contrast.

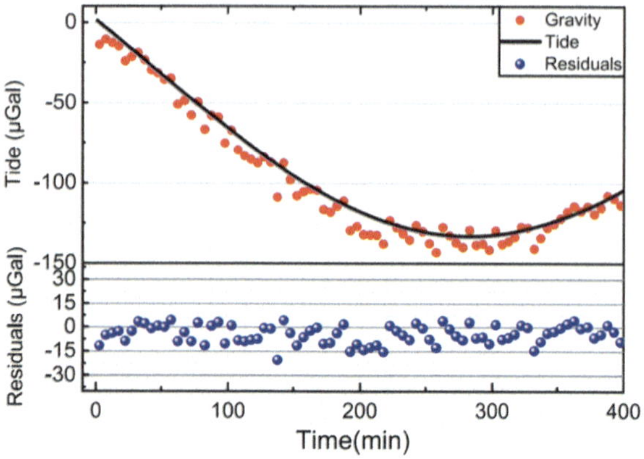

Fig. 7.15 6.5 h of continuous gravity measurement. Each point in the figure is the result of a 5-min average. Black is the theoretical trend; red is the experimental data; blue is the residual error of the experimental data and the theoretical trend

The research group further applied the FA-OSSB system to the gravity measurement of the ^{87}Rb atomic gravimeter. Figure 7.15 is a continuous measurement from 22:00 on November 17, 2023, to 4:30 on November 18. The black line is the theoretical tidal model, the red dots are the experimental model data, and the blue line is the residual after subtracting the experimental data from the theoretical model. The study designed an OSSB Raman laser system based on FADOF. This system has signal-to-noise ratio better than 50 dB and it can completely eliminate the ASB effect, too. Compared with other OSSBs, the FA-OSSB Raman scheme has the characteristics of simple control and high signal-to-noise ratio.

7.2.3 Prospects for the Application of Faraday Lasers in Atomic Gravimeters

As mentioned in Sect. 7.2.2, the internationally pioneering dual-frequency Faraday atomic filter developed by Peking University has been successfully applied to the Raman laser system of the National Institute of Metrology, China [52]. Atomic filter with transmission spectra corresponding to rubidium atoms $5S_{1/2}$ $(F = 1) \rightarrow 5P_{3/2}$ and $5S_{1/2}$ $(F = 2) \rightarrow 5P_{3/2}$ is utilized as filter device. The two transmission narrow bands are less than 1 GHz and correspond to the quantum transition frequency, which realizes the harmonic suppression of the Raman laser, and is successfully applied to the NIM-AGRb-2 type atomic absolute gravimeter system developed by the National Institute of Metrology, China.

Using a dual-frequency Faraday atomic filter, it can effectively suppress the sidebands generated by laser phase modulation, directly produce the core component required by the atomic gravimeter system, i.e., the Raman laser. Compared with the traditional two-laser phase-locking scheme, the system is greatly simplified, and the phase noise is lower. Compared to traditional filtering devices, the use of a dual-frequency Faraday atomic optical filter allows for the direct extraction of frequency components corresponding to atomic spectral lines, with a signal-to-noise ratio superior to 50 dB. The dual-frequency Faraday atomic optical filter can effectively simplify the system complexity of atomic gravimeters and enhance the performance of gravity measurements.

As shown in Fig. 7.10a, the laser used in the above experiment is a 1560 nm fiber laser, which first obtains a first-order sideband through electro-optic phase modulation, and the main frequency and the first-order sideband realize a wavelength of 780 nm laser output through the frequency doubling effect of the crystal. This method of generating Raman lasers increases the complexity of the system, and the action of the electro-optic phase modulation crystal and the frequency doubling crystal increases the loss of the laser, introduces system noise, thereby reducing the quality of the Raman laser. So, if we directly use a 780 nm dual-frequency Faraday laser as a Raman laser, can we further improve the performance of the atomic gravimeter?

We know that in the process of atomic interference, the laser plays a crucial role, playing an important role in key steps such as cooling, trapping, and detecting atoms. The performance of the atomic gravimeter largely depends on the laser, including the linewidth, frequency stability, frequency drift, locking performance, and power stability of the laser.

Considering the impact of the phase noise of the Raman laser on the sensitivity of the atomic interferometer, several potential factors need to be considered. First, the phase noise of the laser itself is an important factor. The stability and accuracy of the laser directly determine the phase characteristics of the emitted laser, and this characteristic is transmitted to the atomic interference process, affecting the stability and accuracy of the atomic interference phase. In addition, the electronic circuit in the Raman laser system may also introduce phase noise, directly affecting the stability of the laser frequency. The phase noise introduced by the reference frequency source is also a factor to consider, because any instability of the reference frequency directly affect the frequency of the Raman light, thereby affecting the interference process of the atoms.

Based on the above analysis, we conclude that the Faraday laser is the ideal choice for generating Raman lasers. First, the Faraday laser, combined with a fully automatic frequency stabilization laser system, locks the center wavelength of the output laser to the atomic transition line through modulation transfer spectroscopy, with narrow linewidth and excellent laser frequency stability. Furthermore, the Faraday laser and its frequency stabilization system adopt an integrated design, making it compact in size, stable in structure, and suitable for various working environments. Moreover, the matching fully automatic frequency stabilization laser system has a one-key automatic locking function, which can quickly restore the locked state in the event of accidental unlocking, ensuring the continuous and reliable operation of the instrument.

In the test, the linewidth of the Faraday laser after locking is less than 100 kHz, and the output power reaches 90 mW, showing excellent performance. According to the calculation of the gravimeter indicators, the residual frequency fluctuation of the laser does not exceed 50 kHz, and the repeated locking frequency error is within 100 kHz.

It meets the requirements of atomic interferometers for laser performance. The application of Faraday lasers in atomic gravimeters inevitably promote the further development of the field of atomic interference measurement, providing reliable tools and methods for research and application in the field of quantum precision measurement.

7.3 Application of Faraday Lasers in Atomic Magnetometers

Magnetism is one of the earliest physical phenomena recognized by humans, with a long history of study and extremely important scientific value. Around 5000 years ago, humans discovered natural magnets (Fe_3O_4). During the Warring States period in China, people creatively invented the earliest compass—Sinan, using the Earth's magnetic field to guide direction. In the development history of magnetism, in 1785, French physicist Coulomb obtained Coulomb's law describing the interaction between charges and magnetic poles through a torsion balance experiment. In 1820, Danish physicist Oersted discovered the phenomenon of current-induced magnetism. In 1831, British physicist Faraday discovered the phenomenon of electromagnetic induction and first introduced the basic physical quantity of magnetic induction intensity. Faraday's proposal marked the beginning of quantitative research on the properties of magnetic fields, laying a solid foundation for subsequent magnetism research. The introduction of this important concept has promoted the further development of magnetism and provided a powerful tool for scientists to understand and explain magnetic phenomena more deeply.

Magnetism has a wide range of applications in contemporary science and engineering fields. For example, in the field of biomedicine, the currently hotly researched Brain-Computer Interface (BCI) is closely related to magnetic technology. BCI is a technology that directly connects the human brain and external devices, realizing direct interaction between the human brain and the computer by recording and interpreting brain activity. Among them, to obtain spatial distribution information of brain activity and decode brain activity, Magnetic Resonance Imaging (MRI) technology is used; to activate or inhibit specific brain areas, change brain activity, and control brain activity, magnetic stimulation technology is needed. In the field of geological exploration, magnetic technology is used for the detection of geomagnetic field anomalies and mutations [53]. For example, volcanic and earthquake activities can cause short-term changes in the intensity and direction of the geomagnetic field, resulting in geomagnetic pulsations. These changes can be detected using a magnetometer; geological structures (such as fault zones, fold zones, etc.) can cause changes in the magnetism of underground rocks, resulting in geomagnetic pulsations. Magnetometer arrays can study the distribution and properties of these geological structures. In non-destructive testing technology, by measuring the leakage magnetic field near the surface of the magnetized workpiece, the defect location can be quickly and accurately located [54]. In addition, magnetic memory detection is an emerging magnetic technology. It applies a magnetic field to the surface of the object to be tested, then records the changes in the magnetic field. When the change in the magnetic field exceeds a certain threshold, it can be determined that there are defects in the object being tested. In the field of military defense, magnetic field detection technology is mainly used in anti-submarine, unexploded bomb detection, underwater geomagnetic navigation, etc. These applications help to improve the efficiency and safety of military operations [55, 56].

Instruments that accurately measure magnetic fields (including the direction and magnitude of magnetic induction) are usually referred to as magnetometers. Depending on the measurement principle, magnetometers can be divided into fluxgate magnetometers [57], proton precession magnetometers [58], SQUID (Superconducting Quantum Interference Device) magnetometers [59], and atomic magnetometers [60], etc. Among them, the atomic magnetometer is a magnetic field measuring instrument based on the atomic energy level structure and magnetic resonance principle. Similar to Nuclear Magnetic Resonance (NMR) and Electron Spin Resonance (ESR) technologies, atomic magnetometers use the energy level transition characteristics of atoms in an external magnetic field to measure the strength and direction of the magnetic field, with very high accuracy and sensitivity.

7.3.1 Principle of Atomic Magnetometer

As a scientific instrument for high-sensitivity precision measurement of magnetic fields, the atomic magnetometer is realized by polarizing the atomic ensemble through a pump laser, and then detecting the atomic ensemble undergoing the Zeeman splitting effect in the magnetic field through a probe laser, thereby achieving high-sensitivity magnetic field measurement. Specifically, the optical pumping technology is used to achieve the angular momentum transfer between the light field and the atoms, preparing the atomic ensemble to a non-thermal equilibrium state, changing the population distribution of the ground state atoms. Usually, the frequency of the pump laser is chosen to be the resonance frequency of the atoms, which can effectively excite the internal energy levels of the atoms. During the optical pumping process, the photons of the pump laser are absorbed by the atoms, causing the electrons of the atoms to transit from the ground state to the excited state. This energy level transition causes the electron cloud of the atom to change in state, thereby polarizing the atomic ensemble. This polarized state of the atomic ensemble is very sensitive to the external magnetic field, so it can be used to measure very weak magnetic fields.

In the presence of a magnetic field, the energy levels of the atoms undergo Zeeman splitting. This is because the magnetic field exerts a torque on the magnetic moment of the atom, thereby changing the energy level of the atom. This Zeeman splitting causes the energy levels of the atoms to split in the magnetic field, resulting in Zeeman sub-energy levels with different magnetic quantum numbers m.

The spin polarization of atoms in a magnetic field can be approximated as classical Larmor precession, and the Larmor frequency of the atom ω_L is related to the magnetic field B and satisfies the relationship $\frac{d\mathbf{M}}{dt} = \omega_L \times \mathbf{B}$, where $\omega_L = \gamma \mathbf{B}$. γ is the gyromagnetic ratio. For a specific atom in an external magnetic field, the Larmor frequency of the atom is directly proportional to the magnetic induction intensity of the external magnetic field. Furthermore, by using another specific wavelength of linearly polarized light as the probe light incident on the atomic cell, the magnetic field strength to be measured can be inferred by detecting the oscillation frequency of the polarization plane of the linearly polarized light.

In the development history of atomic magnetometers, the choice of light source plays a crucial role. The performance of the light source directly affects the energy level transition of atoms and the absorption process of photons, affecting the measurement accuracy (uncertainty) and sensitivity of the magnetic field. Therefore, researchers have been looking for and developing better light sources. In the early days, people mainly used arc discharge lamps (electrodeless discharge lamp) as the light source for magnetometers. For example, in 1961, Bell and Bloom used an arc discharge lamp (electrodeless discharge lamp) as the light source to make a magnetometer that uses modulated light instead of a radio frequency field to achieve magnetic resonance [61]. Since the frequency of light is usually much higher than the frequency of the radio frequency field, higher accuracy and sensitivity can be theoretically obtained. However, the spectral range of the induction lamp is wide, and the pumping efficiency is low. Since the birth of lasers, lasers have gradually replaced induction lamps and other gas discharge lamps as the main light source for magnetometers due to their excellent monochromaticity, directivity, and coherence. Lasers can provide highly monochromatic light, which means that magnetometers can select specific spectral lines for excitation. In addition, lasers can also improve the coherence of the atomic ensemble, making the measurement results more reliable. Taking the ^{4}He atomic magnetometer as an instance, in 1987, McGregor and others used a 1083 nm narrow linewidth laser as the pump laser for the helium magnetometer, and used a larger atomic cell (2 in. in diameter and 6 in. in length) to significantly improve the sensitivity and accuracy of the magnetometer [62]. In 2000, the American Polatomic Company applied a distributed Bragg reflector laser to the P-2000 type laser-pumped ^{4}He atomic magnetometer. In the frequency band range of 0.03 ~ 1 Hz, the sensitivity reached 0.3 pT/$\sqrt{\text{Hz}}$ [63]. During the same period, the French Atomic Energy Commission's Laboratory for Electronics and Information Technology (CEA-Leti) applied a 1083 nm narrow linewidth fiber laser to a laser-pumped ^{4}He atomic magnetometer, with a scalar measurement sensitivity (NSD) of 1 pT/$\sqrt{\text{Hz}}$/10 ~ 100 Hz [64].

In 1976, the Beijing Geological Instrument Factory developed China's first self-excited lamp-pumped cesium atomic magnetometer [65]. The laser-pumped ^{4}He atomic magnetometer developed by Peking University has a sensitivity of 0.28 pT/$\sqrt{\text{Hz}}$/1 Hz [60]. In 2021, Peking University developed a high-sensitivity potassium atomic magnetometer based on DAVLL frequency stabilization [66]. The research team stabilized the diode laser on the potassium atom D_1 transition line using DAVLL frequency stabilization technology, providing pump and probe light for the potassium atomic magnetometer. After optimizing the experimental parameters, the potassium atomic magnetometer achieved a peak-to-peak magnetic field fluctuation of 1.89 pT under an open-loop operation with a magnetic field strength of 50,000 T (1 T = 10^5 G), and the noise at 1 Hz is about 0.1 pT/$\sqrt{\text{Hz}}$ performance indicator.

7.3.2 Prospects for the Application of Faraday Lasers in Atomic Magnetometers

As a high-sensitivity magnetic field measuring instrument based on the interaction of light and atoms, the quality of the light source directly affects the performance of the atomic magnetometer. On the one hand, the light frequency shift during the interaction of light and atoms directly affects the frequency of the atomic ground state magnetic sublevel, thereby affecting the magnetic field measurement value [66]. The optical frequency shift effect refers to the shift in energy levels of atoms under the influence of an optical field. This occurs due to the imperfect matching between the frequency of the optical field and the resonant frequency of the atoms, resulting in the transfer of photon energy to the atoms and consequently causing a shift in their energy levels. This movement affects the splitting of atomic energy levels and affect the results of magnetic field measurements. On the other hand, the intensity, spectral linewidth, and coherence of the pump light and the probe light also affects the accuracy of magnetic field measurement by affecting parameters such as the amplitude and linewidth of the magnetic resonance curve [66]. In order to improve the sensitivity of the atomic magnetometer, it is usually necessary to select the frequency of the laser to a specific value to maximize the absorption cross-section of the atom. At the same time, in order to reduce the impact of light shift and to improve the sensitivity of the atomic magnetometer, it is necessary to select the frequency and keep it as stable as possible at a certain value. Therefore, the stability of the laser frequency is of great significance to the atomic magnetometer.

The Faraday laser is a new type of diode laser that uses the Faraday effect for frequency selection. With its advantages of narrow linewidth, stable frequency, strong robustness to current and temperature fluctuations of laser diode, and a variety of selectable wavelengths, it is an ideal choice for the light source of an atomic magnetometer. First, the Faraday laser has a narrow linewidth, which allows the Faraday laser to provide highly monochromatic light, thereby improving the accuracy of the measurement. At the same time, the frequency of the Faraday laser is stable, which allows the Faraday laser to maintain a stable value for a long time, thereby improving the stability of the magnetometer measurement. In addition, the Faraday laser has very strong robustness to current and temperature fluctuations of laser diode, which allows the Faraday laser to work stably under various environmental conditions, thereby improving the environmental adaptability of the magnetometer. At the same time, the Faraday laser can provide lasers of various wavelengths, which allows the

Faraday laser to meet different experimental needs and expands the application range of the magnetometer.

The latest developed potassium Faraday laser uses natural potassium-based FADOF as the frequency selection element. Under optimal working parameters, the transmission bandwidth of potassium FADOF is > 2 GHz, the transmittance is > 60%, and it can cover the potassium atom D_1 and D_2 transitions, simultaneously. Besides, through changing the vapor temperature and the magnetic field, the line-shape and the transmittance can be altered. And it has been demonstrated that the vapor temperature of FADOF can be set as a key to achieve wavelength switching between 770 and 766 nm. That is, when the vapor temperature of FADOF is set at 105 °C, the Faraday laser operates at 770 nm. Once changing the vapor temperature to 95 °C, the wavelength switches to 766 nm, naturally. Using the atomic transition as an absolutely frequency reference, the potassium Faraday laser is immune to changes in temperature and current of the laser diode within a certain range, demonstrating excellent performance. By using modulation transfer spectroscopy to lock the potassium atom Faraday laser, an ultra-stable potassium Faraday laser with a frequency stability of 1.57×10^{-12}/s for 770 nm and 8.77×10^{-13}/s for 766 nm measured by optical heterodyne between two identical systems is achieved.

The application of the potassium Faraday laser in atomic magnetometers is progressing in an orderly manner. The experimental system scheme is shown in Fig. 7.16. The entire magnetometer system can be divided into laser source module, probe module, and control module. The 770 nm potassium atom Faraday laser corresponds to the potassium D_1 transition line, serving as the pump light for the atomic magnetometer, and the 766 nm potassium Faraday laser corresponds to the potassium D_2 transition line, serving as the detection light for the atomic magnetometer. First, the power of the two Faraday lasers is stabilized through AOM, and the stabilized laser is injected into the probe part of the magnetometer through a polarization-maintaining fiber. The signal output by the probe is received by the detector through a multimode fiber. When the magnetometer system is operating in open loop, the amplitude of the 770 nm potassium Faraday laser (pump light) is modulated by the signal generator to maximize the amplitude of the magnetic resonance signal. The modulation frequency ω_m is kept constant. When the external magnetic field undergoes a slight change, the corresponding Larmor precession frequency ω_L changes. This change is reflected in the slight fluctuations appearing near the zero point of the dispersive magnetic resonance curve, that is, the Y value of the phase-locked amplifier (the Y value of the phase-locked amplifier refers to the amplitude value of the output signal of the phase-locked amplifier, usually used to represent the phase difference between the measured signal and the reference signal). By calculating the peak-to-peak value and linewidth of the dispersive line shape, the slope at the zero point can be obtained. Then, by dividing the fluctuation of the Y value by the slope, the frequency fluctuation $\Delta\omega = |\omega_1 - \omega_0|$ can be obtained. Furthermore, through calculation and conversion, the fluctuation relationship of the magnetic field can be obtained as $\Delta B = \Delta\omega/\gamma$. When the magnetometer system is in closed-loop operation, the error signal of frequency fluctuation is fed back through PID2, which

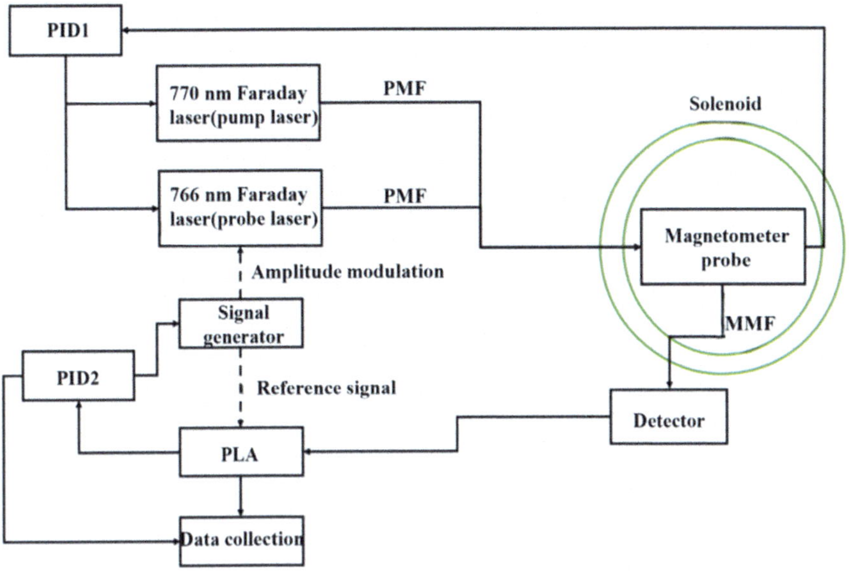

Fig. 7.16 Faraday laser applied to potassium atom magnetometer system scheme

stabilizes the Y value of the phase-locked amplifier at zero. At this time, the modulation frequency of the pump light is locked at the Larmor precession frequency ω_L, which can achieve real-time response to the actual Larmor frequency. When the external magnetic field changes and causes ω_L to change, the amplitude of the change is proportional to the output signal of PID2. Therefore, the fluctuation of the external magnetic field can be detected by measuring the voltage signal output by PID2.

Compared with the diode laser based on DAVLL frequency stabilization, the potassium Faraday laser undoubtedly has superior performance. Firstly, the potassium Faraday laser has the capability to freely switch its wavelength between the potassium D_1 and D_2 lines, which can not only serve as the pump laser for potassium magnetometer, but can also act as the probe laser, greatly simplifying the experimental complexity and cost. Besides, the linewidth of the potassium Faraday laser are also fascinating. In addition, using the MTS signal to stabilize the laser to the ^{39}K $4S_{1/2}F = (1, 2) \rightarrow 4P_{1/2}F' = 2$ and $4S_{1/2}F = (1, 2) \rightarrow 4P_{3/2}F'$ transitions, operating at 770.108 nm and 766.701 nm, respectively, the potassium Faraday laser has excellent frequency stability. Taking the cesium atom Faraday laser as an example, the in-loop frequency stability is $5.8 \times 10^{-15}/\sqrt{\tau}$, which drops to 2×10^{-15}/50 s [66]. The in-loop frequency stability of the Faraday laser is 2.2×10^{-14}/s for 770 nm and 7.0×10^{-15}/s for 766 nm. And the frequency stability measured by optical heterodyne between two identical systems is 1.57×10^{-12}/s for 770 nm and 8.77×10^{-13}/s for 766 nm. This means that the magnetometer based on the Faraday laser has better sensitivity.

7.4 Applications of Faraday Lasers in Optical Communication

Optical communication uses light waves as the information carrier for information transmission, which has the advantages of large transmission capacity, flexible construction methods, strong anti-electromagnetic interference, and strong confidentiality. Because lasers have good monochromaticity, directionality, and coherence, these characteristics ensure the stability and reliability of signal transmission, so they are widely used as the transmission light source in optical communication systems.

7.4.1 Progress in the Study of Lasers as Light Sources in Optical Communication

In 1962, the first generation of diode lasers—GaAs homojunction injection diode lasers were successfully developed in the United States [67]. Subsequently, with the advantages of lower thresholds and better single longitudinal mode characteristics of diode lasers, they were first applied to optical communication systems in 1978. For different application scenarios in the field of optical communication, various types of diode lasers are currently mainly used, such as Fabry–Pérot laser diode (FP-LD), Distributed Feedback Laser LD (DFB-LD), and Vertical Cavity Surface Emitting Laser (VCSEL), etc.

In the field of diode lasers, different types of diode lasers have their unique structures and characteristics. For example, the FP diode laser adopts the FP cavity structure, which can achieve lower threshold current and higher emission power. FP-LD has good divergence characteristics and is suitable for medium and long distance transmission, such as long-distance transmission in optical communication and optical fiber communication networks. The Distributed Feedback (DFB) Laser adopts a distributed feedback structure, which enables it to achieve single-mode output and narrow linewidth. DFB Laser has high frequency stability and low phase noise characteristics, and therefore, performs well in scenarios requiring high precision modulation and long distance transmission, such as high-speed data transmission and fiber optic sensing applications in optical communication systems. The VCSEL is a new type of surface emitting laser, which has the characteristics of simple manufacturing process, high degree of automation, and good emission beam. VCSELs are widely used in short-distance communication and fiber optic sensing fields, such as internal communication in data centers and fiber optic sensing networks. The advent of these new diode lasers provides strong support for the performance improvement and application expansion of optical communication systems. They have promoted the continuous progress and development of optical communication technology, laying a solid foundation for achieving faster, more stable, and more reliable optical communication networks.

This section briefly introduces the research progress of diode lasers as light sources in optical communication, represented by VCSEL lasers. The VCSEL laser was first proposed by Dr. Iga of Japan and others in 1977, and the first VCSEL laser based on GaAs material was developed in 1979 [68]. Due to the short cavity length and high reliability of the VCSEL laser, and its advantages of high modulation rate, it is often used as a light source in the field of optical communication. In order to adapt to the needs of large-capacity communication, VCSEL is gradually developing towards higher modulation speeds and larger modulation bandwidths. For example, in 1997, Lear and others achieved an 850 nm VCSEL, which can achieve a large signal modulation rate of 12 Gbps, with a bit error rate as low as 10^{-13} [69]. In 2008, Finisar Corporation in the United States used non-strained gallium arsenide quantum well technology to achieve a short-wave communication VCSEL with a rate of 30 Gbps at room temperature [70]. In 2015, Kuchta and others from IBM in the United States achieved a non-return-to-zero modulated 850 nm vertical cavity surface emitting laser, with a room temperature error-free data transmission of 71 and 50 Gbps at 90 °C [71]. In 2020, Lavrencik from the Georgia Institute of Technology in the United States achieved an 850 nm VCSEL with a transmission of over 100 Gbps [72].

In recent years, in order to expand the application range of VCSEL in optical communication, such as using VCSEL for long-wavelength communication, including 1310 and 1550 nm, researchers have optimized and innovated in device structure, including improving the refractive index and thermal conductivity of the device, forming effective current limiting structures, etc. These optimization measures aim to improve the performance and stability of long-wavelength VCSELs to meet the needs of longer-distance communication. In 2016, Spiga and others from the Technical University of Munich in Germany realized a single-mode 1.5 μm wavelength VCSEL based on InP material. The cavity length of this laser is 1.5λ, and a data transmission of up to 50 Gbps is achieved through non-return-to-zero modulation and other technologies, with a bit error rate slightly higher than 10^{-12}, showing its excellent dynamic characteristics [73]. In 2019, Blokhin and others from the National Research University Higher School of Economics in Russia studied the dynamic characteristics of a 1.55 μm single-mode VCSEL [74]. By reducing the photon lifetime to increase optical loss, the VCSEL laser achieved a modulation bandwidth increase from 9.2 to 11.5 GHz at 20 °C, and a modulation bandwidth of 8.5 GHz at 85 °C. In 2021, researchers such as Zhang from the Changchun Institute of Optics, Fine Mechanics and Physics of the Chinese Academy of Sciences successfully realized a 1550 nm wavelength single transverse mode VCSEL laser with an output power reaching milliwatt level. At a working temperature of 15 °C, the maximum side mode suppression ratio reached 35 dB [75]. In 2022, the same team from the Changchun Institute of Optics, Fine Mechanics and Physics of the Chinese Academy of Sciences, in collaboration with the Babichev team, used wafer fusion technology to successfully achieve a VCSEL at wavelength of 1550 nm with a modulation rate of 37 Gbps at room temperature [76]. This research achievement marks a further breakthrough in the field of high-speed communication for long-wavelength VCSELs, providing new possibilities for faster and longer-distance optical communication.

7.4.2 Faraday Lasers in Optical Communication Systems

Underwater wireless optical communication (UWOC) is a current research hotspot in the field of wireless optical communication. Compared to underwater acoustic wireless communication, it has the characteristics of high speed and low latency, and its transceiver equipment is smaller and consumes less energy. However, the underwater environment and underwater channels are complex, and there are often interferences such as underwater stray light and other environmental light noise, which can degrade the performance of UWOC. In 2022, a research team from Peking University proposed using Faraday atomic filters to filter out underwater environmental noise interference, and in cooperation with Beijing University of Posts and Telecommunications, developed a UWOC system resistant to broadband background noise [77]. The schematic diagram and experimental setup of this system are shown in Figs. 7.17 and 7.18, respectively.

At the transmitting end, the signal light carrier is generated by an 852 nm Faraday laser, which can reach a maximum output power of 20 dBm. The free-space beam output from the Faraday laser is coupled into a multimode fiber through a lens, and the other end of the multimode fiber is fused with a single-mode fiber. The single-mode fiber is used for subsequent external optical modulation, encoding the information to be transmitted onto the optical carrier. After modulation, the modulated Faraday laser signal light and the unmodulated broadband environmental background noise

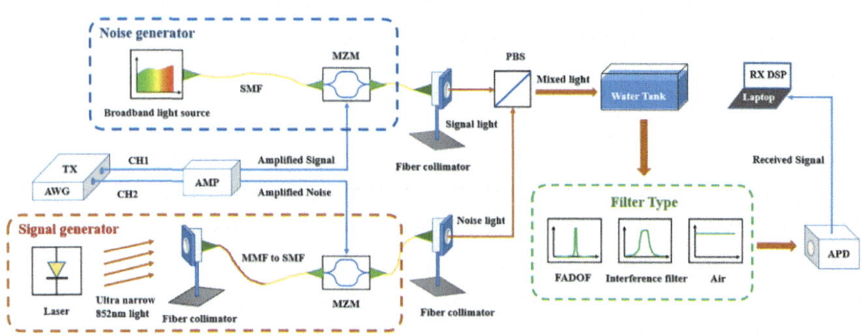

Fig. 7.17 Schematic diagram of the underwater wireless optical communication system structure resistant to background light interference based on FADOF

Fig. 7.18 Experimental setup of the underwater wireless optical communication system based on FADOF

light are combined into the same optical path through a polarization beam splitter and incident together into a 2 m long water tank. In the water tank, the signal light and background light mix and propagate underwater, with an adjustable propagation distance. In the study, it is set to 0.4 m, which simulates the coupling and propagation process of the optical signal and noise.

The electrical signal and simulated electrical noise are generated by a dual-port arbitrary waveform generator. The electrical signal generated by the arbitrary waveform generator is first amplified by an amplifier, and then through an 850 nm Mach–Zehnder modulator with a bandwidth of 2.5 GHz, used to intensity modulate the narrowband optical signal output from the Faraday laser. To simulate broadband background noise, the simulated broadband background noise is first emitted by the arbitrary waveform generator, amplified and then through another Mach–Zehnder modulator of the same type, modulating a broadband light source with a center wavelength of 850 nm and a spectral range of 20 nm, thus loading this broadband background noise onto the broadband light source. In the study, M-ary Quadrature Amplitude Modulation (M-QAM) and Orthogonal Frequency-Division Multiplexing (OFDM) are used as digital signals. The time-domain waveform and Probability Density Function (PDF) of the simulated digital noise are shown in Fig. 7.19, where the PDF follows a Gaussian distribution.

To explore the impact of different filtering schemes on system performance, three filtering schemes were set up at the receiving end in the study: using FADOF filtering, using interference filter filtering, and not using any filtering. Theoretically, FADOF has the strongest filtering capability, while using interference filters, due to their much larger passband than FADOF, their ability to suppress background noise is relatively weak. Without filtering, the signal light and background light directly enter the subsequent receiver circuit, and the interference from the background noise light shows the greatest impact.

Figure 7.20 shows the process of offline digital signal processing. At the transmitting end, a Pseudo-Random Binary Sequence (PRBS) is used as the digital signal data, which is converted into parallel binary streams through Serial to Parallel (S/P) conversion. Next, these parallel binary data are mapped into M-QAM symbols,

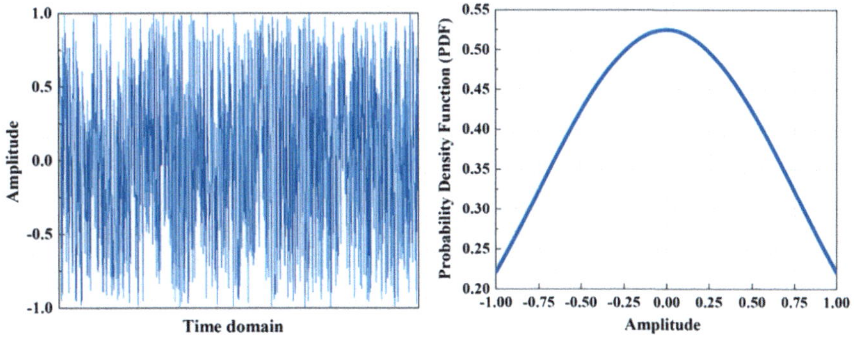

Fig. 7.19 Time-domain simulation of digital noise waveform and its probability density function

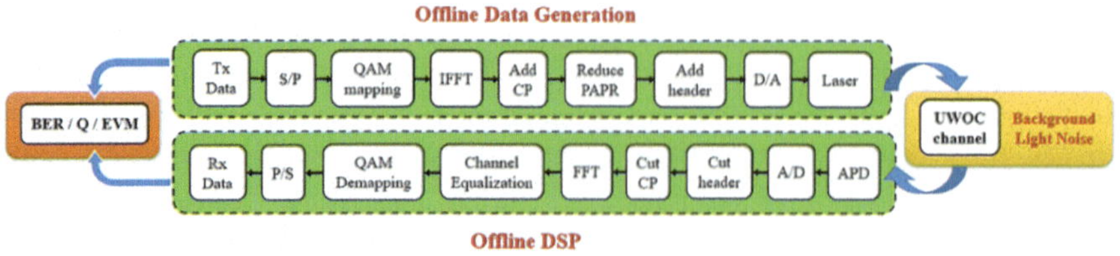

Fig. 7.20 Time domain simulation of digital noise waveform and its probability density function

and by performing a 256-point Inverse Fast Fourier Transform (IFFT) operation on the M-QAM symbol sequence. They can be modulated onto the subcarriers of the OFDM signal, forming a complete OFDM symbol. 300 OFDM symbols form a complete data frame, and a synchronization header composed of 800 data points is added at the beginning of each frame for frame synchronization at the receiving end. At the same time, to alleviate system performance degradation caused by Inter Symbol Interference (ISI), a 16-sample point Cyclic Prefix (CP) is added in front of each OFDM symbol. To obtain a real-value OFDM signal, Hermitian symmetry is applied in the IFFT operation. At the receiving end, the arriving optical signal is optically converted and amplified, and then the oscilloscope samples the electrical signal. After sampling, it goes through frame synchronization, CP removal, Fast Fourier Transform (FFT), channel equalization, QAM demapping, and a series of digital signal processing steps such as parallel to serial (Parallel to Serial, P/S) conversion, finally restoring the original bit sequence. In order to evaluate the transmission quality and reliability of the system, key performance indicators such as bit error rate, Q factor, and error vector magnitude (Error Vector Magnitude, EVM) are calculated.

As shown in Fig. 7.21, the study selected four common M-QAM modulation methods: QPSK, 16QAM, 32QAM, and 64QAM. During the testing process, the signal power at the output end of the polarization beam splitter is fixed at − 10 dBm, while the output power of the broadband light source simulating solar background radiation is continuously changed to simulate different intensities of background noise interference conditions. It can be seen from the figure that after using FADOF filtering, the transmission quality of the system has been significantly improved, especially in the case of strong background noise, the performance improvement is more prominent. When the background noise power is − 13.67 dBm, the Q factor of the FADOF filtering system is 19.2 dB higher than that of the system without FADOF filtering. For 16QAM and 32QAM modulation formats, when the background noise power is − 16.2 dBm and − 16.8 dBm, the Q factor improvements brought by FADOF are 14.8 dB and 12.9 dB respectively. This indicates that the higher the modulation order, the higher the system's requirement for signal-to-noise ratio. Although the gain brought by FADOF has slightly decreased, there is still a significant improvement in the performance.

From the research results, it can be observed that without using any filtering measures, the transmission quality of the system deteriorates as the background

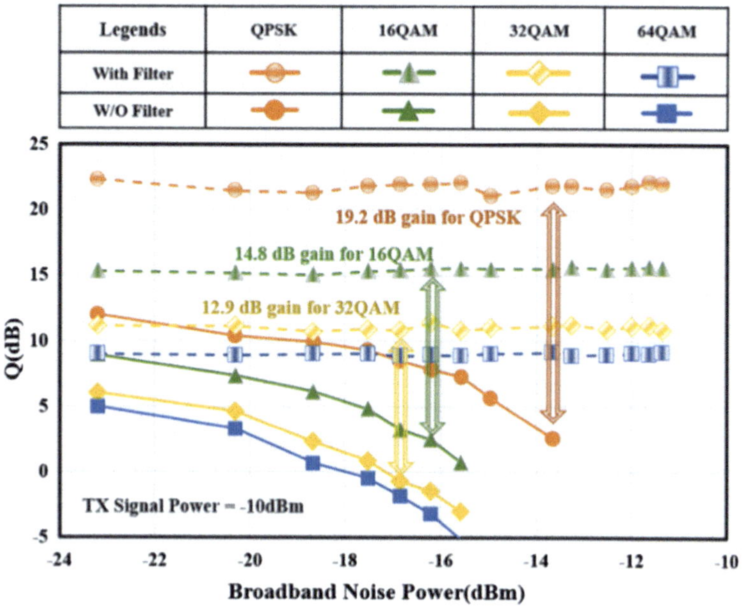

Fig. 7.21 Relationship between Q factor and broadband noise optical power under different filtering conditions and modulation types

noise power increases. When the noise power is greater than − 13.67 dBm, the Q factor without FADOF filtering is unacceptably low and communication is no longer possible. In contrast, when using FADOF for filtering, the degradation of the Q factor is not obvious. For QPSK, 16QAM, and 32QAM modulation formats, the minimum Q factors measured are 21.8, 15.6, and 10.9 dB, all above the usable threshold.

In addition, the study also used a narrow-band interference filter (NBIF) to compare the effects of filtering on communication quality. The NBIF has a bandwidth of 5 nm at a center wavelength of 852 nm, with a peak transmittance of 95%. As shown in Figs. 7.22 and 7.23, experimental and numerical results are obtained for FADOF-based filtering, NBIF-based filtering, and no filtering. The horizontal axis "signal light power/noise light power (dB)" represents the power ratio of the signal light to the background environmental light noise before filtering. "Exp" represents experimental results. "Theo" represents theoretical results. When QPSK is modulated, the optical signal power at the transmitter is − 24 dBm; when 16QAM is modulated, the optical signal power at the transmitter is − 21 dBm.

From Figs. 7.22 and 7.23, it can be seen that for QPSK and 16QAM signals, the experimental results are quite consistent with the numerical results. When the power ratio is high (such as the power ratio of signal light to noise light is greater than − 5 dB), the system performance of FADOF filtering and NBIF filtering is very close, and the difference in Q factor is generally less than 0.4 dB (experimental value) or 0.3 dB (theoretical value). This indicates that in situations where the background

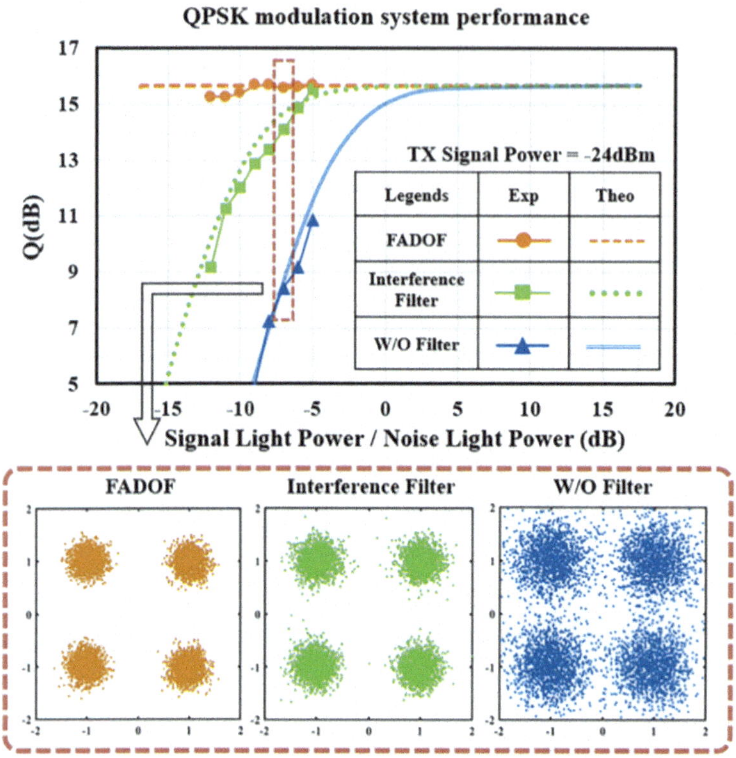

Fig. 7.22 Variation of Q factor performance with the ratio of signal light to noise light power under QPSK modulation format

noise interference is relatively weak, the gain advantage of FADOF filtering is not obvious, and is comparable to the gain effect of traditional interference filtering. However, as the power ratio decreases (such as when the power ratio of the signal light to the noise light is less than − 10 dB), the advantage of FADOF filtering begins to stand out. The Q-factor improvement of QPSK and 16QAM modulation systems can reach 3.37 dB and 3.94 dB, respectively. In addition, compared to the results without using a filter, after using FADOF filtering, the Q-factor of QPSK is increased by 14.67 dB, and the Q-factor of 16QAM is increased by 10.51 dB.

7.5 Application of Faraday Lasers in Laser Wavelength Standards

Metrology is a discipline that studies measurements, units of measurement, and measurement systems. Length is an important part of metrology and is a physical quantity used to describe the extent of an object's extension in a certain direction.

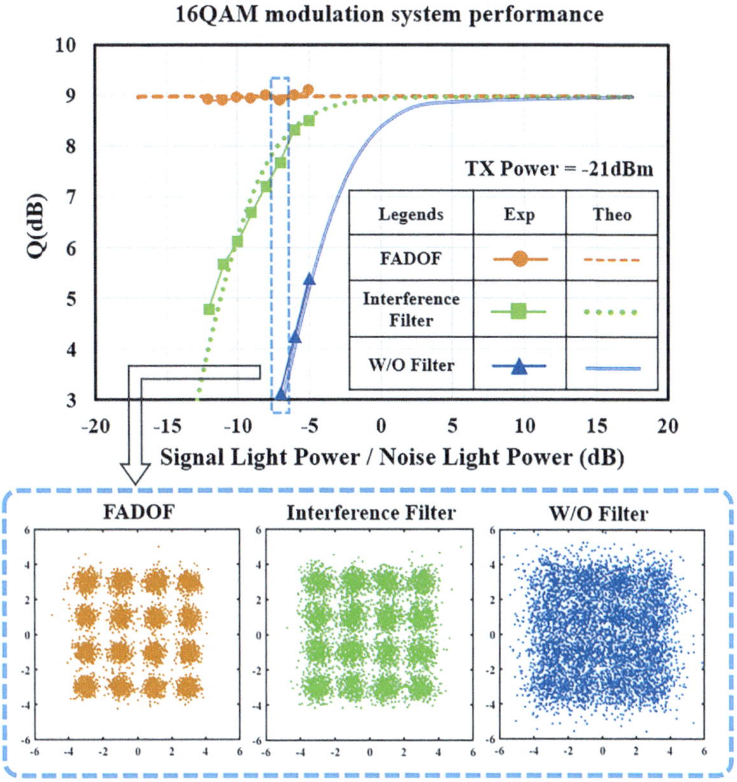

Fig. 7.23 Variation of Q factor performance with the ratio of signal light to noise light power under 16QAM modulation format

Early on, humans used body parts, objects, or natural phenomena as benchmarks for length, such as fingers, steps, sticks, etc., which are relatively simple and intuitive ways of measuring. Simple measuring tools such as rulers and levels appeared in ancient China, Egypt, Greece, and Rome, marking the beginning of systematic exploration of length measurement by humans. With the development of society and the rise of commercial trade, to avoid misunderstandings and confusion caused by differences in length measurement units, people began to establish standard and unified length measurement units. For example, the ancient Egyptian cubit, the Roman pace, and the unified weights and measures of the Qin Dynasty in China. These measures promoted the development of commerce and trade and laid the foundation for the exchange and development of civilization. With the advancement of technology, the technology in the field of metrology has also greatly improved. From traditional measuring tools, such as rulers and calipers, to modern laser rangefinders and other high-precision equipment, the accuracy of measurement continues to improve, providing a reliable basis for scientific research and engineering practice.

Laser wavelength standards use specific wavelengths of laser beams as high-precision length standards. Laser wavelengths have high stability and repeatability and are widely used in precision measurement and calibration. Laser wavelength standards are usually achieved by locking the frequency of the laser to specific quantum transition spectral lines. These spectral lines have very stable wavelengths and can be regarded as a constant length standard. For example, the wavelength of the laser emitted by Iodine molecules stabilized He–Ne laser at 632.8 nm is a common laser wavelength standard. In precision measurement, laser wavelength standards are often linked with interferometric measurement or frequency counting techniques. By comparing the length to be measured with the laser wavelength standard, very high-precision measurements can be achieved.

Laser wavelength standards are widely used in scientific research, engineering technology, geographic measurement, laser communication, and astronomical observation. For example, a laser tracker is a high-precision measuring device that measures the position, distance, and motion of a target by emitting a laser onto the target and receiving the reflected laser, and the laser wavelength standard plays a crucial role in it. High-precision laser wavelength standards can improve the accuracy and reliability of laser trackers, ensuring that they can provide consistent measurement results under various environmental conditions.

7.5.1 Research Progress in Laser Wavelength Standards

Before the advent of lasers, people made several attempts to define the unit of length. In 1875, at an international conference held in Paris, 20 countries including France, the United States, and Russia jointly established the "Metric Convention", defining the "meter" as the basic unit of length based on the platinum-iridium alloy bar meter prototype. With the advancement of science and technology, people realized that using the wavelength of light to define the unit of length might be more accurate. In 1895, the second International Metrology Conference passed a proposal to define the "meter" using the value of the light wavelength. In the following International Metrology Conferences, scientists determined that the light spectral lines are suitable as the basis for the unit of length. On October 14, 1960, at the eleventh International Metrology Conference, after extensive discussion and research, the new definition of the length unit "meter" was officially determined. According to this definition, one meter is equal to 1,650,763.73 times the vacuum wavelength value of the unperturbed transition radiation spectral line of the 2p10–5d5 energy level of the krypton-86 atom [78]. This definition takes advantage of the stability of the krypton-86 atom and the accuracy of the spectral lines, providing a more accurate and reliable standard for length measurement worldwide.

In 1960, the birth of the first laser—the ruby laser, laid the foundation for the new definition of the length unit "meter". However, in the early days, the frequency stability and reproducibility of lasers were not satisfactory, so early laser technology was not widely used. With the continuous development of laser frequency stabilization technology, the linewidth, stability, and reproducibility of laser output have been significantly improved. For example, in 1972, the National Institute of Standards and Technology (NIST) in the United States successfully stabilized the He–Ne laser to the transition spectral line of methane, and through precise measurement, determined the value of the speed of light as c = 299,792,458.7(1.1) m/s, which greatly improved the accuracy of length measurement [79].

With the invention of the optical frequency comb, a direct connection was established between optical frequency and microwave frequency, enabling people to gradually achieve accurate measurement of laser wavelength. The invention of the femtosecond optical comb greatly improved the measurement accuracy of optical frequency, thereby promoting the development of reproducing the length unit "meter" using the frequency values of atomic and molecular spectral lines.

The microscopic references commonly used in current laser wavelength standards include iodine molecules, methane, and acetylene. For laser wavelength standards realized using iodine absorption spectral lines, there are many wavelengths. For example, in 2014, the research team of Zang at the National Institute of Metrology of China conducted research on the miniaturized iodine frequency-stabilized 532 nm laser wavelength standard. The frequency stability of the 532 nm laser wavelength standard they achieved was 2.4×10^{-12}/s, the absolute frequency value was 563,260,223,436 kHz, and the frequency uncertainty was 52 kHz (k = 2) [80]. In 2021, Sanjuan and others at the German Aerospace Center simultaneously used PDH and modulation transfer spectroscopy frequency stabilization technology to lock the laser wavelength on the iodine molecule R(56)32-0's a10 absorption line. The stability of the laser wavelength standard achieved is close to the 10^{-15} level per second, and the hundred-second stability is at the 10^{-14} level [81]. In 2022, Ikeda and others at Yokohama National University in Japan used $^{127}I_2$ to achieve the uncertainty of the absolute frequency of the iodine molecule laser wavelength standard is 5.4 kHz (relative uncertainty is 9.3×10^{-12}) [82]. Currently, the International Committee for Weights and Measures recommends a light communication band calibration reference wavelength based on acetylene frequency stabilization at 1542 nm, which sparked more intense research in this field [83]. In 2017, Yoshii and others from Yokohama National University in Japan achieved an acetylene frequency stability of 3.6×10^{-12} [84] using MTS technique. In 2020, NIST reported a chip-level acetylene wavelength reference standard, integrating acetylene with an optical waveguide chip to construct a Molecular Coating Waveguide (MCWG), which achieved a 1.5 μm laser wavelength lock, with a frequency jitter of less than 400 kHz in 34 s [85].

7.5.2 Faraday Lasers as Laser Wavelength Standards

Long-term stable and accurate laser wavelength standards play an important role in length metrology, optical communication, remote sensing, radar, and information communication networks. The current laser wavelength standards inevitably drift in laser frequency or wavelength under long-term continuous operation, which is obviously difficult to meet the needs of applications that require long-distance and long-term work. On the other hand, the laser reference sources output by current commercial products are almost all single-wavelength and work independently under near-infrared and communication band conditions. There is an urgent need for a full-band, long-term stable and accurate laser wavelength standard.

Currently, standard devices such as atomic clocks all need to use laser sources corresponding to atomic spectral lines, such as the 420, 421, 780, 795, 1529 nm corresponding to the Rb atomic spectral line, or the 455, 459, 852, 894, 1470 nm corresponding to the Cs atomic spectral line. The frequency stability of the laser source is an important indicator of its performance. Traditional methods all require the laser to be locked to the transition wavelength of Rb, Cs atoms, or acetylene molecules through a complex frequency stabilization system. However, these systems inevitably drifts in frequency or wavelength under long-term continuous operation due to small signal amplitude, mechanical structure, and environmental changes. Each wavelength needs to be checked for its accuracy through calibration or periodic measurement, so the role of multi-wavelength laser wavelength standards is very critical.

Faraday lasers use FADOF as frequency selectors, referencing the output laser source frequency to the atomic transition central frequency. This laser reference source has good immunity to fluctuations in external environment, laser operating temperature, and operating current, and can ensure its long-term continuous operation while corresponding to the atomic transition central frequency. It is an important method to achieve long-term stable and accurate laser reference sources corresponding to the transition frequencies of Rb and Cs atoms. In 2017, Chang et al. from Peking University proposed a scheme using Faraday lasers to achieve wavelength standards, including a communication band 1.5 μm laser wavelength standard immune to temperature and current noise by radio frequency excitation of a Rb vapor cell, and a multi-wavelength standard integrating multiple Faraday laser wavelength standards in one device.

1. Laser wavelength standard at 1.5 μm

In recent years, to cope with the growing communication demand and the pursuit of communication speed, the working band of communication has gradually developed towards the 1.5 μm band. This band has lower atmospheric attenuation and better transparency, which is suitable for long-distance, high-speed communication transmission. In the current international communication field, 1.5 μm band lasers mostly use fiber lasers, which have advantages such as narrow linewidth and small temperature drift coefficient. However, it is quite challenging to achieve the absolute frequency reference standard of the 1.5 μm laser wavelength. The structure diagram of the 1.5 μm laser wavelength standard generation device in the communication band is shown in Fig. 7.24a. To achieve this goal, researchers internationally explore methods of locking the laser frequency on the excited state of acetylene molecules or Rb atoms. However, using acetylene molecules as frequency references have problems of small signal amplitude and complex frequency locking process, which brings

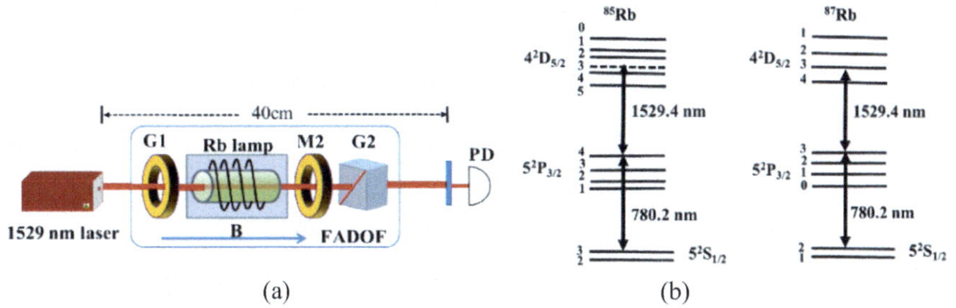

Fig. 7.24 **a** Structure diagram of the 1.5 μm laser wavelength standard generation device in the communication band [87]. **b** Energy level diagram of the ground state 5S to the excited state 5P and 4D of the two isotopes of Rb, ^{85}Rb and ^{87}Rb

certain difficulties to achieve stable frequency locking effect. In contrast, the excited state of Rb atoms (such as the 5P-4D excited state transition of Rb atom at 1529 nm, as shown in Fig. 7.24b) is generally considered a more feasible choice. However, the conventional Rb atom excited state frequency stabilization system internationally requires additional expensive frequency stabilization pump lasers and complex servo loops, making the system costly and bulky.

To solve this problem, in 2017, Peking University proposed a method for generating laser wavelength standards at the 1.5 μm communication band based on a Faraday laser [86]. The idea is to use a radio frequency excited Rb atom excited state filter technology, passing the light beam emitted by the ARLD through a radio frequency excited Rb vapor cell (electrodeless discharge lamp) at a temperature of 120 °C. After mode selection under the action of magnetic field rotation, it is fed back to the ARLD. In the resonant cavity composed of the back end face of the ARLD and the cavity mirror, it oscillates and amplifies to exceed the laser oscillation threshold, making ARLD directly form a laser wavelength standard on the 1529 nm 5P-4D transition wavelength corresponding to the Rb atom filter.

2. Integrated multi-band laser wavelength standard device based on Faraday laser

Currently, there are multiple laser wavelength standards used internationally, with the most common ones being six: 531 nm (diode laser, iodine molecule reference), 532 nm (Nd solid-state laser, iodine molecule reference), 633 nm (helium–neon gas laser, iodine molecule reference), 778 nm (diode laser, rubidium atom reference), 1.5 μm (fiber laser, acetylene molecule reference), and 3.99 μm (helium–neon gas laser, methane molecule reference). These wavelength standards are primarily applied in optical frequency metrology, precision measurement, and optical communications. Typically, each wavelength has an independent standard system. However, if multiple wavelength standards could be generated simultaneously from a single system, it would offer significant convenience and economic benefits. The Faraday laser is precisely advantageous in this regard.

A multi-wavelength laser can generate multiple laser wavelengths in one device. Taking a dual-wavelength laser as an example, when the dual-wavelength laser device operates at two close wavelengths, it can generate a beat signal at microwave frequencies, which can be used as a microwave signal source in electronic signal processing systems. The microwave signal is directly obtained by beating the optical frequency signal, eliminating the need for an optical comb and greatly reducing system complexity. Theoretical analysis shows that the stability of the beat frequency of two adjacent longitudinal modes of the laser can potentially reach the same as the stability of the laser frequency. Based on the correspondence between the laser frequency and the interval of the longitudinal modes, the beat frequency of the two adjacent longitudinal modes is locked on the RF frequency standard through precise phase-locked control technology to control the length of the laser resonator cavity, achieving the purpose of locking the laser frequency.

In 2020, Peking University first proposed the dual-wavelength Faraday laser [88], using it to achieve dual-wavelength standards. Because the FAODF used by dual-wavelength Faraday laser contains two transmission peaks with similar transmittance under specific temperature and magnetic field, it can select two laser modes in the diode laser cavity, ultimately achieving dual-wavelength output of the laser. One of the laser frequencies is stabilized by MTS technique, thereby simultaneously improving the frequency stability of the two wavelengths of the laser, obtaining a highly stable dual-frequency laser wavelength standard.

This new type of wavelength standard has a simple structure, and its frequency stability can be better than 10^{-14}/s. And when this FADOF operates under specific temperature and magnetic field, the transmission spectrum contains two stable transmission peaks with similar transmittance, ensuring that the wavelengths of the dual-wavelength Faraday laser correspond to the dual transmission peaks, and can ensure long-term stable operation. The ARLD has no competition for intra-cavity modes, and its output frequency has good immunity to noise fluctuations caused by external environmental factors, the operating temperature and current of the ARLD, etc. The dual-wavelength laser can work continuously for a long time at the frequencies corresponding to the dual transmission peaks of the FADOF. When the temperature and magnetic field of the vapor cell in the FADOF are changed, the frequencies corresponding to the two transmission peaks of the FADOF can also be changed, thereby changing the two laser modes and achieving the tunability of the output wavelength of the dual-wavelength Faraday laser.

7.5.3 Faraday Lasers Applied to Multi-wavelength Standards for Mixed Atoms

Current commercial laser wavelength standards are mainly single-wavelength, generally used independently in the near-infrared and communication bands. However, in recent years, the demand for long-term stable and accurate laser wavelength standards across all bands has been increasing in the fields of optical communication and atomic clocks. Multi-wavelength output laser reference sources not only expand the frequency coverage range but also improve the system's stability and reliability, providing more flexible and reliable tools for various applications.

In 2023, Peking University proposed an integrated multi-wavelength laser wavelength standard device based on Faraday lasers with mixed atoms [89]. The device integrates various techniques such as atomic filtering technology, static magnetic field regulation, radio frequency excitation, and temperature control, achieving precise control and calibration of multi-wavelength laser wavelengths.

First, Rb and Cs atoms are mixed in an atomic vapor cell. Using the characteristics of different atoms' ground states and excited states corresponding to different wavelengths, filtering conditions are provided for lasers of different wavelengths. The application of a static magnetic field causes changes in the atomic energy level structure. By adjusting the size and direction of the static magnetic field, the frequency of lasers of different wavelengths can be fine-tuned, thereby affecting their interaction with the incident laser and achieving frequency control of lasers of different wavelengths. Secondly, an external radio frequency field can be introduced into the atomic vapor cell. By controlling the frequency and intensity of the radio frequency field, the atomic energy level structure and transition conditions can be further regulated, achieving more refined adjustment of the laser wavelength. At the same time, temperature control is also one of the important factors to ensure the selection effect of lasers of different wavelengths. By controlling the temperature, the motion state and energy level distribution of atoms in the atomic vapor cell can be regulated, thereby affecting the selective transmission of the laser through specific wavelengths. By integrating the above technical means, the device achieves high integration and precise calibration of multi-wavelength laser wavelengths. Finally, by making lasers of different wavelengths resonate with atomic transitions, modes that resonate with atomic transitions are selected, ensuring the stability and accuracy of lasers of different wavelengths, providing reliable laser wavelength standards for applications in fields such as length metrology, optical communication, remote sensing, radar, and information communication networks.

7.5.4 Faraday Lasers Applied to Cold Atom Wavelength Standards

The realization of precise optical atomic clocks has validated the motto of Nobel laureate in Physics, Arthur Schawlow: "Apart from frequency, no other physical quantity needs to be measured." By high-resolution laser spectroscopy measuring atomic energy level transition lines, and stabilizing this spectrum laser frequency, an optical frequency standard (optical clock) can be obtained, achieving time/frequency measurement accuracy of one second error in 30 billion years. The precision that time/frequency measurement can achieve is unparalleled by any other physical quantity, and the future trend of precision measurement is to convert other physical quantities into frequency or time for measurement as much as possible. The International

Committee for Weights and Measures (CIPM) recommends mainly seven types [90] for use as laser wavelength standards: 531 nm diode laser with $^{127}I_2$ as the frequency reference; with $^{127}I_2$ as a frequency reference for 532 nm Nd:YAG laser [91–94]; Rb atoms as a frequency reference for 778/780 nm diode laser [95]; $^{127}I_2$ as a frequency reference for 633 nm He–Ne laser [96]; acetylene as a reference for 1.5 μm laser; and CH_4 as a frequency reference for 3.99 μm He–Ne laser. These wavelength standards have significant application in quantum precision wavelength measurement and mid-infrared communication fields.

However, the above wavelength standard schemes are all based on thermal atoms/molecules. Although these schemes are compact, they are affected by the collision frequency shift and Doppler frequency shift brought by hot atoms, and the uncertainty needs to be further improved. Therefore, it is necessary to use cold atoms as quantum references. Common cold atom schemes include optical lattices, ion traps, MOTs, etc. Although these schemes can achieve temperatures reaching mK, μK or even lower, and generally have shortcomings such as large set up volume, complex system, and relatively low signal-to-noise ratio. Therefore, it is urgent to find new schemes that can overcome the impact of hot atom collision frequency shift, Doppler frequency shift, and solve the problems of limited number of atoms and insufficient compactness of the cold atom system in traditional laser cooling schemes such as optical lattices and ion traps.

To meet the above requirements, Peking University first proposed the application of Faraday lasers to cold atom wavelength standards. Using a thermal atom Faraday optical frequency standard based on high signal-to-noise ratio MTS technique as the system light source, using a long rod-shaped cold Rb atom cluster captured by diffuse reflection laser cooling as a frequency reference, and detecting a high signal-to-noise ratio cold atom transition spectrum. The cold atom transition spectrum is locked in a closed loop, and the remaining error is fed back to correct the drift of the output frequency of the hot atom Faraday optical frequency standard, realizing the suppression effect of the cold atom ensemble on the Doppler frequency shift and collision frequency shift, and the long-term frequency stability of the optical frequency standard continuously decreases according to the trend of $1/\sqrt{\tau}$.

On the system, a home-made diode laser is used as the cooling and repumping laser source, and its output frequency is stabilized on the ^{87}Rb atom hyperfine transition and then the laser power is amplified. The amplified laser is divided into two beams, one as a cooling laser, and the other is frequency modulated by a broadband EOM at 6.58 GHz corresponding to the repumping laser transition, realizing a single laser system directly meeting the preparation requirements of cold atoms, as shown in Fig. 7.25. A diffuse reflective coating with high reflectivity to the cooling laser and repumping laser is applied to the outer surface of the long rod-shaped vacuum glass tube, and the fiber-coupled output cooling laser and repumping laser are shot in from the small holes reserved on the tube wall, and multiple reflections occur inside the tube. Under the radiation pressure of the cooling laser, the ^{87}Rb atom vapor in the tube is cooled, and the entire long rod-shaped vacuum glass tube is filled with ^{87}Rb cold atoms. By optimizing the cooling laser power and detuning, and the repumping laser power and detuning, we expect to achieve an optical density of cold atoms ≥ 2,

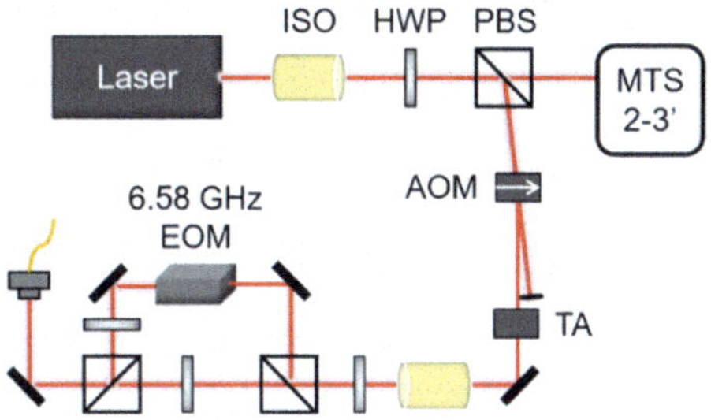

Fig. 7.25 The structure of the cold atom preparation optical path system. *ISO* isolator, *HWP* half-wave plate, *PBS* polarizing beam splitter, *AOM* acousto-optic modulator, *TA* amplifier, *EOM* electro-optic modulator

the number of cold atoms $\geq 2 \times 10^9$, and the temperature of cold atoms $\leq 50\ \mu K$. This overcomes the international issues of limited number of cold atoms and low optical density, providing a reliable guarantee for obtaining cold atom wavelength standards with high signal-to-noise ratio spectra.

Through the above diffuse cooling, cold atoms can be prepared as quantum references. In terms of wavelength standards, a self-developed Faraday optical frequency standard is used as the system light source to construct a modulation transfer spectroscopy system based on a large number of diffuse cooling atom groups. The high stability locking of the Faraday laser based on thermal atoms is first achieved through the modulation transfer spectroscopy. On this basis, the length of the elongated vacuum glass tube is compressed by using an optical path folding scheme, significantly reducing the system volume while ensuring that the laser can fully interact with the cold atoms, as shown in Fig. 7.26.

First, the Faraday laser is locked to thermal atom reference through MTS technique, which serves as the system light source, then the output laser is divided into two paths. One is used as the cold atom laser wavelength standard through the AOM,

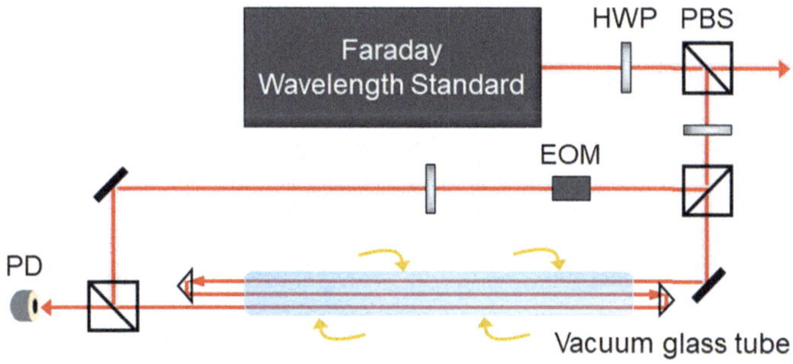

Fig. 7.26 The structure of the high signal-to-noise ratio MTS system based on a large number of cold atoms. *HWP* half-wave plate, *PBS* polarizing beam splitter, *EOM* electro-optic modulator, *PD* photodetector

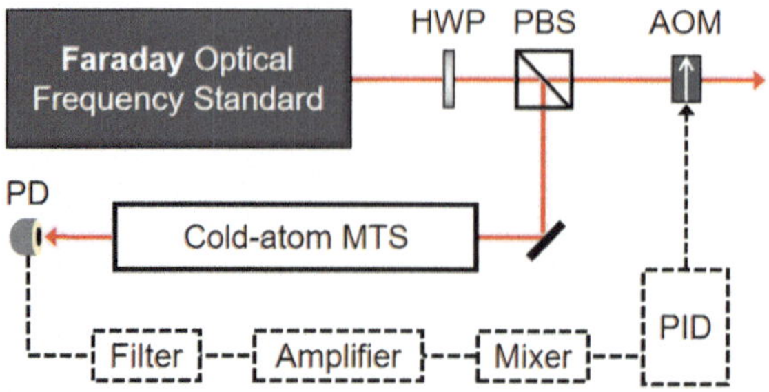

Fig. 7.27 The structure of the Faraday-laser-based cold atom optical frequency standard. *HWP* half-wave plate, *PBS* polarizing beam splitter, *AOM* acousto-optic modulator, *PID* proportional integral derivative feedback circuit, *PD* photodetector

and the other is used for error feedback of the cold atom modulation transfer spectroscopy. The residual error signal of the cold atom modulation transfer spectroscopy after in-loop locking is fed back to the AOM at the user end to calibrate the final output frequency drift of the system (as shown in Fig. 7.27), thereby realizing the suppression effect of the cold atom ensemble on Doppler frequency shift and collision frequency shift, and making the long-term stability of the cold atom laser wavelength standard continuously decline according to the trend of $1/\sqrt{\tau}$. It is expected to solve the international problem of long-term frequency stability deterioration caused by collision frequency shift and Doppler frequency shift currently faced by wavelength standards based on thermal atoms/molecules.

The existing high-accuracy, high-stability cold atom clock devices mostly use PDH ultra-stable optical resonator, magneto-optical trap, and optical lattice to confine atoms, which are essentially laser frequency standards. Taking the PDH ultra-high precision optical resonator as an example, it must rely on ultra-high vacuum maintenance and ultra-high precision temperature control, which undoubtedly makes the system structure complex, expensive, and hard to transport. Compared with these technologies, the advantages of using diffuse reflection laser cooling scheme are firstly that it is all-optical cooling, without the large magnetic field required for magneto-optical trap to capture cold atoms; secondly, diffuse reflection laser cooling is not limited by the geometric shape of the cavity (chamber), with small volume, light weight, low power consumption, and simple structure. Based on this, the realization of a cold atom laser wavelength metrology standard device with original features, high accuracy, and high stability can provide a standard for one of the seven international basic units of length, with more accurate and stable operation capabilities. At the same time, it can also lay a solid technical foundation developing of military standard-length generation, high-speed communication, and cooperative combat.

7.6 Application of Faraday Laser in Ultra-narrow Band Weak Light Coherent Amplification

FADOF [97, 98] has ultra-narrow bandwidth [99], high transmittance, and high signal-to-noise ratio [100, 101], making it widely used as a frequency selector in optical signal processing [102–105] and weak light communication fields, such as free-space optical communication [106] and underwater optical communication [107]. For typical applications such as free-space quantum key distribution [108, 109] and radar long-distance detection systems [110–113], narrow linewidth Faraday detectors can be used to suppress out-of-band noise, thereby reducing the bit error rate and achieving strong background detection. In these systems, the ability of atomic filters to extract weak signals from strong background noise mainly depends on the bandwidth of the filter, and the overall signal transmission efficiency is directly proportional to the transmittance of the Faraday atomic filter. Therefore, to achieve longer communication distances and higher accuracy, traditional Faraday lasers have always been developing towards higher transmittance and narrower bandwidth.

Since the proposal of the Faraday atomic filter, experiments based on different atomic transitions have been conducted, with the obtained transmittance mostly between 40 and 100%, and the equivalent transmission bandwidth around 1 GHz. The relevant transition lines include sodium atom 589 nm (90%, 5 GHz) [114], Rb atom 780 nm (83%, 2.6 GHz) [115], Rb atom 795 nm (70%, 1.2 GHz) [116], Cs atom 459 nm (98%, 1.2 GHz) [117], Cs atom 852 nm (88%, 0.56 GHz) [118], Cs atom 894 nm (77%, 0.96 GHz) [119], strontium atom 461 nm (63%, 1.19 GHz) [120] etc. In 2012, a research group from Peking University achieved a 6.2 MHz ultra-narrowband atomic filter [121], but due to its low transmittance (only 9.7%), its application was limited. To break through the limitation of transmittance, the Faraday atomic filter can be used in series with the Raman optical amplifier [122–124]. This scheme can increase the intensity of the optical signal by 85 times compared to the unamplified situation, thereby expanding the application range of the Faraday atomic filter. However, for extremely weak signals, such an amplification factor cannot meet the requirements, and the noise suppression effect of this filter is limited by the bandwidth of the Faraday filter (0.6 GHz), which is still very limited. However, for extremely weak signals, such amplification coefficients are insufficient. Additionally, the suppression effect of this optical filter on background noise is still quite limited due to the restriction imposed by the bandwidth (0.6 GHz) of the FADOF.

For long-distance optical communication applications facing extremely weak signals, in 2018, Peking University's Duo Pan proposed an ultra-narrowband weak light coherent amplifier based on atomic stimulated amplification and Faraday atomic filtering [125]. It simultaneously achieves coherent amplification of 1470 nm weak light signals [126] and narrowband noise suppression of atomic filters in a single Cs atomic vapor cell. Its amplification principle uses atomic population inversion and coherent stimulated radiation. The gain factor measured in the experiment is greater than 25,000 (44 dB), the gain bandwidth is 13 MHz, and the out-of-band noise suppression ratio is better than 10^5. Compared with traditional atomic filters, this amplifier can more effectively support weak signal extraction under strong backgrounds and has good application prospects in optical communication and weak signal detection.

7.6.1 Scheme of Faraday Atomic Filter Applied to Narrowband Weak Light Coherent Amplification

The experimental setup of this scheme is shown in Fig. 7.28a, where SAS is the saturated absorption spectroscopy system, OC is the optical chopper, AS is the attenuator, and LIA is the lock-in amplifier. M1 is a 459 nm high reflector, M2 is a 459 nm partial reflector, M3 is a 459 nm high reflectance and 1470 nm high transmittance lens, G1 and G2 are a pair of mutually orthogonal Glan–Taylor prisms, and PD is a photodetector. The 459 nm laser is frequency stabilized by SAS technique, and then the Cs atoms are pumped from the ground state $6S_{1/2}$ (F = 4) to the $7P_{1/2}$ (F = 3) state. This transition is a weak transition. After pumping, the atoms undergo spontaneous radiation between multiple intermediate states [127–130], and then form a particle number inversion between the $7S_{1/2}$ (F = 4) and $6P_{3/2}$ (F = 5) states [131]. At this time, under the action of 1470 nm signal light (coinciding with the pump light), the atoms undergo stimulated radiation between two energy levels, and the signal light is coherently amplified. At the same time, a magnetic field is applied to the Cs atom vapor cell, and the vapor cell is placed between a pair of mutually orthogonal Glan–Taylor prisms. When the signal light interacts with the atoms, its polarization direction rotated due to the Faraday effect and passes through the second Glan–Taylor prism. The scattered light that has not rotated is filtered out. The extinction ratio of the Glan–Taylor prism is 10^5, which also determines the out-of-band noise suppression ratio of this system. The optical chopper and lock-in amplifier in the experimental system are used to eliminate the fluorescence background caused by superradiation.

Considering the stimulated radiation amplification process, stimulated radiation can coherently amplify the light signal, and the amplification coefficient is G_{SA}. Considering the action of two orthogonal Glan–Taylor prisms, the gain coefficient of the Faraday atomic filter amplified by stimulated radiation is:

$$G = G_{SA} \times \sin^2 \varphi \tag{7.1}$$

where the rotation angle is represented as:

$$\begin{aligned} \varphi &= \frac{\pi l}{\lambda}(n_+ - n_-) = \frac{\pi l}{2\lambda}\mathrm{Re}(\chi_+ - \chi_-) \\ &= \frac{3N\Gamma\lambda^2 l}{8\pi}\frac{g_F \mu_B B/\hbar}{(g_F \mu_B B/\hbar)^2 + (\Gamma/2)^2} \end{aligned} \tag{7.2}$$

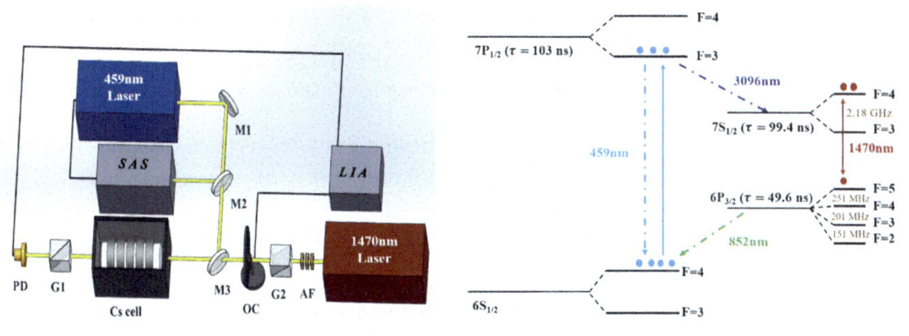

Fig. 7.28 **a** Weak light coherent amplifier experimental setup. *SAS* saturated absorption spectroscopy system, *OC* optical chopper, *AS* attenuation sheet, *LIA* lock-in amplifier, *M1* 459 nm high reflection mirror, *M2* 459 nm partial reflection mirror, *M3* 459 nm high reflection/1470 nm high transmission lens, *G1, G2* a pair of mutually orthogonal Glan–Taylor prisms, *PD* photodetector; **b** Cs atom related energy level

For the relevant energy levels of Cs atoms, the relaxation rate $\Gamma = 55$ MHz, considering the combined relaxation effect of the spontaneous radiation process and the Doppler broadening process brought about by the saturation effect of the pump light [131]. In the formula (7.2), λ represents the wavelength of the light beam, g_F is the Landé g-factor, μ_B is the Bohr magneton, B is the magnetic field size. For $\varphi \leq \pi/2$, when $g_F \mu_B B/\hbar = \Gamma/2$ when $\sin^2 \varphi$ has a maximum value, for $B \approx 7.8$ G. In the experiment, the transmission signal has a maximum value when $B \approx 8$ G, and the maximum transmittance is close to 100% at 135 °C, that is, at this time, the gain coefficient is basically the same without the Glan–Taylor prism. For simplicity, the magnetic field is kept at the optimal value of 8 G in the subsequent calculations and experimental process, and it is assumed at 135 °C that $G = G_{SA}$, for the interaction between the two-level atom and the radiation field, the transition probability can be expressed as $W(t) = |c(t)|^2$, where

$$c(t) = -i\frac{\Omega}{\sqrt{\Omega^2 + \Delta\omega^2}} \sin\left(\frac{\sqrt{\Omega^2 + \Delta\omega^2}}{2} t\right) \exp\left[-i\frac{\Delta\omega}{2} t\right] \tag{7.3}$$

where Ω and $\Delta\omega$ represent the Rabi frequency and the frequency detuning, respectively. Therefore, for the resonant radiation field, the transition probability can be expressed as:

$$W(t) = \sin^2 \frac{\Omega t}{2} \tag{7.4}$$

For an atom with an average life span of, its interaction time distribution function with the radiation field is expressed as: $f(t) = \frac{1}{\tau} e^{-t/\tau}$. Therefore, the transition probability expression can be written as:

$$\langle W \rangle = \int_0^\infty f(t)W(t)dt = \frac{1}{2}\frac{\Omega^2}{\Omega^2 + \Gamma^2} \tag{7.5}$$

Similarly, Γ is the relaxation rate considering spontaneous radiation and Doppler broadening. To match the theoretical calculation with experimental parameters, considering the length of the Cs vapor cell L and the radius of the 1470 nm signal light waist w_0, then the distance the signal light travels in the vapor cell is the cycle time, $\Delta\rho$ is the change in density, w_0 is the initial width, hν is the Planck constant times the frequency, Ω is the Rabi frequency, and Γ is the decay rate. The change in the post-signal power dP is given by:

$$dP = \frac{1}{2}\eta\Delta\rho\pi w_0^2 h\nu \times \frac{\Omega^2}{\Omega^2 + \Gamma^2}dL \tag{7.6}$$

where $\eta = 1/\tau_{cyc}$ is the pump rate, and τ_{cyc} is the cycle time. For the cyclic transition time of atoms. For Cs atoms, we have:

$$\tau_{cyc} = \frac{1}{\Omega} + \frac{1}{\Gamma_{23} + \Gamma_{24}} + \frac{1}{\Gamma_{35} + \Gamma_{36}} + \frac{1}{\Gamma_{51}} \tag{7.7}$$

where Γ_{23}, Γ_{24}, Γ_{35}, Γ_{36}, Γ_{51} correspond to the pumping rates for $7P_{1/2} - 7S_{1/2}$, $7P_{1/2} - 5D_{3/2}$, $7S_{1/2} - 6P_{3/2}$, and $7S_{1/2} - 6P_{1/2}$ transitions. Therefore, the calculated value of η is 3.6×10^6/s. At 135 °C, the Cs atom density determined by the thermal distribution is 1.0×10^{20}, and the population inversion probability is approximately 0.03 [132]. Therefore, the number of atoms between the states $7S_{1/2}$ $(F = 4)$ and $6P_{3/2}$ $(F = 5)$ is 3.0×10^{18}. Due to the Doppler effect, only atoms at the Doppler frequency of 1470 nm signal light width (about 300 kHz) can effectively amplify the signal light. Therefore, the final effective atomic number density $\Delta\rho$ is $3.4 \times 10^{15}/\text{m}^3$.

Therefore, by integrating over the interaction length L between the atoms and the signal light, and utilizing the relation $\Omega^2 = \frac{\Gamma^2}{2} \times \frac{I}{I_s}$ [133], the final result can be obtained as

$$P - P_0 + 2\pi w_0^2 I_s \times \ln\frac{P}{P_0} = \frac{1}{2}\eta\Delta\rho\pi w_0^2 h\nu L \tag{7.8}$$

where $I_s = h\pi c\Gamma/3\lambda^3$, is the saturated light intensity, P_0 is the intensity of incident signal. Since $G = P/P_0$, we can get

$$P_0 \times (G - 1) + 2\pi w_0^2 I_s \times \ln G = \frac{1}{2}\eta\Delta\rho\pi w_0^2 h\nu L \tag{7.9}$$

Through Eqs. (7.7) and (7.8), the changes in transmitted light intensity and gain coefficient with the change in detection light intensity can be obtained, which is demonstrated and analyzed later.

7.6.2 *The Effect of Faraday Atomic Filter Applied to Weak Light Coherent Amplification*

1. Super-radiation background

Due to the multi-atom coherence feature of steady-state super-radiation [134, 135], the atomic ensemble with population inversion between the $7S_{1/2}$ ($F = 4$) state and the $6P_{3/2}$ ($F = 5$) state is faster and stronger than the spontaneous radiation of a single atom, and has a definite directionality. At the same time, the intensity of steady-state super-radiation fluorescence changes with the pump light power and temperature, and this fluorescence overlaps with the signal light in space and frequency, and cannot be filtered out by optical filtering means. Therefore, when the pump light power and system temperature are stable, this fluorescence appears as a background in the detection signal, and the fluorescence background changes with the pump light power and temperature as shown in Fig. 7.29a. This fluorescence background effect can be eliminated by synchronous modulation. In the experiment, an optical chopper is used to pre-modulate the signal light, with a modulation frequency of 1.5 kHz. The detected gain signal is then demodulated by a phase-locked amplifier, with the reference signal of the phase-locked amplifier synchronized with the modulation signal of the optical chopper. In this way, the gain signal amplified by the signal light can be separated from the fluorescence background for separate detection. The transmission signal before and after the signal light modulation is shown in Fig. 7.29b. The signal after demodulation is as shown in the lower half, at which time the fluorescence background is effectively suppressed. This method can significantly improve the signal-to-noise ratio of the system, and can also be used in other atomic filters based on atomic spectral lamps to eliminate the background caused by the non-negligible fluorescence effect.

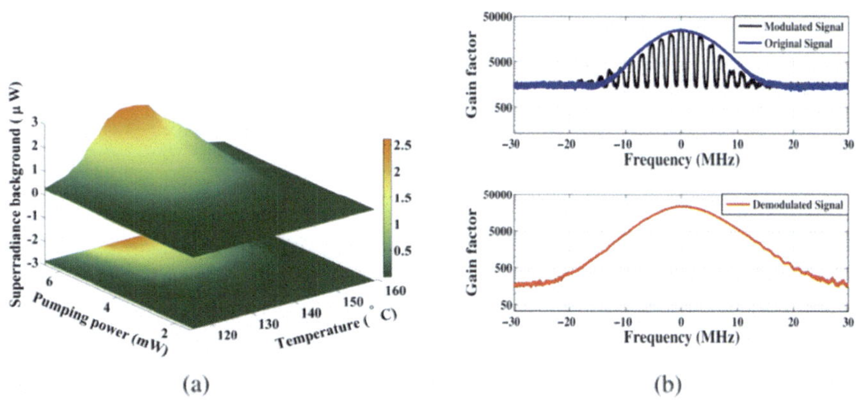

Fig. 7.29 **a** The change in super-radiation background with pump light power and atomic temperature. **b** Top: transmission signal before and after modulation. Bottom: demodulated transmission signal

2. Gain coefficient

In a weak light detection environment, it is particularly important to achieve longer communication distances and higher transmission accuracy. This requires the filter to have a high transmission to suppress loss, or ideally, the filter should have a higher gain.

The purple dashed line in Fig. 7.30a is calculated by formula (7.7) showing the change in transmitted light power with detection light power at resonance, with the temperature of the Cs vapor cell at 135 °C and the 459 nm pump laser power at 3.5 mW. At the same time, the red solid line in the figure is the gain coefficient calculated by formula (7.8) showing its change with the detection light power. In the figure, we can see that as the signal light power increases, the transmitted light power quickly tends to a saturation value, which is due to the limited amplification ability of the atomic system. After the output power tends to the saturation value, the gain coefficient is inversely proportional to the detection light power. In the experiment, the measured output power and gain coefficient are shown as the green dashed line and blue solid line in the figure, respectively. For the case where the detection light power is relatively large, the measured results match well with the theoretical values. However, in the case of extremely weak detection light power, the measured gain coefficient curve shows a significant downward trend compared to the theoretical value, which may be due to the relatively strong super-radiation effect under extremely weak detection light conditions, and the phase-locking effect is not enough to peel the detection photons from it. The gain coefficient has a maximum value when the detection light power is 50 pW, which can reach more than 25,000 (44 dB).

For different detection powers and Cs vapor cell temperatures, the measured gain spectra are shown in Fig. 7.30b, c. Figure 7.30b was measured at a temperature maintained at 135 °C, while the pump laser power used in Fig. 7.30c is 50 pW. At lower temperatures, due to the smaller density of Cs atoms, the gain gradually increases with increasing temperature. At higher temperatures, the enhanced atomic collision effect shortens the coherent time of the $7S_{1/2}$ state, thereby reducing the gain coefficient. The optimal atomic vapor cell temperature is around 135 °C. Similar effects have been reported in active hydrogen lasers [136] and in atomic systems such as Rb and Cs [131, 134, 137].

3. Gain bandwidth

For atomic filters, considering the need to suppress background noise, gain bandwidth is an important indicator. In narrowband coherent amplifiers, since the pump laser is frequency-stabilized by the SAS technique, zero-speed atoms can be selected to participate in the amplification, and the gain bandwidth is close to the natural linewidth of the corresponding transition. In the actual working process, as the pump power increases, the saturation effect of the pump laser causes the atoms pumped to the $7P_{1/2}$ state to have a rate broadening. These atoms relax to the $7S_{1/2}$ and participate in the stimulated radiation process, which ultimately leads to the broadening of the amplifier gain linewidth through the Doppler effect. Figure 7.30d shows the changes

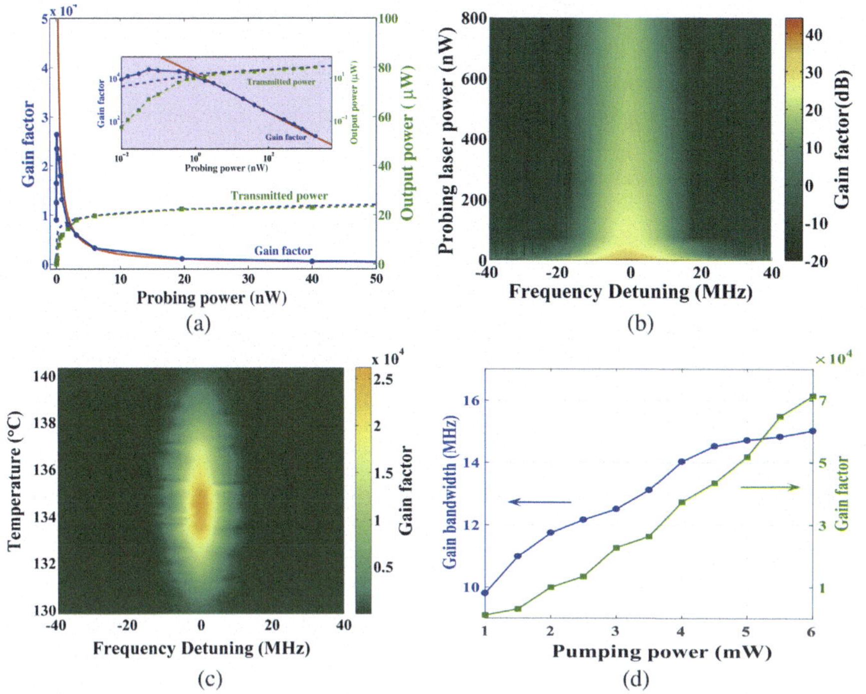

Fig. 7.30 **a** Theoretical and measured curves of transmission power and gain coefficient under different detection powers. **b** Gain spectrum lines under different detection powers. **c** Gain spectrum lines at different temperatures. **d** Changes in gain coefficient and gain bandwidth with pump light power

in gain coefficient and gain bandwidth with pump power at the atomic vapor-cell temperature of 135 °C. It can be seen that as the pump power continues to increase, the gain coefficient increases, and the gain bandwidth also widens. This means that there is an optimal pump light power for different applications, which is determined by the actual situation.

This section describes the research on ultra-narrowband weak light coherent amplifiers by the research group at Peking University. This work has realized a Faraday atomic filter based on stimulated emission, that is, an ultra-narrowband weak light coherent amplifier, based on a Cs atomic vapor, operating at 1470 nm. The maximum gain coefficient that this amplifier can achieve is 25,000 (44 dB), and it has an ultra-narrow bandwidth of 13 MHz, which has potential applications in the field of weak light detection and weak light communication.

In order to eliminate the fluorescence background caused by the super-radiation effect of Cs atoms, this work proposes a synchronous modulation method, which suppresses background noise in the experiment. This method can be further extended to other atomic filters based on discharge lamps to eliminate the fluorescence background of atoms. In the experiment, the characteristics of the gain coefficient and gain bandwidth of the Faraday atomic filter based on stimulated emission under different pump powers, detection powers, and working temperatures are studied. The gain coefficient decreases approximately exponentially with the detection power and increases with the pump power. The gain bandwidth mainly increases with the pump power, so in practical applications, a balance needs to be made between the gain coefficient and the gain bandwidth.

7.6.3 High-Power Coherent Amplification of Faraday Lasers

Coherent amplification in gas lasers usually refers to the amplification of light through energy level transitions of particles (atoms or molecules) in the gas medium. In gas lasers, particles in the gas medium are excited to a high energy level by an external excitation source (such as an electric field or light field). When these particles transit come back to a lower energy level, they emit coherent light, i.e., laser. Coherent amplification is one of the key processes to achieve laser output, ensuring the monochromaticity and coherence of the laser.

Based on the principle of coherent amplification, Peking University first proposed the use of Faraday lasers to achieve high-power coherent amplification. Specifically, a 780 nm laser is used as a pump source to provide energy for the energy level transition of Rb atoms, as shown in Fig. 7.31. In order to generate a population inversion and gain of particles undergoing the D_1 transition ($5P_{1/2} - 5S_{1/2}$) at 795 nm in Rb atom vapor, a 780 nm Faraday laser corresponding to the wavelength of the D_2 line ($5S_{1/2} - 5P_{3/2}$) is used as the pump source to excite Rb atoms to the $5P_{3/2}$ state, as shown in Fig. 7.32. Using a buffer gas to transfer Rb atoms from the $5P_{3/2}$ state to the $5P_{1/2}$ state provides a large gain medium for the coherent amplification of the 795 nm laser.

The front surface of the 795 nm laser diode is coated with an antireflection film, and the back surface forms an optical resonant cavity with the back mirror, forming a separate Faraday laser. The 795 nm laser is coherently amplified through the gain medium. The coherence of the laser comes from the atomic coherence of population inversion stimulated radiation. By precisely controlling the pump process design,

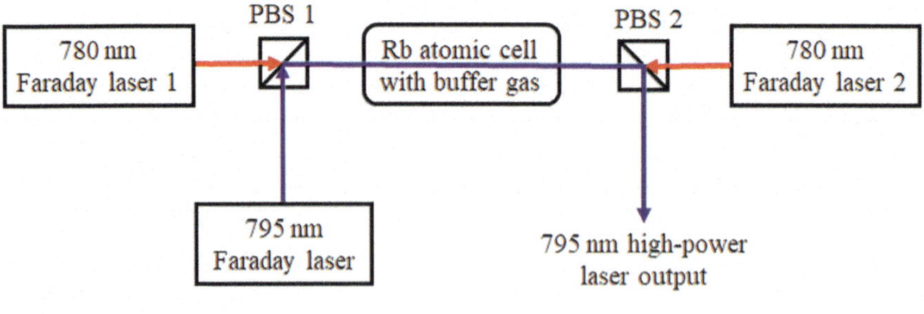

Fig. 7.31 Schematic diagram of the principle of high-power coherent amplification of Faraday lasers

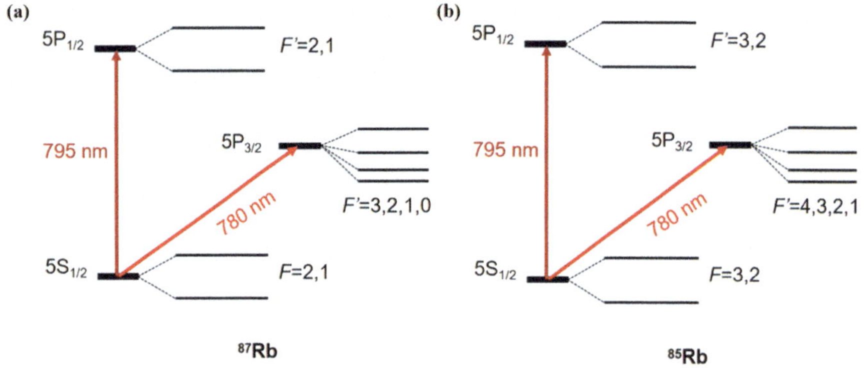

Fig. 7.32 **a** Energy level transition diagram of ^{87}Rb atoms. **b** Energy level transition diagram of ^{85}Rb atoms

high-coherence laser amplification output can be achieved. High-performance high-power laser sources can further be widely used in research, industry, and military fields, such as for spectral analysis, laser cutting, laser radar, etc.

7.7 Application of Faraday Lasers in Raman Spectroscopy Detection

Raman spectroscopy technology [138, 139] is an indispensable technology in the fields of material property research and chemical substance detection. The detection of low Raman frequency shift (GHz–THz) spectral signals can measure lower energy transitions, such as temperature measurement [140, 141], polycrystalline structure detection [142], semiconductor material analysis [143], nanomaterial analysis [144], analysis of macromolecular materials [145, 146] and two-dimensional material analysis [147]. The current commercial equipment for detecting low Raman frequency shift signals is a cascade of multiple monochromators, which can measure down to below 10 wavenumbers [148, 149], but the system is complex and expensive. Experimental research groups have used narrow-band Bragg filters [105] and holographic gratings [150] for low-frequency Raman signal detection studies. Commercial systems based on holographic gratings can measure Raman signals of about 5 wavenumbers. However, there are still difficulties in measuring Raman scattering spectra below 5 wavenumbers, mainly due to the wider linewidth of the pump laser itself and the wider bandwidth of the filter device used to filter the pump laser during Raman signal detection.

By using the absorption or dispersion effect of atoms or molecules near the resonance transition line, specific frequency light signals can be selected. This method provides a new idea for obtaining ultra-low frequency shift Raman spectral signals. In 2014, the research team of Lin from East Carolina University in the United States used the absorption spectrum of a thermal rubidium atomic vapor cell to study the Stokes and anti-Stokes Raman spectra of single living cells and tiny particles, measuring Raman signals below 10 wavenumbers [105]. The experimental setup is shown in Fig. 7.33, the diode laser (TEC-420-0785-1500, Sacher Lasertechnik) emits a laser wavelength of 780.2 nm, corresponding to the Rb $5S_{1/2} \rightarrow 5P_{3/2}$ transition, after a reflective Bragg bandpass filter (BPF-785, OptiGrate Corp, Oviedo, Florida), the background noise is reduced to below − 70 dB. The laser then reflects through a notch filter (BNF-785, OptiGrate Corp, Oviedo, Florida) with a bandwidth of 5 cm^{-1}, enters an inverted microscope with an objective lens (100×, NA 1.30) through a dichroic mirror (DM), forming an optical trap [151]. The particle to be observed is at the focus of the laser, its bright field image is observed through a CCD camera, its Raman scattered light transmits along the original path through the notch filter, then enters a 5 cm^{-1} long rubidium atomic cell through a confocal pinhole, and finally is received by a CCD spectrometer for obtaining the Raman spectrum.

The experimental results are shown in Fig. 7.34. Figure 7.33a is the curve of the transmittance of the Rb cell at temperatures of 25 °C (blue line) and 130 °C (red line) as a function of laser frequency. Figure 7.33b shows the optical density (OD) at different cell temperatures, the frequency corresponds to the arrow shown in Fig. 7.34a. As can be seen from the figure, the absorption bandwidth at nm is 10 GHz, with a maximum optical density (OD) close to 5.0.

Figure 7.34c is the Raman spectrum of the optically trapped 2 μm polystyrene bead, which tested its Raman spectral intensity above the glass surface (curve a) and very close to the glass surface (curve b). Curve c is the background spectrum, mainly from the Rayleigh scattering of the bottom glass surface. As can be seen from the figure, this setup can obtain a Raman spectral signal below 10 cm^{-1}.

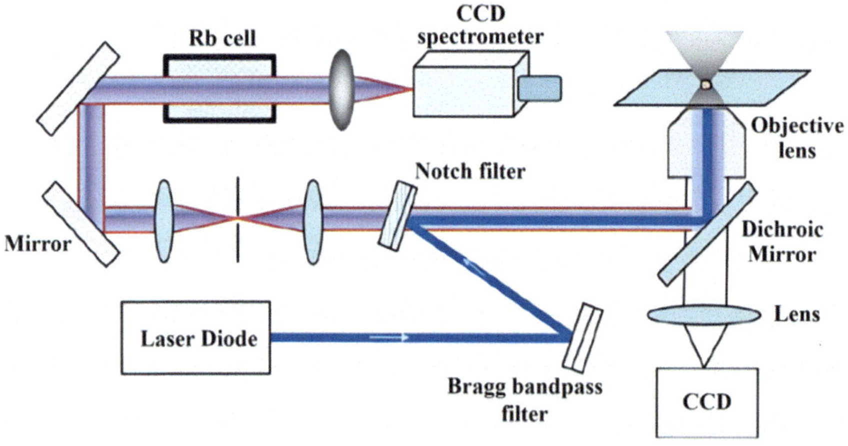

Fig. 7.33 Experimental scheme for ultra-low frequency Stokes and anti-Stokes Raman spectroscopy analysis of single living cells and microparticles using a thermal Rb atomic vapor cell [105]

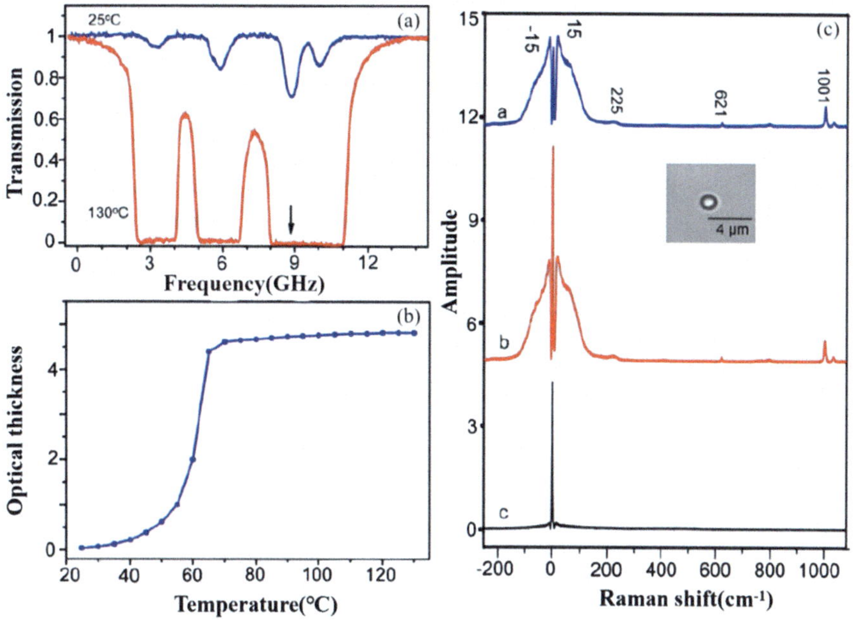

Fig. 7.34 Research results [105]: **a** transmission curve of the Rb cell at 25 °C and 130 °C; **b** optical density of the Rb cell at different temperatures; **c** Raman spectrum of the optically trapped 2 μm polystyrene sphere

Based on the above experiments, in 2016, Peking University and Pennsylvania State University used FADOF as a filter for the pump laser to filter out the broadband scattered background light, and used the atomic absorption bubble as the absorption medium for the pump light, to obtain ultra-low frequency shift Raman spectral signals.

7.7.1 *Faraday Atomic Filter Applied to Raman Spectroscopy Detection Scheme*

The experimental setup of the FADOF applied to the Raman spectroscopy detection scheme is shown in Fig. 7.35 [103]. In the experiment, the atomic vapor cell in the FADOF is a 5 cm natural Rb cell, and the FADOF contains a pair of ring permanent magnets and a pair of Glan–Taylor prisms, with the transmission polarization directions of the two prisms perpendicular to each other. The 780 nm pump laser frequency is adjusted to the maximum transmission peak of the Rb-FADOF. After filtering by the FADOF, it is coupled into a multimode silicon fiber about 2 km-long through a 10× microscope objective, with a core diameter of 50 μm. The Raman signal of the silica material is generated inside the fiber, and the emitted light from

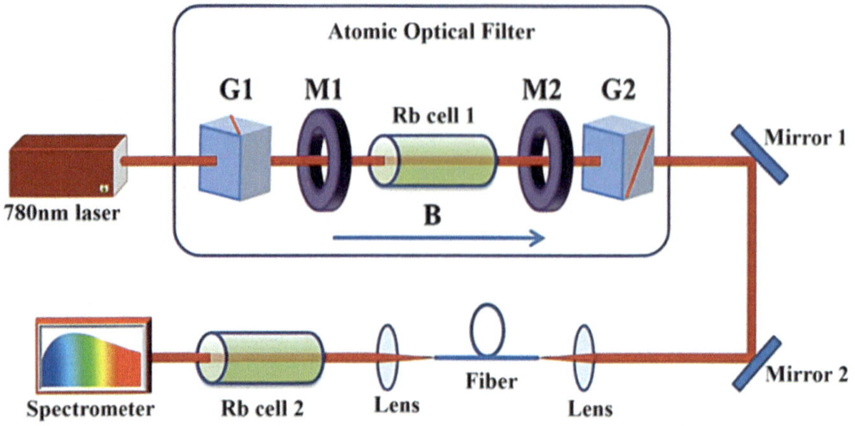

Fig. 7.35 Raman spectroscopy measurement experimental structure

the fiber is collimated by the microscope objective and absorbed by the Rb atoms. The final Raman signal is detected by a spectrometer. This research group measured the Raman scattering signal in the fiber in the experiment, which is down to 0.7 cm^{-1}, achieving Raman spectroscopy detection of less than 1 wavenumber.

In the experiment, a 780 nm laser is used as the pump laser to generate the Raman signal, the FADOF is used to filter the broadband scattered background in the pump laser, and the atomic vapor cell is used to absorb the pump laser. This requires the transmission peak of the FADOF and the absorption peak frequency of the atomic vapor cell to be consistent, and the temperature and magnetic field of the atomic vapor cell to be controlled. Adjust the temperature and magnetic field to adjust the shape of the transmission spectrum, so that the transmission peak corresponding to the atomic transition line frequency is the highest, to minimize the components of other transmission peaks. The FADOF operates at 112 °C. As shown in Fig. 7.36, the bottom curve is the transmission spectrum of the FADOF, the top green curve is the absorption spectrum of the Rb vapor cell at room temperature, and the middle red curve is the absorption spectrum of the Rb vapor cell at 120 °C. From the figure, it can be seen that the highest transmission peak of the FADOF corresponds to the strongest absorption of the $5S_{1/2}$ $(F = 3) \rightarrow 5P_{3/2}$ $(F = 2, 3, 4)$ transition line, so the laser frequency is stabilized at the maximum transmission peak frequency.

In the experiment, the highest peak transmittance of the atomic filter is 42%, which included the loss of all internal components, and the equivalent noise bandwidth (ENBW) of the highest peak is about 2.3 GHz. The power of the incident laser in this system is 6.2 mW, and the laser power after the atomic filter is 2.6 mW. The suppression ratio outside the sideband is 5.5×10^5, which is mainly limited by the suppression ratio of the orthogonal Glan–Taylor prism. Here, the suppression ratio outside the sideband can be improved by using a Glan–Taylor prism with a higher suppressed ratio. In the experiment, by tuning the frequency of the ECDL to the leftmost transmission peak of the atomic filter, it corresponds to the position where

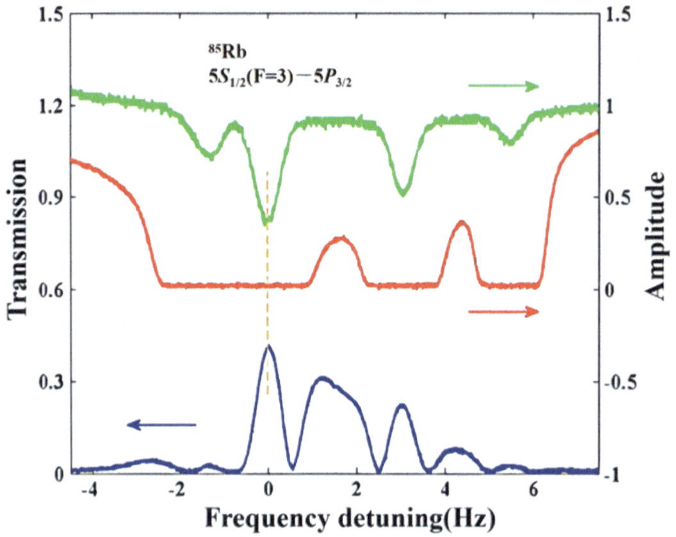

Fig. 7.36 Atomic filter transmitted spectrum and absorption spectrum

the frequency detuning is zero in Fig. 7.36. Therefore, the atomic filter can be used as an ultra-narrow laser filter to eliminate the spontaneous radiation background in the laser cavity.

The structure of the Raman spectroscopy measurement experiment is shown in Fig. 7.35. The laser used is an ECDL, G1 and G2 are two Glan–Taylor prisms that transmit light in perpendicular polarization directions, M1 and M2 are ring permanent magnets, both vapor cell 1 and 2 are filled with natural Rb atoms, the fiber length is about 2 km, and the core diameter is 50 μm. In order to detect low-frequency Raman signals, an additional atomic cell is added in the experiment as a notch filter to eliminate the residual pump laser background. The green curve in the upper part of Fig. 7.36 shows the absorption spectrum of the vapor cell at room temperature. The Rb cell contains natural Rb atoms, of which 27.8% are ^{85}Rb atoms and 72.2% are ^{85}Rb atoms. The outer two peaks correspond to ^{87}Rb atoms, and the inner two peaks correspond to the transitions of ^{85}Rb atoms. As shown in Fig. 7.36, at the frequency detuning of 1.8 and 4.5 GHz, the atomic filter and the atomic notch filter show obvious transmission, which limits the suppression effect of the spontaneous radiation background near these bands. If a pure ^{85}Rb atom or ^{85}Rb atom vapor cell is used, and the bias magnetic field and temperature of the vapor cell are further optimized to improve the response of the filter, suppress the spontaneous radiation background. When the temperature is low, due to the low density of atoms and saturation effects, the number of atoms at higher energy levels is more, making the absorption of the atomic notch filter limited. However, as the temperature increases, the atomic density significantly increases, the absorption is enhanced, so the absorption can be increased by heating the atomic vapor cell. The

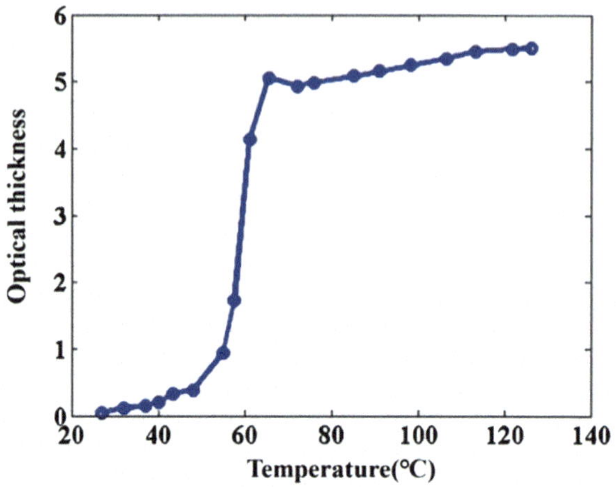

Fig. 7.37 The change in optical density (OD) of Rb atom vapor cells with temperature

red curve in the middle of Fig. 7.36 is the absorption spectrum at 112 °C. At an incident power of 0.5 mW and a temperature of 112 °C, the optical density is 5.5. The relationship between the optical thickness of the atomic vapor cell and temperature is shown in Fig. 7.37.

7.7.2 The Effect of Faraday Atomic Filter Applied to Raman Spectroscopy Detection

Figure 7.37 shows the experimental measurement of the optical thickness with temperature change curve. At temperature below 50 °C, as the temperature rises, the optical thickness slowly increases, the optical thickness rapidly increases with the rise in temperature between 50 and 70 °C. When the temperature is large than 70 °C, the growth slows down again and tends to stabilize. The particle density in the atomic vapor cell increases exponentially with the rise in temperature. Due to the strong pump laser, which is far greater than the saturation intensity, the saturation effect is significant at lower temperatures, and the number of particles in the excited state of the atom forms a population, so the absorption of the pump light is limited. As the temperature rises, the collision effect between atoms strengthens, and the atoms in the excited state can quickly return to the ground state through the collision effect, weakening the saturation effect of the pump light, so the absorption of the pump laser rapidly increases. At higher temperatures, due to the limitation of the residual spontaneous emission fluorescence of the laser, the optical thickness no longer increases.

To verify the effectiveness of the atomic filter and Rb vapor cells, a spectrometer can be used to measure the transmission spectrum of a single vapor cell and the cascade atomic filters. When there is no atomic filter in the optical path, the pump laser passes through the Rb cell and is detected by the spectrometer, obtaining a

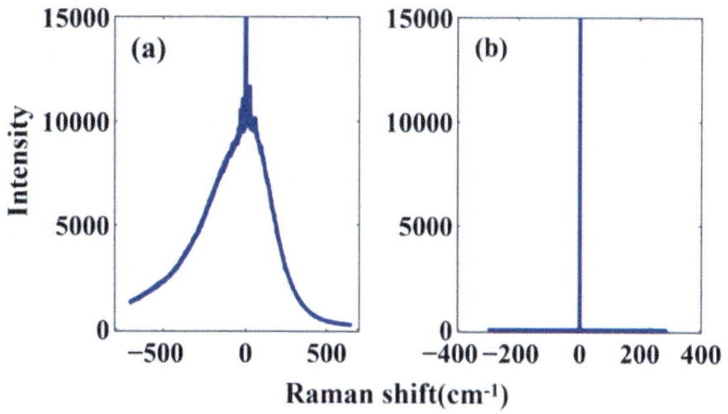

Fig. 7.38 **a** Fluorescence background without Faraday atomic filter; **b** fluorescence background with Faraday atomic filter

spectrum as shown in Fig. 7.38a. It can be seen that the 780 nm laser intensity is highly absorbed, the laser fluorescence background is very obvious, and the spectral coverage exceeds 1000 cm^{-1} (approximately 57 nm). When the atomic filter is placed in the optical path, the laser passes through the Faraday atomic filter and the vapor cell in sequence, and the spectrum obtained by the spectrometer detection is shown in Fig. 7.38b. The figure only has 780 nm pump laser components, and the laser fluorescence background is completely filtered out by the Faraday atomic filter. A multimode fiber of about 2 km in length is placed in the optical path, with a fiber core diameter of 50 μm, to measure the Raman scattering signal in the fiber. Using a 10× objective lens, the spatial light is coupled into the fiber, and another objective lens is used to collimate the output light of the fiber. The laser power of the incident objective lens is 2.6 mW, and the light power of the output objective lens is 0.5 mW. Considering that each objective lens has a power loss of 30%, the coupling efficiency is 39%.

Using a single-stage spectrometer (Spectro Pro 2500i) to measure the Raman spectrum, select the high-resolution grating of 1200/mm in the spectrometer, and obtain the spectrum as shown in Fig. 7.39. The zero Raman frequency shift is the 780 nm pump laser, the positive frequency shift part is the Stokes line, and the negative frequency shift part is the anti-Stokes line. The middle pump laser part is magnified, and the spectrum is shown in the inset of Fig. 7.39. The spectral half-width is 1.4 cm^{-1}, so it can be considered that the minimum measurable Raman frequency shift is 0.7 cm^{-1}, while most of the previous research groups could only measure the spectral components of large Raman frequency shifts (tens to 100 cm^{-1} and above). The experimental results show that this set of devices can obtain Raman spectral signals below 1 cm^{-1}. From Fig. 7.39, it can also be seen that a strong signal peak is present near the Raman frequency shift, and the signal weakens near zero frequency shift. To further verify the effectiveness of FADOF, we removed the FADOF from the optical path in the presence of optical fibers, and the spectral measurements obtained are presented in Fig. 7.40. It can be seen that the fluorescence background of the laser drowned out the Raman light signal.

Using the atomic filter to measure the low-frequency Raman spectrum, this study has proven that the Faraday atomic filter can be used to measure Raman frequency shifts as low as 5 cm^{-1}, with a linewidth of several mHz, and can further carry out

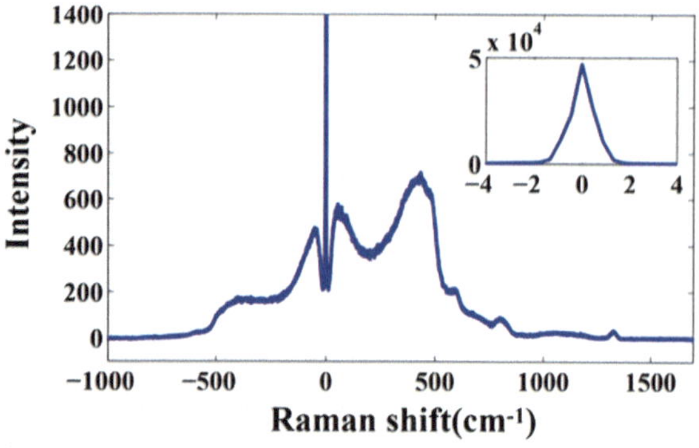

Fig. 7.39 Raman scattering spectrum in multimode fiber

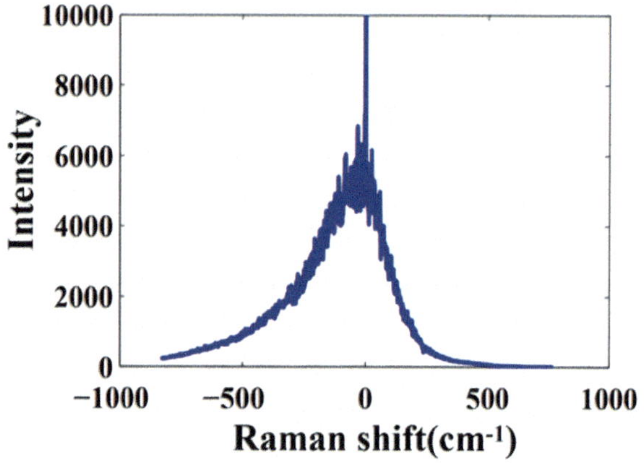

Fig. 7.40 Spectrum after multimode fiber and atomic vapor cell without adding atomic filter

cm^{-1} level Raman measurements in the future. This work provides a new scheme for detecting low-frequency Raman and Brillouin scattering.

References

1. R. Schmeissner, P. Favard, A. Douahi et al., Optically pumped cs space clock development, in *2017 Joint Conference of the European Frequency and Time Forum and IEEE International Frequency Control Symposium (EFTF/IFCS)* (2017), pp. 136–137
2. F.K.P. Berthoud, M. Haldimann, V. Dolgovskiy, High performance industrial cesium beam clock, in *Proceedings of the 4th IFSA Frequency & Time Conference (IFTC 2022)* (2023), pp. 5–6
3. H. Shi, X. Qin, H. Chen, Y. Yan, Z. Lu, Z. Wang, Z. Liu, X. Guan, Q. Wei, T. Shi, J. Chen, Faraday laser pumped cesium beam clock. arXiv (2024)
4. S. Kraemer, J. Moens, M. Athanasakis-Kaklamanakis, Observation of the radiative decay of the ^{229}Th nuclear clock isomer. Nature **617**, 706–710 (2023)
5. C. Zhang, T. Ooi, J.S. Higgins, J.F. Doyle, L. von der Wense, K. Beeks, A. Leitner, G.A. Kazakov, P. Li, P.G. Thirolf, T. Schumm, J. Ye, Frequency ratio of the ^{229m}Th nuclear isomeric transition and the ^{87}Sr atomic clock. Nature **633**, 63–70 (2024)

6. F. Riehle, P. Gill, F. Arias et al., The CIPM list of recommended frequency standard values: guidelines and procedures. Metrologia **55**, 188 (2018)
7. N. Dimarcq, M. Gertsvolf, G. Mileti et al., Roadmap towards the redefinition of the second. Metrologiae **61**, 012001 (2024)
8. T. Shi, Q. Wei, X. Qin, Z. Liu, K. Chen, S. Cao, H. Shi, Z. Liu, J. Chen, Dual-frequency optical-microwave atomic clocks based on cesium atoms. Photonics Res. **12**(9), 1972–1980 (2024)
9. S.L. Campbell, R.B. Hutson, G.E. Marti et al., A fermi-degenerate three-dimensional optical lattice clock. Science **358**, 90–94 (2017)
10. W.F. McGrew, X. Zhang, R.J. Fasano et al., Atomic clock performance enabling geodesy below the centimetre level. Nature **564**, 87–90 (2018)
11. Q. Shen, J.Y. Guan, J.G. Ren et al., Free-space dissemination of time and frequency with 10^{-19} instability over 113 km. Nature **610**, 661–666 (2022)
12. J. Zhang, T. Shi, J. Miao et al., The development of active optical clock. AAPPS Bull. **33**, 1–17 (2023)
13. M. Viteau, M. Reveillard, L. Kime et al., Ion microscopy based on laser-cooled cesium atoms. Ultramicroscopy **164**, 70–77 (2016)
14. B. Bloom, T. Nicholson, J. Williams et al., An optical lattice clock with accuracy and stability at the 10^{-18} level. Nature **506**, 71–75 (2014)
15. B. Young, F. Cruz, W. Itano et al., Visible lasers with subhertz linewidths. Phys. Rev. Lett. **82**(19), 3799 (1999)
16. J. Chen, Active optical clock. Chin. Sci. Bull. **54**(3), 348–352 (2009)
17. K. Debnath, Y. Zhang, K. Mølmer, Lasing in the superradiant crossover regime. Phys. Rev. A **98**(6), 063837 (2018)
18. T. Shi, D. Pan, J. Chen, Realization of phase locking in good-bad-cavity active optical clock. Opt. Express **27**(16), 22040–22052 (2019)
19. J.Y. Wan, X. Wang, X. Zhang et al., Quasi-one-dimensional diffuse laser cooling of atoms. Phys. Rev. A **105**, 033110 (2022)
20. Z.C. Xu, D. Pan, W. Zhuang et al., Experimental scheme of 633 nm and 1359 nm good-bad cavity dual-wavelength active optical frequency standard. Chin. Phys. Lett. **32**, 083201 (2015)
21. I. Marson, J. Faller, G-the acceleration of gravity: its measurement and its importance. J. Phys. E Sci. Instrum. **19**, 22–32 (1986)
22. Z.K. Qiao et al., Research on aeromagnetic compensation of a multi-rotor UAV based on robust principal component analysis. J. Appl. Geophys. **206**, 104791 (2022)
23. J.E. Faller, Thirty years of progress in absolute gravimetry: a scientific capability implemented by technological advances. Metrologia **39**(5), 425 (2002)
24. R.L. Steiner, E.R. Williams, D.B. Newell et al., Towards an electronic kilogram: an improved measurement of the Planck constant and electron mass. Metrologia **42**(5), 431 (2005)
25. N. De Courtenay, O. Darrigol, O. Schlaudt, *The Reform of the International System of Units (SI)* (Routledge, London, 2019)
26. G. Budetta, D. Carbone, Potential application of the Scintrex CG-3M gravimeter for monitoring volcanic activity: results of field trials on Mt. Etna, Sicily. J. Volcanol. Geotherm. Res. **76**, 199–214 (1997)
27. *Superconductor gravimeter igrav* (2019). http://www.gwrinstruments.com/igrav-gravity-sensors.html
28. M.A. Zumberge, R.L. Rinker, J.E. Faller, A portable apparatus for absolute measurements of the earth's gravity. Metrologia **18**(3), 145 (1982)
29. F. Klopping, *FG5-X Absolute Gravity Meters*. http://microglacoste.com/product/fg5-x-absolute-gravimeter/. Accessed 24 July 2020
30. M. Kasevich, S. Chu, Atomic interferometry using stimulated Raman transitions. Phys. Rev. Lett. **67**(2), 181 (1991)
31. M. Kasevich, S. Chu, Measurement of the gravitational acceleration of an atom with a light-pulse atom interferometer. Appl. Phys. B **54**, 321–332 (1992)

32. A. Peters, K.Y. Chung, S. Chu, Measurement of gravitational acceleration by dropping atoms. Nature **400**(6747), 849–852 (1999)
33. P. Cheinet, F.P. Dos Santos, A. Clairon et al., Cold atom absolute gravimeter. J. Phys. IV **119**, 153–154 (2004)
34. J. Le Gouët, T.E. Mehlstäubler, J. Kim et al., Limits to the sensitivity of a low noise compact atomic gravimeter. Appl. Phys. B **92**, 133–144 (2008)
35. Y. Bidel, O. Carraz, R. Charrière et al., Compact cold atom gravimeter for field applications. Appl. Phys. Lett. **102**(14), 144107 (2013)
36. Y. Bidel, N. Zahzam, C. Blanchard et al., Absolute marine gravimetry with matter-wave interferometry. Nat. Commun. **9**(1), 627 (2018)
37. Y. Bidel, N. Zahzam, A. Bresson et al., Absolute airborne gravimetry with a cold atom sensor. J. Geodesy **94**, 1–9 (2020)
38. X. Wu, Z. Pagel, B.S. Malek et al., Gravity surveys using a mobile atom interferometer. Sci. Adv. **5**(9), eaax0800 (2019)
39. L. Zhou, Z.Y. Xiong, W. Yang et al., Measurement of local gravity via a cold atom interferometer. Chin. Phys. Lett. **28**, 013701 (2011)
40. Z.K. Hu, B.L. Sun, X.C. Duan et al., Demonstration of an ultrahigh-sensitivity atom-interferometry absolute gravimeter. Phys. Rev. A **88**, 043610 (2013)
41. B. Wu, Z.Y. Wang, B. Cheng et al., The investigation of a μGal-level cold atom gravimeter for field applications. Metrologia **51**, 452–458 (2014)
42. Z. Fu, B. Wu, B. Cheng et al., A new type of compact gravimeter for long-term absolute gravity monitoring. Metrologia **56**(2), 025001 (2019)
43. S.K. Wang, Shift evaluation of the atomic gravimeter NIM-AGRb-1 and its comparison with FG5X. Metrologia **55**(3), 360 (2018)
44. S. Merlet, L. Volodimer, M. Lours et al., A simple laser system for atom interferometry. Appl. Phys. B **117**, 749–754 (2014)
45. R. Caldani et al., A prototype industrial laser system for cold atom inertial sensing in space. Eur. Phys. J. D **73**, 1–9 (2019)
46. O. Carraz et al., Compact and robust laser system for onboard atom interferometry. Appl. Phys. B **97**, 405–411 (2019)
47. T. Lévèque et al., A laser setup for rubidium cooling dedicated to space applications. Appl. Phys. B **116**, 997–1004 (2014)
48. F. Theron et al., Narrow linewidth single laser source system for onboard atom interferometry. Appl. Phys. B **118**, 1–5 (2015)
49. L.X. Zhu et al., Application of optical single-sideband laser in Raman atom interferometry. Opt. Express **26**(6), 6542–6553 (2018)
50. Q. Luo et al., Eliminating the phase shifts arising from additional sidebands in an atom gravimeter with a phase-modulated Raman laser. Opt. Lett. **47**(1), 114–117 (2022)
51. G.C. Wang et al., Robust single-sideband-modulated Raman light generation for atom interferometry by FBG-based optical rectangular filtration. Opt. Express **30**(16), 28658–28667 (2022)
52. Y. Cong, Y. Zhao, W. Zhuang et al., Optical single-sideband (OSSB) Raman laser based on atomic filter for atom gravimeters. Opt. Lett. **49**(10), 2745–2748 (2023)
53. S. Xu, C.W. Crawford, S. Rochester et al., Submillimeter-resolution magnetic resonance imaging at the earth's magnetic field with an atomic magnetometer. Phys. Rev. A **78**(1), 013404 (2008)
54. C. Bonavolonta, M. Valentino, G. Peluso et al., Nondestructive evaluation of advanced composite materials for aerospace application using HTS SQUIDs. IEEE Trans. Appl. Supercond.Supercond. **17**(2), 772–775 (2007)
55. J. Tang, Y. Gui, N. Jiang, Review of submarine demagnetization systems. Ship Electr. Technol. **06**, 1–3 (2005)
56. W. Shi, *Development of a Six-Sensor Vehicle-Mounted Magnetic Field Measurement System* (Jilin University, 2020)

57. G.E. Moore Jr., P.A. Turner, K.L. Tai, Current density limitations in permalloy magnetic detectors, in *AIP Conference Proceedings* (1973), pp. 217–221
58. Z. Chen, Proton precession magnetometer. Seismol. Res. **4**, 126–144 (1982)
59. E. Trabaldo, C. Pfeiffer, E. Andersson et al., Grooved Dayem nanobridges as building blocks of high-performance $YBa_2Cu_3O_{7-\delta}$ SQUID magnetometers. Nano Lett. **19**(3), 1902–1907 (2019)
60. X. Peng, H. Guo, Optically pumped atomic magnetometer technology. Navig. Control **21**(Z2), 101–121+198 (2022)
61. W.E. Bell, A.L. Bloom, Optically driven spin precession method and apparatus. US Patent 3,173,082, 1965
62. D. McGregor, High-sensitivity helium resonance magnetometers. Rev. Sci. Instrum. **58**(6), 1067–1076 (1987)
63. R.E. Slocum, G. Kuhlman, L. Ryan et al., Polatomic advances in magnetic detection, in *OCEANS MTS/IEEE*, vol. 2 (2002), pp. 945–951
64. I. Fratter, J.M. Léger, F. Bertrand et al., Swarm absolute scalar magnetometers first in-orbit results. Acta Astronaut. Astronaut. **121**, 76–87 (2016)
65. H. Ding, S. Liu, Progress in weak magnetic measurement research in China. Chin. J. Geophys. **40**(S1), 238–248 (1997)
66. C. Zheng, *Potassium Atom DAVLL Frequency Stabilization and Its Application in High Sensitivity Atomic Magnetometers* (Peking University, 2021)
67. R.N. Hall, G.E. Fenner, J.D. Kingsley et al., Coherent light emission from GaAs junctions. Phys. Rev. Lett. **9**(9), 366 (1962)
68. H. Soda, K. Iga, C. Kitahara et al., GaInAsP/InP surface emitting injection lasers. Jpn. J. Appl. Phys.. J. Appl. Phys. **18**(12), 2329–2330 (1979)
69. V.M. Hietala, K.L. Lear, M.G. Armendariz et al., Electrical characterization and application of very high speed vertical cavity surface emitting lasers (VCSELs), in *1997 IEEE MTT-S International Microwave Symposium Digest*, vol. 1 (IEEE, 1997), pp. 355–358
70. R.H. Johnson, D.K. Serkland, 17 G directly modulated datacom VCSELs, in *2008 Conference on Lasers and Electro-Optics and 2008 Conference on Quantum Electronics and Laser Science*, San Jose, CA, 4–9 Mar 2008 (IEEE, Piscataway, NJ, 2008)
71. D.M. Kuchta, A.V. Rylyakov, F.E. Doany et al., A 71-Gb/s NRZ modulated 850-nm VCSEL-based optical link. IEEE Photonics Technol. Lett. **27**(6), 577–580 (2015)
72. J. Lavrencik et al., 168 Gbps PAM-4 multimode fiber transmission through 50 m using 28 GHz 850 nm multimode VCSELs, in *Optical Fiber Communication Conference* (Optica Publishing Group, 2020)
73. S. Spiga et al., Single-mode high-speed 1.5-μm VCSELs. J. Lightwave Technol. **35**(4), 727–733 (2016)
74. S.A. Blokhin et al., Influence of output optical losses on the dynamic characteristics of 1.55-μm wafer-fused vertical-cavity surface-emitting lasers. Semiconductors **53**, 1104–1109 (2019)
75. J. Zhang, X. Zhang, Y. Zhou et al., Milliwatt-level single transverse mode 1550 nm vertical cavity surface emitting diode laser. Acta Phys. Sin. **71**(6), 346–352 (2022)
76. S. Han, S. Tian, H. Xu et al., Research progress on high-speed 1550 m vertical cavity surface emitting lasers. J. Lumin.Lumin. **43**(5), 736–744 (2022)
77. J. Zhang, G. Gao, B. Wang et al., Background noise resistant underwater wireless optical communication using Faraday atomic line laser and filter. J. Lightwave Technol. **40**(1), 63–73 (2022)
78. J.J. Li, J. Wang, Exploring the definition, development, and reproduction methods of length units. Electron. World **18**, 59–60 (2012)
79. K.M. Evenson, J.S. Wells, F.R. Petersen et al., Speed of light from direct frequency and wavelength measurements of the methane-stabilized laser. Phys. Rev. Lett. **29**(19), 1346–1349 (1972)
80. B. Lin, S. Cao, Y. Zhao et al., Miniaturized iodine-stabilized 532 nm solid-state laser. Chin. Laser **41**(09), 14–17 (2014)

81. J. Sanjuan et al., Simultaneous laser frequency stabilization to an optical cavity and an iodine frequency reference. Opt. Lett. **46**(2), 360–363 (2021)
82. K. Ikeda et al., Hyperfine structure and absolute frequency of $^{127}I_2$ transitions at 514 nm for wavelength standards at 1542 nm. J. Opt. Soc. Am. B **39**(8), 2264–2271 (2022)
83. H.P. Yang, J.B. Wang, B.W. Zhang et al., Research progress on acetylene stable frequency laser technology in the 1.5 μm fiber optic communication band. Acta Metrol. Sin. **44**(01), 62–72 (2023)
84. K. Yoshii, T. Inamura, H. Sagawa et al., Modulation free frequency stabilized laser at 1.5 μm using a narrow linewidth diode laser, in *CLEO: Applications and Technology* (California, USA, 2017)
85. R. Zektzer et al., A chip-scale optical frequency reference for the telecommunication band based on acetylene. Laser Photonics Rev. **14**(6), 1900414 (2020)
86. J. Chen, H. Guo, B. Luo et al., A method and device for generating a 1.5-micron laser wavelength standard in the communication band, CN201710690811.2
87. P. Chang, T. Shi, S. Zhang, H. Shang, D. Pan, J. Chen, Faraday laser at Rb 1529 nm transition for optical communication systems. Chin. Opt. Lett. **15**(12), 121401 (2017)
88. T. Shi et al., A dual-frequency faraday laser. IEEE Photonics J. **12**(4), 1–11 (2020). Art no. 1503211
89. Z. Ge, Z. Wang, P. Chang et al., An integrated multi-band laser wavelength standard device, CN 107370016 A
90. F.L. Hong, Optical frequency standards for time and length applications. Meas. Sci. Technol. **28**, 012002 (2017)
91. T. Schuldt, K. Döringshoff, E.V. Kovalchuk et al., Development of a compact optical absolute frequency reference for space with 10^{-15} instability. Appl. Opt. **56**, 1101–1106 (2017)
92. Y. Tanabe, Y. Sakamoto, T. Kohno et al., Frequency references based on molecular iodine for the study of Yb atoms using the $1S_0$–$3P_1$ intercombination transition at 556 nm. Opt. Express **30**, 46487–46500 (2022)
93. B. Lin, S. Cao, Y. Zhao et al., Miniaturized iodine-stabilized 532 nm solid-state laser. Chin. Laser **41**(9) (2014)
94. F. Cheng, N. Jin, F. Zhang et al., A 532 nm molecular iodine optical frequency standard based on modulation transfer spectroscopy. Chin. Phys. B **30**, 050603 (2021)
95. K.W. Martin, G. Phelps, N.D. Lemke et al., Compact optical atomic clock based on a two-photon transition in rubidium. Phys. Rev. Appl. **9**, 014019 (2018)
96. J. Wang, C. Yin, C. Shi et al., High-power iodine-stabilized He–Ne laser wavelength reference source. Infrared Laser Eng. **50**(4) (2021)
97. Y. Öhman, On some new auxiliary instruments in astrophysical research VI. A tentative monochromator for solar work based on the principle of selective magnetic rotation. Stockholms Obs. Ann. **19**, 9–11 (1956)
98. P.P. Sorokin, J.R. Lankard, V.L. Moruzzi, A. Lurio, Frequency-locking of organic dye lasers to atomic resonance lines. Appl. Phys. Lett. **15**, 179–181 (1969)
99. J. Menders et al., Ultranarrow line filtering using a Cs Faraday filter at 852 nm. Opt. Lett. **11**, 846–848 (1991)
100. D.J. Dick, T.M. Shay, Ultrahigh-noise rejection optical filter. Opt. Lett. **11**, 867–869 (1991)
101. L. Weller et al., Optical isolator using an atomic vapor in the hyperfine Paschen–Back regime. Opt. Lett. **37**, 3405–3407 (2012)
102. L. Abel, U. Krohn, P. Siddons, I. Hughes, C. Adams, Faraday dichroic beam splitter for Raman light. Opt. Lett. **34**, 3071–3703 (2009)
103. X. Xue, C. Janisch, Y. Chen, Z. Liu, J. Chen, Low-frequency shift Raman spectroscopy using atomic filters. Opt. Lett. **41**, 5397–5400 (2016)
104. S.L. Portalupi et al., Simultaneous Faraday filtering of the Mollow triplet sidebands with the Cs-D_1 clock transition. Nat. Commun. **7**, 13632 (2016)
105. J. Lin, Y. Li, Ultralow frequency Stokes and anti-Stokes Raman spectroscopy of single living cells and microparticles using a hot rubidium vapor filter. Opt. Lett. **39**, 108–110 (2014)

106. J. Tang et al., Experimental study of a model digital space optical communication system with new quantum devices. Appl. Opt. **34**, 2619–2622 (1995)
107. R.C. Smith, J.E. Tyler, Optical properties of clear natural water. J. Opt. Soc. Am. **57**, 589 (1967)
108. X. Shan, X. Sun, J. Luo, Z. Tan, M. Zhan, Free-space quantum key distribution with Rb vapor filters. Appl. Phys. Lett. **89**, 191121 (2006)
109. W.T. Buttler et al., Practical free-space quantum key distribution over 1 km. Phys. Rev. Lett. **81**, 3283–3286 (1998)
110. A. Popescu, T. Walther, On an ESFADOF edge-filter for a range resolved Brillouin-lidar: the high vapor density and high pump intensity regime. Appl. Phys. B **98**, 667–675 (2010)
111. Y. Yang et al., A flat spectral Faraday filter for sodium lidar. Opt. Lett. **32**, 1302–1304 (2011)
112. F.Q. Li et al., A Doppler lidar with atomic Faraday devices frequency stabilization and discrimination. Opt. Laser Technol. **44**(6), 1982–1986 (2012)
113. A. Rudolf, T. Walther, Laboratory demonstration of a Brillouin lidar to remotely measure temperature profiles of the ocean. Opt. Eng. **53**, 051407 (2014)
114. W. Kiefer, R. Low, J. Wrachtrup, I. Gerhardt, The Na-Faraday filter: the optimum point. Sci. Rep. **4**, 06552 (2014)
115. M.A. Zentile et al., Optimization of atomic Faraday filters in the presence of homogeneous line broadening. Opt. Soc. Am. B Mol. Opt. Phys. **48**, 185001 (2015)
116. J.A. Zielinska, F.A. Beduini, N. Godbout, M.W. Mitchell, Ultranarrow Faraday rotation filter at the Rb D_1 line. Opt. Lett. **37**, 524–526 (2012)
117. X. Xue et al., Faraday anomalous dispersion optical filter at ^{133}Cs weak 459 nm transition. Photon. Res. **3**(5), 275–278 (2015)
118. M.D. Rotondaro, B.V. Zhdanov, R.J. Knize, Generalized treatment of magneto-optical transmission filters. Opt. Soc. Am. B **34**(12), 2507–2513 (2015)
119. M.A. Zentile et al., Atomic Faraday filter with equivalent noise bandwidth less than 1 GHz. Opt. Lett. **40**, 2000–2003 (2015)
120. D. Pan et al., Hollow cathode lamp based Faraday anomalous dispersion optical filter. Sci. Rep. **6**, 29882 (2015)
121. Y. Wang et al., Nonlinear optical filter with ultranarrow bandwidth approaching the natural linewidth. Opt. Lett. **37**, 4059–4061 (2012)
122. X. Shan, X. Sun, J. Luo, M. Zhan, Ultranarrow-bandwidth atomic filter with Raman light amplification. Opt. Lett. **33**, 1842 (2008)
123. W. Zhang, Y. Peng, Transmission characteristics of a Raman-amplified atomic optical filter in rubidium at 780 nm. J. Opt. Technol. **81**, 174–181 (2014)
124. X. Zhao et al., Atomic filter based on stimulated Raman transition at the rubidium D_1 line. Opt. Exp. **23**, 17988 (2015)
125. D. Pan, T. Shi, B. Luo, J. Chen, H. Guo, Atomic optical stimulated amplifier with optical filtering of ultra-narrow bandwidth. Sci. Rep. **8**, 6567 (2018)
126. X. Qi et al., Study of phase coherence degradation induced by a tapered semiconductor amplifier with frequency-modulated continuous-wave and pulsed seed lasers. Appl. Opt. **22**, 4370 (2009)
127. J. Carvalho, A. Laliotis, M. Chevrollier, M. Oria, D. Bloch, Backward-emitted sub-Doppler fluorescence from an optically thick atomic vapor. Phys. Rev. A **96**, 043405 (2017)
128. M. Auzinsh et al., Cascade coherence transfer and magneto-optical resonances at 455 nm excitation of cesium. Opt. Commun. **284**, 2863–2871 (2011)
129. F. Li, H. Li, H. Lu, Realization of a tunable 455.5 nm laser with low intensity noise by intracavity frequency doubled Ti: sapphire laser. IEEE. J. Quantum Electron. **52**, 1700106 (2016)
130. S. Pustelny et al., Nonlinear magneto-optical rotation in rubidium vapor excited with blue light. Phys. Rev. A **92**, 053410 (2015)
131. D. Pan, Z. Xu, X. Xue, W. Zhuang, J. Chen, Lasing of cesium active optical clock with 459 nm laser pumping. Paper presented at 2014 Joint Conference of the IEEE International Frequency Control Symposium, in *Proceedings of 2014 IEEE International Frequency Control Symposium*, Taipei, 19–22 May 2014, pp. 242–245

132. H.J. Metcalf, P.V.D. Straten, in *Laser Cooling and Trapping*, ed. by R.S. Berry et al. (Verlag, 1999), pp. 274–275
133. Q. Sun, Y. Hong, W. Zhuang, Z. Liu, J. Chen, Demonstration of an excited-state Faraday anomalous dispersion optical filter at 1529 nm by use of an electrodeless discharge rubidium vapor lamp. Appl. Phys. Lett. **101**, 211102 (2012)
134. X. Xue, D. Pan, J. Chen, A cavityless laser using cesium cell with 459 nm laser pumping. Paper presented at 2015 Joint Conference of the IEEE International Frequency Control Symposium & European Frequency and Time Forum, in *Proceedings of 2015 IEEE International Frequency Control Symposium*, Denver, 12–16 Apr 2015, pp. 614–616
135. A.V. Andreev, V.I. Emel'yanov, I.A. Il'inskir, Collective spontaneous emission (Dicke superradiance). Usp. Fiz. Nauk. **131**, 653–694 (1980)
136. J. Vanier, C. Audoin, in *The Quantum Physics of Atomic Frequency Standards*, ed. by A.E. Bailey et al. (IOP Publishing Ltd, 1989), pp. 1006–1009
137. A. Sharma, N.D. Bhaskar, Y.Q. Lu, W. Happer, Continuous-wave mirrorless lasing in optically pumped atomic Cs and Rb vapors. Appl. Phys. Lett. **39**, 209 (1981)
138. A.C. Ferrari, D.M. Basko, Raman spectroscopy as a versatile tool for studying the properties of graphene. Nat. Nanotech. **8**, 235–246 (2013)
139. E. Duval, A. Boukenter, B. Champagnon, Vibration eigenmodes and size of microcrystallites in glass: observation by very-low-frequency Raman scattering. Phys. Rev. Lett. **56**, 2052–2055 (1986)
140. K.K. Mon, Y.J. Chabal, A.J. Sievers, Temperature dependence of the far-infrared absorption spectrum in amorphous dielectrics. Phys. Rev. Lett. **35**, 1352–1355 (1975)
141. R.H. Stolen, M.A. Bösch, Low-frequency and low-temperature Raman scattering in silica fibers. Phys. Rev. Lett. **48**, 805–808 (1982)
142. P. Ranzieri, A. Girlando, S. Tavazzi, M. Campione, L. Raimondo, I. Bilotti, A. Brillante, R.G. Della Valle, E. Venuti, Polymorphism and phonon dynamics of α-quaterthiophene. ChemPhysChem **10**, 657–663 (2009)
143. P.H. Tan, W.P. Han, W.J. Zhao, Z.H. Wu, K. Chang, H. Wang, Y.F. Wang, N. Bonini, N. Marzari, N. Pugno, G. Savini, A. Lombardo, A.C. Ferrari, The shear mode of multilayer graphene. Nat. Mater. **11**, 294–300 (2012)
144. M. Ivanda, K. Furic, S. Music, M. Ristic, M. Gotic, D. Ristic, A.M. Tonejc, I. Djerdj, M. Mattarelli, M. Montagna, F. Rossi, M. Ferrari, A. Chiasera, Y. Jestin, G.C. Righini, W. Kiefer, R.R. Goncalves, Low wavenumber Raman scattering of nanoparticles and nanocomposite materials. J. Raman Spectrosc. **38**, 647–659 (2007)
145. P. Pakhomov, S. Khizhnyak, V. Galitsyn, E. Rogova, B. Hartmann, A. Tshmel, Application of the low frequency Raman spectroscopy for studying ultra-high molecular weight polyethylenes. Macromol. Symp.. Symp. **305**, 63–72 (2011)
146. V.N. Sankaranarayanan, R.T. Bailey, A.J. Hyde, Low frequency Raman scattering from polystyrene. Spectrochim. Acta A **34**, 387–389 (1978)
147. C. Janisch, N. Mehta, D. Ma, A.L. Elías, N.P. López, M. Terrones, Z. Liu, Ultrashort optical pulse characterization using WS_2 monolayers. Opt. Lett. **39**, 383–385 (2014)
148. http://www.princetoninstruments.com/products/spec/trivista/
149. X.L. Wu, S.J. Xiong, Y.M. Yang, J.F. Gong, H.T. Chen, J. Zhu, J.C. Shen, P.K. Chu, Nanocrystal-induced line narrowing of surface acoustic phonons in the Raman spectra of embedded Ge_xSi_{1-x} alloy nanocrystals. Phys. Rev. B **78**, 165319 (2008)
150. J.T. Carriere, F. Havermeyer, Ultra-low frequency Stokes and anti-Stokes Raman spectroscopy at 785 nm with volume holographic grating filters, in *Proceedings of SPIE, 8219, Biomedical Vibrational Spectroscopy V: Advances in Research and Industry* (2012), p. 821905
151. C. Xie, A. Mumtaz et al., Near-infrared Raman spectroscopy of single optically trapped biological cells. Opt. Lett. (2002)

Chapter 8
Future Trends

Faraday lasers use the atomic transition lines in atomic filters as the frequency selection references. The output laser frequency can intrinsically align with the atomic transition frequency, thus achieving more reliable frequency stability. With the advantages of narrow linewidth, wavelength intrinsically corresponding to atomic transitions, and immunity to fluctuations of laser diode temperature and current, it has gradually become a research hotspot globally. In the future, Faraday lasers are expected to become the frontier high-end instrument in frequency-stabilized diode lasers, with applications involving quantum precision measurements and basic research of time, frequency, gravity, magnetic field, length, angle, and other parameters. The previous chapters of this book have detailed the research process of Faraday lasers, including basic principles, implementation methods, current research status, and applications. This chapter mainly introduces the latest research directions and future development strategies of Faraday lasers for specific potential innovative technology solutions, functions, and application scenarios.

8.1 Overview of Future Development Prospects of Faraday Lasers

Faraday lasers, based on the quantum frequency selection effect, have advantages such as wavelengths corresponding to atomic transition lines, stable frequencies, strong robustness to fluctuations of laser diode current and temperature, multiple wavelength options, and narrow linewidths. As the output wavelength of the Faraday laser directly corresponds to the atomic transitions, it could work within the passband of the atomic filter for a long time. That's why, it has good adaptability to the environment and does not need to manually search for the wavelength corresponding to the atomic transitions with the help of a wavelength meter. As an instrument or component, it has unique advantages in specific fields such as aviation, satellite, and

J. Chen et al., *Faraday Laser*, https://doi.org/10.1007/978-981-97-8023-5_8

field detection. This high-performance Faraday laser can be applied to fields such as atomic clocks, atomic gravimeters, atomic magnetometers, atomic radars, optical communications, navigation timing, etc.

In addition to the currently realized applications, Faraday lasers also have the potential to achieve miniaturization and intelligent integration by integration technology, thereby improving system integration and greatly expanding the application scenarios. Moreover, high-power, narrow-linewidth diode lasers are crucial light sources in interstellar laser communication, large-scale laser ranging, and laser weapons. Using high-power diode laser and FADOF parameter matching, it is possible to achieve high-power, narrow-linewidth Faraday lasers with output power at the watt or even the hundred-watt level, and use them in related fields. In addition, by setting FADOF parameters, we can realize multi-transmission peak atomic filters, and then demonstrate multi-frequency Faraday lasers, which have the potential to be used in multi-frequency laser spectroscopy, coherent population trapping atomic clocks, optically generated microwave sources, precision length and angle measurement and so on. Furthermore, a series of Faraday lasers based on new principles and technologies have been developed, including Faraday laser frequency combs based on alkali metal atomic nonlinear effects in FADOF, Faraday–Michelson lasers for precision length measurement with traceability to laser frequency, cold atom Faraday lasers based on laser-cooled atoms, Faraday active optical clocks using an active optical clock scheme, and ultra-long cavity narrow-linewidth Faraday lasers based on optical fibers. The latest technological advances based on Faraday lasers have forward-looking guidance for the future development of Faraday lasers, and this chapter will discusses these novel schemes in detail.

8.2 Miniaturized Chip Faraday Lasers

As the core light source for many applications, narrow-linewidth diode lasers have high research and application value. In the future, with the continuous development of photonic integration technology, high-performance narrow-linewidth diode lasers will inevitably tend to be miniaturized and fully automated, thereby stimulating more application potential, especially in sensors and the Internet of Things, driving the extensive and deep development of diode lasers and their downstream applications. Currently, the frontier direction is to miniaturize Faraday lasers by developing micro FADOF or FADOF in hollow fibers, combining with integrated technology to reduce the overall volume, thus improving their stability and reliability.

For example, by utilizing MEMS vapor cells and more compact packaging methods to realize miniaturized Faraday lasers, the length of the laser cavity is effectively shortened. Therefore, we can get a larger free spectral range (FSR), which meets the demand for larger mode-hop-free intervals in specific scenarios, expanding the wavelength range of Faraday lasers. At the same time, the volume of the laser is reduced, thus improving the mechanical stability and frequency stability, making it easier for large-scale mass production and application with lower production costs. Moreover, using fusion technology to develop hollow fiber FADOF and other related laser components to realize all-fiber Faraday lasers, we can further improve the system's robustness to mechanical vibrations, narrow the laser linewidth, and improve system stability. Figure 8.1 shows the recently developed micro FADOF based on a micro atomic vapor cell, whose size is comparable to a coin. The concept diagram of the miniaturized Faraday laser built based on this micro FADOF is shown in Fig. 8.2, and the physical picture of the laser is shown in Fig. 8.3. The miniaturized Faraday laser has the advantages of small size, strong robustness to mechanical vibrations, the wide transmission spectrum bandwidth of the micro FADOF, and the large FSR ensuring the single-mode operation of the laser. This work effectively improves the performance of the Faraday laser, reduces the system volume, improves the reliability of the Faraday laser in satellites, field use, and long-term use, and lays a solid foundation for the later development of microchip Faraday lasers.

This miniaturized, narrow linewidth Faraday laser can be applied to quantum sensors, laser cooling, high-resolution spectroscopy, laser sensing, optical communication, laser radar, space applications, and other fields, becoming a high-quality

Fig. 8.1 Design of micro Faraday atomic filter

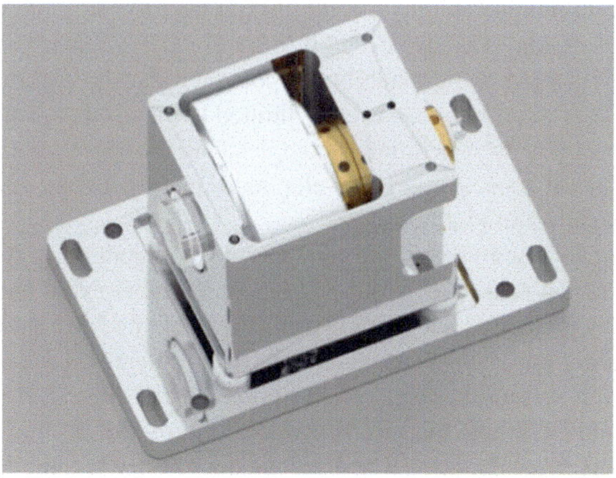

Fig. 8.2 Design of micro Faraday laser

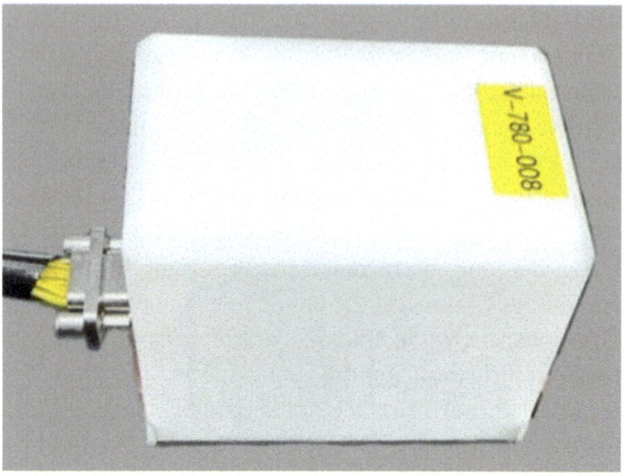

Fig. 8.3 Physical picture of micro Faraday laser

narrow linewidth laser source. Currently, it has started to be applied in large-scale precision measurement devices such as reference atomic clocks and atomic gravimeters, reducing system volume and enhancing system performance.

8.3 Intelligent Integrated Faraday Laser

At present, the existing commercial diode laser frequency stabilization systems have the following problems in application: the modules are scattered and are not easy to integrate; they are mainly designed for laboratory scientific research applications, the interfaces are complex, and multiple signal/power supply lines are required to connect between modules, which to some extent affects system reliability and application convenience; in order to achieve the universality of the equipment, there is a lack of targeted development, and the functions are all but not refined; laser application units need to spend a lot of effort to study the working principle of the laser itself, and it is also difficult to accurately evaluate the laser frequency stabilization effect; the overall stability is poor, sensitive to external temperature, vibration and other interference, basically limited to the use in scientific research laboratory environment; the operation is complex, and the operation interface is not friendly. Therefore, it is necessary to develop more intelligent and integrated diode laser frequency stabilization systems.

In recent years, through the development of high-stability low-noise constant current source technology, laser-specific precision temperature control technology, integrated MTS technology, and the integrated design of laser frequency stabilization algorithms and systems, an intelligent integrated Faraday laser frequency stabilization system that can be used immediately upon startup, automatically searches for and locks onto the spectrum upon startup, and automatically relocks after losing lock has been realized. The implementation of this intelligent integrated laser system ensures the long-term stable operation of the laser, reduces the complexity of the operation, and can be used in different applications such as atomic clocks, atomic interferometers, atomic magnetometers, atomic gyroscopes, and high-precision laser spectroscopy measurements.

8.4 High-Power Faraday Lasers

In specific fields such as interstellar laser communication, large-scale laser ranging, military applications, and quantum optics such as diode laser-pumped alkali metal gas lasers, spin-exchange optical pumping, and metastable rare gas laser pumping, high-power diode lasers with wavelength corresponding to atomic transition lines and narrow linewidth are needed. High-power diode lasers often use arrays or stacks to achieve high power. The laser consists of multiple light-emitting units, which increases the difficulty of narrowing the output laser spectrum. Currently, the spectrum narrowing of high-power diode lasers is usually achieved by using external cavity schematics and utilizing frequency selection components such as gratings to realize frequency selection.

Traditional high-power narrow-linewidth diode lasers are mostly grating-type external cavity diode lasers, where the gratings serve as the feedback and frequency selection components of diode lasers to narrow the linewidth of the laser. Among them, frequency selection devices often use surface diffraction gratings and volume Bragg gratings. The former uses the angular dispersion of the diffraction grating to achieve spectrum narrowing. Still, this scheme has a more complex beam collimation system, low power and efficiency, and the directionality of the emitted light is related to the frequency, so it is inconvenient to use. Volume Bragg gratings also have selectivity for the angle and wavelength of the incident light, equivalent to a narrow-linewidth filter, and the directionality of the emitted light is better than that of surface diffraction gratings. In 2010, Podvyaznyy and colleagues [1] used a volume grating to obtain a narrow-linewidth output with a central wavelength of 780 nm. The output power is 250 W, and the laser linewidth is 10 GHz (0.02 nm). This scheme has a compact structure, simple adjustment, and a better linewidth narrowing effect, so it is often used as the pump light source in alkali metal gas lasers. However, because the wavelength-temperature drift coefficient of the volume grating laser is large (8 ~ 10 pm/K), the output laser wavelength does not intrinsically correspond to the atomic spectrum line. Therefore, one must use the complex temperature control system to ensure the required central wavelength.

In 2018, Rotondaro and colleagues. locked the wavelength precisely on the Cs-D_2 transition line using Cs-FADOF on a vertical diode laser array successfully. They narrowed the linewidth from 3 THz to 10 GHz, with an output power of 518 W and an external cavity efficiency of 80% [2].

This work has realized a diode laser based on FADOF with output power ranging from hundreds of milliwatts to hundreds of watts, demonstrating the feasibility of high-power Faraday lasers. However, in this scheme, the power of the laser diode array is all injected into the FADOF, which may cause thermal deposition problems. In 2021, the National University of Defense Technology reported a high-power Faraday laser based on Rb-FADOF [3]. This structure allows a small portion of the laser to enter the alkali metal atom vapor cell, effectively reducing thermal deposition. The center wavelength of the laser is precisely locked on the Rb-D_2 transition line, with a linewidth of 1.2 GHz, a continuous output power of 18 W, and an external cavity efficiency of 80%.

High-power Faraday lasers can effectively overcome the problems of large center frequency drift, frequency sensitivity to laser diode current, temperature, and mechanical vibration, as well as wide linewidth in high-power external cavity diode lasers, which effectively improve the overall application performance of high-power external cavity diode lasers. By adding buffer gas in the atomic vapor cell of FADOF, the sidebands of the transmission spectrum can be effectively suppressed, thus developing a high-power single-peak and high-transmission FADOF. Using the FADOF as a frequency selection component, we can effectively limit the frequency of the output laser to the atom transition line. One scheme is to set up an optical path structure so that only the laser power for optical feedback passes through the high-power single-peak high-transmission FADOF. The rest of the laser is directly reflected and emitted at the polarization beam splitter in front of FADOF, thus effectively avoiding the thermal deposition effect of the atomic vapor cell and increasing the power of the output laser. In addition, by coating an anti-reflection film on the inner and outer surfaces of the atomic vapor cell, we get a higher transmittance of the atomic vapor cell, which enhances the optical feedback and suppresses the side modes better, reducing the equivalent noise bandwidth. Finally, we get a high-power narrow-linewidth Faraday laser with stable frequency.

8.5 Multi-frequency Faraday Lasers

The Faraday lasers reported so far work in single-frequency mode mostly. If the single-frequency Faraday laser could be extended to dual-frequency or even multi-frequency mode, the research and application range of Faraday lasers can be further expanded. Dual-frequency lasers are widely applied in various fields such as laser frequency stabilization and high-resolution spectroscopy. For instance, the commonly used dual-frequency He–Ne laser can stabilize the microwave frequency difference between two resonant laser modes through phase locking [4]. This enhances the frequency stability of the two laser modes operating in the optical frequency range by stabilizing the cavity length. In addition, the Tm:YAG laser using a standard as the intra-cavity frequency selection component [5], the frequency difference adjustment range between the two laser modes is up to several GHz, and the linewidth of each laser mode is less than 1 MHz, promoting the development of precision spectroscopy experiments. Experiments have proven that a saturation absorption spectroscopy (SAS) with high contrast and enhanced amplitude can be achieved with dual-frequency lasers [6, 7], which could optimize the frequency stability of diode lasers by nearly an order of magnitude. By using MTS technology, one of the laser frequencies can be locked on the atomic transition frequency, and the beat linewidth of the two laser modes can also be narrowed, which is expected to serve as an optically generated microwave laser source [8].

The transmission spectrum of FADOF is changeable by adjusting the applied magnetic field, temperature, and the length of the atomic vapor cell. By setting the FADOF parameters experimentally, we can achieve a high-transmission and dual-peak or even multi-peak transmission spectrum. If the FSR of the laser can cover two transmission peaks, reducing the impact of mode competition, a dual-frequency or even multi-frequency Faraday laser can be realized. In 2020, Shi and others from Peking University reported an 852 nm dual-frequency Faraday laser using a cesium atom FADOF with a dual-peak transmission spectrum as a frequency selector. Its FADOF transmission spectrum is shown in Fig. 8.4 [9]. This dual-frequency Faraday laser can tune the frequency difference between the two longitudinal modes with the temperature of the atomic vapor cell, and beat with a single-mode interference filter laser as a reference, with the mode distribution shown in Fig. 8.5. The coherence between the two modes of this laser is excellent, and the most probable linewidth measured by beating at different atomic vapor cell temperatures is 903 Hz. Since the output frequency difference of the dual-frequency Faraday laser is related to the hyperfine energy level of the ground state of alkali metal atoms, this dual-frequency laser can be used to achieve high-contrast sub-Doppler spectra as well as the application of coherent layout number trapped atomic clocks.

In 2022, Chang et al. from Peking University reported on the MTS frequency stabilization based on the Cs atom 852 nm dual-frequency Faraday laser [8], when

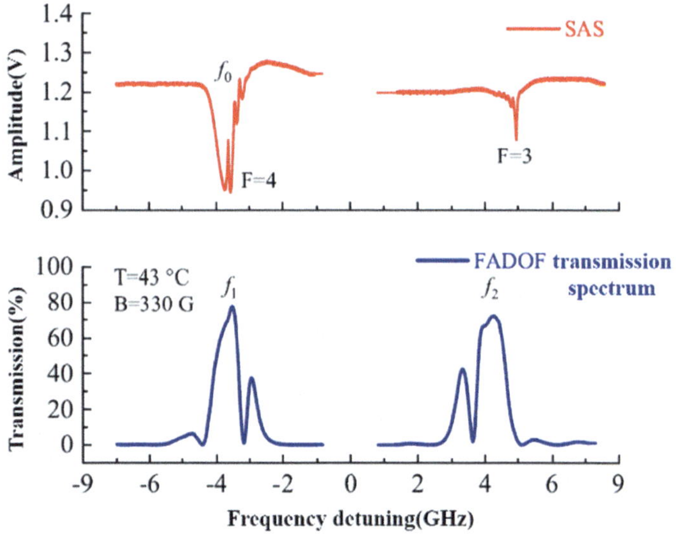

Fig. 8.4 Cs D_2 line SAS (top) and dual-frequency FADOF transmission spectrum (bottom) [9]

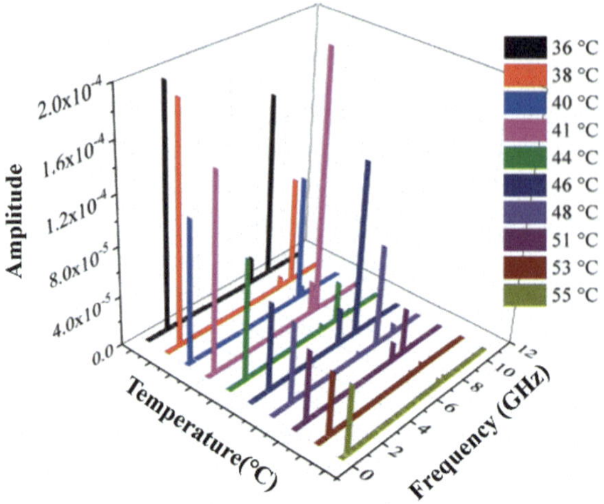

Fig. 8.5 Beat power spectrum of 852 nm dual-frequency Faraday laser and reference laser at different atomic vapor cell temperatures [9]

one of the laser modes is locked on the cesium atom $6^2S_{1/2}F = 3 \rightarrow 6^2P_{3/2}F' = 2$ transition, the beat linewidth of the two laser modes can be narrowed to 85 Hz, as shown in Fig. 8.6. This narrow linewidth beat signal has expanded the application scenarios of the dual-frequency Faraday laser.

In addition to the Cs dual-frequency Faraday laser mentioned above, the research team from Peking University also realized a dual-frequency Faraday laser based on Rb atoms, with the dual-frequency difference tunable with PZT, as shown in Fig. 8.7. The two modes have good coherence, and the Lorentz fit linewidth of the beat signal is 649 Hz, as shown in Fig. 8.8. At the same time, we can also use other alkali metal atoms with two fine energy levels in the ground state to realize dual-frequency

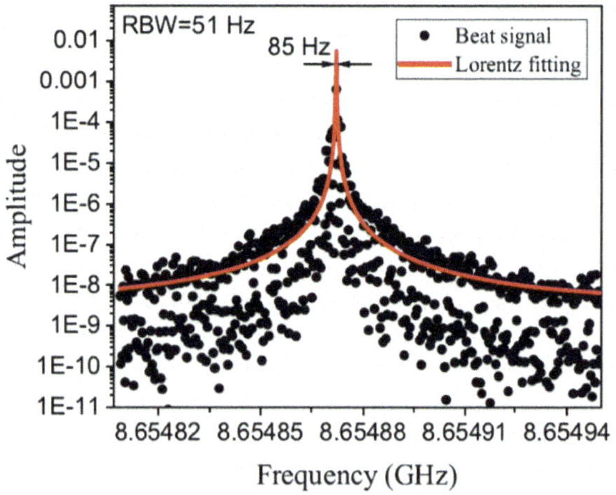

Fig. 8.6 Lorentz fit of the beat signal between the two modes of the locked 852 nm dual-frequency Faraday laser [8]

Faraday lasers, expanding the application range of Faraday lasers. This type of dual-frequency Faraday laser can be applied in atomic physics, such as high-contrast sub-Doppler spectroscopy. Compared with traditional single-frequency SAS, the SAS realized by dual-frequency laser fields can more fully and effectively utilize atoms of different velocity groups, optimizing the frequency stability of diode lasers by nearly an order of magnitude. By precisely designing and adjusting the cavity length of the dual-frequency Faraday laser, if the interval between adjacent longitudinal modes of the laser is equal to an integer multiple of the hyperfine energy level interval of the cesium atom and finely adjusted, it can be used to realize coherent population trapping atomic clocks. The dual-frequency Faraday laser can generate a microwave signal with high coherence, providing a new technical solution to promote the development of optically generated microwave sources. Analogous to the frequency stabilization method of dual-frequency helium–neon lasers, by phase-locking, the two laser modes output by the dual-frequency Faraday laser can be locked to an external reference source, further enhancing the coherence between the two laser modes, and then using phase-locking to narrow the linewidth of each laser mode.

In addition, using the characteristic of FADOF transmission spectrum changeable with light intensity, in 2024, Qin et al. from Peking University realized a dual-frequency single-longitudinal-mode Faraday laser with output laser wavelength switchable with laser diode current [10]. Its experimental structure and corresponding energy level structure are shown in Fig. 8.9. Under appropriate FADOF working parameters, the output wavelength of the laser is switchable between ^{85}Rb $5^2S_{1/2}$ $(F = 3) \rightarrow 5^2P_{3/2}$ and ^{87}Rb $5^2S_{1/2}$ $(F = 2) \rightarrow 5^2P_{3/2}$ with the most probable linewidth of the free-running laser mode being 18 kHz, as shown in Fig. 8.10. This 780 nm switchable Faraday laser with a wavelength corresponding to the D_2 transition of two isotopes of Rb, and its immunity to fluctuations in laser diode current and temperature is usable in laser cooling atoms, atomic gravimeters, atomic magnetometers, small optical

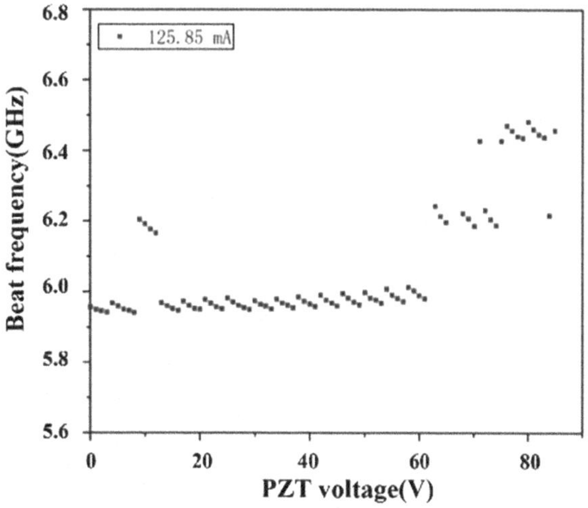

Fig. 8.7 Frequency difference of Rb 780 nm dual-frequency Faraday laser with PZT tuning

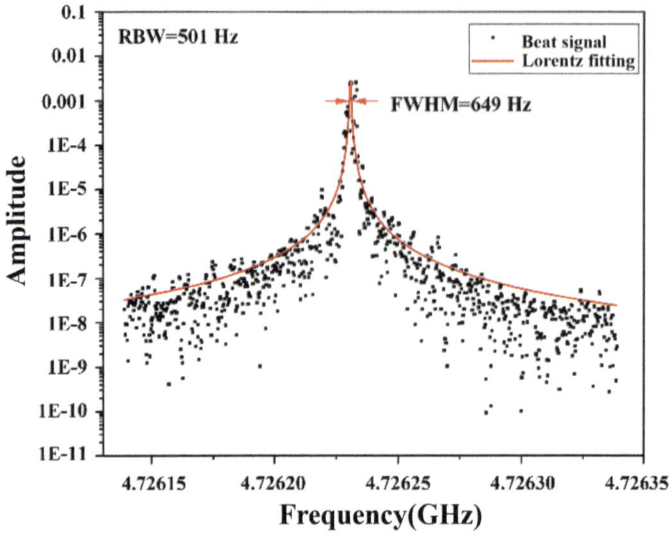

Fig. 8.8 Lorentz fit of beat signal of Rb 780 nm dual-frequency Faraday laser

clocks, and other atomic physics applications, as a high-quality, high-performance diode laser light source that is ready to use.

8.6 Quantum Mode-Locked Faraday Laser Frequency Comb

Diode laser mode locking is a technique that allows lasers to produce extremely short coherent pulses, with durations ranging from picoseconds to femtoseconds. Mode-locked lasers have applications in various fields, including quantum precision measurement, optical communication, laser ranging, biomedical imaging, gas detection, etc. In diode lasers, mode locking is realizable by using a mode-locker in the laser resonator-active components (optical modulators) or nonlinear passive components (saturable absorbers), including passive mode-locking, active mode-locking, and hybrid mode-locking.

Passive mode-locking automatically generates short pulses using nonlinear components (such as saturable absorbing mirrors or saturable absorbers) inside the laser cavity. A saturable absorber is a nonlinear absorbing medium, the absorption or loss of the laser in the resonator changes with the intensity of the light. When the light intensity increases, the absorption of the light decreases. This scheme is technically simple and stable, does not require an external modulation signal, and can produce shorter pulses.

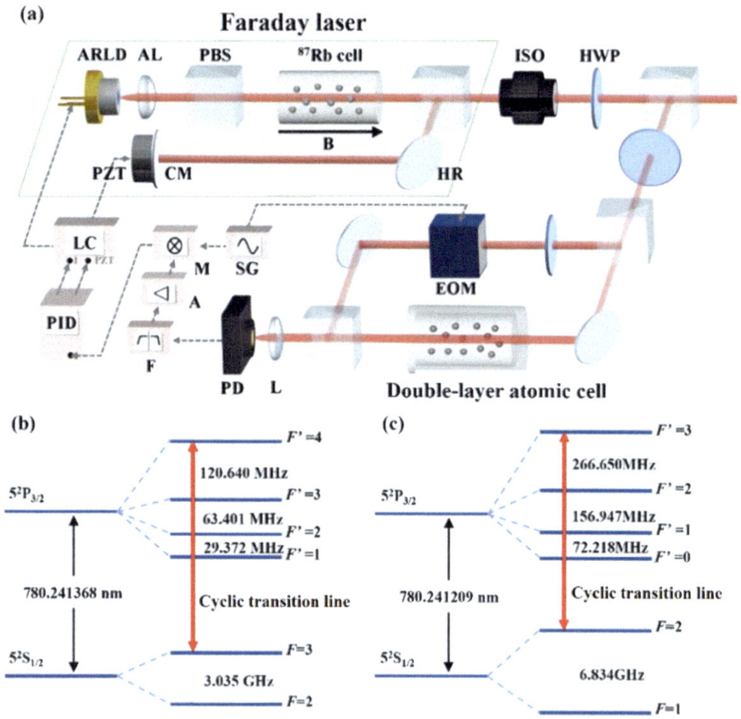

Fig. 8.9 Rb 780 nm dual-frequency switchable Faraday laser based on MTS locking technique. **a** Schematic diagram of the experimental system; **b** related energy levels of ^{85}Rb-D_2 line transition; **c** related energy levels of ^{87}Rb-D_2 line transition [10]

Active mode-locking involves periodic modulation of resonator loss or round-trip phase changes, which can be achieved through various methods such as electro-optic modulators, acousto-optic modulators, or semiconductor electro-absorption modulators. When the modulation signal is synchronized with the round-trip period of the laser resonant cavity, ultrashort pulses can be generated in the form of pulse sequences. In diode lasers, active mode-locking can be achieved by directly modulating the driving current, by periodically changing the gain of the gain medium, thereby generating a series of pulses.

Hybrid mode-locking combines the features of passive and active mode-locking, using an external modulator to stabilize the pulse repetition frequency while using a saturable absorber or other nonlinear effects to optimize the shape and width of the pulse. This method can achieve shorter pulse durations like passive mode-locking, and maintain good repetition frequency stability simultaneously.

In 2018, Luo from the Beijing University of Posts and Telecommunications [11] and Xiong from Peking University [12] found that the transmittance of FADOF is related to the incident light intensity. When FADOF is located in the laser cavity as a frequency selection element, the intensity in the cavity increases when the injection

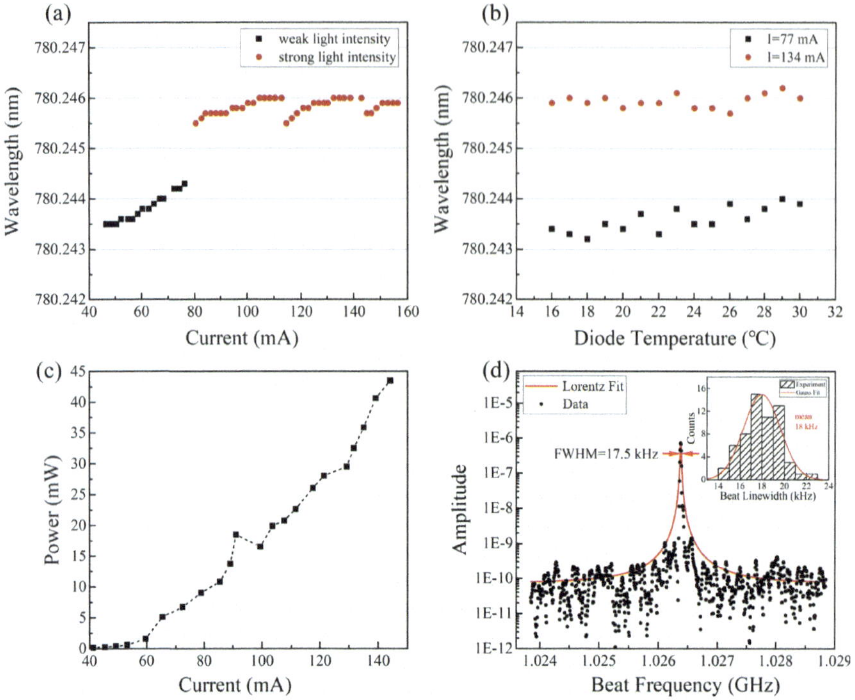

Fig. 8.10 The characteristics of switchable Faraday laser including **a** the output laser wavelength versus drive current, where the black squares and red dots indicate the wavelength under weak and strong laser intensities, respectively; **b** the output laser wavelength versus diode temperature with laser current I = 77 mA (black squares) and 134 mA (red dots), respectively; **c** the output power versus drive current; **d** typical beating signal of two 780 nm free-running Faraday lasers with frequencies tuned to transition 1 and transition 2 respectively. The Lorentz-fitted FWHM of the beating linewidth is 17.5 kHz. Inset: the linewidth distribution of 60 groups of beat-note spectrum with a mean linewidth of 18 kHz. For all the measurement data, the span bandwidth is 5 MHz, and the resolution bandwidth (RBW) is 750 Hz [10]

current of the laser diode increases. Therefore, the transmission spectrum of FADOF also changes. This fact indicates that FADOF may introduce loss related to intensity, similar to the working principle of a saturable absorber. From another perspective, alkali metal atom vapor (such as Rb) has a third-order nonlinear refractive index (Kerr coefficient) of 2×10^{-8} cm^2/W and a self-focusing effect [13, 14], which is of great significance for achieving self-mode-locking operation at a lower threshold.

In 2024, Gao et al. from Peking University observed the quantum mode-locking phenomenon in a 780 nm Faraday laser. The time-domain pulse sequence detected by the oscilloscope is shown in Fig. 8.11, with a pulse width of 252 ps and a period of 3.8 ns. The frequency-domain signal detected by the spectrum analyzer is shown in Fig. 8.12, with a repetition rate of 260 MHz. This laser uses ^{85}Rb-FADOF as a saturable absorber, mixed with 5-torr Ar buffer gas, with a modulation depth of 10%, and has narrow spectral transmission characteristics resonant with atomic transitions.

In addition, due to the selection function of FADOF, the Faraday laser can switch to a single-frequency mode by adjusting the injection current of the laser diode. The output wavelength of the laser under single-frequency and mode-locking corresponds to the ^{87}Rb $5^2S_{1/2}F = 2 \rightarrow 5^2P_{3/2}$ Doppler line. The realization of the 780 nm quantum mode-locked Faraday laser will expand its application in rubidium atomic physics experiments, especially in quantum precision spectroscopy technology using multi-frequency lasers. In addition, this study provides a new method for realizing narrow-spectrum mode-locked lasers with pulse spectrum solely covering the atomic transitions.

The quantum self-mode-locking principle of this Faraday laser is worth further in-depth study and exploration. One possible explanation is that the nonlinear effects in the atomic vapor cell (such as the saturable absorption effect) promote the appearance of mode-locking under certain conditions. The cell acts like a saturable absorber,

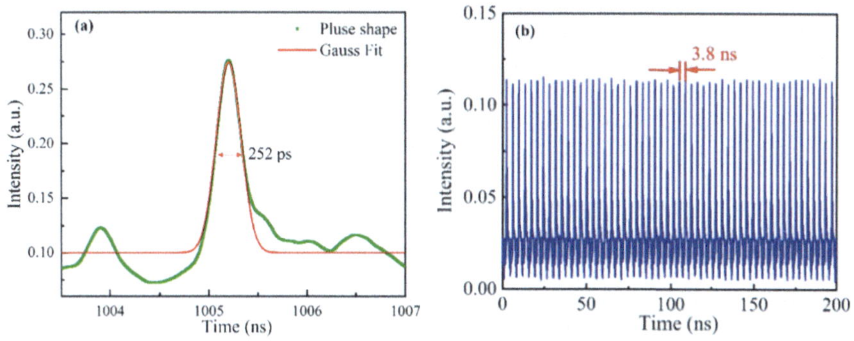

Fig. 8.11 Time-domain distribution of quantum mode-locked Faraday laser frequency comb. **a** Single pulse waveform; **b** time-domain pulse sequence within 200 ns

Fig. 8.12 Spectrum of quantum mode-locked Faraday laser frequency comb within a 2 GHz frequency range

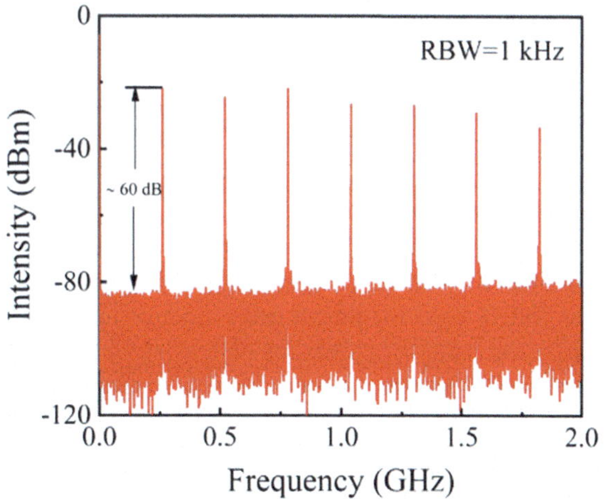

especially when the light intensity in the laser cavity is high enough to saturate atomic quantum transitions. The quantum mode-locking mechanism appears as the high-intensity light passing while low-intensity light blocking.

The discovery of the quantum mode-locked Faraday laser frequency comb has pioneered a new direction for Faraday lasers. In the future, it is necessary to study the physical mechanism further to realize a simple structure with a mode-locked diode laser with a repetition rate of several hundred MHz. It's expected to expand applications in precision laser spectroscopy, new atomic clocks, etc.

8.7 Faraday–Michelson Laser

Precision displacement measurement technology with high resolution is indispensable in major frontier scientific experiments such as gravitational wave detection, microscopic process observation in cell biology and virology at the atomic scale, and ultra-precision equipment fields such as lithography machines, precision machine tools, and IC equipment. Laser interference precision measurement, with its simple structure, high measurement resolution, traceability to laser wavelength, and wide measurement range, has been widely used in scientific research and production applications.

The commonly used laser interferometers include single-frequency laser interferometers (also known as zero-difference laser interferometers) and dual-frequency laser interferometers (also known as heterodyne laser interferometers). Among them, the heterodyne interferometer has a high anti-interference ability, and efforts have been made to achieve 10 pm level displacement measurement with it [15], but this requires the use of fast phase-detection methods, such as pulse counting/zero crossing, phase-locked amplifiers, single bin discrete Fourier transform, dual phase-locked loops, etc. However, due to the electronic noise and environmental noise in the interferometer, it is not easy to measure mechanical displacements with a resolution of 10 pm or less, and it requires the use of complex algorithms or vacuum devices to minimize these noises. Moreover, this scheme is not easy to operate, and the resolution of displacement measurement is limited to this order of magnitude.

The Michelson interferometer is a precision optical instrument invented and manufactured in 1881 by Michelson, the American Nobel laureate in physics, in collaboration with Morley in 1887. It uses the amplitude splitting method to achieve dual-beam interference, which was used to verify the ether drift experiment and finally confirmed the non-existence of ether. After improvements and development, it was later widely used for length and refractive index measurements. For his contributions to the creation of precision optical interferometers, spectroscopy and metrology research, and accurate measurement of the speed of light, Michelson was awarded the Nobel Prize in Physics in 1907. The Michelson interferometer achieves interference measurement through an external light source, such as a mercury or sodium lamp. With the birth of the He–Ne laser in 1961, the He–Ne laser began to serve as the light source for the Michelson interferometer, thus promoting the rapid development and widespread application of laser interferometers. Henceforth, laser interferometers with simple structure, high measurement accuracy, and traceability to the laser wavelength have been widely used in basic scientific research and high-end instrument production fields.

Based on this, Chen et al. at Peking University [16] propose and demonstrate a new displacement measurement scheme named the Michelson laser. The Michelson laser divides the laser cavity into two paths. Due to the different optical lengths of the two paths, it can generate dual-mode or multi-mode lasers whose frequency depends on the sub-cavity length, as shown in Fig. 8.13a. Since frequency measurement can currently achieve more than a high precision of 16 digits, this scheme converts displacement measurement into frequency measurement by measuring the beat signal of the output dual-mode or multi- mode laser, and it is expected to achieve precise displacement measurement with a resolution of pm, or even finer. This system is essentially a new type of laser, and its optical path structure draws on the shape structure of the Michelson interferometer. In honor of the contribution of Michelson, the new laser provided by this invention is named as the Michelson laser. It is worth mentioning that the traditional Michelson interferometer, whether its light source is a laser source or a spectral lamp, is essentially a dual-optical-path device that produces a light field bright and dark interference effect. However, the Michelson laser is a high-end precision instrument that achieves precise displacement measurement by directly measuring frequency, which is fundamentally different from the Michelson interferometer.

The dual-mode or multi-mode Michelson laser can convert the displacement change of the cavity mirror into a frequency change for measurement, thereby achieving precise displacement measurement. Specifically, according to the relationship between the cavity length and laser frequency in the laser resonator $\Delta\nu/\nu = dl/L$, the change in the resonator length L of the laser resonator dl will cause a change in the laser frequency ν, $\Delta\nu$, which means that a slight change in the laser cavity length brings about a significant change in the laser frequency. In the Michelson laser, one of the laser cavities is stable as a reference arm, and the other optical path is moveable as a measurement arm. The change in the position of the corresponding cavity mirror brings about a change in the laser frequency. By testing the change in the beat signal of the dual mode Michelson laser, we can infer the displacement change of the measurement arm cavity mirror. Since the light frequency is of the order of 10^{14} Hz, the current relative measurement resolution of the beat signal $\Delta\nu/\nu$ is much better than 10^{-12}, according to the proportional relationship between the intra-cavity laser frequency and the cavity length $\Delta\nu/\nu = dl/L$ and hence, the displacement measurement resolution of the Michelson laser can theoretically be far better than 10^{-12} m. Under special conditions of vacuum thermal insulation and vibration isolation, we can reduce the interference of factors such as air refractive index, mechanical vibration, and temperature changes, which improves the displacement measurement accuracy of this system.

Due to the presence of a beam splitter in the laser, the loss in the cavity increases, the gain of the He–Ne laser diode is not large enough, and it is hard to adjust the system to oscillate. In addition, the laser modes of this composite cavity are prone

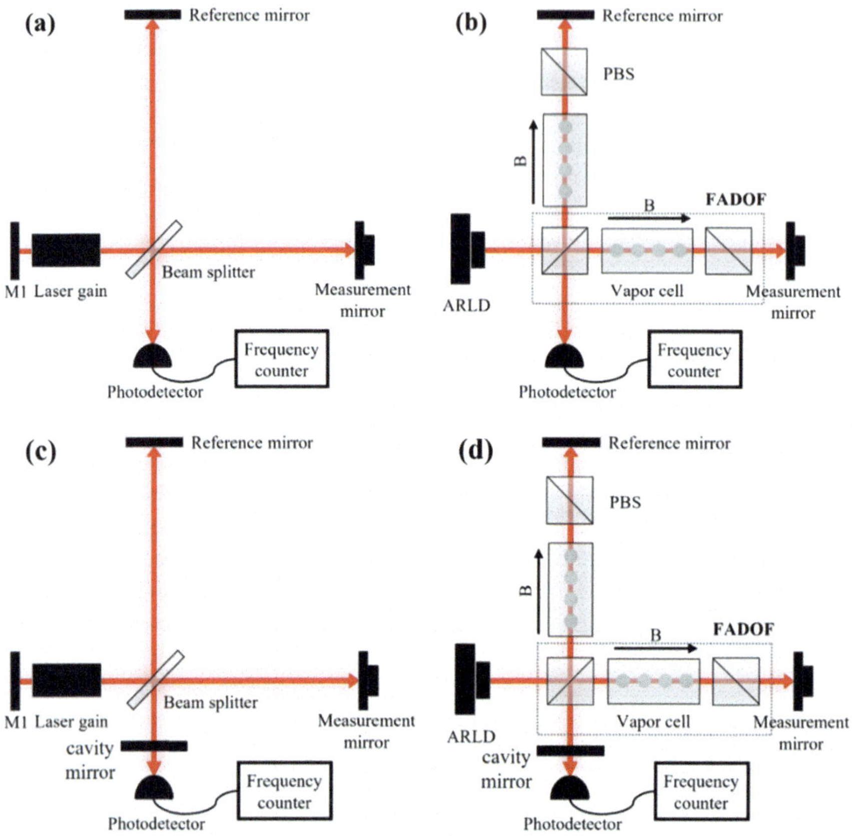

Fig. 8.13 **a** Structural diagram of Michelson laser; **b** structural diagram of Faraday–Michelson laser; **c** structural diagram of multi-mode Michelson laser; **d** structural diagram of multi-mode Faraday–Michelson laser

to compete, which further increases the loss and is inconvenient to use. Professor Chen's team from Peking University proposed a new type of Michelson laser based on FADOF (Faraday–Michelson laser), which replaces the He–Ne laser diode with a laser diode, or Faraday laser pumped alkali metal gas laser, to provide higher gain. FADOF is placed in the two sub-cavities to select the frequency of the two lasers. By setting the appropriate temperature and magnetic field parameters for the filter, the transmission peaks of the two filters produce a specific frequency difference (GHz level), preventing mode competition between the laser modes, and finally achieving dual-frequency laser output of different frequencies in the two arms.

Figure 8.13b shows the structure of this Faraday–Michelson laser. The Faraday–Michelson laser uses a Michelson interferometer-style composite cavity structure, and a FADOF is placed in each of the two sub-cavities for frequency selection. By setting appropriate FADOF parameters, the transmission peaks of the two FADOFs correspond to different atomic transition frequencies, preventing mode competition when the two modes are too close. The gain medium is a laser diode with a high reflection film coated on the back end, which can effectively avoid the inability to oscillate due to excessive loss in the cavity. In addition, a 780 nm Faraday laser can be used as the pump source to achieve a Michelson laser output of 795 nm from the Rb gas laser. The gain medium of alkali metal gas has a higher laser gain and can achieve a higher power Michelson laser output. The designs of a multi-mode Michelson and Faraday–Michelson laser are shown in Fig. 8.13c, d. A high-reflection mirror is set at the laser output end to fully utilize the laser gain to achieve oscillation while ensuring continuous operation of multiple longitudinal modes and the stable operation of the Michelson laser. The proof of principle of a Faraday–Michelson laser utilizing the Rb narrowband FADOF as the frequency selection elements, with simultaneous two laser modes oscillation has been demonstrated experimentally.

Unlike the traditional laser interferometer scheme dividing bright and dark interference fringes of light intensity for displacement measurement, the Michelson laser traces the displacement measurement back to the laser frequency measurement, effectively utilizing the high resolution of frequency measurement. It is expected to improve the measurement resolution further to the pm level or even much finer and is usable in applications such as atomic interference gravimeters[16]. Figure 8.14 shows the schematic diagram of atomic gravimeter based on the Michelson laser [17].

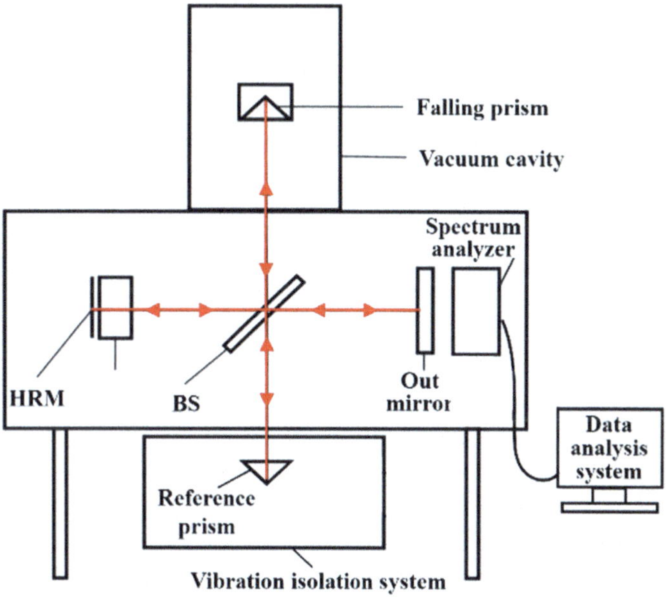

Fig. 8.14 Schematic diagram of the atomic gravimeter based on the Michelson laser

8.8 Cold Atom Faraday Laser

Faraday lasers have the advantages of output laser frequency directly corresponding to atomic transition lines, insensitivity to temperature and current fluctuations of laser diodes, etc. Narrow linewidth lasers of kHz level have been realized combined with heterodyne MTS technology with high SNR. Furthermore, we can use laser cooling technology (such as magneto-optical traps (MOT), diffuse reflection laser cooling, etc.) to suppress the influence of the Doppler effect. This section takes the cooling of ^{87}Rb atoms in a MOT as an example, first introducing the Faraday cold atom filter that rotates light using the principle of magneto-optic birefringence, and the induced birefringence cold atom filter that rotates light using the principle of photo-induced birefringence; then, based on the ultra-narrow band transmission of the atomic filter, the implementation method of the cold atom Faraday laser is given.

8.8.1 Ultra-narrow Bandwidth Faraday Atomic Filter Based on Cold Atoms

The difference in structure between the cold atom Faraday filter and the hot atom Faraday filter is essentially the replacement of the hot atom vapor cell with a cold atom ensemble captured by a MOT. The overall framework design is shown in Fig. 8.15. The cold atom Faraday filter mainly consists of three parts: a MOT system for capturing cold atoms, a detection laser system for detecting the light rotation, and the polarization components and magnetic field required for the Faraday filter.

The MTS technique is used for precise locking of the detection laser frequency. The detection laser passes through an attenuator (for adjusting the power of the detection laser), a Glan–Taylor prism, and an ensemble of cold atoms trapped in the MOT, and is then reflected by a 0° mirror along the original direction. Here, the purpose of returning the detection laser along the original path is to reduce the force

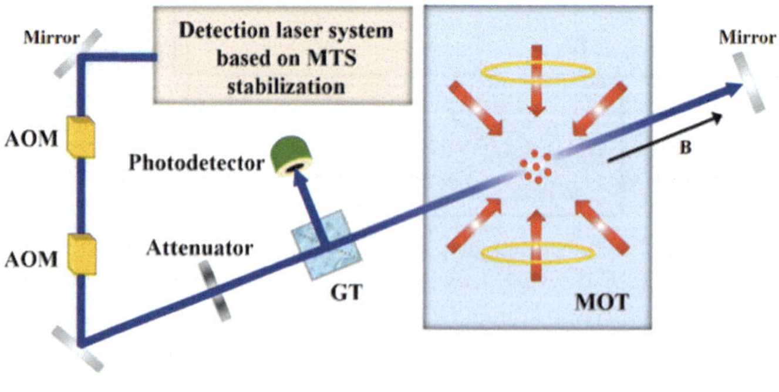

Fig. 8.15 Overall framework of the cold atom Faraday filter

exerted by the one-way detection laser on the cold atom group. Still, the incident and reflected detection lasers should be adjusted to overlap completely. The transmission signal of the cold atom Faraday filter is received by a high-speed photodetector at the other end of the Glan–Taylor prism. The uniform magnetic field for light rotation is generated by a pair of energized coils, whose direction is parallel with the direction of the incident detection laser.

Unlike continuous working hot atom Faraday filters, cold atom Faraday filters require careful consideration and repeated practice of the timing relationship between the closure of the MOT and the detection of the Faraday rotation effect. On the one hand, the detection laser and the Faraday rotation magnetic field may affect the capture of atoms by the MOT; on the other hand, the gradient magnetic field of the MOT may affect the atomic energy levels of the Faraday rotation. Therefore, a sequence is necessary to capture atoms and detect the Faraday rotation effect. After capturing enough cold atoms, the MOT is closed, the cold atoms are released, and then the detection laser and the Faraday rotation magnetic field are turned on to measure the transmission signal of the cold atom Faraday filter. This sequence is repeated continuously. Various factors should be considered to determine the optimal timing interval and opening time. For example, if the gradient magnetic field of the MOT is closed too slowly, the remaining magnetic field may affect the atomic energy levels of Faraday rotation; if the Faraday rotation signal is not detected for a long time after the MOT is closed, the cold atom group may diffuse, resulting in a scarcity of atoms; if the detection laser and the Faraday rotation magnetic field are turned on for too long, they may exert force on the cold atom group, accelerating the diffusion of the cold atom group. Through multiple observations and measurements, the timing relationship of the cold atom ultra-narrow bandwidth Faraday atomic filter is shown in Fig. 8.16.

Figure 8.17 shows the transmission spectrum of a 780 nm cold atom Faraday filter working on the ^{87}Rb atom D_2 line transition (natural linewidth: 6.1 MHz). The frequency of the detection laser is locked to the $5^2S_{1/2}F = 2$ to $5^2P_{3/2}F' = 3$

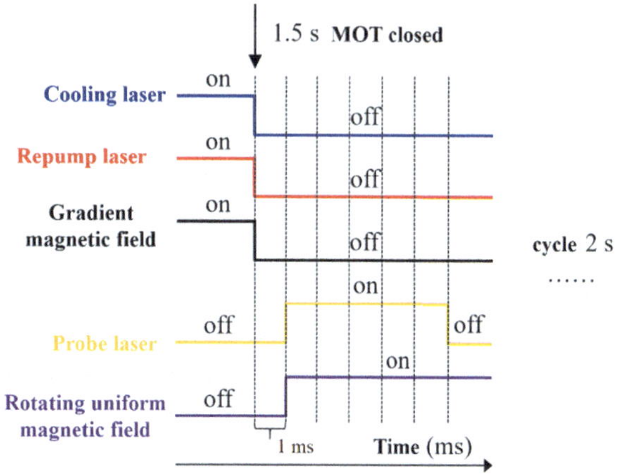

Fig. 8.16 Timing relationship between the shutdown of the MOT and the detection of the Faraday rotation effect

transition, with a power of 10 μW. The intensity of the Faraday uniform magnetic field is 3 G. The Lorentz-fitted transmission bandwidth (red curve) of the filter is 7.1(8) MHz, close to the natural linewidth of the atomic transition (6.1 MHz), and the peak transmission is 2.6%. Similarly, the transmission spectrum of a 420 nm cold atom Faraday filter working on the ^{87}Rb atom $5^2S_{1/2}F = 2$ to $6^2P_{3/2}F' = 3$ transition is shown in Fig. 8.18, with a transmission bandwidth of 2.7(2) MHz and a peak transmission of 3.2%, which is the narrowest bandwidth atomic filter known to date.

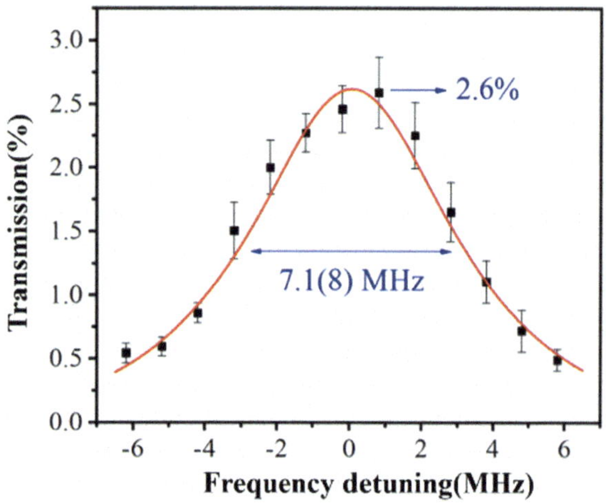

Fig. 8.17 Transmission spectrum of cold atom Faraday filter working on ^{87}Rb atom 780 nm $5^2S_{1/2}$ (F = 2) to $5^2P_{3/2}$ ($F' = 3$) transition

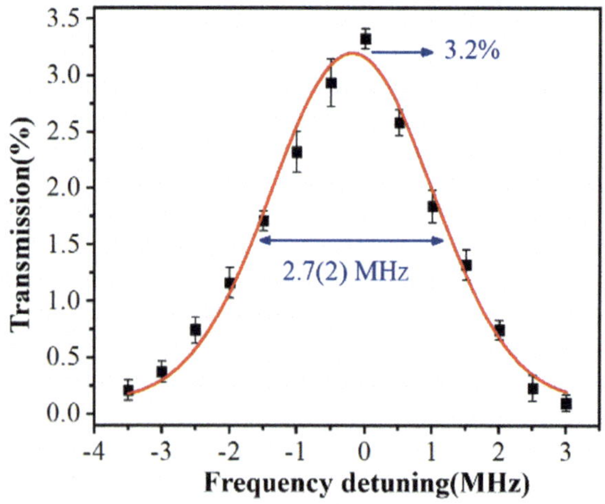

Fig. 8.18 Transmission spectrum of cold atom Faraday filter working on ^{87}Rb atom 420 nm $5^2S_{1/2}$ (F = 2) to $6^2P_{3/2}$ ($F' = 3$) transition

8.8.2 Cold Atom-Based Induced Birefringence Atomic Filter

Taking the 780 nm transition of ^{87}Rb atoms as an example (as shown in Fig. 8.19), according to the selection rules of the corresponding $5^2S_{1/2}F = 2$ to $5^2P_{3/2}F' = 2$ transition, circularly polarized pumping laser gradually pumps atoms from the lower energy level of the ground state to the higher energy level, making the atoms accumulating at the $5^2S_{1/2}$ ($F = 2$, $m_F = +2$) energy level. The atoms located at the ground state $5^2S_{1/2}$ ($F = 2$, $m_F = +2$) energy level can only absorb the left circularly polarized component of the linearly polarized detection laser and thus transition to the $5^2P_{3/2}$ ($F' = 2, m_{F'} = +1$) energy level, the right circularly polarized component is hardly absorbed. This asymmetric absorption causes a big refractive index difference between the left and right circularly polarized light components in the linearly polarized detection laser, because the rotation angle of the filter is proportional to the refractive index difference between the left and right circularly polarized light components, so a larger angle and higher transmission of rotation are formed.

The schematic diagram of the bidirectional color type cold atom filter is shown in Fig. 8.20. The stabilized detection laser is first transmitted to the attenuator to adjust the laser power, then passes through the cold atom group trapped in the vacuum system, a pair of Glan–Taylor prisms with orthogonal polarization directions, and then enters the photodetector, where the photodetector measures the transmission signal of the cold atom filter. At the same time, a beam is split from the stabilized cooling laser to serve as the pumping laser. This pumping laser sequentially passes through a half-wave plate (HWP), a polarizing beam splitter, an expanding lens group, and a quarter-wave plate (QWP), and is then reflected by a mirror with a T:R = 9:1 ratio. The reflected pumping and detection laser overlap in the opposite directions and pass through the cold atom group trapped in the vacuum system. Here, the HWP and the polarizing beam splitter are used to adjust the optical power of the pumping laser and make the pumping laser into pure linearly polarized light. The expanding lens group is used to expand the pumping laser to fully encompass the cold atom group when passing through the vacuum system, thereby improving the efficiency

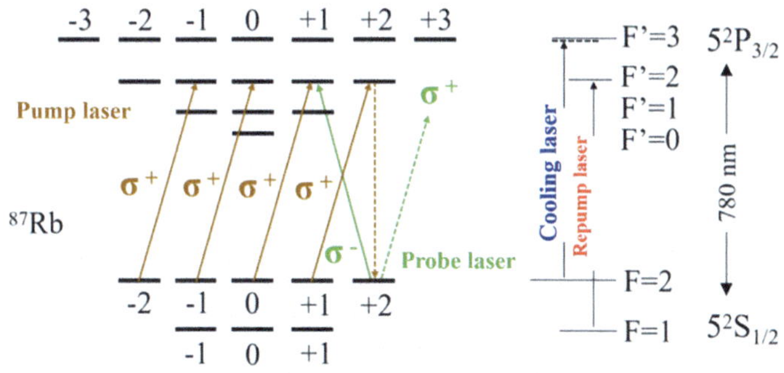

Fig. 8.19 Energy level diagram related to induced birefringence cold atom filter

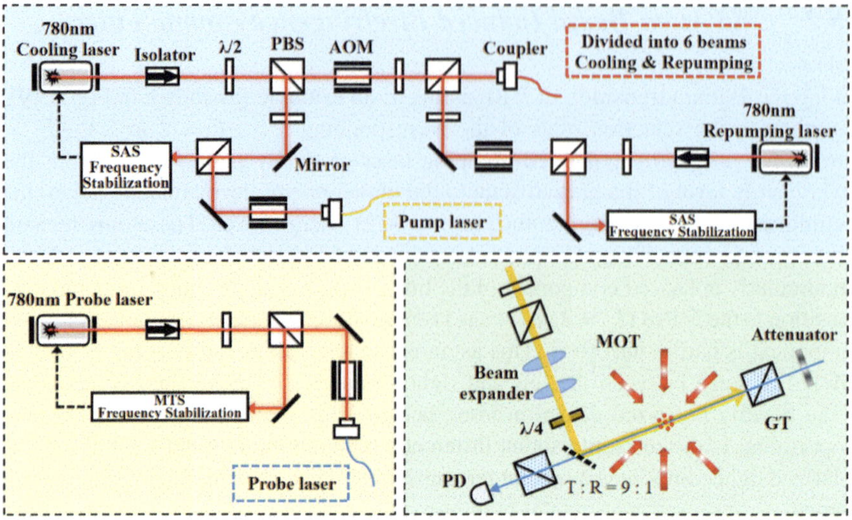

Fig. 8.20 Schematic diagram of the induced birefringence cold atom Faraday filter

of optical pumping atoms. The QWP is to convert the linearly polarized pumping laser into a standard circularly polarized pumping laser.

It's worth noting that atoms pumped to the $5^2P_{3/2}F' = 2$ energy level by the circularly polarized laser may fall to the ground state $5^2S_{1/2}F = 1$ energy level by the process of spontaneous radiation, causing atom accumulation and resulting in losses. In this case, the repumping laser during the stage of capturing cold atoms in the MOT keeps on, allowing atoms on the ground state $5^2S_{1/2}F = 1$ energy level to return to the $5^2S_{1/2}F = 2$ energy level. In addition, the time setting between the pumping laser and the detection laser is also crucial. It is necessary not only to ensure that the cold atoms are effectively pumped to the desired Zeeman sublevel but also to avoid the pumping laser having a significant impact on the cold atom group, thereby affecting detection. After exploration and improvement, the timing relationship of the induced birefringence color type cold atom filter can be referred to as shown in Fig. 8.21.

Figure 8.22 shows the operation at ^{87}Rb atom 780 nm $5^2S_{1/2}F = 2$ to $5^2P_{3/2}F' = 2$ transition of the induced bichromatic cold atomic filter's transmission spectrum. The Lorentz-fitted bandwidth of the transmission spectrum is 6.6(4) MHz, and the peak transmission is 15.6%.

8.8.3 Cold Atom Faraday Laser

Diffuse reflection laser cooling of atoms does not require a magnetic field, which avoids the influence of the gradient magnetic field of the MOT on the atomic energy level. In addition, the long tube-shaped diffuse reflection laser cooling system can

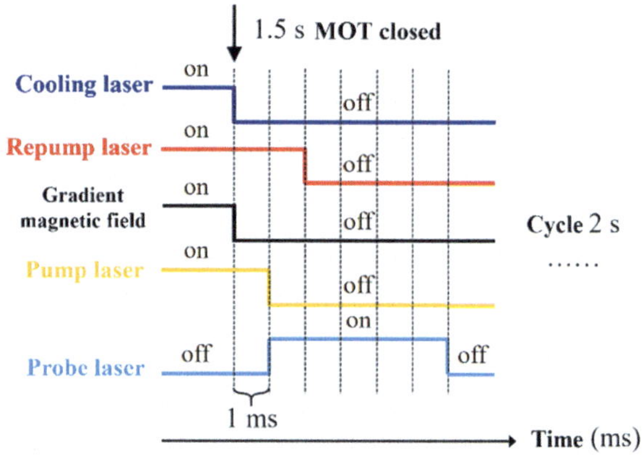

Fig. 8.21 Timing diagram of the induced birefringence cold atom Faraday filter

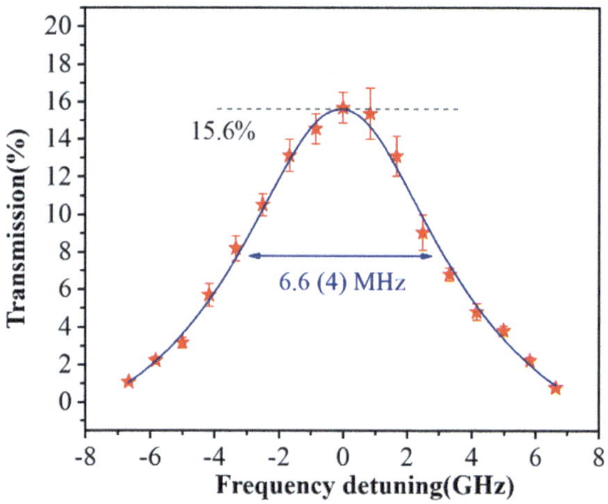

Fig. 8.22 Transmission spectrum of the induced birefringence type cold atomic filter working at ^{87}Rb 780 nm $5^2S_{1/2}F = 2$ to $5^2P_{3/2}F' = 2$ transition

achieve a higher optical density compared to the spherical cavity. This section gives the implementation method of the cold atom Faraday laser with the long glass tube diffuse reflection laser cooling system as an example.

As shown in Fig. 8.23, the stabilized cooling laser and the repump laser are combined and evenly divided into 6 beams coupled into the fiber and then shot into the long glass tube vacuum cavity from the side. This vacuum glass cavity contains a thin ^{87}Rb atomic vapor and ^{87}Rb cold atom clusters.

A coil is wound around the outside of the long glass tube vacuum cavity to apply a magnetic field to the atoms, constructing a Faraday atomic filter. Under the effect of the Faraday magneto-optical effect, the output polarization plane rotates, forming an ultra-narrow band transmission corresponding to the atomic hyperfine transition. By adjusting the angle of the coated external cavity mirror with a certain transmittance, the oscillation forms between the back surface of the laser diode and the cavity mirror, and stimulated radiation is emitted within the ultra-narrow band

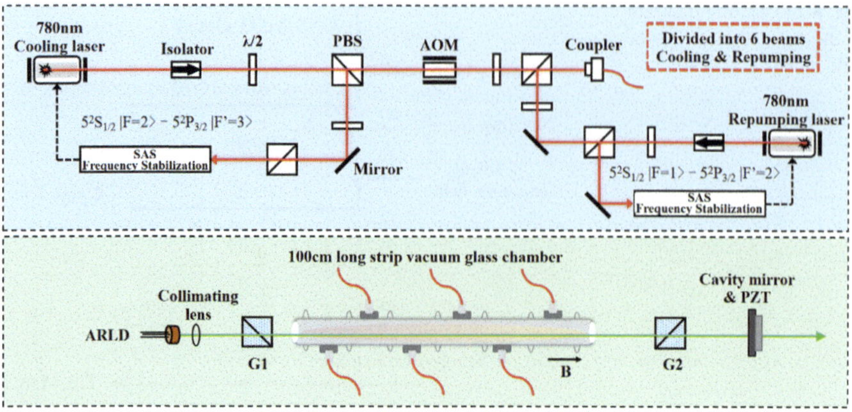

Fig. 8.23 Basic structure of the cold atom Faraday laser

transmission bandwidth, thus realizing the cold atom Faraday laser. It's worth noting that although the cold atom Faraday laser can achieve a narrower laser linewidth, the transmission of the cold atom Faraday filter is much lower than that of hot atoms, which may make it difficult for Faraday lasers to oscillate.

8.9 Faraday Active Optical Clock

The traditional optical clock locks the frequency-stabilized local oscillator laser onto the quantum transition spectrum line through frequency discrimination and phase discrimination technology, which is a passive optical clock. Passive optical clocks usually need to use PDH technology to narrow the laser linewidth and obtain ultra-stable local oscillator lasers. However, the Johnson–Nyquist noise, also known as thermal noise, exists in the PDH cavity. It's the electronic noise when the charge carriers (usually electrons) inside the conductor reach equilibrium due to heat. It is necessary to reduce thermal noise through an ultra-low temperature environment. The PDH technology implemented with ultra-low temperature single-crystal silicon has achieved a laser output with a linewidth of 8 mHz. However, this scheme is complex, costly, and hard to maintain. To break through the high requirements and restrictions of the passive optical frequency standard on the super cavity, in 2005, Chen et al. from Peking University [18–20] proposed a new principle and technology for active optical clocks in principle, placing atoms with ultra-narrow linewidth transition spectrum lines in a low-finesse optical resonator, using the stimulated emission output of atoms in the bad cavity directly as a stable optical frequency standard. In this scheme, the output signal depends on the atom itself, and hence, the performance of the optical clock is expected to be further improved.

8.9.1 Basic Principle of Active Optical Clock

The active optical clock is a new concept based on the coherent stimulated radiation of multiple atoms in a bad cavity, which is comparable to the active hydrogen clock in the microwave band. The principle of the active hydrogen clock is based on the stimulated radiation of hydrogen atoms. The excited hydrogen atoms enter the storage bubble and interact with the microwave resonator to produce stimulated radiation. The output microwave signal serves as the frequency standard signal. Because the frequency signal generated by this stimulated radiation has a purer spectrum and narrower linewidth, the short-term stability of the active hydrogen clock is usually higher than that of the Rb and Cs clock. In the optical band, the working principles of the active and passive optical clock are compared, as shown in Fig. 8.24.

The traditional passive optical clock uses an electronic servo loop to stabilize the PDH-stabilized local oscillator laser on the clock transition line and electrical feedback to improve long-term stability. In contrast, the active optical clock does not need a local oscillator laser. It forms multi-atom coherent stimulated radiation between atomic transition energy levels through weak feedback in the bad cavity, and the emitted laser can be directly used as the clock laser signal. In addition, the active optical clock works in the bad cavity, where the atomic gain linewidth is much smaller than the resonator cavity mode linewidth, and the output clock laser frequency depends on the quantum transition frequency rather than the external reference cavity. The cavity pulling coefficient P demonstrates that the stimulated radiation output frequency ν varies with the rate of change of cavity mode frequency ν_{cavity} as [22],

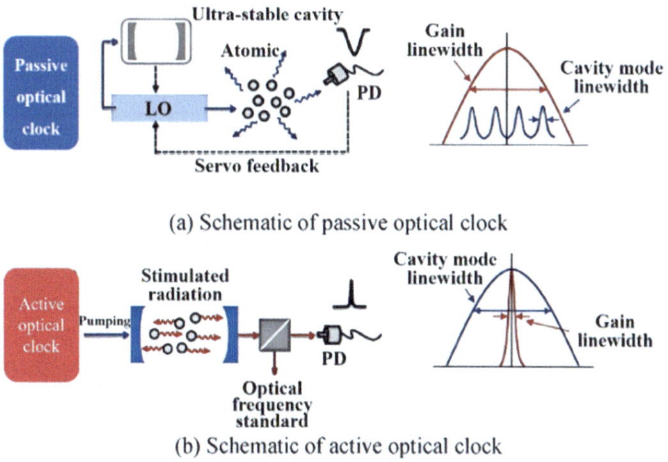

Fig. 8.24 Schematic diagram of the working principles of **a** passive optical clock and **b** active optical clock [21]

$$P = \frac{d\nu}{d\nu_{cavity}} = \frac{\Gamma}{\kappa + \Gamma} = \frac{1}{1 + a} \tag{8.1}$$

where Γ is the atomic gain linewidth, κ is the cavity mode linewidth. For bad cavity lasers, the bad cavity coefficient $a = \kappa/\Gamma \gg 1$, so the cavity pulling effect can be suppressed effectively. Secondly, Schawlow–Townes laser linewidth formula can be modified for under bad cavity conditions as [23],

$$\Delta\nu_{correction} = \frac{h\nu_0}{4\pi} \frac{\kappa^2}{P_{Output}} N_{sp} \left(\frac{\Gamma}{\kappa + \Gamma}\right)^2 \tag{8.2}$$

Here it is assumed that the cavity mode center frequency coincides with the atomic transition center frequency ν_0, where P_{output} is the power of the output signal and N_{sp} is the spontaneous emission factor. The first two terms of the formula represent the quantum-limited linewidth determined by spontaneous emission. However, the active optical clock can narrow the quantum-limited linewidth to $1/(1 + a)^2$, breaking the spontaneous radiation quantum noise limit linewidth.

Therefore, the active optical clock has two main advantages: (1) a narrow quantum limit linewidth; (2) cavity-pulling suppression. Compared with the traditional passive optical clock, the signal output of the active optical clock depends on the atomic transition spectrum line, which is used as the frequency standard signal directly. In principle, it effectively solves the problem of the stability limitation caused by the thermal noise of the resonant cavity in the passive optical clock system. Its output frequency linewidth is expected to reach the millihertz level, which has important application prospects and research significance in quantum precision measurement.

8.9.2 Faraday Active Optical Clock Scheme

The Faraday active optical clock uses the Faraday atomic filter as the intra-cavity frequency reference. Semiconductors, solids, dyes, and other broadband gain media are optional to provide intra-cavity light gain, achieving a stable stimulated radiation output signal. The principle is to use FADOF as the frequency selector to realize the independence of the gain and the quantum reference, thereby reducing the noise impact on the clock laser frequency stability. Considering that the bandwidth of the FADOF is usually widened due to the Doppler effect, it can generally reach the GHz level. For the active optical clock, to achieve a bad cavity coefficient much greater than 1, a narrow-bandwidth atomic filter is necessary as the gain medium. The velocity selection effect can produce ultra-narrow band atomic filters called induced-dichroism-excited atomic line filters (IDEALF) [24]. The basic principle is to use the laser and incident light to enter the FADOF in the opposite directions, using the saturated absorption effect to select atoms with a velocity close to 0 for filtering. The Doppler broadening of these atoms is very small, so the atomic filter made by this method can achieve MHz sub-Doppler bandwidth. However, this filter only selects a part of the atoms, and its transmittance is significantly lower than that of FADOF, which increases the single-pass loss of light, making it hard to achieve the stimulated radiation output of the Faraday active optical clock.

In 2014, Zhuang et al. from Peking University [25] realized the thermal atomic Faraday active optical clock based on the new principle of the active optical clock, combining the characteristics of FADOF and IDEALF. The experiment uses a strong pump light to enter the atomic vapor cell in the opposite direction to select the velocity of the atoms. Different from the above IDEALF, they added a weak magnetic field in the axial direction of the atomic vapor cell. The strength of this magnetic field is much smaller than the magnetic field strength of FADOF, making the Zeeman splitting of the atoms approximately equal to the width of the SAS. The bandwidth of the atomic filter obtained from this can be narrower than that of IDEALF, and the transmittance is significantly improved compared to it.

The experimental setup of the Faraday active optical clock, as shown in Fig. 8.25, can be divided into four main parts: the main system of the Faraday active optical clock, the detection laser system, the pump laser system, and the performance testing system. The main system consists of a laser diode and a FADOF placed in an external cavity structure, with its stimulated radiation output serving as the clock signal. The detection laser system is composed of an 852 nm ECDL, used to detect the performance of the atomic filter. The pump laser system consists of another 852 nm ECDL and a SAS locking system, used to pump the FADOF. The performance testing system consists of two beat frequency optical paths and corresponding high-speed detectors and spectrometers, used for the beat frequency of the two output optical beams of the Faraday active optical clock and the beat frequency of one of the optical beams with the pump light.

The following introduces the Faraday active optical clock system and its working principle, where the two orthogonally polarized elements Glan–Taylor prisms GT1 and GT2 have an isolation ratio of up to 10^5:1, and the transmission of the 852 nm laser reaches 85%. The cesium vapor cell Cs_1, with a length of 15 cm and a diameter of 2.54 cm, is temperature-controlled at room temperature. The magnetic field coil generates a weak magnetic field along the axis of the cesium vapor cell when energized. The temperature control system includes a heating wire wrapped around the outside of the cesium vapor cell, a thermistor, and a temperature control circuit. The magnetic shielding box, with a volume of 20 cm $\times$ 10 cm $\times$ 10 cm, is to isolate the external magnetic field's effect on the cesium vapor cell. It has two light holes with a diameter of 4 cm at both ends for the incident light and reverse pump light to pass through. The pump light system consists of an 852 nm ECDL, with an output power of up to 50 mW. Part of its output light is reflected by the beam splitter BS_2 to the SAS device cesium vapor cell Cs_2, and the pump light frequency is locked to the SAS using a phase-locked amplifier and PID circuit. The other part of the output light passes through an expander, with the beam diameter enlarged three times to cover the incident light of the atomic filter. Then, the power of the pump light is adjusted using a neutral density filter (NDF), and the pump light is changed from linearly to circularly polarized light using a QWP. The detection light system consists of another 852 nm external cavity diode laser, with an output power of less than 1 mW

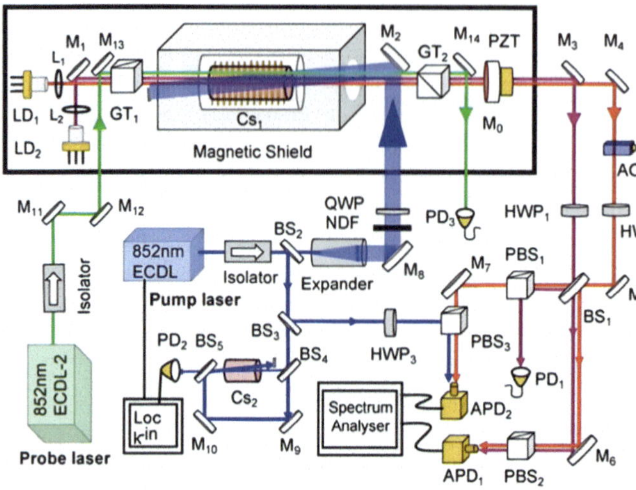

Fig. 8.25 Experimental scheme of Faraday active optical clock [25]. *LD* laser diode, *L* lens, *M* mirror, *GT* Glan–Taylor prism, *PZT* piezoelectric ceramic, *AOM* acousto-optic modulator, *HWP* half-wave plate, *QWP* quarter-wave plate, *NDF* neutral density filter, *ECDL* external cavity diode laser, *PD* photodiode, *APD* avalanche photodiode, *PBS* polarizing beam splitter, *BS* beam splitter

and a beam diameter of 0.4 mm. It is incident on the atomic filter, and its frequency is tunable within 10 GHz. The power of the detection light after passing through the Faraday atomic filter is detected using a photodetector PD_3, thereby obtaining the transmission spectrum of the filter. To test the performance of the Faraday active optical clock, two laser diodes of the same model, LD_1 and LD_2, are used. After the output light passes through the collimating lens, it enters the Faraday atomic filter placed in the resonant cavity to form stimulated radiation, and finally emits the clock signal from the cavity mirror M_0 with a reflectivity of 80%. The external cavity feedback of LD_1 and LD_2 can be directly achieved by adjusting the angle of M_0 and M_1, respectively. After adjustment, the two beams of light are parallel and can achieve feedback simultaneously. The two light paths share a Faraday atomic filter and an external cavity feedback mirror to minimize the impact of external vibration noise.

The experimental results are shown in Fig. 8.26, where the red curve represents the atomic transmission spectrum of the ^{133}Cs Faraday filter. The center of the transmission spectrum corresponds to the SAS of the black curve ^{133}Cs atom $6^2S_{1/2}$ ($F = 4$) to $6^2P_{3/2}$ ($F' = 4$ and 5) crossover peak. The filter has a transmission of 25% and a bandwidth of 45 MHz. The blue curve represents the cavity mode linewidth theoretically calculated considering various actual losses in the external cavity, is found to be 298.3 MHz, corresponding to the bad cavity coefficient 6.6, and the cavity pulling coefficient P of 0.13. The frequency linewidth of the output light is hundreds of Hz through the beat frequency of the two sets of devices, which is limited by the vibration noise of the cavity mode. In addition, by changing the cavity mode frequency, the cavity pulling coefficient can be obtained after calculating the change of the

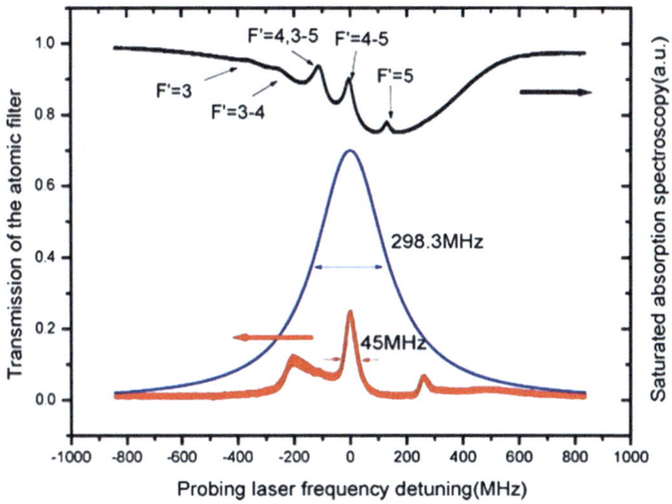

Fig. 8.26 The Faraday filter transmission spectrum of the crossover peak of ^{133}Cs atom $6^2S_{1/2}$ ($F = 4$) to $6^2P_{3/2}$ ($F' = 4$ and 5) [25]

beat signal. The final experimental measurement value is 0.11, which fits well to the theoretical calculation value of 0.13. This result indicates that due to the suppression of the cavity-pulling effect, the Faraday active optical clock system obtains a narrower linewidth than ordinary lasers. Since the system does not use vibration isolation facilities, although the vibration noise still has a significant impact under the suppression of cavity pulling, the current linewidth measurement value still has a gap as compared to the theoretical value of 0.33 Hz [25].

8.9.3 Future Prospects of Faraday Active Optical Clock

However, for Faraday active optical clocks realized using thermal atoms, it is difficult to narrow the transmission bandwidth of the atomic filter to the level of the natural linewidth, which makes it challenging to further enhance the bad-cavity coefficient. Consequently, the experimentally achieved laser linewidth has not yet reached the level predicted by theoretical calculations. In the future, cold atom clusters can be used as quantum frequency reference selection devices to suppress the Doppler effect, reduce the transmission bandwidth, and improve the cavity pulling suppression coefficient, thereby narrowing the laser linewidth. Peking University has also proposed an optical lattice active optical clock, which uses the Cs atomic ensemble captured in the optical lattice to provide a continuous super-radiation laser signal; the magic wavelength of its optical lattice is calculated, and the atoms are confined in the Lamb–Dick region, which can avoid the influence of optical frequency shift. Theoretically, the frequency uncertainty is at the 10^{-15} level; the linewidth is at the Hz level. To further reduce the impact of hot atoms on the active optical frequency standard signal, such as collision frequency shift, collision broadening, etc., researchers at Peking University proposed cold atom active optical clocks, which can benchmark the EU's iq Clock project. Two types of Faraday active optical clock schemes based on cold atoms have also been proposed [26], the specific experimental principal diagrams of which, are shown in Figs. 8.27 and 8.28.

The active optical clock, as a new concept of optical clock, essentially uses the correlation of atoms to achieve a highly coherent bad cavity stimulated radiation light field, where the frequency of the light field is referenced to quantum transitions. The natural immunity of the active optical clock frequency to the thermal noise of the resonant cavity length allows it to break through the thermal noise limit of the resonant cavity, achieving ultra-narrow linewidth.

Regardless of how the optical clock develops in the future, the narrow linewidth local oscillator laser, which is insensitive to environmental noise based on different principles, is one of the most critical technical bottlenecks of the optical clock. By further researching and developing the cold atom Faraday active optical clock, in addition to realizing a new type of optical clock, it is likely to use its bad cavity effect

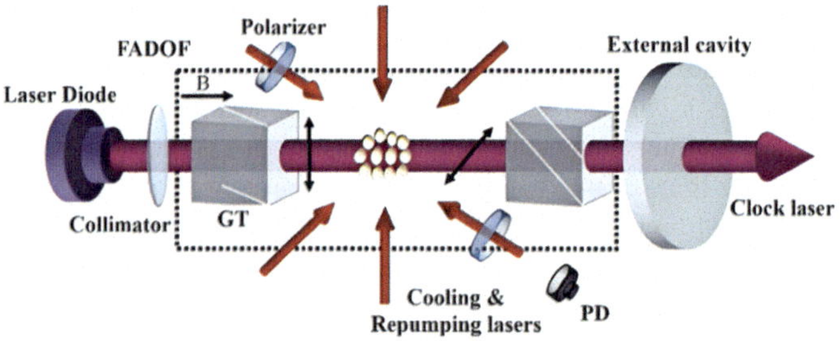

Fig. 8.27 Schematic diagram of the two-level cold atom Faraday active optical clock based on gray optical clump [26]

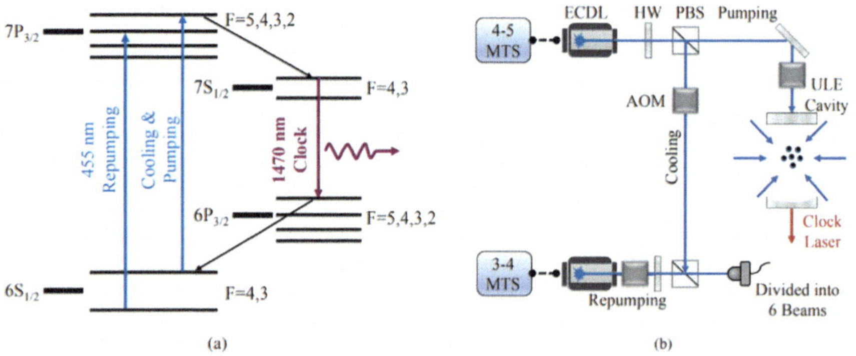

Fig. 8.28 Schematic diagram of the four-level cold atom Faraday active optical clock based on blue light MOT: **a** related energy level; **b** experimental principle diagram [26]

to bring immunity to external noise, providing a more stable, narrower linewidth local oscillator laser for future optical clocks, thereby promoting the progress of quantum precision measurement.

8.10 Optical Fiber and Ultra-long Cavity Faraday Laser

For the mode hopping of Faraday lasers, even if mode hopping occurs, the output frequency remains within the transmission bandwidth of the FADOF. There are two methods to limit the significant frequency hopping of Faraday lasers further. One is to use a short resonant cavity, while increasing the transmission spectrum width of the FADOF, ensuring that the FSR of the Faraday laser is wider than the transmission spectrum of the FADOF. Therefore, only one cavity mode within the transmission spectrum of the FADOF can form laser output. This method can increase the tunable range of the Faraday laser and avoid mode hopping by changing the cavity length to adjust the laser frequency. The other is to enlarge the laser cavity length to reduce the FSR, so that at least one cavity mode is always closes to the peak of the FADOF transmission spectrum, preventing significant laser frequency hopping. If the laser cavity is long enough, the FSR decreases to the kHz level, and even if mode hopping occurs, the frequency change also reduces to the kHz level, reducing the probability of significant frequency hopping.

The first method has already been implemented by the research team at Peking University. For the second method, a Faraday laser using a 1.2 km fiber as an extended cavity with narrow FSR of 83 kHz at ^{87}Rb 780 nm has been demonstrated at Peking University in 2016. There are 350 cavity modes within the FADOF transmitted peak competing with each other. Even for long-term running with the length free drift of 1.2 km fiber, the stability of 6.0×10^{-11} can be kept [27]. Considering practical applications, the fiber resonant cavity is one of the best choices to increase the length of the Faraday laser resonant cavity. Compared to the free-space laser, an all-fiber structure could improve mechanical stability, overall volume, and portability. When the cavity length increases to the 10 km level, the FSR decreases to the kHz level, reducing the frequency range of mode hopping compared to free-space light.

The key to achieve all-fiber Faraday laser output is the fiber atomic filter. As shown in Fig. 8.29, this is a feasible fiber atomic vapor cell process. The fiber collimator is installed at both ends of the vapor cell throughout the process, and high-precision collimation is performed to ensure that the two beams of light have good collimation and overlap effects to ensure its performance.

In addition to the above process, the University of Stuttgart in Germany has another method for the fiber atomic vapor cell, allowing atoms to enter the fiber [28–31].

Like free-space light, the fiber atomic filter mainly plays the role of laser mode selection, including atomic filters based on principles such as Faraday anomalous dispersion and Voigt anomalous dispersion. Except for permanent magnets, other devices like PBSs and atomic vapor cells are all fiber devices, as shown in Fig. 8.29. To ensure that the all-fiber atomic filter filters effectively, the perfect cooperation of

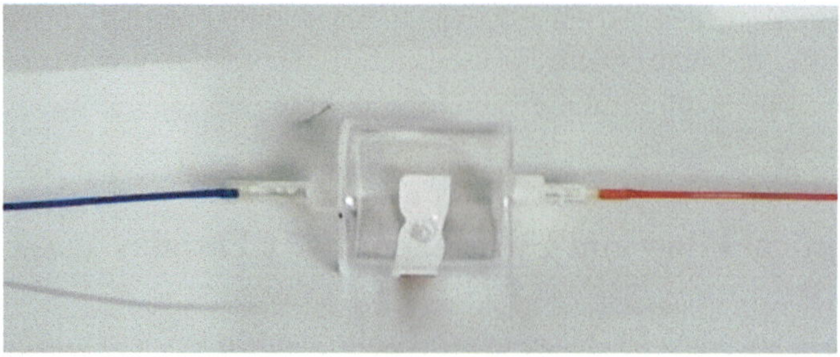

Fig. 8.29 Fiber atomic vapor cell

two fiber PBSs is required. The ARLD is utilized as the gain medium, outputting through a collimated fiber. Generally speaking, in an atomic filter, the fiber PBS in front of the vapor cell acts as a polarizer, ensuring that the light incident on the fiber cell is linearly polarized light, which is then decomposed into left and right circularly polarized light. Similar to a spatial light Faraday laser, in an all-fiber structure, due to the presence of an external magnetic field, the energy levels of atoms in the fiber cell generates Zeeman splitting, causing the left and right circularly polarized light to correspond to different transition frequencies, resulting in corresponding splitting of their absorption and dispersion curves. The dispersion curve is directly related to the refractive index of light, whose splitting causes the left and right circularly polarized light to have different refractive indexes. When passing through the same path, they have different phase changes, causing the polarization direction of linearly polarized light to deviate, where the optical rotation effect occurs. The optical rotation effect is directly related to the incident light intensity, the size of the external magnetic field, the length of the fiber cell, and the atomic density (temperature of the atomic cell) in the fiber cell. When the incident light intensity is determined, the rotation angle is determined by the parameters mentioned above. In an all-fiber atomic filter, the two fiber PBSs can transmit light with polarization directions perpendicular to each other. Therefore, only light with a rotation angle close to 90° can pass through the all-fiber atomic filter, incident on the fiber reflection cavity mirror, and then return to the ARLD, thus forming a laser oscillation output. The above optical rotation effect only occurs for light close to the atomic transition frequency, so the output frequency of the Faraday laser is always near the quantum transition frequency, which is immune to changes in the driving current and working temperature of the ARLD.

The all-fiber atomic filter can select a specific frequency to pass through, forming laser oscillation and producing a single longitudinal mode laser output. At the same time, it can also make two specific laser modes pass through by changing the parameters that affect the transmission spectrum, forming laser oscillation, and producing a single cavity dual longitudinal mode laser output. The single and dual longitudinal mode laser outputs are switchable according to actual application conditions.

The output laser frequency can be locked through atomic laser spectrum stabilization, such as MTS, SAS, polarization spectroscopy, etc. The laser frequency is

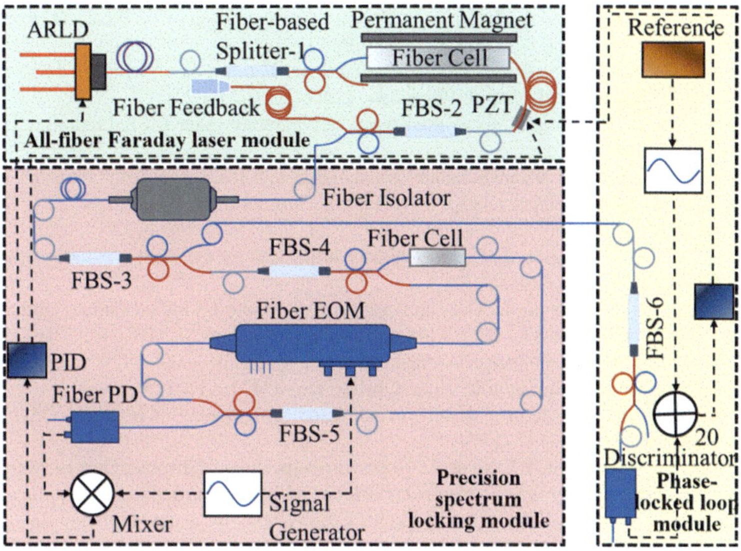

Fig. 8.30 Schematic diagram of the structure of an all-fiber Faraday laser

locked to the corresponding atomic transition line, obtaining a traceable absolute frequency reference.

The all-fiber Faraday laser can output a single or dual longitudinal mode laser as needed, as shown in Fig. 8.30. In the case of dual longitudinal mode output, one of the modes can be locked to the quantum transition line through the precision spectrum locking module. Then, the beat frequency between the two modes is locked through the phase-locked loop module, ensuring the ultra-high stability and low noise of the beat note between the two laser modes, so that both longitudinal modes can be locked to the corresponding quantum reference.

References

1. A. Podvyaznyy, G. Venus, V. Smirnov et al., 250 W diode laser for low pressure Rb vapor pumping. Proc. SPIE **7583**, 758313 (2010)
2. M.D. Rotondaro, B.V. Zhdanov, M.K. Shaffer et al., Narrowband diode laser pump module for pumping alkali vapors. Opt. Express **26**(8), 9792–9797 (2018)
3. H. Tang, H. Zhao, R. Wang et al., 18 W ultra-narrow diode laser absolutely locked to the Rb D_2 line. Opt. Express **29**(3), 38728–38736 (2021)
4. S. Yokoyama, T. Araki, N. Suzuki, Intermode beat stabilized laser with frequency pulling. Appl. Opt. **33**(3), 358–363 (1994)
5. G. Quehl, J. Grinert, V. Elman et al., A tunable dual frequency Tm: YAG laser. Opt. Commun. **190**(1–6), 303–307 (2001)
6. M.A. Hafiz, D. Brazhnikov, G. Coget et al., High-contrast sub-Doppler absorption spikes in a hot atomic vapor cell exposed to a dual-frequency laser field. New J. Phys. **19**(7), 073028 (2017)

7. M.A. Hafiz, G. Coget, E. De Clercq et al., Doppler-free spectroscopy on the Cs D_1 line with a dual-frequency laser. Opt. Lett. **41**(13), 2982–2985 (2016)
8. P. Chang, T. Shi, J. Miao et al., Frequency-stabilized Faraday laser with 10^{-14} short-term instability for atomic clocks. Appl. Phys. Lett. **120**(14), 1102 (2022)
9. T. Shi, X. Guan, P. Chang et al., A dual-frequency Faraday laser. IEEE Photonics J. **12**(4), 1503211 (2020)
10. X. Qin, Z. Liu, H. Shi et al., Switchable Faraday laser with frequencies of ^{85}Rb and ^{87}Rb 780 nm transitions using a single isotope ^{87}Rb Faraday atomic filter. Appl. Phys. Lett. **124**(15), 161104 (2024)
11. B. Luo, L. Yin, J. Xiong et al., Signal intensity influences on the atomic Faraday filter. Opt. Lett. **43**(11), 2458–2461 (2018)
12. J. Xiong, B. Luo, L. Yin et al., The characteristics of Ar and Cs mixed faraday optical filter under different signal powers. IEEE Photonics Technol. Lett. **30**(8), 716–719 (2018)
13. V.A. Sautenkov, M.N. Shneider, S.A. Saakyan et al., A self-focusing of CW laser beam with variable radius in rubidium atomic vapor. Opt. Commun. **431**, 131–135 (2019)
14. J.E. Bjorkholm, A.A. Ashkin, CW self-focusing and self-trapping of light in sodium vapor. Phys. Rev. Lett. **32**, 129–132 (1974)
15. T.D. Nguyen, M. Higuchi, T.T. Vu et al., 10-pm-order mechanical displacement measurements using heterodyne interferometry. Appl. Opt. **59**(27), 8478–8485 (2020)
16. X. Qin et al., High resolution displacement measurement by a novel Michelson laser, in *Proceedings of the 37th European Frequency and Time Forum*, Neuchatel, Switzerland, 25–27 June 2024
17. J. Chen, X. Qin, T. Shi et al., A Michelson laser and its implementation method, displacement measurement method. Patent number: ZL202210393703.X, 2 May 2023
18. J.B. Chen, X.Z. Chen, Optical lattice laser, in *Proceedings of the 2005 IEEE International Frequency Control Symposium and Exposition* (2005), pp. 608–610
19. J. Chen, Active optical clock. Chin. Sci. Bull. **54**, 348–352 (2009)
20. X. Zhang et al., in *2015 Joint Conference of the IEEE IFCS & EFTF*, Denver, CO, USA (2015), pp. 618–621
21. J. Zhang, T. Shi, J. Miao et al., Research progress of active optical clocks. Meas. Technol. **43**(3), 1–16 (2023)
22. J. Bohnet, Z. Chen, J. Weiner et al., A steady-state superradiant laser with less than one intracavity photon. Nat. Photonics **484**, 78–81 (2012)
23. S.J.M. Kuppens, M.P. van Exter, J.P. Woerdman, Quantum-limited linewidth of a bad-cavity laser. Phys. Rev. Lett. **72**(24), 3815–3818 (1994)
24. S.K. Gayen, R.I. Billmers, V.M. Contarino et al., Induced-dichroism-excited atomic line filter at 532 nm. Opt. Lett. **20**(12), 1427–1429 (1995)
25. W. Zhuang, J.B. Chen, Active Faraday optical frequency standard. Opt. Lett. **39**(21), 6339–6342 (2014)
26. T. Shi, D. Pan, W. Zhuang, X. Guan, J. Miao, J. Zhang, J. Chen, Active optical clock based on laser cooling of alkali-metal atoms, in *2021 Joint Conference of the European Frequency and Time Forum and IEEE International Frequency Control Symposium (EFTF/IFCS)* (2021), pp. 1–3
27. Z. Tao, X. Zhang, D. Pan, M. Chen, C. Zhu, J. Chen, Faraday laser using 1.2 km fiber as an extended cavity. J. Phys. B At. Mol. Opt. Phys. **49**, 13LT01 (2016)
28. I. Caltzidis, H. Kibler, T. Pfau et al., Atomic Faraday beam splitter for light generated from pump-degenerate four-wave mixing in a hollow-core photonic crystal fiber. Phys. Rev. A **103**(4), 043501 (2021)
29. D.R. Häupl, D. Weller, R. Löw et al., Spatially resolved spectroscopy of alkali metal vapour diffusing inside hollow-core photonic crystal fibers. New J. Phys. **24**(11), 113017 (2022)
30. D. Weller, A. Yilmaz, H. Kübler et al., High vacuum compatible fiber feedthrough for hot alkali vapor cells. Appl. Opt. **56**(5), 1546–1549 (2017)
31. J. Gutekunst, D. Weller, H. Kübler et al., Fiber-integrated spectroscopy device for hot alkali vapor. Appl. Opt. **56**(21), 5898–5902 (2017)

Epilogue

The publication of "Faraday Laser" represents not only the first academic monograph in the field of atomic frequency-stabilized Faraday diode lasers, but also a compilation of our team's years of research for the Faraday lasers. This book aims to introduce the technological achievements of Faraday lasers, enhance their promotion and application, facilitate exchanges and learning within related scientific research fields, and contribute to support the development of scientific instruments, quantum precision measurement, and other domains. Here, we briefly share the original intentions behind writing this book.

Firstly, it is a tribute to the renowned British physicist, Michael Faraday.

Advancements in scientific research stem from the diverse contributions of scientists within the scientific community. In this process, genius scientists often leave behind prominent, individualistic, and indelible achievements. These genius scientists can be broadly categorized into two types. One type possesses a solid foundation in mathematics and physics, enabling them to carry out their work with relatively clear mathematical logic as support and guidance. Newton, Maxwell, and Einstein are prominent examples of this type. The other type comprises profound experimental scientific geniuses, with Michael Faraday (9/22/1791–8/25/1867) being a typical representative. Limited by his early living conditions, Faraday's education only equipped him with basic mathematical knowledge. Through diligent and assiduous experimental research, he continually expanded and deepened his knowledge, pioneered innovative explorations, and achieved extraordinary scientific accomplishments unparalleled in human history. His contributions, particularly in electromagnetism and optics, laid the foundation for Maxwell's subsequent theoretical work and provided essential support for significant transformative progress in quantum mechanics. Maxwell deemed Faraday "in reality a mathematician of a very high order". Einstein also hung Faraday's portrait on the wall of his Berlin office. Among Faraday's brilliant scientific achievements, the Faraday magneto-optical effect stands out as a particularly noteworthy accomplishment.

Faraday held a steadfast belief that there must be a connection between electricity, magnetism, and light. This belief was deeply rooted in his heart. To nurture this seed, Faraday conducted countless experiments over decades. The intrinsic

J. Chen et al., *Faraday Laser*, https://doi.org/10.1007/978-981-97-8023-5

link between light and magnetism, from a scientific perspective, is a significant issue. Whether one can persevere tirelessly around this major scientific question is a testament to the character and spiritual strength of a genius scientist. Over several decades, Faraday attempted experiments using various transparent materials under magnetic field conditions, including rock salt, quartz, alum, and fluorspar, but without success. It wasn't until September 13, 1845, when Faraday happened to recall a special lead borate glass he had developed 20 years earlier, a significant achievement from his five-year project aimed at "improving the optical properties of glass". Using this special lead borate glass, Faraday finally succeeded in discovering what later became known as the Faraday effect—the magneto-optical effect. Shortly after, he borrowed a powerful magnet from the Royal Military Academy in Woolwich (now the Royal Military Academy Sandhurst in the UK). On September 18, Faraday joyfully confirmed the magneto-optical effect, successfully validating his belief in the connection between magnetism and light. That day, he eagerly tested various materials and excitedly wrote in his diary, "does an excellent day's work" "There was an effect produced on the polarized ray and thus magnetic force and light were proved to have relations to each other", adding, "Have got enough for today". On November 5, 1845, Faraday submitted a paper titled "On the Magnetization of Light and the Luminescence of Magnetic Lines of Force" to the Royal Society of London. This marked the first discovery of the interconnection between light and magnetism in human history, signifying the birth of magneto-optics. Faraday's long-cherished wish had finally been fulfilled.

"Faraday Laser" is a book based on Faraday's magneto-optical effect discovered over 170 years ago, realizing a frequency-stabilized diode laser, which we have named the Faraday laser, in homage to Faraday.

Secondly, it is a continuation of the scientific research and innovative exploration of Professor Yiqiu Wang and Professor Donghai Yang in the field of laser-pumped cesium beam atomic clock.

President Yiqiu Wang, the former Executive Vice President of Peking University, was the first in China to embark upon the research of laser-pumped cesium beam atomic clock. In the laser-pumped cesium beam atomic clock system, a high-performance 852 nm cesium-stabilized laser capable of long-term stable operation is required. For decades, the commercialization of diode laser-pumped cesium beam atomic clock has remained a significant international challenge. Through the relentless efforts of President Yiqiu Wang and Professor Donghai Yang over several decades, coupled with vigorous promotion from various sources, China has finally achieved a breakthrough in this field, reaching world-leading standards. To further fulfill President Yiqiu Wang's aspirations in laser-pumped cesium beam atomic clocks, we have specifically developed an 852 nm cesium-stabilized Faraday diode laser and successfully created a high-performance Faraday laser pumped cesium beam atomic clocks. Additionally, under the guidance of Prof. Lemin Zheng, the Institute of Quantum Electronics at Peking University has been conducting research on Faraday filters for optical communication and laser frequency stabilization since the 1980s, laying a solid foundation for the development of Faraday lasers.

Finally, this book aims to continuously promote technological innovation and multi-domain application of Faraday lasers.

There has been considerable research in China and internationally on Faraday filters and their application in laser frequency stabilization. However, our core contribution lies in integrating Faraday filters directly within the laser system as frequency-selective devices, successfully developing atomic frequency-selective Faraday lasers. This effectively addresses the international challenges faced by traditional frequency-stabilized lasers, which are susceptible to temperature and current disturbances, as well as long term drift, and enables widespread application in fields such as quantum precision measurement. In this process, we have also significantly expanded the range of Faraday filter types and available atomic spectral lines. Looking ahead, our goal is to develop ultra-high-stability Faraday lasers that can be instantly activated and directly output wavelengths corresponding to atomic spectral lines, facilitating industrialized application in quantum precision measurement, time–frequency communication, precision spectroscopy, and other domains.

Lastly, we would like to express our heartfelt gratitude to all those who have provided assistance and support during the compilation of this book. Our sincere thanks go out to our family, friends, colleagues, classmates, as well as the editors and staff of the Springer. Without your support and help, this book would not have been possible. While the completion of this book signifies a significant milestone, it is by no means the end of our exploration and research in the field of Faraday lasers. In the days to come, we will continue to delve deeper into related research areas, and we hope this book will inspire more researchers to develop an interest and enthusiasm for Faraday lasers, jointly advancing the development and progress of this field.

September 29, 2024

Index

J. Chen et al., *Faraday Laser*, https://doi.org/10.1007/978-981-97-8023-5

The manufacturer's authorised representative in the EU is Springer Nature Customer Service Centre GmbH, Europaplatz 3, 69115 Heidelberg, Germany. If you have any concerns regarding our products, please contact ProductSafety@springernature.com

Printed and bound by CPI Group (UK) Ltd, Croydon, CR0 4YY
01/10/2026
02225969-0017